STRUCTURE REPORTS

for 1977

Volume 43 A

Structure Reports is prepared under the guidance of a Commission of the International Union of Crystallography. The members of the Commission sometime concerned with the preparation of this volume are listed below.

STRUCTURE REPORTS

for 1977

Volume 43A

METALS AND INORGANIC SECTIONS

General editor

J. Trotter

Section editors

L. D. Calvert *and* J. Trotter

Springer Science+Business Media, B.V.

First published in 1979

ISBN 978-94-017-3141-6   ISBN 978-94-017-3139-3 (eBook)
DOI 10.1007/978-94-017-3139-3

TABLE OF CONTENTS

Introduction, VI

Metals, 1

Inorganic Compounds, 107

Subject Index, 359

Formula Index, 371

Author Index, 382

Corrigenda, 393

INTRODUCTION

The present volume continues the aim of Structure Reports to present critical accounts of all crystallographic structure determinations. The increase in the number of crystal structure papers (now about 2000 per year in the Organic Section) has prompted a minor change in format, which increases the information density per page, hopefully without any loss of clarity.

Details of the arrangement in the volumes, symbols used etc. are given in previous volumes (e.g. 41B or 42A, pages vi-viii).

University of British Columbia   J. TROTTER
Vancouver, Canada

15 July 1978

STRUCTURE REPORTS

SECTION I

METALS

Edited by

L. D. Calvert

(National Research Council of Canada)

with the assistance of

J. K. Byron
F. L. Lee
Yu Wang

ARRANGEMENT

As in previous volumes the arrangement in the Metals section is approximately, but not strictly, alphabetical, and to find particular substances the subject index or formula index should be used.

ALKALINE-EARTH ANTIMONIDES and BISMUTHIDES
A_2B, A_5B_3 (A = Ba, Sr; B = Bi, Sb)

B. EISENMANN and K. DELLER, 1975. Z. Naturf., 30B, 66-72.

A_2B. Tetragonal, La_2Sb type (1), I4/mmm, Z = 4. Mo radiation, photographic data.
Atoms are placed A(1) in 4(c): 0,1/2,0; A(2) and B in 4(e): 0,0,z.

A_5B_3. Hexagonal, Mn_5Si_3 type (2), $P6_3$/mcm, Z = 2. Mo radiation, diffractometer
data. Atoms are placed A(1) in 4(d): 1/3,2/3,0; A(2) and B in 6(g): x,0,1/4.

	Sr_2Bi	Ba_2Sb	Sr_5Bi_3	Ba_5Sb_3	Ba_5Bi_3
a (Å)	5.01	5.22	9.63	9.97	10.13
c	17.68	18.46	7.63	7.73	7.79
c/a	3.53	3.54	0.792	0.776	0.769
D_x	5.75	5.23	5.77	5.25	6.31
R	0.147	0.146	0.098	0.103	0.105
Refl.	138	149	368	368	387
z(A(2))	0.329	0.327	-	-	-
z(B)	0.138	0.136	-	-	-
x(A(2))	-	-	0.7492	0.7497	0.7513
x(B)	-	-	0.3942	0.3918	0.3926

Interatomic distances (Å)

A_2B		Sr_2Bi	Ba_2Sb
A(1)	- 8 A	3.54, 3.93	3.69, 4.13
	- 4 B	3.50	3.62
A(2)	- 8 A	3.93, 4.51	4.13, 4.65
	- 5 B	3.38, 3.59	3.52, 3.76
B	- 9 A	3.38 - 3.59	3.52 - 3.76

A_5B_3		Sr_5Bi_3	Ba_5Sb_3	Ba_5Bi_3
A(1)	- 8 A	3.82, 4.14	3.87, 4.27	3.89, 4.34
	- 6 B	3.52	3.63	3.68
A(2)	- 10 A	4.14 - 4.52	4.27 - 4.60	4.34 - 4.64
	- 5 B	3.33 - 4.06	3.43 - 4.11	3.48 - 4.16
B	- 9 A	3.33 - 4.06	3.43 - 4.11	3.48 - 4.16

1. Structure Reports, 35A, 12; 39A, 6; 40A, 8, 30; 41A, 12.
2. Strukturbericht, 4, 24, 137, 246; Structure Reports, 24, 78; 32A, 102; 34A, 34.

ALKALINE-EARTH BERYLLIUM SILICIDES and GERMANIDES
$ABe_{0.75}X_{1.25}$ (A = Ba, Ca, Sr; X = Ge, Si)

N. MAY, W. MÜLLER and H. SCHÄFER, 1974. Z. Naturf., 29B, 325-327.

Hexagonal, AlB_2 type (1), P6/mmm, Z = 1. Mo radiation, photographic data. P6m2 was
rejected. Atoms are placed A in 1(a): 0,0,0; Be/X in 2(d): 1/3,2/3,1/2. Occupancy
of Be/X = 0.375Be, 0.625X.

A	X	a(Å)	c(Å)	c/a	D_X	R
Ca	Si	3.94	4.38	1.11	2.29 (D_m)	0.125
Sr	Si	4.03	4.68	1.16	3.27	0.102
Ba	Si	4.05	5.05	1.25	4.15	0.067
Ca	Ge	4.00	4.34	1.09	3.80	0.131
Sr	Ge	4.08	4.66	1.14	4.58	0.103

Interatomic distances (Å)

A	X	A - 8 A	A - 12 Be/X	Be/X - 3 Be/X	Be/X - 6 A
Ca	Si	3.94, 4.38	3.16	2.27	3.16
Sr	Si	4.03, 4.68	3.30	2.33	3.30
Ba	Si	4.05, 5.05	3.43	2.34	3.43
Ca	Ge	4.00, 4.34	3.16	2.30	3.16
Sr	Ge	4.08, 4.66	3.34	2.36	3.34

1. Strukturbericht, 3, 28, 311; 4, 121; Structure Reports, 20, 51.

ALKALINE-EARTH PNICTIDES
AMg_2X_2 (A = Ba, Ca, Sr; X = As, Bi, Sb)

K. DELLER and B. EISENMANN, 1977. Z. Naturf., 32B, 612-616.

Trigonal, Al_2CaSi_2 type (1), P$\bar{3}$m1, Z = 1. Mo radiation, diffractometer data. Atoms are placed A in 1(a): 0,0,0: Mg and X in 2(d): 1/3,2/3,z.

A	X	a(Å)	c(Å)	c/a	D_X	R	Refl.	z(Mg)	z(X)
Ca	As	4.34	7.13	1.64	3.40	0.033	235	0.6342	0.2507
Ca	Sb	4.66	7.58	1.63	3.95*	0.039	366	0.6315	0.2460
Sr	Sb	4.70	7.83	1.66	4.20	0.051	325	0.6286	0.2571
Ba	Sb	4.77	8.10	1.70	4.47	0.106	212	0.6247	0.2680
Ca	Bi	4.73	7.68	1.62	5.80*	0.062	200	0.6299	0.2422
Ba	Bi	4.86	8.22	1.69	5.95	0.064	366	0.6274	0.2641

* D_m

Interatomic distances (Å)

		Ca/As	Ca/Sb	Sr/Sb	Ba/Sb	Ca/Bi	Ba/Bi
A	- 6 X	3.08	3.27	3.38	3.51	3.30	3.55
	- 6 Mg	3.62	3.88	3.98	4.10	3.94	4.16
	- 6 A	4.43	4.66	4.70	4.77	4.73	4.86
		[4.34]					
Mg	- 3 X	2.64	2.85	2.86	2.89	2.90	2.95
	- 1 X	2.73	2.92	2.91	2.89	2.98	2.99
	- 3 Mg	3.15	3.35	3.38	3.42	3.38	3.50
	- 3 A	3.62	3.88	3.98	4.10	3.94	4.16
X	- 3 Mg	2.64	2.85	2.86	2.89	2.90	2.95
	- 1 Mg	2.73	2.92	2.91	2.89	2.98	2.99
	- 3 A	3.08	3.27	3.38	3.51	3.30	3.55

1. Structure Reports, 12, 174; 32A, 5.

ALUMINUM BARIUM GERMANIUM
$Al_3Ba_{10}Ge_7$

A. WIDERA and H. SCHÄFER, 1977. Z. Naturf., 32B, 619-624.

Hexagonal, P6$_3$/mcm, a = 9.749, c = 16.47 Å, c/a = 1.69, D$_m$ = 4.77, Z = 2. Mo radiation, R = 0.071 for 1119 reflexions, diffractometer data.

Atomic positions

			x	y	z
Ba(1)	in	4(c)	1/3	2/3	1/4
Ba(2)	in	4(d)	1/3	2/3	0
Ba(3)	in	12(k)	0	0.2522	0.4018
Ge(1)	in	12(k)	0.3955	0	0.8832
Ge(2)	in	2(a)	0	0	1/4
Al	in	6(g)	0	0.727	1/4

Interatomic distances (Å)

Ba(1)	- 8 Ba	4.12, 4.47		Ge(1)	- 8 Ba	3.40 - 3.81
	- 6 Ge	3.71			- 1 Al	2.50
	- 3 Al	3.58				
				Ge(2)	- 6 Ba	3.51
Ba(2)	- 8 Ba	4.05, 4.12			- 3 Al	2.66
	- 6 Ge	3.56				
				Al	- 6 Ba	3.58
Ba(3)	- 8 Ba	4.05 - 4.47			- 3 Ge	2.50, 2.66
	- 5 Ge	3.40 - 3.81				
	- 2 Al	3.58				

The structure contains propeller-like Al$_3$Ge$_7$ units (Fig. 1) with $\bar{6}$m2 symmetry.

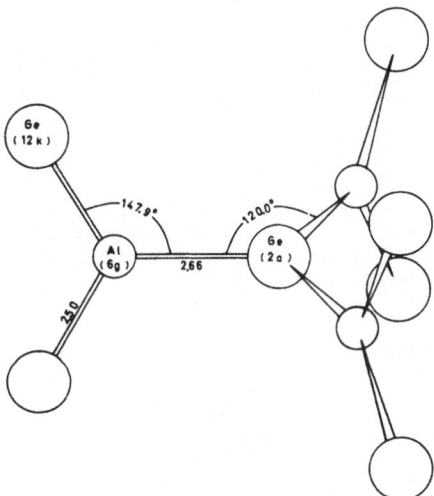

Fig. 1.   The characteristic Al$_3$Ge$_7$ units in Al$_3$Ba$_{10}$Ge$_7$.

ALUMINUM BARIUM IRON SULPHUR   BARIUM IRON SULPHUR

$(Al_{0.4}Ba_{0.6})Ba_3Fe_2S_6[S_{0.6}(S_2)_{0.4}]$ (I) $Ba_4Fe_2S_6[S_{2/3}(S_2)_{1/3}]$ (II)

J.T. HOGGINS, L.E. RENDON-DIAZMIRON and H. STEINFINK, 1977. J. Solid State Chem., 21, 79-90.

I. Triclinic, P$\bar{1}$, a = 8.993, b = 6.780, c = 12.869 Å, α = 106.38, β = 104.17, γ = 90.90°, Z = 2, D$_x$ = 3.90 for P-cell; a' = a, b' = b, c' ∿ 2c, α = 91.11, β = 105.04, γ = 90.90° for A-centred cell. Mo radiation, diffractometer data, R = 0.082 for

3589 reflexions, 17 site-sets for I.

II. Triclinic, $P\bar{1}$, a = 9.002, b = 6.7086, c = 12.861 Å, α = 106.61, β = 104.27, γ = 90.74°, Z = 2, D_X = 4.15 for P-cell; a' = a, b' = b, c' $\sim$ 2c, α = 91.49, β = 105.10, γ = 90.74° for A-centred cell. Mo radiation, diffractometer data, R = 0.047 for 1682 reflexions, 15 site-sets for II.

BaS_6 trigonal prisms share edges to form distorted hexagonal rings, which form one-dimensional chains leaving two free lateral edges. The chains link in a stair-step manner. These stairsteps join in a complicated manner to form a three-dimensional network. Fe ions are in two sites forming isolated FeS_4 tetrahedra (Fe-S   2.21-2.27 Å) and isolated Fe_2S_6 (Fe-S 2.14-2.16 Å) dimers by edge-sharing tetrahedra. The Al substitution occurs in the BaS_6 trigonal prisms which have free edges with Al replacing Ba, but occupying one of the two free square faces on a statistical basis instead of the centre; thus the Ba^{2+} has been replaced by $(Al,e)^{2+}$ with Fe^{3+} reduced to maintain charge balance; Ba-S distances range from 2.88 to 3.65 Å and average about 3.3 Å. The Fe-Fe distances across the shared edge are 2.79 Å in the Ba-Fe-S compound and 2.86 Å in the Al-substituted one. The S_2 disulphide ion replaces one S atom around a Ba on a statistical basis; Al-S distances range from 2.24 to 2.88 Å. Mossbauer spectra and valence calculations indicate that Fe in edge-sharing tetrahedra are reduced ($Fe^{2.5+}$) while Fe in the isolated tetrahedron is unchanged.

ALUMINUM BORON
α-AlB_{12}

I.   I. HIGASHI, T. SAKURAI and T. ATODA, 1977. J. Solid State Chem., 20, 67-77.
II. J.S. KASPER, M. VLASSE and R. NASLAIN, 1977. Ibid., 20, 281-285.

I. Tetragonal, $P4_32_12$, a = 10.158, c = 14.270 Å, D_m = 2.55. Mo radiation, diffract-ometer data, R = 0.03 for 1478 reflexions.

II. Tetragonal, $P4_32_12$, a = 10.161, c = 14.283 Å, D_m = 2.65. Mo radiation, diffract-ometer data, R = 0.026 for 2393 reflexions. 28 site-sets with occupancies. See also 1 and 2. There is excellent agreement between the two studies. [In II the z coordinate for Al(2) should be 0.3080, not 0.3030.] Both I and II give 176 B atoms in the cell; I gives 12.95 Al, II gives 13.21 Al atoms.

Summarized interatomic distances (Å)

Type	Range	Average
Within B_{12} unit	1.74-1.86	1.805
Between B_{12} units	1.63-1.81	1.709
Within B_{19} unit	1.74-1.91 or	1.814
	1.74-1.98	1.820
Between B_{12} and B_{19} units	1.66-1.87	1.745
Between B_{19} and B_{19} units	1.715	-
B_{23}^{*} - B_{12}	-	1.874
B_{23}^{*} - B_{19}	-	1.792
Al - B (first B neighbours)	2.02-2.98	-
Al - Al†	2.15-2.72	-

* B_{23} is the unique B atom (Fig. 2)
† Short Al-Al distances (0.55-1.22 Å) occur between sites with total occupancies
   less than 100%.

The structure contains 8 B_{12} icosahedra and 4 B_{19} units (Fig. 1), linked by
single B atoms (Fig. 2). All B atom sites are fully occupied. The Al atoms occupy
5 sites (Fig. 2) partially. One Al site is only 2% occupied and II suggests that
B might occupy this site. The B_{19} group is a new group and can be described as a
twinned icosahedron, with a triangular face as the composition plane and two atoms
missing. The possible relationships between AlB_{12} and BeB_6, β-tetragonal B (3) and
LiB_6 (4) was not confirmed when powder patterns were calculated (II).

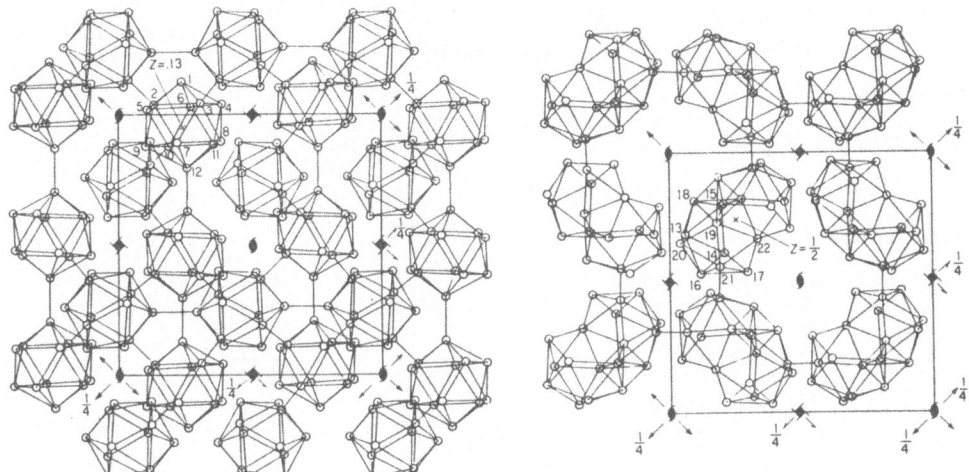

Fig. 1.   The B_{12} icosahedra (left) and B_{19} units (right) in α-AlB_{12}, projected
           onto (001).

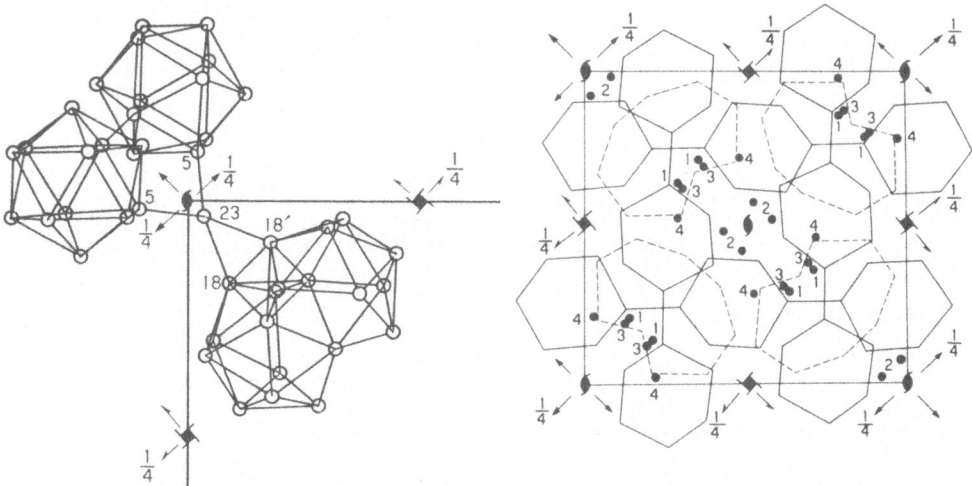

Fig. 2.   The unique B atom, B_{23}, linking B_{12} and B_{19} units in α-AlB_{12}, as
           projected onto (001) (left), and the Al sites, projected onto (001)
           (right, the B_{12} and B_{19} units are outlined with solid and dashed
           lines, respectively; the Al(5) site (0.02 occupied) lies between Al(1)
           and Al(3) and is omitted).

1. Strukturbericht, 7, 110.
2. Structure Reports, 22, 6; 26, 4; 29, 4.
3. Ibid., 24, 249; 26, 57.
4. D.R. SECRIST, 1967. J. Amer. Ceram. Soc., 50, 520.

ALUMINUM CERIUM   ALUMINUM PRASEODYMIUM
AlCe   AlPr

H. ASMAT, B. BARBARA and D. GIGNOUX, 1977. J. Solid State Chem., 22, 179-184.

Orthorhombic, AlCe type (1), Cmcm, a = 9.267, 9.146, b = 7.680, 7.625, c = 5.746, 5.698 Å, D_X = 5.43, 5.61, for AlCe, AlPr, respectively, Z = 8. Neutron data.

Atomic positions

			x	y	z
Ce or Pr	in	8(g)	0.179	0.161	1/4
Al(1)	in	4(c)	0	0.790	1/4
Al(2)	in	4(b)	0	1/2	0

[Interatomic distances (Å)]

		AlCe	AlPr
Ce or Pr	- 7 Ce or Pr	3.32 - 4.06	3.27 - 4.03
	- 8 Al	3.14 - 3.53	3.10 - 3.49
Al(1)	- 8 Ce or Pr	3.14 - 3.34	3.10 - 3.31
	- 2 Al	2.65	2.63
Al(2)	- 8 Ce or Pr	3.40, 3.53	3.38, 3.49
	- 4 Al	2.65, 2.87	2.63, 2.85

1. Structure Reports, 26, 174; 32A, 9.

ALUMINUM CHROMIUM   ALUMINUM VANADIUM
γ-Al$_8$Cr$_5$ γ-Al$_8$V$_5$

J.K. BRANDON, W.B. PEARSON, P.W. RILEY, C. CHIEH and R. STOKHUYZEN, 1977.
Acta Cryst., B33, 1088-1095.

Al$_8$Cr$_5$, rhombohedral, R3m, a = 7.811 Å, α = 109.12°, A $\sim$ 26 atoms per cell (a = 9.057 Å, α = 89.28°, body-centred cell, $\sim$ 52 atoms), D_m = 4.22. Mo radiation, R = 0.049 for 263 reflexions, diffractometer data. See also 1. [The wrong cell is given by mistake on p. 11 of 1; see 2 for details.] The crystal used was quenched from 1215°C, corresponding to the γ_1 of 1.

Atomic positions

Site label*	CN	Occupancy	Site set	x	y	z
IT1	13	1/2Cr	1(a)xxx	0.1795	-	-
IT3	13	3/2Cr+3/2Al	3(b)xxz	0.0055	-	-0.2096
OT1	13	1Cr	1(a)xxx	-0.3071	-	-
OT3	13	3Cr	3(b)xxz	0†	-	0.3451
OH3+	13	2Cr+1Al	3(b)xxz	0.3699	-	0.0306
OH3-	15	3Al	3(b)xxz	-0.3608	-	-0.0095
CO3+	13	3/2Cr+3/2Al	3(b)xxz	0.3608	-	0.5720
CO3-	15	3Al	3(b)xxz	-0.2709	-	-0.6270
CO6	15	6Al	6(c)xyz	-0.3179	0.2982	0.0533

* See 3 for nomenclature. Here + and - denote positions + or - along [111] from
the centre of the 26-atom cluster. The numbers give the multiplicity of the
site-set in R3m.
† Set at 0 to fix origin.

Averaged interatomic distances (Å)

Atom pair	Average distance	Distances averaged
Cr - Cr	2.59	3
Cr - Al/Cr	2.62	30
Cr - Al	2.66	24
Al/Cr - Al/Cr	2.67	18
Al/Cr - Al	2.70	48
Al - Al	3.15	54

γ-Cr_5Al_8 belongs to a group of γ-brass like structures with rhombohedral cells
and α < 90° (for the 52-atom body-centred cell). In body-centred cubic brasses
(3), the 26 atom clusters are roughly spherical in shape but in rhombohedral
γ-Cr_5Al_8, where there is only one threefold symmetry axis [111], the smaller Cr
atoms predominate in that half of the cluster along the positive [111] direction.
The larger Al atoms predominate in the other half; thus the 26-atom clusters in
Cr_5Al_8 show a conical distortion resulting in α < 90°. These results agree well
with 1 except for the occupancy and coordinates of sites IT1 and OT1. In 1 these
are fully occupied by Cr and Al, respectively, and thus located somewhat further
from the centre of the cluster. Finally, because the two studies agree so well
either the crystal used decomposed to the γ₂ of 1 on quenching or the structures
differ mainly in IT1 and OT1 sites.

Al_8V_5, Cubic, Cu_5Zn_8 type (3), I43m, a = 9.234 Å, D_m = 3.97, Z = 4. Mo radiation,
diffractometer data, R = 0.049 for 140 reflexions. See also 4.

Atomic positions

Site label*	Occupancy	Site set	x	z
IT	4Al+4V	8(c)xxx	0.1062	-
OT	8V	8(c)xxx	0.8300	-
OH	4Al+8V	12(e)00z	-	0.3522
CO	24Al	24(g)xyz	0.3087	0.0394

* For nomenclature and detailed description see 3.

Interatomic distances (Å)

Site	Neighbours	Number	Distance
IT(1/2Al+1/2V)	IT	3	2.77
	OT	3	2.68
	OH	3	2.66
	CO	3	2.72
OT(V)	IT	3	2.68
	OH	3	2.79
	CO	3	2.65
	CO	3	2.70
OH(1/3Al+2/3V)	IT	2	2.66
	OT	2	2.79
	OH	1	2.73
	CO	4	2.90
	CO	2	2.69
	CO	2	3.04

Site	Neighbours	Number	Distance
CO(A1)	IT	1	2.72
	OT	1	2.65
	OT	1	2.70
	OH	2	2.90
	OH	1	2.69
	OH	1	3.04
	CO	2	3.52
	CO	4	2.77
	CO	2	3.61

It is supposed that, in the γ-brasses with rhombohedral cells and α < 90° (for the 52-atom cell) which all appear to have valence-electron concentrations in excess of 88 or 89 per 52-atom cell (say about 95), the structures are stabilized by band-structure energy. Since V_5Al_8 has the cubic rather than the rhombohedral structure it appears that V must absorb some electrons from the conduction band.

1. Strukturbericht, 5, 11, 65.
2. J.W. VISSER, 1977. Acta Cryst., B33, 316.
3. Strukturbericht, 1, 499, 533; 2, 693; Structure Reports, 33A, 72.
4. Structure Reports, 18, 29; 19, 41.

ALUMINUM COBALT LANTHANUM
Al_4CoLa

R.M. RYKHAL, O.S. ZAREČNJUK and Ja.P. JARMOLJUK, 1977. Dop. Akad. Nauk Ukr., No. 3, 265-268.

Orthorhombic, Pmma, a = 7.701, b = 4.082, c = 7.023 Å, D_m = 4.61, Z = 2. Cu radiation, R = 0.147, photographic data.

Atomic positions

			x	y	z
La	in	2(e)	1/4	0	0.388
Co	in	2(e)	1/4	0	0.813
Al(1)	in	4(j)	0.061	1/2	0.708
Al(2)	in	2(f)	1/4	1/2	0.022
Al(3)	in	2(a)	0	0	0

Interatomic distances (Å)*

La	- 4 La	4.07, 4.16		Al(2) -	2 La	3.30
	- 12 Al	3.21 - 3.38			- 8 Al	2.67 - 3.01
	- 4 Co	3.02 - 4.14			- 2 Co	2.52
Co	- 1 La	3.02		Al(3) -	2 La	3.36
	- 8 Al	2.34 - 2.61			- 8 Al	2.81, 2.92
					- 2 Co	2.33
Al(1) -	4 La	3.21, 3.38				
	- 6 Al	2.67 - 3.11		* [Calculated values differ slightly]		
	- 2 Co	2.61				

ALUMINUM IRIDIUM NIOBIUM
AlIrNb - A' phase

R. HORYŃ, 1977. J. Less-Common Metals, 56, 103-111.

Hexagonal, P6$_3$/mcm, a = 8.840, c = 8.220 Å, c/a = 0.930, A = 36, D$_X$ = 9.36, for a composition of Al$_{0.4227}$Ir$_{0.2222}$Nb$_{0.3550}$. Mo radiation, R = 0.088 for 240 reflexions, diffractometer data. See also 1.

Atomic positions

			x	y	z
Nb	in	12(k)	0.6674	0	0.0645
AlIr*	in	12(j)	0.8366	0.3326	1/4
Al	in	6(g)	0.1704	0	1/4
AlNb†	in	4(d)	1/3	2/3	0
Ir	in	2(b)	0	0	0

* Al$_{0.5}$Ir$_{0.5}$ † Al$_{0.805}$Nb$_{0.195}$

Interatomic distances (Å)

Nb	- 4 Nb	3.05 - 3.14		Al	- 6 Nb	2.96, 2.97
	- 6 AlIr	2.96 - 2.98			- 2 AlIr	2.49
	- 3 Al	2.96, 2.97			- 2 Al	2.61
	- 2 AlNb	3.00			- 2 Ir	2.55
	- 1 Ir	2.99				
				AlNb	- 6 Nb	3.00
					- 6 AlIr	2.55
AlIr	- 6 Nb	2.96 - 2.98				
	- 3 AlIr	2.50, 2.61				
	- 1 Al	2.49		Ir	- 6 Nb	2.99
	- 2 AlNb	2.55			- 6 Al	2.55

A superstructure of the MgZn$_2$ type (1).

1. Structure Reports, 42A, 129.

ALUMINUM MAGNESIUM SELENIUM
Al$_2$Mg$_5$Se$_8$

K.-P. DOTZEL and H. SCHÄFER, 1977. Z. Naturf., 32B, 1488-1489.

Orthorhombic, Pna2$_1$, a = 13.30, b = 6.32, c = 7.72 Å, D$_m$ = 4.08, Z = 2. Mo radiation, photographic data, R = 0.120 for 296 reflexions.

Atomic positions

	x	y	z	Occupancy
Se(1)	0.320	0.243	0.049	1
Se(2)	0.080	0.252	0.268	1
Se(3)	0.894	0.726	0.287	1
Se(4)	0.664	0.746	0.018	1
Mg(1)	0.004	0.485	0.061	0.80
Mg(2)	0.238	0.530	0.282	0.35
Mg(3)	0.504	0.530	0.530	0.50
Mg(4)	0.729	0.508	0.309	0.84
Al	0.420	0.405	0.286	1

Mg is octahedrally coordinated by Se (Mg-Se = 2.40-3.08 Å) and Al tetrahedrally by Se (Al-Se = 2.32-2.48 Å); Mg-Al distances range from 2.26 to 3.55 Å. The structure is similar to those of Mg$_2$SiO$_4$ and Li$_3$PO$_4$.

ALUMINUM PALLADIUM
AlPd (at r.t.)

T. MATKOVIĆ and K. SCHUBERT, 1977. J. Less-Common Metals, 55, 45-52.

Rhombohedral, R$\bar{3}$, a = 15.659, c = 5.251 Å, A = 26 x 3, D_X = 7.75. Mo radiation,
R = 0.091 for 509 reflexions, diffractometer data. See also 1; contrary to 2.

Atomic positions

		x	y	z
Pd(1)	in 18(f)	0.1784	0.1110	0.6655
Pd(2)	in 18(f)	0.3196	0.0831	0.3351
Pd(3)	in 3(a)	0	0	0
Al(1)	in 18(f)	0.1799	0.1117	0.1657
Al(2)	in 18(f)	0.3200	0.0829	0.8365
Al(3)	in 3(b)	0	0	1/2

[Interatomic distances (Å)]

Pd(1)	- 6 Pd	2.92 - 3.01	Al(1)	- 7 Pd	2.58 - 2.63
	- 7 Al	2.59 - 2.63		- 6 Al	2.92 - 3.02
Pd(2)	- 6 Pd	2.92 - 3.02	Al(2)	- 7 Pd	2.58 - 2.63
	- 7 Al	2.58 - 2.63		- 6 Al	2.92 - 3.02
Pd(3)	- 6 Pd	3.01	Al(3)	- 8 Pd	2.59, 2.63
	- 8 Al	2.61, 2.63		- 6 Al	3.02

1. Structure Reports, 13, 23; 15, 12; 17, 31; 29, 12.
2. Ibid., 17, 31; 21, 12; 29, 12.

ALUMINUM RHENIUM SILICON
$Al_{1.20}ReSi_{0.8}$

Ju.B. KUZ'MA and V.V. MILJAN, 1977. Izv. Akad. Nauk SSSR, Neorg. Mater., 13, 926-
927 [Inorg. Mater., 13, 760-761].

Tetragonal, $MoSi_2$ type (1), I4/mmm, a = 3.180, c = 8.020 Å, c/a = 2.522, Z = 2,
D_X = 9.87. Cu radiation, photographic data. Re in 2(a): 0,0,0; (2.4Al + 1.6Si)
in 4(e): 0,0,1/3.

[Interatomic distances (Å)]

Re	- 10 AlSi	2.62, 2.67	AlSi	- 9 AlSi	2.62 - 3.18
	- 4 Re	3.18		- 5 Re	2.62, 2.67

1. Strukturbericht, 1, 219, 740, 783; Structure Reports, 8, 102.

AMERICIUM SILICON
$AmSi_X$ (x = 1.88)

F. WEIGEL, F.D. WITTMANN and R. MARQUART, 1977. J. Less-Common Metals, 56, 47-53.

Tetragonal, α-$ThSi_2$ type (1), $I4_1$/amd, a = 4.019, c = 13.688 Å, Z = 4, D_X = 8.92.
Cu radiation, X-ray powder data. 4 Am in 4(a): 0,3/4,1/8; ~7.5 Si in 8(e): 0,3/4,
0.542.

Interatomic distances (Å)

Am	- 8 Am	3.97, 4.02	Si	- 6 Am	3.04, 3.06
	- 12 Si	3.04, 3.06		- 3 Si	2.27, 2.32

1. Structure Reports, 9, 121; 12, 124; 23, 237; 26, 271.

ANTIMONY ARSENIC LEAD SULPHUR   (GEOCRONITE)
$As_{5.5}Pb_{28}S_{46}Sb_{6.5}$

R.W. BIRNIE and C.W. BURNHAM, 1976. Amer. Min., 61, 963-970.

Monoclinic, $As_{12}Pb_{28}S_{46}$ (jordanite) type (1), $P2_1/m$, a = 8.963, b = 31.93, c = 8.500 Å, β = 118.02°, Z = 1. Mo radiation, R = 0.121 for 1392 reflexions, diff-ractometer data; 25 site-sets with occupancies are given. See also 2.

Geocronite is the Sb-containing analogue of jordanite. Each of the three layers in the asymmetric unit (1) contains one As site; in addition one Pb site is half-occupied by As. Only three of these can expand to accommodate Sb (Sb-3S = 2.26-2.54, mean 2.40 Å; the fourth site has As-3S = 2.19-2.27, mean 2.24 Å, and cannot expand because of its Pb neighbour). Thus the limiting Sb content is 8 Sb to 4 As atoms. Pb-S distances range from 2.65 to 3.48 Å with average values Pb-6S = 2.98, Pb-7S = 3.07, Pb-8S = 3.03 Å.

1. Structure Reports, 18, 471; 26, 426; 31A, 164; 40A, 19.
2. Ibid., 18, 473.

ANTIMONY ARSENIC URANIUM
AsSbU

R. TROJKO and Z. DESPOTOVIC, 1977. J. Nucl. Mater., 67, 105-108.

Tetragonal, Cu_2Sb type (1), P4/nmm, a = 4.152, c = 8.463 Å, c/a = 2.037, Z = 2, D_x = 9.90. Cu radiation, R = 0.124 for 17 reflexions, diffractometer data. U in 2(c): 3/4,3/4,0.295; As in 2(c): 3/4,3/4,0.641; Sb in 2(a): 3/4,1/4,0 [origin at centre].

[Interatomic distances (Å)]

U	- 4 U	4.15	Sb	- 4 U	3.25
	- 5 As	2.93, 2.99		- 4 As	3.68
	- 4 Sb	3.25		- 4 Sb	2.94
As	- 5 U	2.93, 2.99			
	- 4 As	3.78			
	- 4 Sb	3.68			

1. Strukturbericht, 2, 742; 3, 33, 288.

ANTIMONY BERYLLIUM
$Be_{13}Sb$

A. HAASE and M. MARTINEZ-RIPOLL, 1977. Acta Cryst., B33, 555-557.

Cubic, NaZn$_{13}$ type (1), Fm3c, a = 10.046 Å, D$_m$ = 3.0, Z = 8. Mo radiation, R = 0.043 for 105 reflexions, diffractometer data.

Atomic positions

			x	y	z
Sb	in	8(a)	1/4	1/4	1/4
Be(1)	in	8(b)	0	0	0
Be(2)	in	96(i)	0	0.1781	0.1157

Interatomic distances (Å)

Be(1) - 12 Be 2.13   Sb - 24 Be 2.94

Be(2) - 10 Be 2.13 - 2.32
      - 2 Sb 2.94

1. Strukturbericht, 6, 8, 157; Structure Reports, 16, 139.

ANTIMONY BERYLLIUM SODIUM   ARSENIC BERYLLIUM SODIUM
BeNaSb   AsBeNa

C. TIBURTIUS and H.U. SCHUSTER, 1977. Z. Naturf., 32B, 1133-1138.

Hexagonal, Ni$_2$In type (1), P6$_3$/mmc, Z = 2. Mo radiation, diffractometer data.

	a(Å)	c(Å)	c/a	D$_m$	R	Reflexions
NaBeSb	4.144	9.320	2.25	3.71	0.096	28
NaBeAs	3.820	8.948	2.34	3.16	0.079	∼185

Atoms are placed Na in 2(a): 0,0,0; Be in 2(d): 1/3,2/3,3/4; As or Sb in 2(c): 1/3,2/3,1/4.

Interatomic distances (Å)

		NaBeAs	NaBeSb
Na - 8 Na		3.82, 4.47	4.14, 4.66
- 6 Be		3.14	3.34
- 6 As or Sb		3.14	3.34
Be - 6 Na		3.14	3.34
- 3 As or Sb		2.21	2.39
As or Sb - 6 Na		3.14	3.34
- 3 Be		2.21	2.39

1. Structure Reports, 9, 91.

ANTIMONY GOLD YTTRIUM
Au$_3$Sb$_4$Y$_3$

A.E. DWIGHT, 1977. Acta Cryst., B33, 1579-1581.

Cubic, I$\bar{4}$3d, a = 9.818 Å, D$_m$ = 9.35, Z = 4. Cu radiation, photographic data. Atoms are placed Y in 12(a): 0,1/4,3/8; Au in 12(b): 7/8,0,1/4; Sb in 16(c): x,x,x; x = 0.088. The lanthanides Nd to Lu (except for Pm, Eu, and Yb) form isotypic compounds but Sc does not.

Interatomic distances (Å)

Y - 4 Au	3.01	Au - 4 Y	3.01	Sb - 3 Y	3.35
- 4 Sb	3.35	- 4 Sb	2.77	- 3 Y	3.45
- 4 Sb	3.45			- 3 Au	2.77
				- 3 Sb	3.62

ANTIMONY LITHIUM
Li$_2$Sb

W. MÜLLER, 1977. Z. Naturf., 32B, 357-359.

Hexagonal, Mg$_2$Ga type (1), P6̄2c, a = 7.946, c = 6.527 Å, D$_m$ = 3.79, Z = 6. Mo radiation, R = 0.058 for 341 reflexions, diffractometer data.

Atomic positions

			x	y	z
Li(1)	in	6(g)	0.294	0	0
Li(2)	in	6(h)	0.622	0.010	1/4
Sb(1)	in	2(b)	0	0	1/4
Sb(2)	in	4(f)	1/3	2/3	0.0225

Interatomic distances (Å)

Sb(1) - 9 Li	2.85, 3.04	Li(1) - 8 Li	3.04 - 3.26
- 2 Sb	3.26	- 4 Sb	2.82, 2.85
Sb(2) - 9 Li	2.82 - 3.00	Li(2) - 8 Li	3.04 - 3.27
- 2 Sb	2.97, 3.56	- 5 Sb	2.94 - 3.04

1. Structure Reports, 35A, 62.

ANTIMONY LITHIUM ZINC
Li$_2$SbZn

G. SCHROEDER and H.-U. SCHUSTER, 1977. Z. anorg. Chem., 431, 217-220.

Cubic, [CuHg$_2$Ti type (1)], F4̄3m, a = 6.472 Å, Z = 4, D$_x$ = 4.93. R = 0.058 for 114 reflexions, diffractometer data. Sb in 4(a): 0,0,0; Zn/Li(1) (2/3:1/3) in 4(c): 1/4,1/4,1/4; Zn/Li(2) (1/3:2/3) in 4(b): 1/2,1/2,1/2; Li in 4(d): 3/4,3/4,3/4.

[Interatomic distances (Å)]

Sb	- 10 Zn/Li	2.80, 3.24	Zn/Li(2) - 6 Sb	3.24
	- 4 Li	2.80	- 4 Zn/Li	2.80
			- 4 Li	2.80
Zn/Li(1) -	4 Sb	2.80		
	- 4 Zn/Li	2.80	Li - 4 Sb	2.80
	- 6 Li	3.24	- 10 Zn/Li	2.80, 3.24

1. Structure Reports, 40A, 56; this volume, p. 52.

ANTIMONY MANGANESE ZINC
MnSbZn

V. JOHNSON and W. JEITSCHKO, 1977.   J. Solid State Chem., 22, 71-75.

Tetragonal, ZrSiSe type (1), P4/nmm, a = 4.17, c = 6.233 Å, Z = 2, D_X = 7.11.   Mo
radiation, R = 0.044 for $\bar{1}$41 reflexions, diffractometer data.   Mn atoms occupy the
tetrahedral sites and Zn the divalent cation site, with square pyramidal coordination by Sb.

Atomic positions

	x	y	z	Occupancy
Mn in 2(a)	1/4	3/4	0	0.889
Zn in 2(c)	1/4	1/4	0.2808	0.942
Sb in 2(c)	1/4	1/4	0.7160	1

Interatomic distances (Å)

Mn - 4 Sb	2.74	Zn - 5 Sb	2.71, 2.95	Sb - 4 Mn	2.74
- 4 Zn	2.72	- 4 Mn	2.72	- 5 Zn	2.71, 2.95
- 4 Mn	2.95				

1. Structure Reports, 29, 38.

ANTIMONY PALLADIUM
$Pd_{20}Sb_7$

W. WOPERSNOW and K. SCHUBERT, 1977.   J. Less-Common Metals, 51, 35-44.

Rhombohedral, R$\bar{3}$ [misprinted as R3], a = 11.734, c = 11.021 Å, Z = 3, D_X = 11.30.
Mo radiation, R = 0.098 for 659 reflexions, diffractometer data.   See also 1.

Atomic positions

		x	y	z
Pd(1) in 6(c)		0	0	0.3774
Pd(2) in 18(f)		0.1505	0.9639	0.1881
Pd(3) in 18(f)		0.2041	0.9517	0.9397
Pd(4) in 18(f)		0.2133	0.9712	0.4481
Sb(1) in 3(a)		0	0	0
Sb(2) in 18(f)		0.1918	0.9613	0.6848

[Interatomic distances (Å)]

Pd(1) - 10 Pd	2.70 - 3.31	Pd(4) - 9 Pd	2.80 - 3.31
- 3 Sb	2.60	- 4 Sb	2.62 - 3.08
Pd(2) - 8 Pd	2.82 - 3.01	Sb(1) - 12 Pd	2.80, 2.89
- 5 Sb	2.65 - 3.56		
		Sb(2) - 12 Pd	2.60 - 3.56
Pd(3) - 9 Pd	2.82 - 3.40		
- 4 Sb	2.64 - 2.82		

1. Structure Reports, 37A, 161, ref. 42.

ANTIMONY RHODIUM   ANTIMONY RUTHENIUM
RhSb   RuSb

K. ENDRESEN, S. FURUSETH, K. SELTE, A. KJEKSHUS, T. RAKKE and A.F. ANDRESEN, 1977.
Acta Chem. Scand., A31, 249-252.

Orthorhombic, MnP type (1), Pnma, a = 5.9718, 5.9608, b = 3.8621, 3.7023, c =
6.3242, 6.5797 Å, for RhSb, RuSb, respectively, Z = 4.   X-ray and neutron diffraction data.   All atoms in 4(c): x,1/4,z.   RhSb, Rh (0.0053,1/4,0.1942), Sb
(0.1949,1/4,0.5915); RuSb, Ru (0.0053,1/4,0.2037), Sb (0.1992,1/4,0.5808). Rh-Sb =
2.589-2.756, mean 2.677 Å; Ru-Sb = 2.614-2.737, mean 2.662 Å.

The NiAs → MnP structure transition is largely due to the formation of 4 T-T
bonds (T = Rh or Ru, Rh-Rh = 3.068, 3.125, Ru-Ru = 3.042, 3.258 Å) and also to the
fact that the structural parameters lead to T-Sb bonds that are approximately equal.

1. Strukturbericht, 3, 17, 264; Structure Reports, 27, 319; 30A, 98.

ANTIMONY SILVER   (DYSCRASITE)
$Ag_{3.15}Sb_{0.85}$

J.D. SCOTT, 1976.   Canad. Miner., 14, 139-142.

Orthorhombic, Pmm2, a = 3.008, b = 4.828, c = 5.214 Å, D_m = 9.712, Z = 1.   Ag radiation, R = 0.042 for 246 reflexions, photographic data.   In agreement with 1, but
contrary to 2.

Atomic positions

			x	y	z
Sb*	in	1(a)	0	0	-0.02488
Ag(1)	in	1(c)	1/2	0	1/2
Ag(2)	in	1(b)	0	1/2	0.64808
Ag(3)	in	1(d)	1/2	1/2	0.16016

* Occupancy = Sb 0.849, Ag 0.151

The structure is essentially a hexagonal close-packing, with all atoms 12-
coordinated; Ag-Sb = 2.90-3.12 Å and Ag-Ag = 2.95-3.01 Å.   It is suggested that
allargentum ($Ag_{1-x}Sb_x$) is a disordered version of the dyscrasite structure.

1. Strukturbericht, 2, 744; Structure Reports, 8, 29; 24, 49.
2. Strukturbericht, 1, 596; Structure Reports, 23, 131.

ANTIMONY THALLIUM
Sb_2Tl_7

R. STOKHUYZEN, C. CHIEH and W.B. PEARSON, 1977.   Canad. J. Chem., 55, 1120-1122.

Cubic, Im3m, a = 11.618 Å, D_m = 10.67, Z = 6.   Mo radiation, R = 0.097 for 106
reflexions, diffractometer data.   See also 1 and 2.

Atomic positions

			x	y	z
Tl(1)	in	2(a)	0	0	0
Tl(2)	in	16(f)	0.1704	0.1704	0.1704
Sb	in	12(e)	0	0	0.3138
Tl(3)	in	24(h)	0.3497	0.3497	0

Interatomic distances (Å)

Tl(1) - 8 Tl 3.43	Tl(3) - 2 Sb 3.28
	- 6 Tl 3.39, 3.49
Tl(2) - 5*Tl 3.21 - 3.55	
- 3 Sb 3.26	Sb - 8 Tl 3.26, 3.28

* [Misprinted as 7]

The structure is a true supercell of the CsCl type structure with 54 atoms in a block of 27 CsCl cells. The mean Sb-Tl distance is 3.27 Å and the mean Tl-Tl distance 3.43 Å; near-neighbour coordination remains 8-fold.

1. Strukturbericht, 1, 600.
2. Ibid., 3, 362.

ANTIMONY TITANIUM   ANTIMONY VANADIUM
Sb_3Ti_5 Sb_2V_3

I. J. STEINMETZ, B. MALAMAN and B. ROQUES, 1977. C.R. Acad. Sci. Paris; C, 284, 499-502.

Sb_3Ti_5, orthorhombic, β-Yb_5Sb_3 type (1), Pnma, a = 10.22, b = 8.354, c = 7.181 Å, Z = 4, D_X = 6.55. R = 0.08. See also 2. Cu radiation, diffractometer data.

Sb_2V_3, rhombohedral, Fe_3Sn_2 type (3), R$\bar{3}$m, a = 5.551, c = 20.35 Å, Z = 6, D_X = 7.27. R = 0.10. See also 4. Cu radiation, diffractometer data.

II. R. BERGER, 1977. Acta Chem. Scand., A31, 889-890.

Sb_3Ti_5, orthorhombic, β-Yb_5Sb_3 type (1), Pnma, a = 10.2173, b = 8.3281, c = 7.1459 Å, Z = 4, D_X = 6.50. R = 0.087 on F^2 for 1205 reflexions, diffractometer data. See also 2.

Atomic positions

Sb_3Ti_5 (II), Ti(1) and Sb(1) in 8(d), remainder in 4(c)

	x	y	z
Ti(1)	0.05705	0.05823	0.20174
Ti(2)	0.23539	1/4	0.81717
Ti(3)	0.28158	1/4	0.33160
Ti(4)	0.00533	1/4	0.52152
Sb(1)	0.32279	0.49075	0.06612
Sb(2)	0.47625	1/4	0.58965

Sb_2V_3 (I)

			x	y	z
V	in	18(h)	0.481	-0.481	0.112
Sb(1)	in	6(c)	0	0	0.106
Sb(2)	in	6(c)	0	0	0.333

Interatomic distances

Sb_3Ti_5 (II)

Ti(1) - 5 Sb 2.71 - 2.94	Ti(3) - 5 Sb 2.71 - 2.93
- [7] Ti 2.84 - [3.43]	- [7] Ti 2.95 - [3.50]
Ti(2) - 6 Sb 2.73 - 2.95	Ti(4) - 5 Sb 2.79 - 2.81
- [5] Ti 2.99 - [3.50]	- 8 Ti 2.84 - 3.41

Sb(1) - [5] Sb 3.75 - [3.87]   Sb(2) - 4 Sb 3.75, 3.86
       - 9 Ti 2.79 - 2.94   - 8 Ti 2.71 - 2.95

Sb_2V_3 (I)
  Sb(1) - 9 V   2.78 - 2.95   V   - 6 V   2.46 - 3.09
        - 9 Sb 3.86 - 4.04   - 5 Sb 2.69 - 2.95

  Sb(2) - 6 V   2.69 - 2.89
        - 9 Sb 3.20 - 3.87

1. Structure Reports, 37A, 8; 41A, 76.
2. Ibid., 27, 44.
3. Ibid., 42A, 103.
4. Ibid., 30A, 166, 173, ref. 40.

ARSENIAN ULLMANNITE
$(As_{0.08}Bi_{0.01}Sb_{0.91})(As_{0.02}S_{0.98})(Co_{0.03}Ni_{0.97})$

P. BAYLISS, 1977. Amer. Min., 62, 369-373.

Triclinic, P1, a = b = c = 5.886 Å, α = β = γ = 90°, Z = 4, D_X = 5.78. Mo radiation, R = 0.034 for 1176 reflexions, diffractometer data; 12 site-sets with occupancies are given. The crystals are pseudo-cubic, contrary to 1-4. The reduction in symmetry from $P2_13$ arises from the ordering of As in one each of the Sb and S sites, confirming the optical observations of 5.

Interatomic distances (Å)

        Ni - 3 S   2.35 - 2.39   S - Sb   2.49, 2.50
           - 3 Sb 2.53 - 2.54

     The structure is essentially the distorted pyrite type described by 3 with the non-metal atoms coordinated to three metal atoms and one non-metal.

1. Strukturbericht, 1, 284.
2. Structure Reports, 11, 257.
3. Ibid., 21, 34.
4. G.B. BOKIJ and L.I. TSENOKEV, 1954. Trudy Inst. Krist. Akad. Nauk, SSSR, 9, 239 (C.A. Abstr., 48, 13553h).
5. D.D. KLEMM, 1962. Neues Jb. Miner., Abh., 97, 337.

ARSENIC BERYLLIUM LITHIUM
AsBeLi

C. TIBURTIUS and H.U. SCHUSTER, 1977. Z. Naturf., 32B, 116-117.

Tetragonal, ZrSiS type (1), P4/nmm, a = 3.749, c = 6.219 Å, c/a = 1.66, D_m = 3.41, Z = 2. R = 0.056 for 419 reflexions, diffractometer data. Occupancies given as Li 1.04, Be 0.98, and As 1.009.

Atomic positions [Origin at centre]

			x	y	z
Li	in	2(c)	-1/4	3/4	0.353
Be	in	2(a)	-1/4	1/4	0
As	in	2(c)	-1/4	3/4	0.7783

Interatomic distances (Å)

Li - 5 As 2.64, 2.77   As - 5 Li 2.64, 2.77
   - 4 Li 3.22   - 4 Be 2.33
   - 4 Be 2.89   - 1 As 3.82

Be - 4 As 2.33
   - 4 Be 2.65
   - 4 Li 2.89

1. Structure Reports, 27, 348; 29, 38, 342.

ARSENIC COBALT IRON SULPHUR   (ALLOCLASITE)
As(Co,Fe)S

J.D. SCOTT and W. NOWACKI, 1976. Canad. Miner., 14, 561-566.

Monoclinic, $P2_1$, a = 4.661, b = 5.602, c = 3.411 Å, β = 90.2°, D_m = 5.95, Z = 2.
Mo radiation, R = 0.029 for 600 reflexions, diffractometer data. See also 1. The
cobaltite of 2 is really alloclasite.

Atomic positions

	x	y	z
1.404 Co + 0.451 Fe	0.24824	0	0.24480
1.967 As + 0.033 S	0.04766	0.37149	0.25547
1.963 S + 0.037 As	0.44104	0.62236	0.24239

Interatomic distances (Å)

Co - 3 As 2.28, 2.31   S - 3 Co 2.30 - 2.37
   - 3 S   2.30 - 2.37   - 1 As 2.31

As - 3 Co 2.28, 2.31
   - 1 S   2.31

     Earlier work gave orthorhombic symmetry (3, 4) due to a structural extinction
for 00ℓ; details of all earlier work and full references are given. The structure
is closely related to that of costibite and this explains the observed alloclasite-
cobaltite transformation. The structure of cobaltite was also refined with use of
published data.

1. Structure Reports, 28, 50.
2. Ibid., 28, 344, 356, ref. 4.
3. Ibid., 30A, 17.
4. P.W. KINGSTON, 1971. Canad. Miner., 10, 838.

ARSENIC EUROPIUM
As_2Eu_2

I. Y. WANG, E.J. GABE, L.D. CALVERT and J.B. TAYLOR, 1977. Acta Cryst., B33,
   131-133.

Hexagonal, Na_2O_2 type (1), P6̄2m, a = 8.154, c = 6.137 Å, D_m = 6.41, Z = 3. Mo radiation, R = 0.061 for 887 reflexions, diffractometer data. See also 2.

Atomic positions

			x	y	z
Eu(1)	in	3(f)	0.3110	0	0
Eu(2)	in	3(g)	0.6461	0	1/2
As(1)	in	2(e)	0	0	0.2992
As(2)	in	4(h)	1/3	2/3	0.2090

The structure has trigonal-prismatic coordination of each Eu atom to six As atoms (Eu-As 3.09-3.18, average 3.14 Å, a distance which indicates divalent Eu) and each As atom is octahedrally coordinated to six Eu plus two As atoms, one at $\sim$ 3.6 and another at $\sim$ 2.5 Å. Thus the As atoms occur in pairs, with the shorter As-As distances (2.46 or 2.56 Å) comparable to that observed in metallic arsenic (2.52 Å). The Na_2O_2 structure, characterized by octahedra sharing faces, allows close approach of the central As atoms, and differs in this way from the rocksalt structure of other lanthanon pnictides, in which octahedra share edges.

As_3Eu

II.   J.F. BRICE and A. COURTOIS, 1976. C.R. Acad. Sci. Paris, C, __283__, 479-481.

Triclinic, a = 5.911, b = 5.626, c = 6.450 Å, α = 120.7, β = 92.3, γ = 104.6°, D_m = 7.08, Z = 2. [This is contrary to __2__, __3__, who give a = 9.500, b = 7.591, c = 5.789 Å, β = 112.62°, C*/* for $EuAs_3$. This monoclinic cell has been confirmed by __4__ for the isomorphous $SrAs_3$.]

__1__. Structure Reports, __21__, 233.
__2__. Ibid., __37A__, 149; __39A__, 21.
__3__. Ibid., __37A__, 164, ref. 11.
__4__. Ibid., __42A__, 30.

ARSENIC MAGNESIUM
As_4Mg

I.   R. GÉRARDIN, J. AUBRY, A. COURTOIS and J. PROTAS, 1977. Acta Cryst., B33, 2091-2094.
II.   R. GÉRARDIN, M. ZANNE, A. COURTOIS and J. AUBRY, 1976. C.R. Acad. Sci. Paris, C, __283__, 135-138.

Tetragonal, $P4_12_12$, a = 5.385, c = 15.798 Å, D_m = 4.62, Z = 4. Mo radiation, R = 0.069 for 532 reflexions, diffractometer data.

Atomic positions

			x	y	z
As(1)	in	8(b)	0.0994	-0.0233	0.3749
As(2)	in	8(b)	0.6754	0.0668	0.2029
Mg	in	4(a)	0.198	0.198	0

The Mg atoms are octahedrally coordinated by 6 As (2.71-2.92 Å), As(1) is tetrahedrally bonded to 2 Mg (2.71 Å) and 2 As (2.41, 2.43 Å) and As(2) is tetrahedrally joined to 1 Mg (2.92 Å) and 3 As (2.41-2.46 Å). The structure (Fig. 1) is similar to that of MgP_4 (__1__).

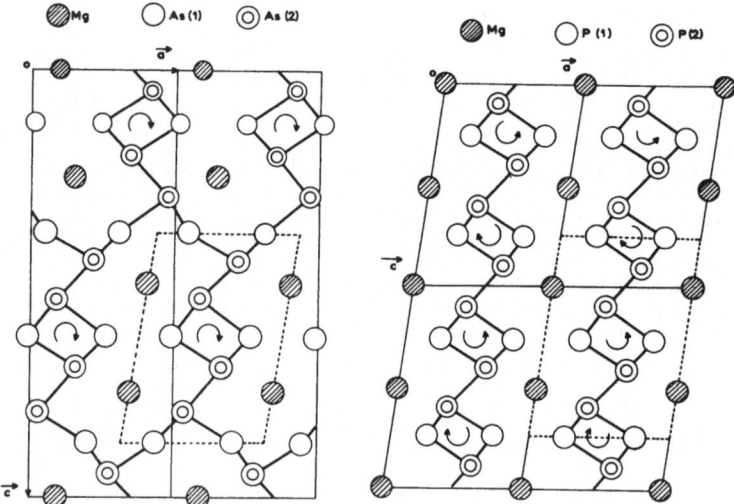

Fig. 1.   The structures of MgAs$_4$ (left) and MgP$_4$ (right) projected onto (010).
           The common unit is outlined.

<u>1</u>.   Structure Reports, <u>40</u>A, 86; <u>41</u>A, 92.

ARSENIC MERCURY SULPHUR THALLIUM   (CHRISTITE)
AsHgS$_3$Tl (synthetic)

I.   K.L. BROWN and F.W. DICKSON, 1976. Z. Kristallogr., <u>144</u>, 367-376.
II.   A.S. RADTKE, F.W. DICKSON, J.F. SLACK and K.L. BROWN, 1977. Amer. Min., <u>62</u>,
      421-425.

Monoclinic, P2$_1$/n, a = 6.113, b = 16.188, c = 6.111 Å, β = 96.71°, D$_m$ = 6.2, Z = 4.
Mo radiation, R = 0.044 for 702 reflexions, diffractometer data.

Atomic positions
    All atoms in 4(e)

	x	y	z
Tl	0.1633	0.45039	0.3366
Hg	0.0793	0.25013	0.9207
As	0.5759	0.3916	0.9254
S(1)	0.2046	0.3917	0.8587
S(2)	0.1419	0.2408	0.3559
S(3)	0.1421	0.1077	0.7962

     The structure contains trigonal AsS$_3$ pyramids (As-S = 2.23-2.26 Å) joined
together by HgS$_4$ tetrahedra (Hg-S = 2.46-2.66 Å) to form sheets parallel to (010).
Tl atoms are situated between these sheets asymmetrically (Tl-7 S = 3.11-3.52 Å);
Tl to 5 S in one sheet are less than 3.45 Å while the remaining two to the other
sheet are 3.51 and 3.52 Å.

ARSENIC MOLYBDENUM NICKEL
As$_3$Mo$_2$Ni$_{0.83}$

R. GUÉRIN, M. POTEL and M. SERGENT, 1975.   Rev. Chim. Minér., 12, 335-346.

Same data are given in 1. The metal atom chains in $Co_2Mo_2S_4$ (Mo-Mo = 2.76-2.96 Å),
Mo_2As_3 (Mo-Mo ∿ 2.95 Å), WP (W-W = 2.86 Å), MoAs (Mo-Mo = 2.93 Å), $NiMo_2As_3$ (Mo-Mo =
3.08, 2.98 Å), and related structures are illustrated and discussed.

1. Structure Reports, 40A, 21.

ARSENIC SODIUM SULPHUR
$AsNa_3S_3$

H. SOMMER and R. HOPPE, 1977. Z. anorg. Chem., 430, 199-210.

Cubic, $P2_13$, a = 8.573 Å, D_m = 2.47, Z = 4. Mo radiation, R = 0.069 for 428
reflexions, diffractometer data. See also 1.

Atomic positions
    S in 12(b), the remainder in 4(a)

	x	y	z
Na(1)	0.5785	0.5785	0.5785
Na(2)	0.3181	0.3181	0.3181
Na(3)	0.8104	0.8104	0.8104
As	0.0281	0.0281	0.0281
S	0.2220	0.6040	0.4981

Interatomic distances (Å)

Na(1) - 8 Na	3.44 - 4.16	
- 3 As	3.52	
- 6 S	3.06, 3.14	
Na(2) - 7 Na	3.37 - 4.07	
- 6 S	2.84, 3.01	
Na(3) - 7 Na	3.37, 4.16	
- 1 As	3.23	
- 6 S	2.85, 3.02	

As - 4 Na	3.23, 3.52
- 3 S	2.25
S - 6 Na	2.84, 3.02
- 1 As	2.25
- 2 S	3.50

The structure contains trigonal pyramidal AsS_3 groups (S-As-S angle 102°); Na
atoms are six-coordinated by S atoms forming irregular octahedra. The structure
can be considered a stuffed variant of the $NaClO_3$ type (2).

1. Structure Reports, 42A, 30.
2. Ibid., 21, 356.

ARSENIC STRONTIUM
As_4Sr_3

K. DELLER and B. EISENMANN, 1977. Z. Naturf., 32B, 1368-1370.

Orthorhombic, Fdd2, [Eu_4As_3 type (1)], a = 14.84, b = 17.89, c = 5.97 Å, Z = 8, D_X =
4.72. Mo radiation, R = 0.057 for 802 reflexions, diffractometer data.

Atomic positions

			x	y	z
Sr(1)	in	8(a)	0	0	1/4
Sr(2)	in	16(h)	0.2520	0.0825	0.9723
As(1)	in	16(h)	0.3361	0.9205	0.9822
As(2)	in	16(h)	0.5858	0.9971	0.2551

Interatomic distances (Å)

As(1)	- 1 As	2.48	
	- 6 Sr	3.14 - 3.31	

Sr(1)	- 8 As	3.14 - 3.31
	- 8 Sr	4.11 - 4.35

As(2)	- 2 As	2.48, 2.55
	- 6 Sr	3.18 - 3.30

Sr(2)	- 8 As	3.16 - 3.31
	- 8 Sr	4.11 - 4.35

The structure is characterized by As_4 chains (As-As-As 116°). The As atoms centre trigonal prisms of Sr atoms. The structure can be formulated as a Zintl-phase $Sr_3^{6+}As_4^{6-}$.

1. M.L. SMART and L.D. CALVERT, 1972. A.C.A. Meeting, Albuquerque. Abstracts, p. 61.

ARSENIC SULPHUR THALLIUM (IMHOFITE)
$As_{15}S_{25.3}Tl_{5.6}$

V. DIVJAKOVIĆ and W. NOWACKI, 1976. Z. Kristallogr., 144, 323-333.

Monoclinic, $P2_1/n$, a = 8.755, b = 24.425, c = 5.739 Å, β = 108.28°, Z = 1, D_x = 4.39. Cu radiation, R = 0.106 for 1376 reflexions, diffractometer data. 15 site-sets with 7 sites partly occupied. See also 1.

Interatomic distances (Å)

			Average				Average
Tl(1)	- 8 S	3.23 - 3.65	3.36	As(2)	- 3 S	2.25, 2.28	2.27
	- 1 As	3.49					
				As(3)	- 3 S	2.37 - 2.52	2.45
Tl(2)	- 7 S	3.08 - 3.78	3.37				
				As(4)	- 3 S	2.16, 2.27	2.23
Tl(3)	- 8 S	3.21 - 3.85	3.53				
				As(5)	- 3 S	2.18 - 2.29	2.25
As(1)	- 3 S	2.20, 2.30	2.27				

Average As - S = 2.29 Å

Tl(1) is coordinated by eight S atoms outlining a deformed bicapped trigonal prism, Tl(2) by seven S atoms making a strongly deformed octahedron with a split corner. The As atoms show a trigonal-pyramidal coordination to the S atoms. The AsS_3 pyramids form isolated $As_{15}S_{25}$ groups. Layers (A) and (B) perpendicular to the b axis differ from one another by their structure and the occupation of the atomic sites. The $Tl(1)S_6$ prisms are in the fully occupied zones (B) and form infinite chains parallel to [101].

1. Structure Reports, 30A, 459.

ARSENIC VANADIUM
α-As$_3$V$_5$

I. R. BERGER, 1977. Acta Chem. Scand., A31, 223-226.

Tetragonal, W$_5$Si$_3$ type (1), I4/mcm, a = 9.5031, c = 4.8255 Å, Z = 4, D$_x$ = 14.07. Cu radiation, R = 0.027, diffractometer data. The crystal used had composition As$_{2.74}$V$_5$. See also 2 and 3.

Atomic positions

			x	y	z
V(1)	in	16(k)	0.07601	0.22405	0
V(2)	in	4(b)	0	1/2	1/4
As(1)	in	8(h)	0.16458	0.66458	0
*As(2)	in	4(a)	0	0	1/4

* occupancy 2.97 As atoms by refinement (2.74 by microprobe)

Interatomic distances (Å)

V(1)	-	9 V	2.69 - 3.18	As(1)	-	10 V	2.52 - 2.77
	-	6 As	2.52 - 2.77		-	2 As	3.33
V(2)	-	10 V	2.41, 2.98	As(2)	-	8 V	2.55
	-	4 As	2.52		-	2 As	2.41

The V(2) and As(2) atoms form straight chains parallel to c, both with distances of 2.41 Å.

As$_2$V$_3$

II. R. BERGER, 1977. Acta Chem. Scand., A31, 287-291.

Tetragonal, P4/m, a = 9.4128, c = 3.3361 Å, Z = 4, D$_x$ = 14.78. Cu radiation, R = 0.023 for 378 reflexions, diffractometer data. See also 2, 3, and 4.

Atomic positions

			x	y	z
V(1)	in	4(k)	0.13335	0.17927	1/2
V(2)	in	4(j)	0.39694	0.28477	0
V(3)	in	2(f)	0	1/2	1/2
V(4)	in	1(d)	1/2	1/2	1/2
V(5)	in	1(a)	0	0	0
As(1)	in	4(k)	0.24604	0.41181	1/2
As(2)	in	4(j)	0.28481	0.03946	0

Interatomic distances (Å)

V(1)	-	5 As	2.43 - 2.56	V(5)	-	4 As	2.71
	-	9 V	2.68 - 3.34		-	10 V	2.68, 3.34
V(2)	-	5 As	2.48 - 2.54	As(1)	-	4 As	3.34, 3.38
	-	10 V	2.80 - 3.34		-	7 V	2.43 - 2.53
V(3)	-	6 As	2.46, 2.65	As(2)	-	4 As	3.34, 3.38
	-	8 V	3.27 - 3.34		-	8 V	2.53 - 2.71
V(4)	-	4 As	2.53				
	-	10 V	2.80, 3.34				

The structure (Fig. 1) can be regarded as a 'filled' Ti_5Te_4 type and is characterized by body-centred cubic units and trigonal prisms.

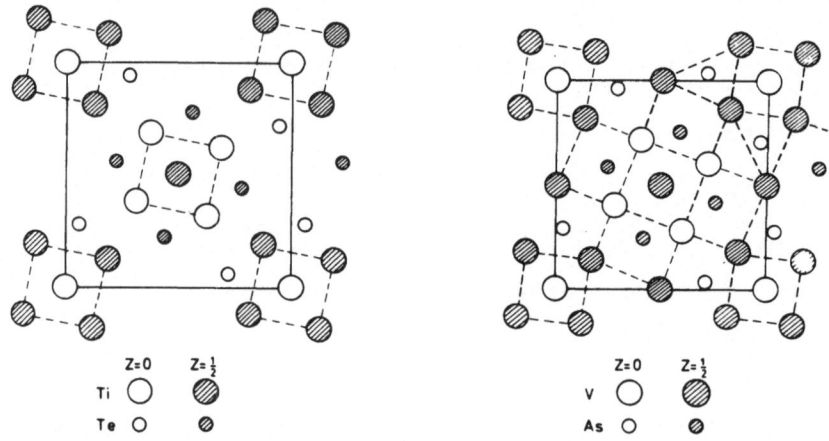

Fig. 1.   The structures of Ti_5Te_4 (left) and As_2V_3 (right) projected onto (001).

1. Structure Reports, 19, 277.
2. H. BOLLER and H. NOWOTNY, 1966. Mh. Chem., 97, 1053.
3. Idem, 1967. Ibid., 98, 2127.
4. Structure Reports, 40A, 26. Details in original text.

BARIUM GERMANIUM
$BaGe_2$ (H.P.)

J. EVERS, G. OEHLINGER and A. WEISS, 1977. Z. Naturf., 32B, 1352-1353.

Tetragonal, α-$ThSi_2$ type (1), $I4_1/amd$, a = 4.755, c = 14.73 Å, D_m = 5.60, Z = 4.
Mo radiation, R = 0.057 for 24 reflexions, photographic data. Ba in 4(a): 0,3/4,1/8;
Ge in 8(e): 0,1/4,0.415 (origin at centre).

[Interatomic distances (Å)]

Ba - 8 Ba   4.38, [4.76]   Ge - 6 Ba   3.40, [3.59]
    - 12 Ge   3.40 - [3.59]   - 3 Ge   2.50, 2.65

1. Structure Reports, 9, 121; 26, 257.

BARIUM IRON SULPHUR
$Ba_9(Fe_2S_4)_8$

J.T. HOGGINS and H. STEINFINK, 1977. Acta Cryst., B33, 673-678.

Tetragonal, P4/mnc, a = 7.7758, c = 44.409 Å, Z = 2, D_x = 3.90. Mo radiation,
R = 0.12 for 679 reflexions, diffractometer data; 14 site-sets. See also 1.

The structure (Fig. 1) consists of FeS_4 tetrahedra (Fe-S 2.21 to 2.31 Å) sharing edges to form chains parallel to c, with the Ba atoms occupying the channels between these chains. The Ba atoms are 8- or 12-coordinated by S atoms which form square antiprisms or face-capped tetragonal prisms (Ba-S 3.07 to 4.13 Å).

This structure belongs to the 'infinitely adaptive' series $Ba_p(Fe_2S_4)_q$ (p and q are integers) in which Ba atoms and Fe_2S_4 groups each define incommensurate subcells. The true supercell space group is the mathematical intersection of the subcell space groups with the proper number of repeat translations; similar supercells occur in $Mn_{15}Si_{26}$ and $Ge_{22}Rh_{17}$ (1).

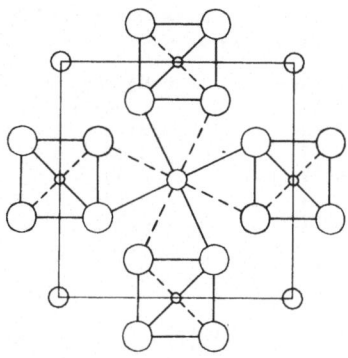

Fig. 1.   An idealized (001) projection of $Ba_9(Fe_2S_4)_8$ with S atom positions averaged.

1. Structure Reports, 32A, 83, 99.

BARIUM LEAD
Ba_3Pb_5 (R.T.)

G. BRUZZONE and E. FRANCESCHI, 1977. J. Less-Common Metals, 52, 211-216.

Orthorhombic, Pu_3Pd_5 type (1), Cmcm, a = 11.148, b = 9.049, c = 11.368 Å, Z = 4, D_x = 8.39. Photographic data. See also 2.

Atomic positions

			x	y	z
Ba(1)	in	4(c)	0	0.615	1/4
Ba(2)	in	8(e)	0.202	0	0
Pb(1)	in	4(c)	0	0.025	1/4
Pb(2)	in	8(f)	0	0.315	0.451
Pb(3)	in	8(g)	0.215	0.295	1/4

Interatomic distances (Å)

Ba(1)	-	4 Ba	4.49		Pb(2)	-	6 Ba	3.46 - 3.76
	-	9 Pb	3.46 - 3.76			-	4 Pb	3.31 - 3.53
Ba(2)	-	5 Ba	4.49 - 4.65		Pb(3)	-	6 Ba	3.52 - 3.90
	-	10 Pb	3.52 - 3.90			-	4 Pb	3.32 - 3.80
Pb(1)	-	5 Ba	3.63, 3.71					
	-	6 Pb	3.42 - 3.80					

1. This volume, p. 90.
2. Structure Reports, 29, 29.

BARIUM LEAD MAGNESIUM   BARIUM TIN ZINC   BERYLLIUM CALCIUM GERMANIUM
AB_2X_2 (A = Ba, Ca; B = Be, Mg, Zn; X = Ge, Pb, Sn)

B. EISENMANN, N. MAY, W. MÜLLER and H. SCHÄFER, 1972.   Z. Naturf., 27B, 1155-1157.

Tetragonal, $CaBe_2Ge_2$ type, P4/nmm, Z = 2. Atoms are placed A in 2(c): 1/4,1/4,z;
B(1) in 2(a): 3/4,1/4,0; B(2) in 2(c): 1/4,1/4,z; X(1) in 2(b): 3/4,1/4,1/2; X(2)
in 2(c): 1/4,1/4,z.

	$CaBe_2Ge_2$	$BaMg_2Pb_2$	$BaZn_2Sn_2$
a(Å)	4.02	5.00	4.69
c	9.92	12.11	11.33
c/a	2.47	2.42	2.42
D_m	4.21 [D_x]	6.50	6.63
R	0.14	0.13	0.17
z(A)	0.249	0.253	0.248
z(B(2))	0.608	0.631	0.614
z(X(2))	0.868	0.867	0.866

Interatomic distances (Å)

		$CaBe_2Ge_2$	$BaMg_2Pb_2$	$BaZn_2Sn_2$
A	- 4 A	4.02	5.00	4.69
	- 9 B	3.18, 3.56	3.80 - 4.58	3.66 - 4.15
	- 9 X	3.07, 3.20	3.82, 3.89	3.56, 3.69
		3.78	4.67	4.33
B(1)	- 4 A	3.18	3.95	3.66
	- 4 X	2.40	2.97	2.79
	- 4 B(1)	-	3.54	3.32
B(2)	- 5 A	3.18, 3.56	3.80, 4.58	3.67, 4.15
	- 5 X	2.28, 2.58	2.86, 2.96	2.68, 2.86
X(1)	- 4 A	3.20	2.90	3.69
	- 4 B	2.28	2.96	2.68
	- 4 X	2.84	3.54	3.32
X(2)	- 5 A	3.07, 3.78	3.82, 4.67	3.56, 4.33
	- 5 B	2.40, 2.58	2.86, 2.97	2.79, 2.86

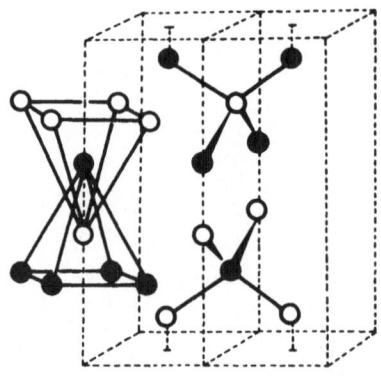

Fig. 1.   The $CaBe_2Ge_2$ type structure. Solid circles Ge atoms, open circles
Be and Ca atoms.

The structure (Fig. 1) is a variant of the $ThCr_2Si_2$ type (1) with a different distribution of atoms, and can be derived from the Al_4Ba type (2) (Fig. 2).

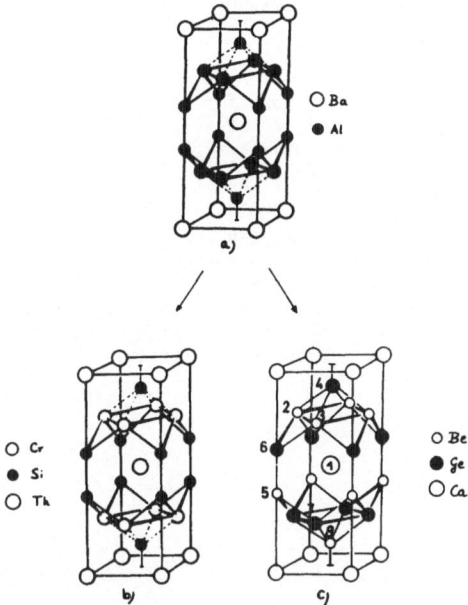

Fig. 2.   The relationship between the (a) Al_4Ba, (b) $ThCr_2Si_2$ and (c) $CaBe_2Ge_2$ types.

1. Structure Reports, 30A, 126; this volume p. 99.
2. Strukturbericht, 3, 45, 330.

BARIUM SILICON
$BaSi_2$ (H.P.)

J. EVERS, G. OEHLINGER and A. WEISS, 1977. Angew. Chem., Int. Ed. Engl., 16, 659-660.

Trigonal, $EuGe_2$ type (1), $P\bar{3}m1$, a = 4.047, c = 5.330 Å, D_m = 4.26, Z = 1. Mo radiation, R = 0.074 for 27 reflexions, photographic data. Ba placed in 1(a): 0,0,0; Si in 2(d): 1/3,2/3,0.432.

Interatomic distances (Å)

Ba -	6 Ba	4.05	Si -	6 Ba	3.28, 3.82
-	12 Si	3.28, 3.82	-	3 Si	2.45

Si-Si-Si = 112°

1. Structure Reports, 29, 50.

BERYLLIUM PHOSPHORUS
BeP_2

P. L'HARIDON, J. DAVID, J. LANG and E. PARTHÉ, 1976.   J. Solid State Chem., 19, 287-297.

Tetragonal, $I4_1/amd$, a = 3.546, c = 15.01 Å, Z = 4, for the substructure; pseudo-tetragonal, a = 7.09, c = 30.02 Å, Z = 32, for the intermediate structure; pseudo-tetragonal, a = 7.09, c = N x 15.01 Å, Z = N x 16, for the true structure, D_m = 2.49. See also 1. For the substructure, R = 0.033 for 215 reflexions, diffracto-meter data. Atoms are placed P in 4(a): 0,3/4,1/8; 4 Be + 4 P in 8(e): 0,1/4,z, z = 0.33453. Differs from 1.

   BeP_2 is one of the $M^{II}P_2$ phosphides, ZnP_2 and CdP_2, with tetrahedral structures, but differs in having an OD structure. In the short range each Be has 4 tetrahedral P neighbours (2.18 Å) while each P has 2 Be (2.18 Å) and 2 P (2.15 Å) neighbours, also tetrahedrally arranged. In the long range, although sites can be specified, their occupancy cannot and thus in the true structure there are two kinds of layer whose stacking is random but not completely arbitrary. The intermediate cell is related to that of 1.

1. Structure Reports, 41A, 123.

BISMUTH CERIUM SULPHUR
$Bi_{3.78}Ce_{1.25}S_8$

R. CÉOLIN, P. TOFFOLI, P. KHODADAD and N. RODIER, 1977.   Acta Cryst., B33, 2804-2806.

Orthorhombic, Pnma, a = 16.55, b = 4.053, c = 21.52 Å, Z = 4 for $(Ce(Ce_{0.25}\square_{0.75})-Bi_3(Bi_{0.78}\square_{0.22}))(S_8)$. Mo radiation, R = 0.047 for 1874 reflexions, diffractometer data. 14 site-sets given, with occupancies for Bi and Ce.

Interatomic distances (Å)

*M(1) - 7 S	2.81 - 3.00	Bi(1) - 6 S	2.55 - 3.61
*M(2) - 8 S	3.24 - 3.51	Bi(2) - 6 S	2.57 - 3.25
Ce - 8 S	2.91 - 3.26	Bi(3) - 6 S	2.67 - 3.02
*M(1) = 0.78 Bi, M(2) = 0.25 Ce			

   The Bi atoms are 6-coordinated by S atoms forming irregular octahedra.

BISMUTH COPPER LEAD SULPHUR
$Bi_{11}CuPb(S,Se)_{18}$ (PEKOITE)

I. W.G. MUMME and J.A. WATTS, 1976.   Canad. Miner., 14, 322-333.

Orthorhombic, $P2_1am$, a = 11.472, b = 11.248 x 3, c = 4.016 Å, Z = 2. Cu radiation, R = 0.121 for 425 reflexions, photographic data; 31 site-sets with occupancies for $Cu_{0.65} Pb_{0.78} Bi_{11.5} S_{14.93} Se_{3.07}$. $CuPbBi_{11}S_{18}$ had a = 11.322, b = 33.504, c = 3.987 Å, and has the ideal formula.

Interatomic distances (Å)

Cu	- 4 S	2.30 - 2.32	Bi(7)	- 7 S	2.55 - 3.37
Bi(1)	- 6 S	2.64 - 3.53	Bi(8)	- 6 S	2.63 - 3.51
Bi(2)	- 7 S	2.63 - 3.40	Bi(9)	- 7 S	2.70 - 3.44
Bi(3)	- 6 S	2.64 - 3.50	Bi(10)	- 6 S	2.74 - 3.28
Bi(4)	- 6 S	2.76 - 3.41	Bi(11)	- 7 S	2.68 - 3.50
Bi(5)	- 6 S	2.71 - 3.39	M*	- 7 S	2.86 - 3.20
Bi(6)	- 7 S	2.61 - 3.31	*M = 0.7 Pb, 0.3 Bi		

The structure of pekoite (Fig. 1) is composed of two bismuthinite ($\underline{1}$) slabs and one krupkaite ($\underline{2}$) slab. The classification of minerals in the bismuthinite (Bi_2S_3) - aikinite ($\overline{CuPbBiS_3}$) series has been discussed ($\underline{3}$). It is proposed that the ordered phases in this series can be represented as $Cu_xPb_xBi_{12-x}S_{18}$, where x = 0 for bismuthinite, x = 1 for pekoite, x = 2 for gladite ($\underline{4}$), and x = 3 for krupkaite, x = 4 for hammarite, and x = 6 for aikinite ($\underline{5}$). Crystal structure studies, detailed below, of synthetic samples prepared at 450-500°C show that Bi_2S_3-$CuPbBiS_3$ form a complete solid solution series, in which Pb occupies Bi sites and extra Cu is added for charge balance. The relationship between the ordered members is summarized in Fig. 1.

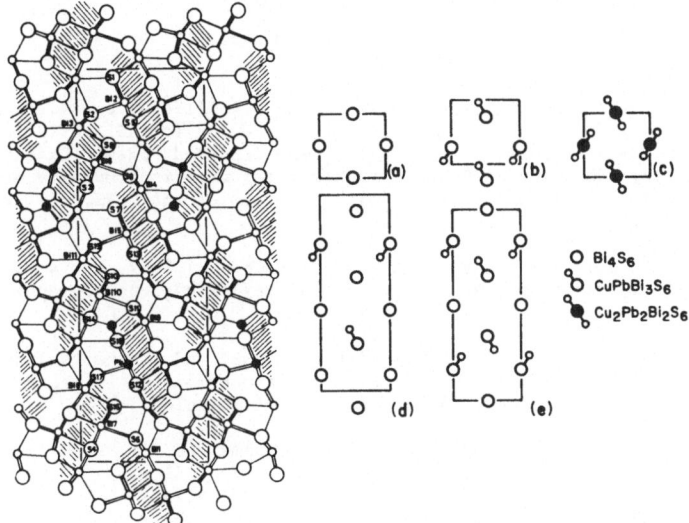

Fig. 1.   The pekoite structure (idealized as $CuPbBi_{11}S_{18}$) projected onto (001), and ribbon models for (a) Bi_2S_3 bismuthinite, (b) krupkaite, (c) aikinite, (d) pekoite, (e) gladite.

$Bi_{1.8}Cu_{0.2}Pb_{0.2}S_3$
$Bi_{1.5}Cu_{0.5}Pb_{0.5}S_3$ (KRUPKAITE)
$BiCuPbS_3$ (AIKINITE)

$Bi_{1.667}Cu_{0.33}Pb_{0.33}S_3$ (GLADITE)
$Bi_{1.25}Cu_{0.75}Pb_{0.75}S_3$

Orthorhombic, Bi_2S_3 type ($\underline{1}$, $\underline{5}$), Pnma, Z = 4.   Cu radiation, photographic data. These structural studies confirm a solid solution between bismuthinite (Bi_2S_3) and aikinite ($CuPbBiS_3$) above 450°C. Refined coordinates are given for each composition. For $BiCuPbS_3$, a = 11.603, b = 4.031, c = 11.301 Å, R = 0.115, with similar values for the others.

Atomic positions
    All atoms in 4(c): x,1/4,z

$BiCuPbS_3$	x	z
Bi	0.0175	0.6802
Pb	0.3342	0.4864
Cu	0.2347	0.2076
S(1)	0.0471	0.1389
S(2)	0.3784	0.0514
S(3)	0.2128	0.8054

Interatomic distances (Å)

	Cu - 4 S	Pb - 7 S	Bi - 7 S
$Bi_{1.8}Cu_{0.2}Pb_{0.2}S_3$	2.22 - 2.38	2.64 - 3.34	2.69 - 3.41
$Bi_{1.667}Cu_{0.33}Pb_{0.33}S_3$	2.23 - 2.42	2.69 - 3.35	2.68 - 3.47
$Bi_{1.5}Cu_{0.5}Pb_{0.5}S_3$	2.25 - 2.38	2.67 - 3.35	2.70 - 3.44
$Bi_{1.25}Cu_{0.75}Pb_{0.75}S_3$	2.31 - 2.43	2.83 - 3.28	2.67 - 3.52
$BiCuPbS_3$	2.31 - 2.43	2.85 - 3.27	2.68 - 3.53

$Bi_7Cu_3Pb_3S_{15}$ (LINDSTRÖMITE) $Bi_4Cu_2Pb_2S_9$ (HAMMARITE)

II.   H. HORIUCHI and B.J. WUENSCH, 1977. Canad. Miner., 15, 527-535.
III.   Idem, 1976. Ibid., 14, 536-539.

$Bi_7Cu_3Pb_3S_{15}$ (II). Orthorhombic, Pbnm, a = 56.115, b = 11.5695, c = 4.001 Å,
(compare 3, 6), Z = 4, D_X = 7.03. Cu radiation, R = 0.163 for 491 reflexions,
photographic data, 28 site-sets. Contrary to 7 who reported $Pb2_1m$. Exsolved
krupkaite masks n-glide absences. The 'lindstromite' of 4 is krupkaite accord-
ing to the approved nomenclature. The complex minerals of the $Cu_2S.Bi_2S_3.PbS$
system are summarized in 3, 8, 9.

$Bi_4Cu_2Pb_2S_9$ (III). Orthorhombic, Pbnm, a = 33.7726, b = 11.5857, c = 4.01 Å,
Z = 4, Dx = 7.05. Cu radiation, R = 0.169 for 181 reflexions, photographic data.
27 site-sets given.

    Both structures contain aikinite-like ($Cu_2Pb_2Bi_2S_6$) and krupkaite-like
($CuPbBi_3S_6$) ribbons, lindstromite in the ratio 1:4 and hammarite in the ratio 1:2.
Both structures have similar coordinations as summarized below:

	lindstromite	hammarite
Bi - 6 S	2.50 - 3.22	2.68 - 3.31
Bi - 7 S	2.57 - 3.37	2.57 - 3.29
Pb - 7 S	2.88 - 3.32	2.84 - 3.12
Cu - 4 S	2.14 - 2.56	2.22 - 2.44

    The crystal-chemical rules in these Bi-Pb-Cu-S minerals can be summarized:
(a) Pb atoms occupy inner sites in ribbons, Bi the outer ones; (b) Pb atoms are
distributed over the maximum number of ribbons; (c) Cu atoms occupy tetrahedral
interstices, coupled to substitution by Pb for a Bi atom adjacent to the Cu site.

1. Strukturbericht, 3, 51, 349; 6, 86; this volume, p. 33.
2. Structure Reports, 42A, 46.
3. D.C. HARRIS and T.T. CHEN, 1976. Canad. Miner., 14, 194.
4. Structure Reports, 42A, 47.
5. Ibid., 17, 451; 18, 470; 35A, 30; 37A, 29.
6. Ibid., 42A, 49, ref. 4.
7. Ibid., 42A, 49, ref. 2.
8. L.L. CHANG and S.H. HODA, 1977. Amer. Min., 62, 346.
9. W.G. MUMME, E. WELIN and B.J. WUENSCH, 1976. Amer. Min., 61, 15.

BISMUTH NICKEL SELENIUM
$Bi_2Ni_3Se_2$

I.   A. CLAUS, 1977. Naturwissenschaften, $\underline{64}$, 145.
II. A. CLAUS and K. WEBER, 1975. Neues Jb. Miner., Mh., 385-395.

Monoclinic, $Bi_2Ni_3S_2$ type ($\underline{1}$), C2/m, a = 11.224, b = 8.188, c = 8.100 Å, β = 133.24°, Z = 4, D_X = 9.21. Ag radiation, R = 0.058 for 400 reflexions, diffracto-meter data. In agreement with $\underline{1}$, but contrary to $\underline{2}$.

Atomic positions

			x	y	z
Bi(1)	in	4(i)	0.2509	0	0.2569
Bi(2)	in	4(i)	0.2177	0	0.7020
Se	in	8(j)	0.0266	0.2831	0.8099
Ni(1)	in	4(f)	1/4	1/4	1/2
Ni(2)	in	4(i)	0.0180	0	0.7904
Ni(3)	in	4(h)	0	0.2737	1/2

[Interatomic distances (Å)]

Bi(1)	- 1 Bi	3.87	Ni(1)	- 4 Bi	2.79, 2.85
	- 6 Se	3.40 - 3.61		- 2 Se	2.31
	- 6 Ni	2.75 - 2.85		- 4 Ni	2.81, 3.01
Bi(2)	- 3 Bi	3.56 - 4.64	Ni(2)	- 4 Bi	2.75 - 2.91
	- 4 Se	3.39, 3.43		- 2 Se	2.32
	- 6 Ni	2.78 - 2.91		- 4 Ni	3.01, 3.15
Se	- 5 Bi	3.39 - 3.61	Ni(3)	- 4 Bi	2.75, 2.86
	- 3 Se	3.52 - 3.71		- 2 Se	2.32
	- 3 Ni	2.31, 2.32		- 4 Ni	2.81, 3.15

The structure has octahedral coordination of Ni by 4 Bi + 2 Se, similar to that in $Ni_3Sn_2S_2$ ($\underline{3}$). This analysis clearly shows that the type structure $Ni_3Bi_2S_2$ (par-kerite) is monoclinic and not orthorhombic ($\underline{2}$).

$\underline{1}$. Structure Reports, 40A, 103.
$\underline{2}$. Ibid., $\underline{9}$, 191; $\underline{13}$, 278; $\underline{39A}$, 33.
$\underline{3}$. Ibid., $\underline{42A}$, 113.

BISMUTH SULPHUR   (BISMUTHINITE)
Bi_2S_3

V. KUPČIK and L. VESELÁ-NOVÁKOVÁ, 1970. Tschermaks Miner. Petrogr. Mitt., $\underline{14}$, 55-59.

Orthorhombic, Pbnm, a = 11.115, b = 11.25, c = 3.97 Å, Z = 4, D_X = 6.88. R = 0.084 for 450 reflexions, photographic data. Confirms $\underline{1}$.

Atomic positions (all atoms in 4(c))

	x	y	z
Bi(1)	0.3257	0.0166	1/4
Bi(2)	0.0341	0.1594	3/4
S(1)	0.3715	0.4517	3/4
S(2)	0.4432	0.1250	3/4
S(3)	0.1938	0.2178	1/4

Interatomic distances (Å)

Bi(1)	- 3 S	2.67, 2.70	Bi(2)	- 3 S	2.56, 2.74
	- 3 S	3.02, 3.05		- 2 S	2.96
	- 1 S	3.37		- 2 S	3.29

The structure resembles Sb_2S_3 but is more regular.

1. Strukturbericht, 3, 51, 349; 6, 86.

BORON (β-Rhombohedral)

B. CALLMER, 1977. Acta Cryst., B33, 1951-1954.

Rhombohedral, β-rhombohedral boron type (1, 2), R$\bar{3}$m, a = 10.9251, c = 23.8143 Å, A = 314.7, D_X = 2.30 (rhombohedral cell, a = 10.139 Å, α = 65.20°). Mo radiation, R = 0.053 for 719 reflexions, diffractometer data. 16 site-sets, with 0.734 occupancy for B(13) and 0.248 for B(16). An accurate refinement on a chemically pure crystal analysed by microprobe.

The structure was described in 1 and 2. Average B-B distances (Å) are compared below between the present refinement and that of 2 (Hoard et al., non-rounded coordinates obtained by private communication):

	Present	Hoard et al. (2)
Overall average	1.803	1.804
Central icosahedron	1.762	1.768
Rhombohedral icosahedron	1.840	1.849
Equatorial icosahedron	1.808	1.805
B_{10} unit	1.827	1.822

1. Structure Reports, 35A, 127.
2. Ibid., 38A, 38.

BORON GERMANIUM MOLYBDENUM
$BGe_{0.3}Mo_{1.7}$

M.A. MARKO, L.S. SAAKJAN and Ju.B. KUZ'MA, 1976. Izv. Akad. Nauk SSSR, Neorg. Mater., 12, 1307-1309 [Inorg. Mater., 12, 1087-1089].

Orthorhombic, Mg_2Cu type (1), Fddd, a = 15.41, b = 7.86, c = 4.57 Å, D_m = 9.32, Z = 16. Cr radiation, photographic data.

Atomic positions (origin at centre)

	x	y	z
*Mo/Ge(1) in 16(e)	0.208	1/8	1/8
*Mo/Ge(2) in 16(f)	1/8	0.458	1/8
B in 16(e)	0.500	1/8	1/8

$*Mo_{0.85}Ge_{0.15}$

[Interatomic distances (Å)]

Mo/Ge(1)	- 11	Mo/Ge	2.56 - 2.93	Mo/Ge(2)	- 11	Mo/Ge	2.63 - 2.93
	- 4	B	2.36, 2.37		- 4	B	2.33
B	- 8	Mo/Ge	2.33 - 2.37				
	- 2	B	2.27				

<u>1</u>. Structure Reports, <u>8</u>, 64.

BORON LITHIUM NICKEL   BORON MAGNESIUM NICKEL
$B_2Li_{1.2}Ni_{2.5}$ $B_2MgNi_{2.5}$

W. JUNG, 1977. Z. Naturf., <u>32B</u>, 1371-1374.

Hexagonal, $P6_222$, a = 4.842, 4.887, c = 8.664, 8.789 Å, D_m = 5.02, 5.34, for Li and
Mg compounds, respectively, Z = 3. Mo radiation, diffractometer data, R = 0.092
and 0.057 for 125 and 195 reflexions.

Atomic positions

			x	y	z	Occupancy
$B_2Li_{1.2}Ni_{2.5}$						
Ni/Li(1)	in	6(f)	1/2	0	0.2092	0.84 + 0.16
Ni/Li(2)	in	3(d)	1/2	0	1/2	0.84 + 0.16
Li	in	3(a)	0	0	0	0.70
B	in	6(i)	0.394	0.788	0	1
$B_2MgNi_{2.5}$						
Ni/Mg(1)	in	6(f)	1/2	0	0.2084	0.83 + 0.098
Ni/Mg(2)	in	3(d)	1/2	0	1/2	0.85 + 0.043
Mg	in	3(a)	0	0	0	0.76
B	in	6(i)	0.393	0.786	0	1

Interatomic distances (Å)

				$B_2Li_{1.2}Ni_{2.5}$	$B_2MgNi_{2.5}$
Ni(1)	-	5	Ni	2.45 - 2.53	2.47 - 2.56
	- 2	[4]	Li (Mg)	2.65, [3.02]	2.68, [3.05]
	-	4	B	2.02	2.04
Li (Mg)	-	12	Ni	2.65 - 3.02	2.68 - 3.05
	-	2	Li (Mg)	2.89	2.93
	-	4	B	2.58	2.60
Ni(2)	-	6	Ni	2.45, 2.52	2.47, 2.56
	-	4	Li (Mg)	2.82	2.85
	-	4	B	2.24	2.26
B	-	6	Ni	2.02, 2.24	2.04, 2.26
	-	2	Li (Mg)	2.58	2.60
	-	1	B	1.77	1.81

There are $[BNi_6]$ trigonal prisms which share edges giving B_2Ni_{10} groups with
B_2 pairs. The structure is related to that of $CeCo_3B_2$ (<u>1</u>).

<u>1</u>. Ju.B. KUZ'MA and P.I. KRIPJAKEVIC, 1969. Akad. Nauk Ukr. RSR, A<u>31</u>, 939.

BORON LUTETIUM RUTHENIUM
B_4LuRu_4

D.C. JOHNSTON, 1977. Solid State Comm., <u>24</u>, 699-702.

Tetragonal, $I4_1/acd$, a = 7.419, c = 14.955 Å, c/a = 2.016, D_m = 10.03, Z = 8. Cu
radiation, R = 0.065, diffractometer data; 27 isostructural compounds are listed in
Table I.

Atomic positions

			x	y	z
Lu	in	8(b)	0	1/4	1/8
Ru	in	32(g)	0.112	0.100	15/16
B	in	32(g)	0.818	0.111	0.961

Interatomic distances (Å)

```
        Ru -  6 Ru  2.71 - 3.12        B - 5 Ru  2.21 - 2.24
           -  3 Lu  2.88 - 3.22          - 1 B   1.54
           -  5 B   2.21 - 2.24

        Lu - 12 Ru  2.88 - 3.22
```

Lu atoms form a face-centred cubic sublattice if c/a = 2; phases with c/a < 2 are superconducting with T_c up to 9.5 K; if c/a > 2 magnetic ordering occurs and superconductivity is less. Ru atoms form an array of Ru_4 tetrahedra linked by Ru-Ru bonds to other Ru_4 tetrahedra to form zigzag Ru chains parallel to <u>a</u> and <u>b</u>. B atoms form pairs located between Ru atoms of adjacent Ru_4 tetrahedra.

BORON MOLYBDENUM THORIUM   BORON THORIUM TUNGSTEN
B_4MoTh B_4ThW

P. ROGL and H. NOWOTNY, 1974. Mh. Chem., <u>105</u>, 1082-1098.

Orthorhombic, Cmmm, a = 7.481, 7.487, b = 9.658, 9.681, c = 3.771, 3.739 Å, for Mo, W compounds, respectively, Z = 4, D_x = 9.05, 11.25. Cu radiation, photographic data, R = 0.12 for 138 and 132 reflexions. Atoms are placed Th in 4(i): 0,y,0, y = 0.3021, 0.3020; Mo or W in 4(g): x,0,0, x = 0.1708, 0.1717; B(1) in 8(q): x,y,1/2, x = 0.234, y = 0.155; B(2) in 4(h): x,0,1/2, x = 0.379; B(3) in 4(j): 0,y,1/2, y = 0.093. Contrary to <u>1</u> who reported a monoclinic cell.

The structure is based on $[BM_6]$ trigonal prisms (Fig. 1) and closely resembles those of $YCrB_4$ and Y_2ReB_6 with planar pentagon-hexagon B nets whose nodes centre the triangular nets of M atoms. The B-B distances range from 1.80 to 1.85 Å; the B-Th distances are 2.76-2.94 Å, while the B-Mo(W) distances are 2.43-2.46 Å in the $[BM_6]$ prisms.

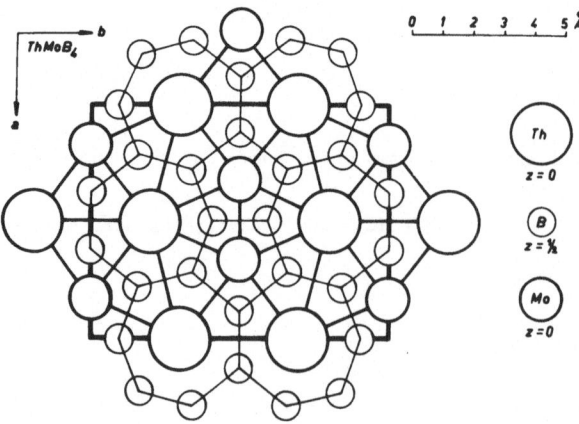

Fig. 1.   The $ThMoB_4$ structure.

<u>1</u>. D.T. PITMAN and D.K. DAS, 1960. J. Electrochem. Soc., <u>107</u>, 763.

BORON RUTHENIUM TUNGSTEN
$B_2Ru_{1.25}W_{1.75}$

P. ROGL and H. NOWOTNY, 1974. Rev. Chim. Minér., 11, 547-555.

Orthorhombic, Mo_2IrB_2 type (1), Pnnm, a = 9.579, b = 7.459, c = 3.072 Å, Z = 4, D_x = 14.21. Cr radiation, 96 reflexions, photographic data.

Atomic positions (all atoms in 4(g))

	x	y	z	Occupancies
W/Ru(1)	0.109	0.122	0	0.15 W + 0.85 Ru
W/Ru(2)	0.368	0.312	0	0.80 W + 0.20 Ru
W/Ru(3)	0.640	0.070	0	0.80 W + 0.20 Ru
B(1)	0.035	0.616	0	1
B(2)	0.228	0.622	0	1

[Interatomic distances (Å)]

W/Ru(1)	- 11 W/Ru	2.77 - 3.07		B(1)	- 7 W/Ru	2.30 - 2.39
	- 3 B	2.19, 2.39			- 2 B	1.85, 1.86
W/Ru(2)	- 11 W/Ru	2.78 - 3.17		B(2)	- 8 W/Ru	2.19 - 2.67
	- 5 B	2.28 - 2.67			- 1 B	1.85
W/Ru(3)	- 10 W/Ru	2.78 - 3.17				
	- 7 B	2.26 - 2.62				

1. Structure Reports, 38A, 49.

BORON SULPHUR
B_2S_3

H. DIERCKS and B. KREBS, 1977. Angew. Chem., Int. Eng. Edition, 16, 313.

Monoclinic, $P2_1/c$, a = 4.039, b = 10.722, c = 18.620 Å, β = 96.23°, D_m = 1.95, Z = 8. R = 0.057, diffractometer data. Contrary to Structure Reports, 38A, 160.

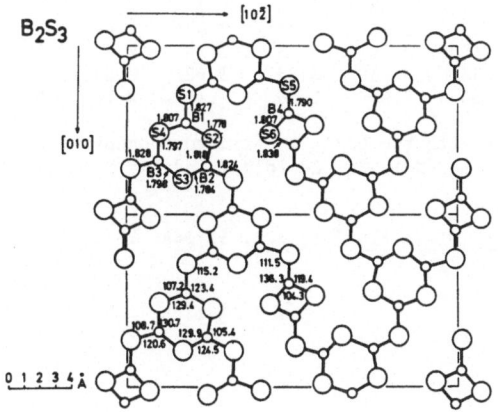

Fig. 1.   Two unit cells of the B_2S_3 structure, projected perpendicular to the plane of the layer.

Atomic positions* (all atoms in 4(e))

	x	y	z
B(1)	0.586	0.4639	0.1798
B(2)	0.535	0.7169	0.2406
B(3)	0.741	0.6858	0.0960
B(4)	0.045	0.3995	0.4902
S(1)	0.567	0.2929	0.1763
S(2)	0.491	0.5501	0.2565
S(3)	0.687	0.7950	0.1656
S(4)	0.694	0.5168	0.0951
S(5)	0.108	0.2339	0.4887
S(6)	0.083	0.5278	0.4291

*[Private communication from authors]

The structure (Fig. 1) is made up of planar B_3S_3 six-membered rings and B_2S_2 four-membered rings linked by S bridges to form almost-planar infinite two-dimensional high-polymer layers.

CaAl$_2$Si$_2$-TYPE COMPOUNDS
AB_2X_2 (A = Ca; B = Cd, Zn; X = As, P)

P. KLÜFERS and A. MEWIS, 1977. Z. Naturf., 32B, 753-756.

Trigonal, CaAl$_2$Si$_2$ type (1), P$\bar{3}$m1, Z = 1. Mo radiation, diffractometer data.
Atoms are placed A in 1(a): 0,0,0; B and X in 2(d): 1/3,2/3,z.

	CaZn$_2$P$_2$	CaCd$_2$P$_2$	CaZn$_2$As$_2$	CaCd$_2$As$_2$
a(Å)	4.038*	4.277	4.162	4.391
c	6.836	7.031	7.010	7.184
c/a	1.693	1.644	1.685	1.636
D$_X$	3.85	4.87	5.06	5.74
R	0.039	0.043	0.039	0.058
Refl.	177	235	158	165
z(B)	0.6318	0.6369	0.6303	0.6331
z(X)	0.2606	0.2405	0.2583	0.2382

* misprinted as 4.083

Interatomic distances (Å)

	A-6B	A-6X	B-3A	B-3B	B-4X	X-3A
CaZn$_2$P$_2$	3.43	2.93	3.43	2.95	2.44, 2.54	2.93
CaCd$_2$P$_2$	3.55	2.99	3.55	3.13	2.62, 2.79	2.99
CaZn$_2$As$_2$	3.53	3.01	3.53	3.02	2.53, 2.61	3.01
CaCd$_2$As$_2$	3.66	3.06	3.66	3.18	2.70, 2.84	3.06

The AB_2X_2 compounds with the CaAl$_2$Si$_2$ structure occur with A = Ca, Sr, Ce, Pr, Zr, Hf, B = Li, Al (d^0), Mn (d^5), Zn, Cd (d^{10}), and X = N, P, As, Sb, Bi, Si, and Ge.

1. Structure Reports, 12, 174; 17, 386; 32A, 5; 34A, 135.

CADMIUM GOLD
AuCd (at r.t.)

I. K.M. ALASAFI and K. SCHUBERT, 1977. J. Less-Common Metals, 55, 1-8.

Trigonal, P31m, a = 8.1047, c = 5.7974 Å, Z = 9, D$_X$ = 14.02. Mo radiation, R = 0.08 for 797 reflexions, diffractometer data. See also 1, 2; contrary to 3, 4, 5, and 6.

Atomic positions

			x	y	z
Au(1)	in	3(c)	0.312	0	0.460
Au(2)	in	3(c)	0.643	0	0.752
Au(3)	in	2(b)	1/3	2/3	0.042
Au(4)	in	1(a)	0	0	0.197
Cd(1)	in	3(c)	0.312	0	0.955
Cd(2)	in	3(c)	0.649	0	0.252
Cd(3)	in	2(b)	1/3	2/3	0.546
Cd(4)	in	1(a)	0	0	0.687

[Interatomic distances (Å)]

Au(1)	- 4 Au	2.95 - 3.21		Cd(1)	- 8 Au	2.84 - 2.97
	- 8 Cd	2.84 - 2.99			- 6 Cd	2.97 - 3.66
Au(2)	- 5 Au	3.11 - 3.21		Cd(2)	- 8 Au	2.86 - 2.99
	- 8 Cd	2.87 - 2.97			- 6 Cd	3.14 - 3.80
Au(3)	- 3 Au	3.11		Cd(3)	- 8 Au	2.84 - 2.92
	- 8 Cd	2.84 - 2.92			- 6 Cd	3.14, 3.66
Au(4)	- 3 Au	2.95		Cd(4)	- 8 Au	2.84 - 2.96
	- 8 Cd	2.84 - 2.96			- 6 Cd	2.97, 3.80

The structure (Fig. 1) is related to that of CsCl.

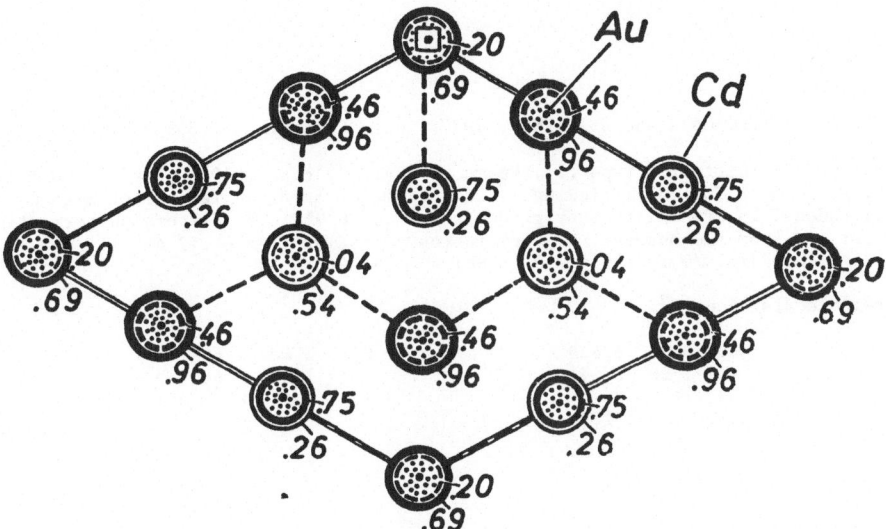

Fig. 1.   The (0001) projection of the AuCd structure.

$AuCd_3$

II. K.M. ALASAFI and K. SCHUBERT, 1977. J. Less-Common Metals, <u>51</u>, 225-233.

Hexagonal, $P6_3cm$, a = 8.147, c = 8.511 Å, Z = 6, D_X = 10.88. Mo radiation, R = 0.12 for 382 reflexions, diffractometer data. This model is an 'average' one; reflexions 0kℓ (ℓ odd) are observed diffuse, but they calculate zero. The model is basically the Cu_3P type (<u>7</u>).

Atomic positions

		x	y	z
Au	in 6(c)	0.332	0	0.75
Cd(1)	in 6(c)	0.292	0	0.078
Cd(2)	in 6(c)	0.363	0	0.420
Cd(3)	in 4(b)	1/3	2/3	0.206
Cd(4)	in 2(a)	0	0	0.331

[Interatomic distances (Å)]

Au	- 11 Cd 2.75 - 3.40		Cd(3)	- 3 Au 2.75
				- 9 Cd 3.10 - 3.56
Cd(1)	- 4 Au 2.81 - 3.40			
	- 8 Cd 2.97 - 3.21		Cd(4)	- 3 Au 2.79
				- 9 Cd 3.05 - 3.21
Cd(2)	- 4 Au 2.82 - 3.17			
	- 9 Cd 2.97 - 3.56			

1. Structure Reports, 21, 12.
2. H.M. LEDBETTER and C.M. WAYMAN, 1972. Acta Met., 20, 19.
3. Strukturbericht, 2, 11, 702.
4. Structure Reports, 11, 56; 15, 25; 23, 73.
5. R.S. TOTH and H. SATO, 1968. Acta Met., 16, 413.
6. W. WALLACE, W.D. HOFF and W.J. KITCHINGMAN, 1968. Acta Cryst., A24, 680.
7. Structure Reports, 38A, 86.

CADMIUM SELENIUM
CdSe

D.K. FREEMAN, S.L. MAIR and Z. BARNEA, 1977. Acta Cryst., A33, 355-359.

Hexagonal, ZnS (wurtzite) type (1), P6$_3$mc, a = 4.299, c = 7.010 Å, Z = 2. Mo radiation, R = 0.0136 for 143 reflexions, diffractometer data. A refinement with use of
two extended-face crystals; average deviation was 0.9%. The dispersion corrections
were refined from the observed Bijvoet ratios, yielding f' = 0.177 and f'' = 2.19 (Se).
Atoms in 2(b): 1/3,2/3,z, z(Cd) = 0, z(Se) = 0.37679.

Interatomic distances (Å)

Cd - 3 Se 2.628		Se - 3 Cd 2.628
- 1 Se 2.641		- 1 Cd 2.641

1. Strukturbericht, 1, 78, 128.

CAESIUM SULPHUR
Cs$_2$S

H. SOMMER and R. HOPPE, 1977. Z. anorg. Chem., 429, 118-130.

Orthorhombic, Pnma, a = 8.571, b = 5.383, c = 10.385 Å, D$_m$ = 4.19, Z = 4. Mo radiation, R = 0.104 for 202 reflexions, diffractometer data.

Atomic positions (all atoms in 4(c))

	x	y	z
Cs(1)	0.0256	1/4	0.1748
Cs(2)	0.1502	1/4	0.5743
S	0.256	1/4	-0.1121*

* [Misprinted as 0.1121]

Interatomic distances (Å)

$$
\begin{array}{ll}
Cs(1) - 10 \text{ Cs} & 4.01 - 4.56 \\
 - 5 \text{ S} & 3.57 - 3.96 \\
\\
Cs(2) - 8 \text{ Cs} & 4.01 - 4.28 \\
 - 4 \text{ S} & 3.38 - 3.41
\end{array}
\qquad
S - 9 \text{ Cs} \quad 3.38 - 3.96
$$

CALCIUM DEUTERIUM
CaD_2

A.F. ANDRESEN, A.J. MAELAND and D. SLOTFELDT-ELLINGSEN, 1977. J. Solid State Chem., 20, 93-101.

Orthorhombic, $PbCl_2$ (Co_2Si) type (1), Pnma, a = 5.925, b = 3.581, c = 6.776 Å, Z = 4, D_X = 1.94. R = 0.0371, neutron diffraction data. See also 2.

Atomic positions (atoms in 4(c))

	x	y	z
Ca	0.2378	1/4	0.1071
D(1)	0.3573	1/4	0.4269
D(2)	0.9737	1/4	0.6766

Interatomic distances (Å)

$$
\begin{array}{ll}
D(1) - 4 \text{ Ca} & 2.24 - 2.28 \\
 - 8 \text{ D} & 2.66 - 2.83 \\
\\
D(2) - 5 \text{ Ca} & 2.38 - 2.63 \\
 - 10 \text{ D} & 2.66 - 3.13
\end{array}
\qquad
Ca - 9 \text{ D} \quad 2.24 - 2.63
$$

1. Structure Reports, 19, 124; 30A, 75.
2. Strukturbericht, 3, 306; Structure Reports, 27, 118.

CALCIUM LUTETIUM SULPHUR
$CaLu_2S_4$

N. RODIER and V. TIEN, 1977. C.R. Acad. Sci. Paris, C, 284, 909-911.

Orthorhombic, Yb_3S_4 type (1), Pnma, a = 12.767, b = 3.829, c = 12.958 Å, Z = 4, D_X = 5.43. Mo radiation, R = 0.048 for 740 reflexions, diffractometer data.

Atomic positions (all atoms in 4(c): x,1/4,z)

	x	y	z
Ca	0.1315	1/4	0.0839
Lu(1)	0.39323	1/4	0.91678
Lu(2)	0.35325	1/4	0.29733
S(1)	0.2816	1/4	0.7376
S(2)	0.2426	1/4	0.4727
S(3)	0.0302	1/4	0.8801
S(4)	0.4651	1/4	0.1181

Interatomic distances (Å)

$$
Ca - 7 \text{ S} \quad 2.85 - 2.98
\qquad
\begin{array}{ll}
Lu(1) - 6 \text{ S} & 2.67 - 2.77 \\
Lu(2) - 6 \text{ S} & 2.65 - 2.73
\end{array}
$$

1. Structure Reports, 32A, 132.

CALCIUM SILICON   EUROPIUM SILICON   STRONTIUM SILICON
MSi_2 (M = Ca, Eu, Sr) (H.P.)

J. EVERS, G. OEHLINGER and A. WEISS, 1977. J. Solid State Chem., 20, 173-181.

Tetragonal, α-ThSi$_2$ type (1), I4$_1$/amd, a = 4.283, 4.304, 4.438, c = 13.52, 13.65,
13.83 Å, for Ca, Eu, Sr compounds, respectively, Z = 4, D$_x$ = 2.58, 5.47, 3.51. Cu
and Mo radiations, photographic data, R = 0.07, 0.08, 0.08. Atoms are placed M in
4(a): 0,0,0; Si in 8(e): 0,0,z, z = 0.4135, 0.414, 0.414. See also 2, 3, and 4.

Interatomic distances (Å)

		CaSi$_2$	EuSi$_2$	SrSi$_2$
M -	8 M	4.00, 4.28	4.03, 4.30	4.11, 4.44
-	12 Si	3.08, 3.25	3.11, 3.26	3.17, 3.36
Si -	6 M	3.08, 3.25	3.11, 3.26	3.17, 3.36
-	3 Si	2.34, 2.38	2.35, 2.40	2.38, 2.47

CaSi$_2$ (2), EuSi$_2$ (3), and SrSi$_2$ (4) have different structures at room-tempera-
ture, each with three-connected Si networks. After high-pressure and high-temperature treatment they have the α-ThSi$_2$ structure with planar three-connected SiSi$_3$
segments interconnected in a manner similar to that of a graphite network with one
corner of each hexagon missing. However the Si-Si bonds are not equal and angles
are not 120°. [In 5 the α-ThSi$_2$ structure is reported for SrSi$_{2-x}$.]

	CaSi$_2$		EuSi$_2$	SrSi$_2$	
	Normal pressure	High pressure	Normal and high pressure	Normal pressure	High pressure
- 1 Si	2.45	2.34	2.39	2.39	2.38 Å
- 2 Si	2.45	2.38	2.40	2.39	2.47
Si-Si-Si	104.2	128.2	127.3	117.8	128.1°
angles	104.2	115.9	116.3	117.8	115.9

1. Structure Reports, 9, 121; 26, 257.
2. Strukturbericht, 1, 175, 218; Structure Reports, 32A, 139; 33A, 51.
3. Structure Reports, 23, 219.
4. Ibid., 30A, 97 (I); 38A, 145.
5. Ibid., 30A, 97 (II).

CALCIUM TIN
Ca$_{31}$Sn$_{20}$

M.L. FORNASINI and E. FRANCESCHI, 1977. Acta Cryst., B33, 3476-3479.

Tetragonal, Pt$_{20}$Pu$_{31}$ type (1), I4/mcm, a = 12.542, c = 40.00 Å, D$_m$ = 3.74, Z = 4. Mo
radiation, R = 0.092 for 274 reflexions, photographic data. 15 site-sets are given.

Interatomic distances (Å)

Ca(1) -	10 Ca	3.77 - 4.25	Ca(4) -	9 Ca	3.60 - 4.52
-	4 Sn	3.07, 3.21	-	6 Sn	3.07 - 3.45
Ca(2) -	9 Ca	3.72 - 4.26	Ca(5) -	10 Ca	3.29 - 4.15
-	6 Sn	3.19 - 3.50	-	4 Sn	3.17, 3.38
Ca(3) -	11 Ca	3.76 - 4.52	Ca(6) -	9 Ca	3.55 - 4.15
-	6 Sn	3.25 - 3.53	-	6 Sn	3.26 - 4.02

Ca(7) - 10 Ca   3.29, 4.04   Sn(5) - 10 Ca   3.35 - 4.02
      - 4 Sn   3.18   - 1 Sn   3.15

Sn(1) - 8 Ca   3.45   Sn(6) - 8 Ca   3.25, 3.29
      [- 2 Sn   3.80]   - 2 Sn   3.07, 3.14

Sn(2) - 8 Ca   3.21 - 3.36   Sn(7) - 10 Ca   3.17 - 3.66
      - 1 Sn   3.05
                                    Sn(8) - 8 Ca   3.26
Sn(3) - 8 Ca   3.07 - 3.52   - 2 Sn   3.07
      - 1 Sn   3.15

Sn(4) - 8 Ca   3.19, 3.34
      - 1 Sn   3.14

The structure is discussed in terms of its resemblance to the W_5Si_3 (2) and Y_3Rh_2 (3) types.

1. This volume, p. 92.
2. Structure Reports, 19, 277.
3. Ibid., 42A, 121.

CARBON COBALT TUNGSTEN

x-$C_{3.4}Co_3W_{10}$

A. HÅRSTA, T. JOHANSSON, S. RUNDQVIST and J.O. THOMAS, 1977. Acta Chem. Scand., A31, 260-264.

Hexagonal, $P6_3/mmc$, a = 7.8304, c = 7.8361 Å, Z = 2, D_X = 16.413. R(I) = 0.078, neutron powder diffraction data. See also 1, 2, and 3.

Atomic positions

			x	y	z
W(1)	in	12(k)	0.2038	0.4076	0.0682
W(2)	in	6(h)	0.5497	0.0994	1/4
W(3)	in	2(a)	0	0	0
Co	in	6(h)	0.8868	0.7736	1/4
C(1)	in	6(g)	1/2	0	0
0.83 C(2)	in	2(c)	1/3	2/3	1/4

Interatomic distances (Å)

W(1) - 3 C   2.12, 2.26   Co   - 10 W   2.49 - 2.79
     - 3 Co   2.78, 2.79   - 2 Co   2.66
     - 6 W   2.82 - 2.96
                                   C(1) - 6 W   2.12, 2.07
W(2) - 2 C   2.07   (octahedral coordination)
     - 2 Co   2.59
     - 6 W   2.75, 2.92   C(2) - 6 W   2.26
                                       (trigonal prismatic)
W(3) - 6 Co   2.49
     - 6 W   2.82

There has been considerable discussion on this phase, first described by 1. The structure proposed by 2, on the basis of powder photographs and geometrical arguments, placed C in 6(g) and 2(c) and (Co + W) in both 6(h) sites; 3, on the basis of many experiments [the full text should be consulted], found that the metal atoms are ordered between the two (h) sites and that the 2(a) site is occupied by a metal atom.

The results here (Fig. 1) agree with this arrangement for the metal atoms. In addition the C-W distances seem reasonable; for the C in 6(g) (octahedral) the W-C distances agree closely with those in W_2C, also octahedral.

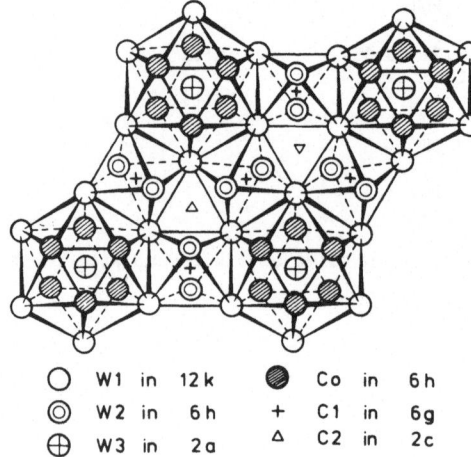

$\bigcirc$	W1	in	12k	$\bullet$	Co	in	6h
$\circledcirc$	W2	in	6h	+	C1	in	6g
$\oplus$	W3	in	2a	$\triangle$	C2	in	2c

Fig. 1.   The structure of $C_{3.4}Co_3W_{10}$ projected onto (0001).

1. Structure Reports, 16, 42.
2. Ibid., 18, 81.
3. Ibid., 39A, 37.

CARBON MOLYBDENUM
α-CMo_2

A.N. CHRISTENSEN, 1977. Acta Chem. Scand., A31, 509-511.

Orthorhombic, Pbcn, a = 4.732, b = 6.037, c = 5.204 Å, Z = 4, D_X = 9.33. R = 0.064 for 87 reflexions, neutron single-crystal data. See also 1. The true cell is in $P2_12_12_1$, with a' = a, b' = 2b, c' = 4c; 301 independent reflexions were obtained but were insufficient for refinement of the 145 parameters; the superstructure is probably due to a modulation of the carbon positions. Contrary to 2.

Atomic positions (subcell)

			x	y	z
	Mo	in 8(d)	0.249	0.130	0.083
0.20	C(1)	in 4(c)	0	0.355	1/4
0.82	C(2)	in 4(c)	0	0.866	1/4

The structure is as given in 1; Mo-6C = 2.00-2.22 Å.

1. Structure Reports, 28, 15.
2. Strukturbericht, 1, 575; 2, 240; Structure Reports, 16, 47; 24, 91; 26, 101.

CERIUM
α'-Ce
α''-Ce

W.H. ZACHARIASEN and F.H. ELLINGER, 1977. Acta Cryst., A33, 155-160.

α'-Ce, orthorhombic, α-U structure type ($\underline{1}$), Cmcm, a = 3.049, b = 5.998, c = 5.215 Å, A = 4, D_X = 9.76, measurements at 58 kbar and room temperature, powder photographs, Mo radiation. The atoms are in 4(c): 0,y,1/4, with y = 0.105 (from $\underline{1}$). See also $\underline{2}$.

Interatomic distances (Å)

Ce - 2 Ce	2.90	Ce-Ce-Ce	128 ± 5°
- 2 Ce	3.05		180
- 4 Ce	3.36		
- 4 Ce	3.49		

The zigzag chains are parallel to $\underline{c}$ and the straight chains parallel to $\underline{a}$.

α''-Ce, monoclinic, I2/m, a = 4.762, b = 3.170, c = 3.169 Å, β = 91.73°, A = 2, D_X = 9.73, measurements at 56 kbar and room temperature; powder photographs, Mo radiation. The atoms are in 2(a): 0,0,0 and 1/2,1/2,1/2.

Interatomic distances (Å)

Ce - 2 Ce	3.169	straight chains parallel to $\underline{c}$
- 2 Ce	3.170	straight chains parallel to $\underline{b}$
- 4 Ce	3.235	
- 4 Ce	3.305	

There has been considerable conflict ($\underline{3\text{-}8}$) on the high pressure phases of Ce and the text gives details of the various reports and compares them with detailed results. The table below summarizes the findings and shows the change in valence from γ-Ce to α''-Ce.

Phase	Pressure (kbar)	Volume (Å)	Radius (Å)	Valence
γ-Ce	0	35.89	1.851	3.00
γ-Ce	0	34.43	1.825	3.15
α-Ce	0	29.00	1.724	3.72
α-Ce	7.5-25	28.52	1.714	3.77
α-Ce	25-40	27.27	1.689	3.90
α-Ce	40-51	26.58	1.675	3.97
α'-Ce	58-95	26.29	1.669	4.00
α''-Ce	56-100	26.26	1.668	4.00

$\underline{1}$. Strukturbericht, $\underline{6}$, 1, 51; Structure Reports, $\underline{18}$, 306; $\underline{24}$, 240; $\underline{26}$, 271; $\underline{27}$, 379; $\underline{28}$, 41.
$\underline{2}$. Structure Reports, $\underline{40A}$, 104.
$\underline{3}$. P.W. BRIDGMAN, 1952. Proc. Amer. Acad. Sci., $\underline{81}$, 169.
$\underline{4}$. E. KING, J.A. LEE, I.R. HARRIS and T.F. SMITH, 1970. Phys. Rev., B, $\underline{1}$, 1380.
$\underline{5}$. D.B. McWHAN, 1970. Phys. Rev., B, $\underline{1}$, 2826.
$\underline{6}$. P.L. SCHAUFELBERGER and H. MERX, 1974. Proc. 4th Internat. Conf. High Pressure, Kyoto (1974), pp. 222-227.
$\underline{7}$. R.A. STAGER and H.G. DRICKAMER, 1964. Phys. Rev., A, $\underline{133}$, 830.
$\underline{8}$. J. WITTIG, 1968. Phys. Rev. Lett., $\underline{21}$, 1250.

CERIUM DYSPROSIUM SULPHUR
$CeDyS_3$

N. RODIER and V. TIEN, 1977. C.R. Acad. Sci. Paris, C, $\underline{285}$, 133-136.

Orthorhombic, Gd_2S_3 type (1), Pnma, a = 7.344, b = 3.948, c = 15.278 Å, Z = 4, D_x = 5.98. Mo radiation, R = 0.05 for 762 reflexions, diffractometer data.

Atomic positions (all atoms in 4(c))

	x	y	z
Ce	0.1424	1/4	0.20264
Dy	0.7677	1/4	0.54342
S(1)	0.0100	1/4	0.3932
S(2)	0.1495	1/4	0.7823
S(3)	0.3744	1/4	0.5649

Interatomic distances

Ce to 8 S atoms at 2.78 - 3.07 Å, average = 2.92 Å
Dy to 7 S atoms at 2.74 - 2.91 Å, average = 2.81 Å

1. Structure Reports, 33A, 60, 82; 34A, 98.

CERIUM RHENIUM SILICON
$CeRe_4Si_2$

O.I. BODAK, E.I. GLADYŠEVSKIJ and V.K. PEČARSKIJ, 1977. Kristallografija, 22, 178-181 [Soviet Physics - Crystallography, 22, 100-103].

Orthorhombic, Cmmm, a = 4.167, b = 14.001, c = 4.145 Å, D_m = 12.87, Z = 2. Cu radiation, R = 0.13 for 70 reflexions, photographic data.

Atomic positions

			x	y	z
Ce	in	2(a)	0	0	0
Re(1)	in	4(j)	0	0.1855	1/2
Re(2)	in	4(i)	0	0.3058	0
Si	in	4(j)	0	0.405	1/2

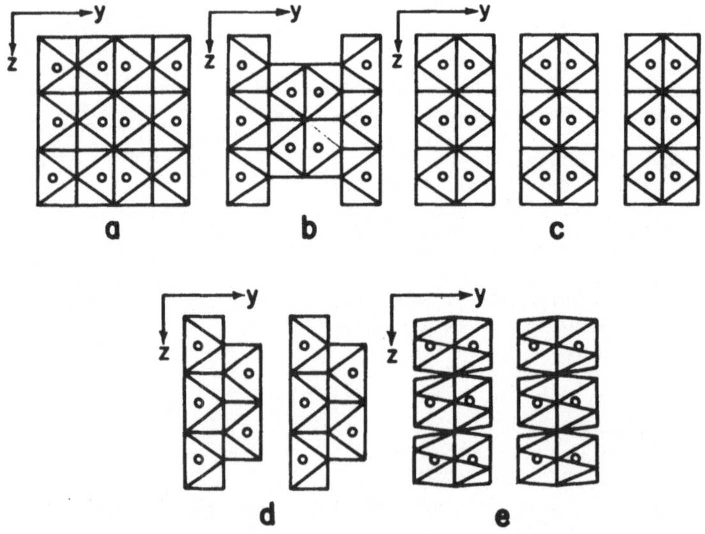

Fig. 1.   Projections of structure types based on tetragonal antiprisms, (a) $CeMg_2Si_2$, (b) $CeAl_2Ga_2$, (c) $CeRe_4Si_2$, (d) CeFeSi, (e) $PuGa_6$.

Interatomic distances (Å)

Ce	- 8 Si	3.23		Re(2)	- 2 Si	2.49,	[2.43]
	- 8 Re	3.32,	3.43		- 8 Re	2.60† - 2.94	
	- 4 Ce	4.14,	4.17		- 2 Ce	3.43	
Re(1)	- 2 Si	2.50,	[2.44]	Si	- 1 Si	2.66	
	- 8 Re	2.67* - 2.94			- 4 Re	2.43,	2.50
	- 2 Ce	3.32				[2.44,	2.49]
					- 4 Ce	3.23	

* [2.76 given as 2.68]
† [given as 2.68]

This structure can be built up from blocks of the CeAl$_2$Ga$_2$ [ThCr$_2$Si$_2$ (1)] and Cu structures. Similar tetragonal antiprisms occur in CeMg$_2$Si$_2$ (2), CeFeSi [PbFCl] (3), and PuGa$_6$ (4) (Fig. 1), and in NiScSi$_3$ (5).

1. Structure Reports, 30A, 129, ref. 152; this volume, p. 99.
2. Ibid., 37A, 64.
3. Ibid., 35A, 44.
4. Ibid., 30A, 51.
5. This volume, p. 83.

CHROMIUM LANTHANUM SULPHUR
CrLaS$_3$

K. KATO and I. KAWADA, 1977. Acta Cryst., B33, 3437-3443.

Triclinic, P1, a = 5.94, b = 17.2, c = 66.2 Å, α = 90.3, β = 95.3, γ = 90.0°, Z = 64 (average cell). Cu radiation, photographic data. The structure is of the OD type with alternate layers of CrS$_2$ with a CdI$_2$ type structure, and LaS layers with the NaCl type structure; in the average cell layers are not strictly commensurate. La-S = 2.92-3.04, Cr-S = 2.39, 2.40 Å.

CHROMIUM NIOBIUM SILICON
Cr$_2$Nb$_4$Si$_5$

J. STEINMETZ and B. ROQUES, 1977. J. Less-Common Metals, 52, 247-258.

Orthorhombic, Cr$_2$Nb$_4$Si$_5$ (Si$_5$V$_6$) type (1, 2), Ibam, a = 16.32, b = 7.800, c = 5.017 Å, Z = 4, D$_x$ = 6.41. Cu radiation, R = 0.05, photographic data.

Atomic positions

			x	y	z
Nb(1)	in	8(j)	0.1423	0.1136	0
Nb(2)	in	8(j)	0.4364	0.2455	0
Cr	in	8(f)	0.3097	0	1/4
Si(1)	in	8(j)	0.2908	0.2912	0
Si(2)	in	8(j)	0.0601	0.4136	0
Si(3)	in	4(a)	0	0	1/4

Interatomic distances (Å)

Nb(1)	- 6 Nb	3.02 - 3.53		Nb(2)	- 6 Nb	3.02 - 3.53
	- 4 Cr	3.13, 3.36			- 2 Cr	3.08
	- 7 Si	2.70 - 2.83			- 7 Si	2.37 - 2.80

Cr	- 6 Nb	3.08 - 3.36		Si(2)	- 5 Nb	2.37 - 2.80
	- 2 Cr	2.51			- 2 Cr	2.56
	- 6 Si	2.56 - 2.63			- 5 Si	2.38 - 3.18
Si(1)	- 5 Nb	2.40 - 2.83		Si(3)	- 8 Nb	2.57, 2.78
	- 4 Cr	2.61, 2.63			- 2 Si	2.51
	- 1 Si	2.91				

In ternary silicides, $(T,T')_6Si_5$, where T = V, Cr, or Mn and T' = Ti, Nb, or Ta, the smallest site (8(f)) is occupied exclusively by a T atom, forming chains (see 1, Fig. 17), and the larger of the 8(j) sites is occupied preferentially by T' atoms. The smaller 8(j) site is occupied by either T or T' atoms.

1. Structure Reports, 33A, 64.
2. Ibid., 38A, 101.

CHROMIUM NITROGEN VANADIUM
$Cr_{0.875}NV_{0.125}$ (at 5°K)

M.N. EDDINE, E.F. BERTAUT, M. ROUBIN and J. PÂRIS, 1977. Acta Cryst., B33, 3010-3013.

Orthorhombic, Pnmm, a = 2.8840, b = 2.9620, c = 4.1314 Å, Z = 2, D_X = 6.20. Diffractometer data. Atoms are placed: 2N in 2(b): 0.26,1/4,3/4; 1.75Cr + 0.25V in 2(a): 0.24,1/4,1/4, giving distances of Cr-6N = 2.07 Å and Cr/V-12Cr/V = 2.88-2.96 Å.

CHROMIUM PHOSPHORUS SULPHUR
$CrPS_4$

R. DIEHL and C.-D. CARPENTIER, 1977. Acta Cryst., B33, 1399-1404.

Monoclinic, C2, a = 10.871, b = 7.254, c = 6.140 Å, β = 91.88°, D_m = 2.88, Z = 4. Mo radiation, R = 0.019 for 810 reflexions, diffractometer data.

Atomic positions (Cr(1) and Cr(2) in 2(a), remainder in 4(c))

	x	y	z
Cr(1)	0	0	0
Cr(2)	0	0.5095	0
P	0.2971	0.2588	0.1655
S(1)	0.1335	-0.0084	0.7007
S(2)	0.1347	0.5248	0.6993
S(3)	0.1062	0.2594	0.1934
S(4)	0.1303	0.7575	0.1452

The structure is characterized by puckered hexagonally close-packed sulphur layers stacked parallel to (100) with Cr in octahedral and P in tetrahedral interstices; Cr-S = 2.38-2.49, mean 2.43, P-S = 2.02-2.09, mean 2.05 Å, S-P-S = 103-114°.

COBALT GERMANIUM SULPHUR
$Co_2Ge_3S_3$

R. KORENSTEIN, S. SOLED, A. WOLD and G. COLLIN, 1977. Inorg. Chem., 16, 2344-2346.

Rhombohedral, R3, a = 8.017 Å, α = 90°, D_m = 5.54, Z = 4 (hexagonal cell has a = 11.338, c = 13.886 Å). Mo radiation, R = 0.115 for 535 reflexions, diffractometer data, 12 site-sets. R = 0.043 for a $CoAs_3$ model and 97 subcell (Im3) reflexions. Co atoms are 6-coordinated, to 3 S [2.16-2.35 Å] and 3 Ge [2.27-2.34 Å]; the As_4 square rings of $CoAs_3$ become -Ge-S-Ge-S- rings [Ge-S 2.33-2.74 Å].

COBALT INDIUM
$CoIn_3$

M.V. KATRIČ, N.N. MATJUŠENKO and Ju.G. TITOV, 1977. Kristallografija, 22, 188-190 [Soviet Physics - Crystallography, 22, 107-108].

Tetragonal, $CoGa_3$ type (1), P$\bar{4}$n2, a = 6.829, c = 7.094 Å, c/a = 1.039, Z = 4, D_x = 8.01. Fe radiation, R = 0.054, diffractometer data. Differs from 2.

Atomic positions

			x	y	z
Co	in	4(f)	0.3500	0.1500	1/4
In(1)	in	4(e)	0	0	1/4
In(2)	in	8(i)	0.1542	0.3458	0

Interatomic distances (Å)

Co	-	1 Co	2.90	In(2)	-	3 Co	2.59, 2.73
	-	8 In	2.59 - 2.73		-	11 In	2.98 - 3.66
In(1)	-	2 Co	2.60				
	-	10 In	3.14, 3.55				

1. Structure Reports, 23, 117.
2. Ibid., 39A, 49.

COBALT IRON YTTRIUM
$(Co_{0.31}Fe_{0.69})_{19.2}Y_2$ (at 20°C)
$(Co_{0.72}Fe_{0.28})_{18.4}Y_2$ (at 20°C)

R.S. PERKINS and P. FISCHER, 1976. Solid State Comm., 20, 1013-1018.

Rhombohedral, Th_2Zn_{17} type (1), R$\bar{3}$m, a = 8.473, 8.389, c = 12.501, 12.348 Å, c/a = 1.4754, 1.4719, for the two compositions, D_x = 7.29, 7.73. R = 0.061 and 0.091 for 43 neutron powder reflexions. See also 2 and 3.

Interatomic distances (Å)

			$(Co_{0.31}Fe_{0.69})_{19.2}Y_2$	$(Co_{0.72}Fe_{0.28})_{18.4}Y_2$
Y	-	4 Y	4.17, 4.89	4.12, 4.84
	-	19 M	2.93 - 3.34	2.90 - 3.22
M(1)	-	1 Y	3.00	2.98
	-	13 M	2.28 - 2.86	2.41 - 2.87
M(2)	-	2 Y	3.19	3.22
	-	14 M	2.42 - 4.89	2.43 - 4.84
M(3)	-	2 Y	2.93	2.90
	-	11 M	2.47 - 3.22	2.44 - 3.18
M(4)	-	3 Y	3.08 - 3.34	3.10 - 3.19
	-	9 M	2.42 - 2.61	2.42 - 2.59

The structures are hyper-stoichiometric (2) and coordinates are consistent with 3; the details of occupancies were not conclusively fixed; Fe atoms occupy 6(c) sites preferentially and avoid the 18(f) site.

1. Structure Reports, 20, 195.
2. Ibid., 38A, 76, 113.
3. Ibid., 33A, 105; 34A, 75.

COBALT NIOBIUM PHOSPHORUS
$CoNb_4P$

Ja.F. PALFIJ and Ju. B. KUZ'MA, 1977. Dop. Akad. Nauk Ukr., No. 3, Ser. A, 262-265.

Tetragonal, Nb_4CoSi type (1), P4/mcc, a = 6.118, c = 4.996 Å, c/a = 0.817, Z = 2, D_X = 8.20. Cr radiation, photographic data.

Atomic positions

			x	y	z
Co	in	2(a)	0	0	1/4
P	in	2(c)	1/2	1/2	1/4
Nb	in	8(m)	0.162	0.662	0

[Interatomic distances (Å)]

Co -	2 Co	2.50		Nb -	2 Co	2.61
	- 8 Nb	2.61			- 2 P	2.61
					- 11 Nb	2.80 - 3.24
P -	2 P	2.50				
	- 8 Nb	2.61				

1. Structure Reports, 30A, 44.

COBALT SILICON URANIUM
$Co_3Si_5U_2$

L.G. AKSEL'RUD, Ja.P. JARMOLJUK and E.I. GLADYŠEVSKIJ, 1977. Kristallografija, 22, 861-863 [Soviet Physics - Crystallography, 22, 492-493].

Orthorhombic, Ibam, a = 9.59, b = 11.13, c = 5.617 Å, Z = 4, D_X = 8.79. Cu radiation, R = 0.122 for 75 hk0 and 0.142 for 55 hk1, photographic data.

Atomic positions

			x	y	z
U	in	8(j)	0.2668	0.1328	0
Co(1)	in	4(a)	0	0	1/4
Co(2)	in	8(j)	0.1150	0.3626	0
Si(1)	in	4(b)	1/2	0	1/4
Si(2)	in	8(g)	0	0.2253	1/4
Si(3)	in	8(j)	0.3467	0.3927	0

Interatomic distances (Å)

U -	4 U	3.85, 4.08	Co(1) -	4 U	3.27
	- 6[7] Co	2.94 - 3.27 [3.34]		- 2 Co	2.81
	- 10 Si	2.89 - 3.09		- 6 Si	2.51 - 2.36

```
Co(2) - 4 [5] U   2.94 - 3.21 [3.34]        Si(2) - 4 U   3.08, 3.09
      - 5   Si  2.25, 2.35                        - 3 Co  2.35, 2.51
                                                  - 5 Si  2.42 - 3.05
Si(1) -   4   U   3.03
      -   4   Co  2.35                       Si(3) - 4 U   2.89 - 3.02
      -   4   Si  2.81, 3.05                      - 3 Co  2.25, 2.36
                                                  - 2 Si  2.42
```

COPPER GALLIUM SELENIUM
$CuGaSe_2$

L. MANDEL, R.D. TOMLINSON and M.J. HAMPSHIRE, 1977. J. Appl. Cryst., 10, 130-131.

Tetragonal, $CuFeS_2$ type (1), I$\bar{4}$2d, a = 5.614, c = 11.022 Å, Z = 4, D_x = 5.57. Cu radiation, diffractometer data. See also 2-4. Atoms are placed Cu in 4(a): 0,0,0; Ga in 4(b): 0,0,1/2; Se in 8(d): 0.259,1/4,1/8, giving distances Cu - 4 Se   2.45, Ga - 4 Se 2.39 Å. The bond angles around Se are Cu-Se-Ga 110.4, Cu-Se-Cu 107.0, and Ga-Se-Cu 111.0°.

1. Strukturbericht, 1, 279, 290; 2, 48, 346; 3, 385; Structure Reports, 10, 122;
    23, 141; 39A, 51.
2. Structure Reports, 17, 19.
3. L.S. LERNER, 1966. J. Phys. Chem. Solids, 27, 1.
4. G.D. BOYD, H.M. KASPER, J.H. McFEE, F.G. STORZ, 1972. IEEE, J. Quantum Electron.,
    8, 900.

COPPER IRON SULPHUR TIN   (MAWSONITE)
$Cu_6Fe_2S_8Sn$

J.T. SZYMAŃSKI, 1976. Canad. Miner., 14, 529-535.

Tetragonal, P$\bar{4}$m2, a = 7.603, c = 5.358 Å, Z = 1, D_x = 4.65. Mo radiation, R = 0.037 for 1004 reflexions, diffractometer data. See also 1.

Atomic positions

			x	y	z
Sn	in	1(a)	0	0	0
Cu(1)	in	2(g)	1/2	0	0.0002
Cu(2)	in	4(i)	0.2463	0.2463	1/2
Fe(1)	in	1(b)	1/2	1/2	0
Fe(2)	in	1(c)	1/2	1/2	1/2
S(1)	in	4(j)	0.2615	0	0.2538
S(2)	in	4(k)	0.2597	1/2	0.2490

Interatomic distances (Å)

```
Sn    - 4 S   2.41                Fe(2) - 4 S   2.27
                                       - 2 Fe  2.68
Cu(1) - 4 S   2.27, 2.38               - 4 Cu  2.73

Cu(2) - 4 S   2.29, 2.35         S(1)  - 1 Sn  2.41
      - 1 Fe  2.73                     - 3 Cu  2.26, 2.29

Fe(1) - 4 S   2.26               S(2)  - 2 Fe  2.26
      - 2 Fe  2.68                     - 3 Cu  2.35, 2.38
```

The structure is based on nearly c.c.p. S atoms with metal atoms in half the tetrahedral holes and an extra Fe atom at 1/2,1/2,1/2 giving Fe chains (Fe-Fe = 2.68 Å) parallel to $\underline{c}$ and providing a possible mechanism for the weak magnetism observed via a one-dimensional -Fe-S-Fe- interaction. The deviations from c.c.p. packing are small and due to the Sn and Fe(1) atoms. The structure can be derived from two sphalerite-like cubes.

1. Structure Reports, 30A, 453.

COPPER MERCURY SULPHUR TIN   (VELIKITE)
$Cu_{3.75}Hg_{1.75}S_8Sn_2$

L.N. KAPLUNNIK, E.A. POBEDIMSKAJA and N.V. BELOV, 1977. Kristallografija, 22, 175-177 [Soviet Physics - Crystallography, 22, 99-100].

Tetragonal, I$\bar{4}$2m, a = 5.542, c = 10.908 Å, D_m = 5.59, Z = 1. Mo radiation, R = 0.077 for 202 reflexions, diffractometer data. Microprobe analysis $(Cu_{3.18}Hg_{1.53}Zn_{0.44})(Sn_{1.90}As_{0.08}Sb_{0.02})S_{9.94}$ [the structural results give only 8 S atoms]. See also 1. Sample from Khaidarkan.

Atomic positions

			x	y	z	occupancy
Hg	in	2(a)	0	0	0	0.875
Sn	in	2(b)	0	0	1/2	1*
Cu	in	4(d)	1/2	0	[1/4]†	0.94
S	in	8(i)	0.2435	0.2435	0.3533	0.99

* [given as 2.02 in 2(b)]   †[given as 0.2498(1)]

A variant of the stannite, Cu_2FeSnS_2, structure (2) with Hg occupying 7/8ths of the Fe positions and tetrahedral coordination of metal atoms by S (Hg-4S 2.57, Sn-4S 2.49, Cu-4S, 2.26 Å).

1. M. FLEISCHER, 1977. Amer. Min., 62, 1260.
2. Strukturbericht, 3, 96, 440.

COPPER MERCURY TITANIUM
$CuHg_2Ti$

M. PUŠELJ and Z. BAN, 1969. Croat. Chem. Acta, 41, 79-83.

Cubic, F$\bar{4}$3m, a = 6.155 Å, D_m = 10.61, Z = 4, D_x = 14.6. Cu radiation, diffractometer data. [Isomorphous compounds have stoichiometric compositions (1).] Measured density corresponds to 30% vacancies on all sites. Atoms are sited thus; Hg(1) 4(a): 0,0,0; Hg(2) 4(d):3/4,3/4,3/4; Ti 4(c):1/4,1/4,1/4; Cu 4(b):1/2,1/2,1/2.

Interatomic distances (Å)

Hg(1) - 6 Cu 3.08   Cu - 4 Ti 2.67
      - 4 Ti 2.67   - 4 Hg 2.67
      - 4 Hg 2.67   - 6 Hg 3.08

Hg(2) - 4 Cu 2.67   Ti - 4 Hg 2.67
      -*6 Ti 3.08   - 4 Cu 2.67
      - 4 Hg 2.67   - 6 Hg 3.08

* [given as 2 Ti]

The structure is an ordered variant of the NaTl type.

1. Structure Reports, 40, 56 [where F$\bar{4}$3m is misprinted as Fd3m.]

COPPER MOLYBDENUM SULPHUR
$Cu_{2-x}Mo_3S_4$ (x = 1.1, 0.62, 0.53, 0.17, at r.t.)

K. YVON, A. PAOLI, R. FLÜKIGER and R. CHEVREL, 1977. Acta Cryst., B33, 3066-3072.

Rhombohedral, $CoMo_3S_4$ type (1), R$\bar{3}$, Z = 2 (rhombohedral cell). R = 0.03 for 613 reflexions, diffractometer data for the 4 samples.

	$Cu_{0.90}Mo_3S_4$	$Cu_{1.38}Mo_3S_4$	$Cu_{1.47}Mo_3S_4$	$Cu_{1.83}Mo_3S_4$
a_{rh}(Å)	6.503	6.560	6.573	6.597
$\alpha°$	94.93	95.51	95.56	95.58
a_{hex}(Å)	9.584	9.713	9.735	9.773
c_{hex}(Å)	10.250	10.213	10.221	10.255
D_x	5.75	5.98	6.01	6.24

Atomic positions (for x = 0.17, similar values for others)
S(2) in 6(c), remainder in 18(f)

$Cu_{1.83}Mo_3S_4$	x	y	z	occupancy
Mo	0.01418	0.16369	0.39282	1
S(1)	0.30683	0.27780	0.40872	1
S(2)	0	0	0.20153	1
Cu(1)	0.72377	0.48270	0.33242	0.23
Cu(2)	0.15164	0.24085	0.89202	0.38

The structure belongs to a group (2) based on the Mo_6Se_8 face-centred 'cubes' of Mo_3Se_4 (3) with general formula $M_xMo_3S_4$ (M = transition or rare earth element, 0 < x < 2). The Mo_6S_8 'cubes' (see figures in 2, 3) have an S atom array with both tetrahedral interstices and a large cubic hole. Both holes are partially occupied by Cu atoms but there is overlap (Cu-Cu 1.1 to 1.3 Å) and a maximum of 4 Cu atoms occupy any given cluster of 12 sites. Both Cu-S distances (2.19-2.73 Å) and Mo-S distances (2.41-2.55 Å) are normal. The Mo-Mo distances (2.66-2.75 Å) are among the shortest for fractional Mo-Mo bonds; Cu-Cu distances between adjacent clusters are 2.1-2.4 Å, allowing the possibility of ionic d.c. conduction. The anisotropic motions of Cu atoms (0.3 Å) are directed toward neighbouring Cu sites. The compounds are generally good superconductors.

1. Structure Reports, 37A, 154.
2. Ibid., 39A, 78, 85; 42A, 67, 106.
3. Ibid., 39A, 85, 86.

COPPER NICKEL TIN   COPPER TIN
Cu_9NiSn_3 δ-$Cu_{41}Sn_{11}$

M.H. BOOTH, J.K. BRANDON, R.Y. BRIZARD, C. CHIEH and W.B. PEARSON, 1977. Acta Cryst., B33, 30-36.

Cubic, $Cu_{41}Sn_{11}$ type (1, 2), F$\bar{4}$3m. Mo radiation, diffractometer data. See also 1 and 2 for $Cu_{41}Sn_{11}$; 3 gave a 9 Å cell for Cu_9NiSn_3.

Phase	a(Å)	D_m	Atoms	R	Reflexions
$Cu_{41}Sn_{11}$	17.980	8.83	412	0.070	310
Cu_9NiSn_3	18.011	8.68	400	0.105	224

Atomic positions

			$Cu_{41}Sn_{11}$ [a]			Cu_9NiSn_3 [b]		
Cluster	Site	Point set	Atoms	x	z	Atoms†	x	z
A	IT	$x\ x\ x$	12 Cu*	0·0573		12 Cu*	0·0577	
B	IT	$x\ x\ x$	16 Cu	0·3005		16 Cu	0·2988	
C	IT	$x\ x\ x$	16 Cu	0·5504		16 Cu	0·5635	
D	IT	$x\ x\ x$	16 Cu	0·8062		16 Cu	0·8046	
A	OT	$x\ x\ x$	16 Sn	−0·0887		16 Sn	−0·0875	
B	OT	$x\ x\ x$	16 Cu	0·1657		16 Cu	0·1638	
C	OT	$x\ x\ x$	16 Cu	0·4166		16 Cu	0·4153	
D	OT	$x\ x\ x$	16 Cu	0·6664		16 Cu	0·6647	
A	OH	$0\ 0\ z$	24 Cu	0·0	0·1763	24 Cu	0·0	0·1759
B	OH	$\frac{1}{4}\ \frac{1}{4}\ z$	24 Cu	0·25	0·4241	24 Cu	0·25	0·4249
C	OH	$\frac{1}{2}\ \frac{1}{2}\ z$	24 Cu	0·5	0·6765	24 Cu	0·5	0·6757
D	OH	$\frac{3}{4}\ \frac{3}{4}\ \text{-}z$	24 Sn	0·75	0·9309	24 Sn	0·75	0·9299
A	CO	$x\ x\ z$	48 Cu	0·1562	0·0186	48 Cu	0·1580 (6)	0·0159
B	CO	$x\ x\ z$	48 Sn	0·4084	0·2680	48 Sn	0·4086 (5)	0·2678
C	CO	$x\ x\ z$	48 Cu	0·6465	0·5278	36 Cu*	0·6510 (9)	0·5255
D	CO	$x\ x\ z$	48 Cu	0·9087	0·7631	48 Cu	0·9106 (7)	0·7600

* occupancy = 75%

a,b	Cu	Sn [a]	Cu	Ni	Sn [b]
From structure refinement	324	88	280.8	32.2	88.0
From chemical analysis	328.3	84.7	278.6	31.0	92.9

† The Ni atom sites are not specified; Cu atom scattering factors were used for all sites except those occupied by Sn.

The classical γ-brass (Cu_5Zn_8 (4)) structure can be referred to clusters of 26 atoms built up from an inner tetrahedron (IT) of four atoms, an outer tetrahedron (OT) of four atoms, then an octahedron (OH) of six atoms, and finally a cubo-octahedron (CO) of 12 atoms. In the cubic γ-brasses with I cells (a ∿ 9 Å) these clusters pack as pseudo-atoms in a b.c.c. array; in those with a P cell (a ∿ 9 Å) two different clusters A and B pack as in the CsCl structure, whereas in the present cases with F cells (a ∿ 18 Å) four different clusters A, B, C, and D pack in that order along <111> and the large Sn atoms occur in the sites OT, CO, none, and OH of four successive clusters. This ordering uniquely prevents any close contact between Sn atoms. This arrangement agrees with 2 and is one which cannot be satisfied in the P or I cells thus accounting for the F cell. A further consequence of this ordering is a maximizing of the number of unlike (Sn-Cu) contacts. Also the phases Ce_5Hg_{21}, Pu_5Hg_{21}, and Pt_5Zn_{21} (with formulae M_5N_{21}) are predicted to have the M atoms in sites, OT, OH, OH, OT... in successive clusters along <111> as this ordering prevents close M-M contacts. The observed Cu-Cu distances range from expected values of 2.56 Å up to 2.91 Å in different clusters and this is correlated with compressions from neighbouring clusters.

1. Strukturbericht, 1, 548; Structure Reports, 20, 103; 21, 116; 29, 112.
2. Structure Reports, 42A, 79.
3. Ibid., 17, 155.
4. Strukturbericht, 1, 497; 2, 693, 699; 3, 589; Structure Reports, 33A, 72.

COPPER POTASSIUM SULPHUR
Cu_3KS_2

C. BURSCHKA and W. BRONGER, 1977. Z. Naturf., 32B, 11-14.

Monoclinic, $CsAg_3S_2$ type (1), C2/m, a = 14.773, b = 3.946, c = 8.182 Å, β = 113.5°, Z = 4, D_x = 4.46. Mo radiation, R = 0.058 for 590 reflexions, diffractometer data.

Atomic positions (all atoms in 4(i))

	x	y	z
K	0.1346	0	0.0433
Cu(1)	0.0964	1/2*	0.6436
Cu(2)	0.1884	1/2*	0.4265
Cu(3)	0.0606	0	0.4021
S(1)	0.1953	0	0.6847
S(2)	0.5204	0	0.2395

* [Given as 4(i): x,0,z; structure requires 1/2]

Interatomic distances (Å)

K - 7 S	3.20 - 3.39	Cu(2) - 4 S	2.24 - 2.86
- 7 Cu	3.52 - 3.77	- 5 Cu	2.63 - 2.68
- 5 K	3.75 - 3.95	- 2 K	3.52
Cu(1) - 3 S	2.28, 2.40	Cu(3) - 3 S	2.32, 2.37
- 6 Cu	2.63 - 2.95	- 7 Cu	2.68 - 2.95
- 3 K	3.67, 3.77	- 2 K	3.52, 3.63
- 1 S	3.04		

Cu(1) and Cu(3) atoms have 3 near S neighbours and Cu(2) atoms 2+2 S near neighbours forming channels and sheets similar to those in $K_2Ag_4S_3$ (2).

1. Structure Reports, 39A, 104.
2. Ibid., 42A, 120.

COPPER SULPHUR TIN
Cu_4S_4Sn

S. JAULMES, J. RIVET and P. LARUELLE, 1977. Acta Cryst., B33, 540-542.

Orthorhombic, Pnma, a = 13.558, b = 7.681, c = 6.412 Å, D_m = 4.86, Z = 4. Mo radiation, R = 0.071 for 536 reflexions, diffractometer data. See also 1.

Atomic positions (Cu(1) and S(1) in 8(d), remainder in 4(c))

	x	y	z
Sn	0.0869	1/4	0.1257
Cu(1)	0.3358	0.9873	0.1272
Cu(2)	0.3423	1/4	0.5580
Cu(3)	0.4340	1/4	0.8878
S(1)	0.1752	0.0011	0.2624
S(2)	0.0838	1/4	0.7502
S(3)	0.4187	1/4	0.2424

Interatomic distances (Å)

Sn - 4 S	2.41 - 2.43	Cu(2) - 3 S	2.27, 2.35
		- 1 Cu	2.45
Cu(1) - 4 S	2.27 - 2.43		
		Cu(3) - 4 S	2.22 - 2.56

Sn, Cu(1), and Cu(3) are tetrahedrally coordinated by S but Cu(3) is almost at the centre of a triangle of S atoms.

1. Structure Reports, 41A, 127.

DEUTERIUM THORIUM
$D_{15}Th_4$ (at r.t. and 90°K)

M.H. MUELLER, R.A. BEYERLEIN, J.D. JORGENSEN, T.O. BRUN, C.B. SATTERTHWAITE and R. CATON, 1977. J. Appl. Cryst., 10, 79-83.

Cubic, [$Cu_{15}Si_4$ type (1)], I$\bar{4}$3d, a = 9.11 Å, Z = 4, D_X = 8.29. R = 0.0368 at r.t. and 0.0381 at 90°K for 155 reflections, neutron data. Atoms are placed Th in 16(c): x,x,x, x = 0.2066 at r.t. and 0.2071 at 90°K; D(1) in 12(a): 3/8,0,1/4; D(2) in 48(e): x,y,z, x = 0.3706, (r.t.), 0.3708 (90°K), y = 0.2155 (r.t.), 0.2169 (90°K), and z = 0.4048 (r.t.), 0.4051 (90°K). See also 2.

Interatomic distances (Å)

Th	- 11 Th	3.84 - 4.10	D(2)	- 3 Th	2.27 - 2.38
	- 12 D	2.27 - 2.46		- 9 D	2.25 - 2.79
D(1)	- 4 Th	2.46			
	- 8 D	2.35, 2.42			

The structure contains chains of Th atoms equally spaced (3.95 Å) along four non-intersecting chains paralled to the body-diagonals. The Th atoms are in the centre of a 12-fold array of D atoms which are arranged on triangles along the body diagonals.

1. Strukturbericht, 3, 62, 366.
2. Structure Reports, 17, 188.

DYSPROSIUM IRON SILICON   IRON SILICON SCANDIUM
$Dy_2Fe_3Si_5$ $Fe_3Sc_2Si_5$

O.I. BODAK, B.Ja. KOTUR, V.I. JAROVEČ and E.I. GLADYŠEVSKIJ, 1977. Kristallografija, 22, 385-388 [Soviet Physics - Crystallography, 22, 217-219].

Tetragonal, P4/mnc, Z = 4. Cu radiation, photographic data. $Dy_2Fe_3Si_5$, a = 10.11, c = 5.492 Å, D_m = 6.81. R = 0.142 for 170 reflections; $Fe_3Sc_2Si_5$, a = 10.05, c = 5.313 Å, D_m = 4.82. R = 0.104 for 229 reflections.

Atomic positions

			x		y		z	
			Dy	Sc	Dy	Sc	Dy	Sc
Dy or Sc	in	8(h)	0.0703	0.0701	0.2385	0.2500	0	0
Fe(1)	in	8(h)	0.3770	0.3790	0.3407	0.3601	0	0
Fe(2)	in	4(d)	0	0	1/2	1/2	1/4	1/4
Si(1)	in	8(g)	0.1840	0.1779	0.6840	0.6779	1/4	1/4
Si(2)	in	4(e)	0	0	0	0	0.2706	0.2528
Si(3)	in	8(h)	0.1779	0.1799	0.4876	0.4761	0	0

Interatomic distances (Å)

		$Dy_2Fe_3Si_5$	$Fe_3Sc_2Si_5$
Dy or Sc	- 4 Dy or Sc	3.55, 3.87	3.69
	- 7 Fe	3.06 - 3.46	2.93 - 3.46
	- 9 Si	2.74 - 3.02	2.53 - 2.95
Fe(1)	- 5 Dy or Sc	3.12 - 3.46	3.04 - 3.46
	- 2 Fe	2.88	2.63
	- 6 Si	2.14 - 2.50	2.28, 2.32

		$Dy_2Fe_3Si_5$	$Fe_3Sc_2Si_5$
Fe(2)	- 4 Dy or Sc	3.06	2.93
	- 2 Fe	2.75	2.66
	- 6 Si	2.27, 2.63	2.26, 2.53
Si(1)	- 4 Dy or Sc	2.89, 3.02	2.92, 2.95
	- 3 Fe	2.19, 2.63	2.33, 2.53
	- 4 Si	2.41, 2.75	2.42, 2.66
Si(2)	- 4 Dy or Sc	2.92	2.93
	- 4 Fe	2.39	2.28
	- 2 Si	2.52, 2.97	2.63, 2.69
Si(3)	- 3 Dy or Sc	2.74, 2.93	2.53, 2.79
	- 4 Fe	2.14 - 2.50	2.26, 2.32
	- 2 Si	2.42	2.42

The structure (Fig. 1) is related to those of W_5Si_3 (1) and U_6Mn (2) in having columns of tetragonal antiprisms centred at 0,0 and 1/2,1/2; the structures differ in the modes of interconnection.

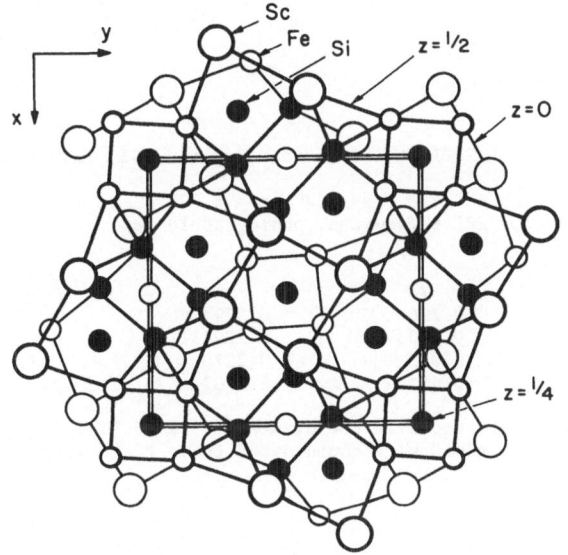

Fig. 1.   The $Fe_3Sc_2Si_5$ structure.

1. Structure Reports, 19, 277.
2. Ibid., 13, 93.

ERBIUM PLATINUM SILICON
$ErPt_2Si_2$

I. MAYER and P.D. YETOR, 1977. J. Less-Common Metals, 55, 171-176.

Tetragonal, $ThCr_2Si_2$ type (1), I4/mmm, a = 4.115, c = 9.822 Å, Z = 2, D_X = 12.25. Mo radiation, R = 0.156 for 24 reflexions, diffractometer data. Atoms were placed Er in 2(a): 0,0,0; Pt/Si(1) in 4(d): 1/2,0,1/4; Pt/Si(2) in 4(e): 0,0,z, z = 0.37.

Interatomic distances (Å) of isostructural compounds

	Ln-8Pt	[Ln-8Si]	Ln-2Si	Pt*-4Pt	[Pt-4Si]	Si-Si†
Sr	3.29	3.28	3.66	3.02	2.44	2.60
La	3.26	3.28	3.63	3.03	2.44	2.55
Nd	3.23	3.24	3.62	2.98	2.41	2.55
Eu	3.25	3.28	3.63	3.02	2.44	2.55
Gd	3.22	3.21	3.62	2.95	2.40	2.54
Dy	3.21	3.20	3.62	2.93	2.38	2.55
Er	3.20	3.18	3.63	2.91	2.37	2.55
Tm	3.21	3.18	3.64	2.91	2.37	2.56
Lu	3.22	3.16	3.68	2.88	2.36	2.59

* 4(d) site   † 4(e) site

The first transition metal found to substitute for X in AM_2X_2 compounds (A =
Th, U, rare earth, or alkaline earth metal; M = 3d or 4d transition metal; X = Si
or Ge) is Pt; this is possibly due to the tetravalency of Pt.

1. Structure Reports, 30A, 126; this volume, p. 99.

ERBIUM SELENIUM SILVER
AgErSe$_2$

M. JULIEN-POUZOL and P. LARUELLE, 1977. Acta Cryst., B33, 1510-1512.

Orthorhombic, $P2_12_12_1$, a = 6.88, b = 13.79, c = 4.175 Å, D_m = 7.32, Z = 4. Mo
radiation, R = 0.074 for 287 reflexions, photographic data.

Atomic positions

	x	y	z
Ag	0.296	0.3794	0.002
Er	0.2903	0.1294	0.2303
Se(1)	0.085	0.2270	0.730
Se(2)	0.4846	0.0267	0.728

The Ag atoms are tetrahedrally coordinated to Se (2.71-2.80, mean 2.74 Å)
while the Er atoms are octahedrally coordinated to Se (2.84-2.87, mean 2.85 Å);
mean Se-Se 3.92 Å.

EUROPIUM SULPHUR TIN
Eu$_5$S$_{12}$Sn$_3$

I. S. JAULMES and M. JULIEN-POUZOL, 1977. Acta Cryst., B33, 1191-1193.

Orthorhombic, $Pm2_1b$, a = 3.924, b = 11.509, c = 20.219 Å, D_m = 5.2, Z = 2. Mo
radiation, R = 0.048 for 1332 reflexions, diffractometer data. 20 atomic positions.

Interatomic distances (Å) (average values in brackets)

Sn(1) - 6 S	2.51 - 2.63	(2.57)	Eu(2) - 8 S	2.87 - 3.19	(2.94)
Sn(2) - 6 S	2.47 - 2.61	(2.57)	Eu(3) - 8 S	2.98 - 3.16	(3.05)
Sn(3) - 5 S	2.41 - 2.54	(2.48)	Eu(4) - 8 S	2.98 - 3.42	(3.06)
Eu(1) - 8 S	2.95 - 3.11*	(3.06)	Eu(5) - 7 S	2.81 - 2.84	(2.82)

* [One value is misprinted as 3.409 for 3.049]

The mean values of Eu-S distances for Eu(2) and Eu(5), 2.94 and 2.82 Å, respectively, are consistent with trivalent Eu and thus the formula can be written $Eu(III)_2Eu(II)_3Sn_3S_{12}$.

$Eu_3S_7Sn_2$

II. S. JAULMES and M. JULIEN-POUZOL, 1977. Acta Cryst., B33, 3898-3901.

Orthorhombic, Pbam, a = 11.542, b = 12.690, c = 3.974 Å, D_m = 5.2, Z = 2. Mo radiation, R = 0.042 for 1126 reflexions, diffractometer data.

Atomic positions

			x	y	z
Eu(1)	in	2(c)	0	1/2	0
Eu(2)	in	4(g)	0.38672	0.64649	0
Sn	in	4(h)	0.2737	0.38827	1/2
S(1)	in	4(g)	0.3482	0.2951	0
S(2)	in	4(h)	0.2951	0.0647	1/2
S(3)	in	4(h)	0.0757	0.3191	1/2
S(4)	in	2(b)	0	0	1/2

Interatomic distances

Sn	- 5 S	2.38 - 2.97
Eu(1)	- 8 S	3.16, 3.20

Eu(2) - 8 S 2.99 - 3.30

GADOLINIUM PHOSPHORUS SULPHUR
GdPS

F. HULLIGER, R. SCHMELCZER and D. SCHWARZENBACH, 1977. J. Solid State Chem., 21, 371-374.

Orthorhombic, Pmnb, a = 5.3620, b = 5.4079, c = 16.742 Å, Z = 8, D_x = 6.03. Mo radiation, R = 0.046 for 726 reflexions, diffractometer data.

Atomic positions (P in 8(d), remainder in 4(c))

	x	y	y
Gd(1)	1/4	0.0160	0.1376
Gd(2)	1/4	-0.0152	0.6469
P	0.0382	0.7046	-0.0025
S(1)	1/4	0.5150	0.1888
S(2)	1/4	0.4883	0.6883

Interatomic distances (Å)

Gd(1)	- 4 P	3.10, 3.13	S(1)	- 4 P	3.55, 3.68
	- 4 Gd	3.81		- 5 Gd	2.77 - 2.91
	- 5 S	2.79 - 2.91		- 4 S	3.39, 3.43
Gd(2)	- 4 P	2.92, 3.11	S(2)	- 2 P	3.58
	- 5 S	2.75 - 2.84		- 5 Gd	2.75 - 2.84
				- 4 S	3.43, 3.51
P	- 4 P	2.25 - 3.22			
	- 4 Gd	2.92 - 3.13			
	- 4 S	3.55 - 3.85			

GdPS (Fig. 1) is non-metallic, with chains based on a distortion of the square net of the PbFCl structure. [Similar chains occur in CaP_3 and $CaAs_3$ (1).]

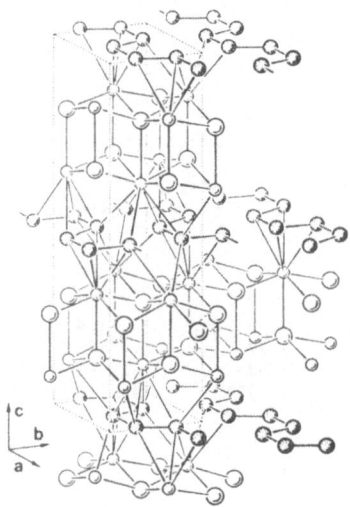

Fig. 1.   The GdPS structure; smallest spheres Gd, largest S, stippled spheres P.
             Some bonds are omitted.

1. Structure Reports, 39A, 42; 42A, 25.

GALLIUM SELENIUM THALLIUM
GaSe$_2$Tl-II (H.P.)

K.J. RANGE, G. MAHLBERG and S. OBENLAND, 1977. Z. Naturf., 32B, 1354-1355.

Tetragonal, TlSe type (1), I4/mcm, a = 8.053, c = 6.417 Å, c/a = 0.797, D$_m$ = 6.72,
Z = 4. R = 0.097, photographic data.

Atomic positions

	x	y	z
Tl in 4(a)	0	0	1/4
Ga in 4(b)	0	1/2	1/4
Se in 8(h)	0.157	0.657	0

[Interatomic distances (Å)]

Tl - 2 Tl 3.21	Se - 4 Tl 3.44
- 4 Ga 4.03	- 2 Ga 2.40
- 8 Se 3.44	- 2 Se 3.84
	Ga - 4 Tl 4.03
	- 2 Ga 3.21
	- 4 Se 2.40

1. Strukturbericht, 7, 6, 77.

GALLIUM TELLURIUM
Ga_2Te_5

M. JULIEN-POUZOL, S. JAULMES and F. ALAPINI, 1977. Acta Cryst., B33, 2270-2272.

Tetragonal, I4/m, a = 7.913, c = 6.848 Å, D_m = 5.85, Z = 2. Mo radiation, R = 0.069 for 475 independent reflexions, diffractometer data. Given as hexagonal $GaTe_3$ by 1.

Atomic positions

			x	y	z
Te(1)	in	2(a)	0	0	0
Te(2)	in	8(h)	0.1904	0.3318	0
Ga	in	4(d)	0	1/2	1/4

Interatomic distances (Å)

Te(1) - 4 Te(2)   3.03   Ga - 4 Te   2.64
                                                   - 2 Ga   3.42
Te(2) - 4 Te(2)   3.78, 3.94
      - 1 Te(1)   3.03

The structure is composed of infinite chains of $[GaTe_4]$ tetrahedra parallel to c. These chains are joined by Te(1) atoms which centre a square of Te(2) atoms; Te-Te bonds are largely covalent.

1. Structure Reports, 26, 149.

GERMANIUM GOLD SODIUM   GOLD SODIUM TIN
$AuGeNa_2$ AuNaSn (L.T.)

G. WROBEL and H.-U. SCHUSTER, 1977. Z. anorg. Chem., 432, 95-100.

Na_2AuGe, orthorhombic, Immm, a = 7.227, b = 7.529, c = 4.417 Å, D_m = 6.35, Z = 3. R = 0.123 for 257 reflexions, diffractometer data.

NaAuSn, orthorhombic, [E-phase, NiSiTi type (1)], $Pna2_1$, a = 7.476, b = 8.088, c = 4.530 Å, D_m = 8.48, Z = 4. R = 0.083 for 639 reflexions, diffractometer data. [Coordinates are centrosymmetric as given, thus Pnam is possible.]

Atomic positions

Na_2AuGe			x	y	z	Occupancies
Na/Ge	in	8(n)	0.322	0.320	0	4 Na + 2 Ge
Ge/Au	in	4(e)	0.202	0	0	1 Ge + 3 Au
Na	in	4(g)	0	0.402	0	2 Na

NaAuSn, all atoms in 4(a) (in $Pna2_1$, as given)

Na	0.489	0.664	0
Au	0.2067	0.3697	0
Sn	0.3420	0.0597	0

Interatomic distances (Å)

Na_2AuGe
Na/Ge - 4 Na/Ge   2.57 - 2.71   Ge/Au - 10 Na   2.56 - 3.36
      - 3 Ge/Au   2.56, 2.60   - 1 Ge/Au   2.92
      - 3 Na   2.41 - 4.05

Na   - 10 Na   3.05 - 4.05
      - 2 Na/Ge   2.41
      - 6 Ge/Au   3.17, 3.36

NaAuSn

Au - 4 Sn	2.70 - 2.79
- 5 Na	3.17 - 3.22

Na - 4 Na	3.49, 3.98
- 5 Au	3.17 - 3.22
- 6 Sn	3.20 - 3.46

Sn - 6 Na	3.20 - 3.46
- 2 Sn	3.41
- 4 Au	2.70 - 2.79

<u>1</u>. Structure Reports, <u>32</u>A, 75.

GERMANIUM IRON LITHIUM

Fe$_6$Ge$_4$Li   Fe$_6$Ge$_5$Li   Fe$_6$Ge$_6$Li

I.   E. WELK and H.-U. SCHUSTER, 1977. Z. Naturf., <u>32</u>B, 749-752.
II.   Idem, 1976. Z. anorg. Chem., <u>424</u>, 193-304.

Fe$_6$Ge$_4$Li and Fe$_6$Ge$_5$Li, monoclinic, C2/m. Mo radiation, R = 0.102 and 0.132 for 350 and 292 reflexions respectively, diffractometer data. Fe$_6$Ge$_6$Li, hexagonal, P6/mmm. Mo radiation, R = 0.133 for 165 reflexions, diffractometer data.

	a(Å)	b(Å)	c(Å)	β(°)	D$_m$	Z	Space group
Fe$_6$Ge$_4$Li	8.739	5.045	7.170	113.94	7.10	2	C2/m
Fe$_6$Ge$_5$Li	8.744	5.045	14.841	101.42	7.16	4	C2/m
Fe$_6$Ge$_6$Li	8.744		8.033		7.21	3	P6/mmm

Atomic positions

Fe$_6$Ge$_4$Li

			x	y	z
Fe(1)	in	4(i)	0.401	0	0.809
Fe(2)	in	8(j)	0.355	0.25	0.190
Ge(1)	in	4(i)	0.126	0	0.122
Ge(2)	in	4(i)	0.333	0	0.5
Li	in	2(c)	0	0	1/2

Fe$_6$Ge$_5$Li   Fe(1) and (2) in 8(j), remainder in 4(i)

	x	y	z
Fe(1)	0.296	0.25	0.138
Fe(2)	0.388	0.25	0.416
Fe(3)	0.365	0	0.583
Fe(4)	0.450	0	0.861
Ge(1)	0.333	0	0
Ge(2)	0.027	0	0.087
Ge(3)	0.425	0	0.276
Ge(4)	0.145	0	0.441
Ge(5)	0.242	0	0.724
Li	0.092	0	0.263

Fe$_6$Ge$_6$Li

Ge(1)	in	6(j)	0.333	0	0
Ge(2)	in	6(k)	0.333	0	1/2
Ge(3)	in	4(h)	1/3	2/3	0.3486
Ge(4)	in	2(e)	0	0	0.1514
Fe(1)	in	12(o)	0.166	0.333	0.246
Fe(2)	in	6(i)	1/2	0	0.25
Li(1)	in	2(c)	1/3	2/3	0
Li(2)	in	1(b)	0	0	1/2

[Interatomic distances (Å)]

Fe_6Ge_4Li

Fe(1)	- 7 Fe	2.47 - 3.18	Ge(1) - 6 + 2 Fe	2.25 - 2.89
	- 3 Ge	2.05, 2.60	- 2 Ge	2.19, 2.58
	- 1 Li	3.30	- 1 Li	3.31

Fe(2)	- 6 Fe	2.47 - 3.18	Ge(2) - 5 Fe	2.05 - 3.01
	- 5 Ge	2.25 - 3.01	- 4 Ge	2.58 - 2.92
	- 1 Li	2.42	- 3 Li	2.91

Li	- 6 Fe	2.42, 3.30
	- 8 Ge	2.91 - 3.31

Fe_6Ge_5Li

Fe(1)	- 4 Fe	2.50 - 2.55	Ge(1) - 6 Fe	2.48 - 2.51
	- 6 Ge	2.48 - 2.64	- 9 Ge	2.91 - 3.19
	- 2 Li	3.09, 3.14		
			Ge(2) - 6 Fe	2.62 - 2.64
Fe(2)	- 6 Fe	2.50 - 2.85	- 7 Ge	2.53 - 3.19
	- 5 Ge	2.50 - 2.56	- 1 Li	2.56
	- 2 Li	3.33, 3.40		
			Ge(3) - 6 Fe	2.49 - 2.51
Fe(3)	- 6 Fe	2.50 - 2.82	- 3 Ge	2.91
	- 5 Ge	2.50 - 2.56	- 3 Li	2.88, 2.94
	- 2 Li	3.37		
			Ge(4) - 6 + 3 Fe	2.53 - 2.56
Fe(4)	- 4 Fe	2.50, 2.55	- 3 Ge	3.35, 3.40
	- 6 Ge	2.44 - 2.63	- 1 Li	2.59
	- 2 Li	3.10		
			Ge(5) - 6 Fe	2.44 - 2.54
Li	- 12 Fe	3.09 - 3.40	- 3 Ge	2.91
	- 8 Ge	2.56 - 2.96	- 3 Li	2.90, 2.96

Fe_6Ge_6Li

Fe(1)	- 4 Fe	2.52	Ge(1) - 6 Fe	2.45, 2.48
	- 6 Ge	2.45 - 2.66	- 5 Ge	2.91 - 3.16
	- 2 Li	3.21, 3.25	- 2 Li	2.92

Fe(2)	- 4 Fe	2.52	Ge(2) - 6 Fe	2.48, 2.51
	- 6 Ge	2.48, 2.65	- 1 + 6 Fe	2.91 - 3.16
	- 2 Li	3.23	- 1 Li	2.91

Li(1)	- 6+2 Ge	2.92, 2.80	Ge(3) - 6 Fe	2.65, 2.66
	- 12 Fe	3.21, 3.23	- 1 + 6 Ge	2.43, 3.16
			- 1 Li	2.80

Li(2)	- 6+2 Ge	2.92, 2.80		
	- 12 Fe	3.25	Ge(4) - 6 Fe	2.64
			- 1 + 6 Ge	2.43, 3.16
			- 1 Li	2.80

These structures are based on trigonal prisms (Fe_6) centred by Ge atoms. Similar prisms occur in Fe_6Ge_5 (1), FeGe (2), $Fe_3Ge_6Mn_4$ (3), and Fe_3Sn_2 (4).

1. Structure Reports, 39A, 64; 40A, 68.
2. Ibid., 32A, 82.
3. Ibid., 42A, 87.
4. Ibid., 42A, 103.

GERMANIUM NICKEL PALLADIUM   GERMANIUM PALLADIUM
GeNiPd $GePd_2$

W. WOPERSNOW and K. SCHUBERT, 1977. J. Less-Common Metals, <u>52</u>, 1-12.

GeNiPd, tetragonal, PbFCl type (<u>1</u>), P4/mmm, a = 3.5952, c = 6.2716 Å, Z = 2, D_x =
9.74. Cu radiation, R = 0.16, photographic data.

Atomic positions [origin at centre]

		x	y	z
Ni in	2(a)	0	0	0
Pd in	2(c)	0	1/2	0.639
Ge in	2(c)	0	1/2	0.248

[Interatomic distances (Å)]

Ni - 4 Ni   2.54   Pd - 4 Pd   3.08
   - 4 Pd   2.89   - 5 Ge   2.45, 2.64
   - 4 Ge   2.38   - 4 Ni   2.89

Ge - 4 Ni   2.38
   - 5 Pd   2.45, 2.64

$GePd_2$, hexagonal, Fe_2P type (<u>2</u>), P$\bar{6}$2m, a = 6.712, c = 3.408 Å, Z = 3, D_x = 10.69.
Cu radiation, R = 0.18, photographic data. See also <u>3</u>.

Atomic positions

		x	y	z
Pd(1) in	3(f)	0.266	0	0
Pd(2) in	3(g)	0.604	0	1/2
Ge(1) in	2(c)	1/3	2/3	0
Ge(2) in	1(a)	0	0	1/2

[Interatomic distances (Å)]

Pd(1) - 10 Pd   2.84 - 3.41   Ge(1) - 9 Pd   2.49, 2.67
      - 4 Ge   2.47, 2.49   - 2 Ge   3.41

Pd(2) - 5 Ge   2.66, 2.67   Ge(2) - 9 Pd   2.47, 2.66
      - 8 Pd   2.84 - 3.41   - 2 Ge   3.41

<u>1</u>. Strukturbericht, <u>2</u>, 45, 362; <u>3</u>, 369.
<u>2</u>. Ibid., <u>2</u>, 15, 284; Structure Reports, <u>23</u>, 68; <u>39A</u>, 76.
<u>3</u>. Structure Reports, <u>16</u>, 95; <u>17</u>, 174.

GERMANIUM PALLADIUM
Ge_8Pd_{21}

P. MATKOVIĆ, W. WOPERSNOW and K. SCHUBERT, 1977. J. Less-Common Metals, <u>56</u>, 69-75.

Tetragonal, Pt_8Al_{21} type (<u>1</u>), $I4_1/a$, a = 13.067, c = 10.033 Å, D_m = 10.92, Z = 4.
Cu radiation, R = 0.133 for 571 reflexions, photographic data.

Atomic positions (Pd(6) in 4(b), remainder in 16(f))

	x	y	z
Pd(1)	0.0834	0.1442	0.846
Pd(2)	0.2205	0.2367	0.614
Pd(3)	0.2213	0.4548	0.611
Pd(4)	0.2523	0.3401	0.176
Pd(5)	0.0887	0.3044	0.022
Pd(6)	0	1/4	5/8
Ge(1)	0.205	0.154	0.072
Ge(2)	0.097	0.125	0.472

[Interatomic distances (Å)]

Pd(1)	-	4 Ge	2.49 - 2.81		Pd(5)	-	3 Ge	2.52 - 2.55
	-	10 Pd	2.74 - 3.18			-	8 Pd	2.68 - 2.94

Pd(2) - 4 Ge 2.52 - 2.60   Pd(6) - 4 Ge 2.58
      - 9 Pd 2.74 - 3.18   - 8 Pd 2.83, 2.89

Pd(3) - 3 Ge 2.51 - 2.88   Ge(1) - 10 Pd 2.49 - 2.77
      - 8 Pd 2.70 - 3.15

                             Ge(2) - 10 Pd 2.51 - 2.88
Pd(4) - 4 Ge 2.51 - 2.72
      - 8 Pd 2.68 - 3.15

1. Structure Reports, 31A, 10.

GERMANIUM PLATINUM SELENIUM
GePtSe

I.   S.C. ABRAHAMS, J.L. BERNSTEIN and E. BUEHLER, 1976. Mater. Res. Bull., 11,
     707-712.
II. S.C. ABRAHAMS and J.L. BERNSTEIN, 1977. Acta Cryst., B33, 301-302.

Orthorhombic, Pca2_1, a = 6.00984, b = 6.06174, c = 5.98187 Å, Z = 4, D_x = 10.57.
A new refinement based on the data of 1. See also 2. The model of 1 (Ge and Se
interchanged) gave R = 0.0863 for 1187 observed reflexions. The new model gave
R = 0.0861 and more equal temperature factors and is thus more likely.

Atomic positions

	x	y	z
Pt in 4(a)	0.0081	0.7422	0
Ge in 4(a)	0.6190	0.3751	0.3830
Se in 4(a)	0.3842	0.1370	0.6151

Interatomic distances (Å)

	Range	Mean	
Pt - 3 Ge	2.453 - 2.490	2.466	octahedral
- 3 Se	2.528 - 2.573	2.555	
Ge - 3 Pt	2.453 - 2.490	2.466	tetrahedral
- 1 Se	2.450	-	
Se - 3 Pt	2.528 - 2.573	2.555	tetrahedral
- 1 Ge	2.450	-	

The present model gives atomic radii, Pt 1.285, Ge 1.181, and Se 1.269 Å, in
better agreement with recent data for PtGe (3) and PtSe$_2$ (4) than the original model
of 1. The crystals are frequently twinned; ferroelastic reorientation under uniaxial

stress can detwin such crystals and is accompanied by atomic displacements as large
as 2 Å. The tetrahedral angles about Ge range from 102 to 116°, apparently reflect-
ing a tendency towards metallic bonding. GePtSe can be regarded as a coupled but
semiconducting ferroelastic-ferroelectric.

1. Structure Reports, 39A, 67.
2. M. EL-BORAGY and K. SCHUBERT, 1971. Z. Metallk., 62, 667.
3. Structure Reports, 39A, 66.
4. Ibid., 30A, 76.

GERMANIUM SILVER SULPHUR   (SYNTHETIC ARGYRODITE)
Ag_8GeS_6

G. EULENBERGER, 1977. Mh. Chem., 108, 901-913.

Orthorhombic, $Pna2_1$, a = 15.149, b = 7.476, c = 10.589 Å, D_m = 6.21, Z = 4. Mo
radiation, R = 0.082 for 3431 reflexions, diffractometer data. 15 site-sets.
See also 1; compare 2.

The structure consists of slightly distorted isolated GeS_4 tetrahedra (mean
Ge-S 2.212 Å, range 2.21-2.23 Å). The GeS_4 tetrahedra and the remaining S atoms
are connected by the Ag atoms to form a three-dimensional framework. Three Ag
positions have a strongly distorted tetrahedral environment, four Ag positions
an approximately planar threefold coordination, while one Ag atom is almost
linearly coordinated by two S atoms. The Ag-S distances are 2.56-2.94 Å, 2.49-
2.76 Å, and 2.42-2.44 Å, respectively. All Ag atoms have Ag neighbours between
2.93 and 3.60 Å, with total coordination numbers of 10 (x 4), 11, 12, or 14 (x 2).

1. Structure Reports, 8, 174.
2. Ibid., 9, 190; 30A, 100.

GOLD INDIUM   INDIUM SILVER
Au_9In_4 Ag_9In_4

J.K. BRANDON, R.Y. BRIZARD, W.B. PEARSON and D.J.N. TOZER, 1977. Acta Cryst.,
B33, 527-537.

Cubic, Al_4Cu_9 type (1), $P\bar{4}3m$, Z = 4. Mo radiation, diffractometer data. Contrary
to 2 who gave a body-centred cell of Cu_5Zn_8 type.

	a(Å)	D_m	R	Reflexions
Au_9In_4	9.829	15.60	0.082	324
Ag_9In_4	9.922	9.90	0.088	146

Atomic positions

Cluster	Site	Point set	Atoms	Au_9In_4 x	y	z	Ag_9In_4* Point set	Atoms	x	z
A	IT	4(e) xxx	4 In	0·1212			4(e) xxx	4 In	0·122	
B	IT	4(e) xxx	4 Au	0·6080			4(e) xxx	4 Ag	0·605	
A	OT	4(e) xxx	4 Au	−0·1659			4(e) xxx	4 Ag	−0·162	
B	OT	4(e) xxx	4 Au	0·3241			4(e) xxx	4 Ag	0·318	
A	OH	6(f)00z	6 Au	0·0	0·0	0·3575	6(f)00z	6 Ag	0·0	0·355
B	OH	6(g)½½z	6 Au	0·5	0·5	0·8566	6(g)½½z	6 Ag	0·5	0·854
A	CO	24(j) xyz	9 Au + 3 In	0·3319	0·3031	0·0307	12(i) xxz	12 Ag	0·320	0·034
B	CO	24(j) xyz	9 In + 3 Au	0·8259	0·8000	0·5361	12(i) xxz	12 In	0·808	0·530

* Approximate atomic coordinates based on refinement of reflexions with h+k+ℓ = 2n.

Interatomic distances (Å)
Au$_9$In$_4$. For distances involving CO atoms, the x and y coordinates have been averaged to place the CO atoms on xxz sites.

Number	Cluster A		Cluster B		Number	Cluster A		Cluster B	
	A IT (In) to:		B IT (Au) to:			A OH (Au) to:		B OH (Au) to:	
3	A IT	3·37	B IT	3·00	2	A IT	2·87	B IT	2·87
3	A OT	2·89	B OT	2·95	2	A OT	2·98	B OT	3·02
3	A OH	2·87	B OH	2·87	1	A OH'	2·80	B OH'	2·82
3	A CO	2·86	B CO	2·93	4	A CO	3·16	B CO	3·13
1	B OT	3·46	A OT	3·85	2	B CO	2·80	A CO	2·77
					2	B CO	3·14	A CO	3·06
	A OT (Au) to:		B OT (Au) to:			A CO ($\frac{3}{4}$Au + $\frac{1}{4}$In) to:		B CO ($\frac{3}{4}$In + $\frac{1}{4}$Au) to:	
3	A IT	2·89	B IT	2·95	1	A IT	2·87	B IT	2·93
3	A OH	2·98	B OH	3·02	1	A OT	2·86	B OT	2·82
3	A CO	2·86	B CO	2·82	1	B OT	2·89	A OT	2·94
3	B CO	2·94	A CO	2·89	2	A OH	3·16	B OH	3·13
1	B IT	3·85	A IT	3·46	1	B OH	2·77	A OH	2·80
					1	B OH	3·06	A OH	3·14
					2	A CO	3·99	B CO	3·85
					2	B CO	2·93	A CO	2·88
					2	B CO	2·88	A CO	2·93
					2	A CO'	3·64	B CO'	3·75

The γ-brass structure can be described as built up of 26-atom clusters, composed of an inner tetrahedron (IT) of four atoms, followed by an outer tetrahedron (OT), an octahedron (OH) of six atoms, and finally a cubo-octahedron (CO) of twelve atoms. γ-Brasses with the I cell are composed of only one type of cluster packed as pseudoatoms in the b.c. cubic arrangement, those with the P cell of two clusters, A and B, packed as pseudoatoms in the CsCl structure arrangement, and those with the F cell (cell edge 2a) of four clusters A, B, C, and D. Factors which determine whether a cubic γ-brass adopts the I, P or F cell and the relative ordering of the atoms are: (i) obtaining a high value of the packing fraction, (ii) maximizing the number of contacts between like atoms, and (iii) avoiding contact between the atoms present in lesser proportion. Which factor dominates depends on the composition of the phase and the relative sizes of the atoms. In addition the electrostatic potential energy, as represented by the Ewald constant, is also a factor.

1. Strukturbericht, 1, 498, 499, 542; 3, 57, 590; Structure Reports, 30A, 3; 33A, 72.
2. Structure Reports, 11, 123; 15, 82.

GROUP VIII PLUTONIUM COMPOUNDS
Pu$_5$Ru$_3$ Pu$_5$Os$_3$ Pu$_5$Ir$_3$ Pu$_5$Rh$_3$ Pu$_5$Pt$_3$

A.V. BEZNOSIKOVA, N.T. ČEBOTAREV, A.S. LUK'JANOV, A.V. ČERNJI and E.A. SMIRNOVA, 1974. Atomn. Energ., 37, 144-148 [Soviet Atomic Energy, 37, 842-846].

Pt$_3$Pu$_5$ is hexagonal, the remainder are tetragonal. Mo radiation, photographic data. See also 1, 2.

	a(Å)	c(Å)	D$_m$	Z	Type	Space group
Pu$_5$Ru$_3$	10.7685	5.7473	14.80	4	W$_5$Si$_3$ (3)	I4/mcm
Pu$_5$Os$_3$	10.8818	5.6645	17.48*	4	W$_5$Si$_3$	I4/mcm
Pu$_5$Ir$_3$	11.0438	5.6115	17.18*	4	W$_5$Si$_3$	I4/mcm
Pu$_5$Rh$_3$	10.941	6.0203	13.4	4	Pu$_5$Rh$_3$	P4/ncc
Pu$_5$Pt$_3$	8.4905	6.0944	15.4	2	Mn$_5$Si$_3$ (4)	P6$_3$/mcm

* D$_x$

Atomic positions

Pu$_5$X$_3$, X = Ru, Os, Ir	x	y	z
Pu(1) in 4(b)	0	1/2	1/4
Pu(2) in 16(k)	0.083	0.219	0
X(1) in 4(a)	0	0	1/4
X(2) in 8(h)	0.157	0.657	0

Pu$_5$Rh$_3$ (origin at centre)			
Pu(1) in 4(b)	1/4	3/4	0
Pu(2) in 16(g)	0.034	0.339	0.889
Rh(1) in 4(c)	1/4	1/4	0.071
Rh(2) in 8(f)	0.407	0.593	1/4

Pt$_3$Pu$_5$			
Pu(1) in 4(d)	1/3	2/3	0
Pu(2) in 6(g)	0.240	0.240	3/4
Pt in 6(g)	0.606	0.606	3/4

Interatomic distances Å

		Pu$_5$Ru$_3$	Pu$_5$Os$_3$	Pu$_5$Ir$_3$
Pu(1) -	10 Pu	2.87, 3.47	2.83, 3.49	2.81, 3.53
-	4 X	2.79	2.80	2.82
Pu(2) -	9 Pu	3.02 - 3.57	3.05 - 3.60	3.09 - 3.66
-	6 X	2.88 - 3.27	2.91 - 3.24	2.94 - 3.23
X(1) -	2 X	2.87	2.83	2.81
-	8 Pu	2.90	2.92	2.94
X(2) -	10 Pu	2.79 - 3.27	2.80 - 3.24	2.82 - 3.23

		Pu$_5$Rh$_3$			Pu$_5$Pt$_3$
Pu(1) -	6 Pu	3.01, 3.40		- 8 Pu	3.05, 3.63
-	4 Rh	2.86		- 6 Pt	3.02
Pu(2) -	8 Pu	3.20 - 3.61		- 10 Pu	3.54 - 3.67
-	5 Rh	2.90 - 3.08		- 5 Pt	2.92 - 3.31
Rh(1) -	2 Rh	3.01	Pt -	9 Pu	2.92 - 3.31
-	8 Pu	2.91, 3.02		- 2 Pt	3.54
Rh(2) -	8 Pu	2.86 - 3.08			

Pu$_5$Rh$_3$ is a more general version of the W$_5$Si$_3$ type structure.

1. Structure Reports, 41A, 103.
2. This volume, p. 71.
3. Structure Reports, 19, 277.
4. Strukturbericht, 4, 24, 137, 246; Structure Reports, 32A, 102.

HAFNIUM NICKEL SILICON
Hf$_3$Ni$_2$Si$_3$

Ja.P. JARMOLJUK, Ju.N. GRIN' and E.I. GLADYŠEVSKIJ, 1977. Kristallografija, 22, 726-730 [Soviet Physics - Crystallography, 22, 416-419].

Orthorhombic, Cmcm, a = 3.831, b = 9.862, c = 13.003 Å, Z = 4, D_x = 9.97. Cu radiation, R = 0.124 for 214 reflexions, photographic data.

Atomic positions   (Hf(2) and Si(2) in 4(c), remainder in 8(f))

	x	y	z
Hf(1)	0	0.4268	0.1177
Hf(2)	0	0.1337	1/4
Ni	0	0.7166	0.0844
Si(1)	0	0.1090	0.0388
Si(2)	0	0.8342	1/4

Interatomic distances (Å)

Hf(1)	-	7	Hf	3.29 - 3.44 [3.83]		Si(1)	-	5 [6]	Hf	2.76, 2.82 [3.30]
	-	4	Ni	2.86, 2.98			-	3	Ni	2.27, 2.35
	- 6 [7]		Si	2.73, 2.82 [3.30]			-	1	Si	2.37
Hf(2)	-	8	Hf	3.29 - 3.83		Si(2)	-	7	Hf	2.73 - 2.78 [2.95]
	-	4	Ni	3.00			-	2	Ni	2.45
	-	5	Si	2.75 - 2.78 [2.94]						
Ni	-	6	Hf	2.86 - 3.00						
	-	2	Ni	2.99						
	-	4	Si	2.27 - 2.45						

The structure is based on $[SiM_6]$ trigonal prisms and can be built up from segments of the Al_2CuMg and TlI [CrB] structures.

HYDROGEN LITHIUM PALLADIUM
$D_{0.70}LiPd$

HYDROGEN LITHIUM PLATINUM
$D_{0.66}LiPt$

B. NACKEN and W. BRONGER, 1977. J. Less-Common Metals, 52, 323-325.

$D_{0.70}LiPd$, tetragonal, P4/mmm, a = 2.798, c = 3.768 Å, Z = 1, D_x = 6.42. R = 0.052, neutron radiation; Pd in 1(a): 0,0,0; Li in 1(d): 1/2,1/2,1/2; 0.7 D in 1(c): 1/2,1/2,0.

$D_{0.66}LiPt$, trigonal, P3m1, a = 2.728, c = 4.266 Å, Z = 1, D_x = 12.36. R = 0.061, neutron radiation; Pt in 1(a): 0,0,0; Li in 1(b): 1/3,2/3,1/2; 1/3 D in 1(a): 0,0,0.37; 1/3 D in 1(b): 1/3,2/3,0.905.

In D_xLiPt, D is in tetrahedral holes (D(1)-3Li 1.67, -1Pt 1.56, plus 1 Pt at 2.66 Å; D(2)-3Pt 1.63, -1Li 1.71, plus 1 Li at 2.51 Å). In D_xLiPd, D is in octahedral holes (D-4Pd 1.98, -2Li 1.88 Å).

HYDROGEN RUTHENIUM YTTERBIUM
H_6RuYb_2

R. LINDSAY, R.O. MOYER, J.S. THOMPSON and D. KUHN, 1976. Inorg. Chem. 15, 3050-3053.

Cubic, H_6RuSr_2 type (1), Fm3m, a = 7.248 Å, D_m = 7.68, Z = 4. R = 0.15, photographic data; Yb in 8(c): 1/4,1/4,1/4; Ru in 4(a): 0,0,0; H in 24(e): 0.223,0,0.

1. Structure Reports, 37A, 83.

INDIUM SAMARIUM SULPHUR
InS_6Sm_3

D. MESSAIN, D. CARRÉ and P. LARUELLE, 1977. Acta Cryst., B33, 2540-2542.

Orthorhombic, [ScS_6U_3 type (1)], Pnnm, a = 16.513, b = 13.632, c = 3.901 Å, Z = 4, D_x = 5.74. Mo radiation, R = 0.029 for 879 reflexions, diffractometer data. 11 site-sets given.

Interatomic distances (Å)

In(1) - 6 S 2.62	Sm(2) - 8 S 2.80 - 3.14
In(2) - 6 S 2.43, 2.72	Sm(3) - 7 S 2.76 - 2.94
Sm(1) - 8 S 2.83 - 2.94	

The In atoms are at the centres of octahedra which form chains along c by edge-sharing. The Sm atoms are in both 7- and 8-fold coordination.

1. Structure Reports, 42A, 123.

INDIUM SELENIUM
InSe

K.C. NAGPAL and S.Z. ALI, 1976. Indian J. Pure Appl. Phys., 14, 434-440.

Rhombohedral, R3m, a = 4.0046, c = 24.960 Å, D_m = 5.59, Z = 6. Cu radiation, R = 0.064, photographic data. See also 1, 2, and 3; contrary to 4 and 5. Atoms are in 3(a): 0,0,z, z = 0.0555, 0.9445, $\overline{0.7727}$, $\overline{0.5607}$ for In(1), In(2), Se(1), Se(2), respectively. Crystals are seldom single, being mostly composed of thin lamellae (25-32 Å thick) of alternate obverse and reverse structures. Reflexions with h-k = 3n are sharp and the others diffuse. [The same structure is also given by 6.]

Interatomic distances (Å)

In - 1 In 2.77	Se - 3 Se 3.81
- 3 Se 2.63	- 3 In 2.63

The distances are more reasonable than those of 5 and agree with 6.

1. W. KLEMM and V. VOGEL, 1934. Z. anorg. Chem., 219, 45.
2. Structure Reports, 18, 176.
3. K.C. NAGPAL and S.Z. ALI, 1975. Indian J. Pure Appl. Phys., 4, 258.
4. S. SUGAIKE, 1957. Mineral. J., 2, 63.
5. Structure Reports, 22, 142.
6. Ibid., 41A, 80.

INDIUM SULPHUR TERBIUM
$In_5S_{12}Tb_3$

D. CARRÉ, 1977. Acta Cryst., B33, 1163-1166.

Monoclinic, $P2_1/m$, a = 10.998, b = 21.259, c = 3.897 Å, γ = 96.36°, D_m = 5.40, Z = 2. Mo radiation, R = 0.037 for 2064 reflexions, diffractometer data, 20 site-sets.

Interatomic distances (Å)

In(1) - 6 S 2.53 - 2.84	Tb(1) - 8 S 2.77 - 3.13
In(2) - 4 S 2.45 - 2.48	Tb(2) - 8 S 2.83 - 2.97
In(3) - 6 S 2.55 - 2.74	Tb(3) - 7 S 2.73 - 2.84
In(4) - 6 S 2.55 - 2.77	
In(5) - 6 S 2.53 - 2.74	

Indium atoms are octahedrally or tetrahedrally coordinated and Tb atoms 7- or 8-coordinated as is usual.

IRIDIUM PLUTONIUM
Ir_3Pu_5

D.T. CROMER, 1977. Acta Cryst., B33, 1996-1997.

Tetragonal, W_5Si_3 (1), I4/mcm, Z = 4. Mo radiation, R = 0.0417, 281 reflexions for approximately stoichiometric alloy, R = 0.0536, 260 reflexions for Ir-rich alloy, diffractometer data. Atoms are placed Pu(1) in 4(b): 0,1/2,1/4; Pu(2) in 16(k): x,y,0; Ir(1) in 4(a): 0,0,1/4; Ir(2) in 8(h): x,x+1/2,0. See also 2, 3.

Ir_3Pu_5	a(Å)	c(Å)	D_m	x(Pu(2))	y(Pu(2))	x(Ir(2))
∼ Stoichiometric	11.012	5.727	16.64	0.0845	0.2184	0.1583
Ir-rich	11.015	5.621	16.99	0.0845	0.2204	0.1590

Interatomic distances (Å)

		Stoichiometric	Ir-rich
Pu(1)	- 10 Pu	2.86, 3.54	2.81, 3.51
	- 4 Ir	2.85	
Pu(2)	- 9 Pu	3.07 - 3.65	3.04 - 3.68
	- 4 Ir	2.91 - 3.27	2.91 - 3.22
Ir(1)	- 8 Pu	2.95	2.96*
	- 2 Ir	2.86	2.81
Ir(2)	- 10 Pu	2.85 - 3.27	2.85 - 3.22
	- 2 Ir	4.05	3.99

* [Misprinted as 2.995]

The site 0,1/2,1/4 (4(b)) occupied by Pu(1) is substituted partly by Ir in the Ir-rich alloy as indicated by larger and more anisotropic thermal motion and also a smaller c axis. This agrees with the situation in $(Pu,Ce)_5Co_3$ (4).

1. Structure Reports, 19, 277.
2. A.V. BEZNOSIKOVA, N.T. ČEBOTAREV, A.S. LUK'JANOV, A.V. ČERNJI and E.A. SMIRNOVA, 1974. Atomn. Energ.,USSR, 37, 144.
3. V.I. KUTAITSEV, N.T. CHEBOTAREV, I.G. LEBEDEV, M.A. ANDRIANOV, V.N. KONEV and T.S. MENSHIKOVA, 1965. Plutonium 1965, Edit. A.E. KAY and M.B. WALDRON, pp. 420-449. London: Chapman and Hall.
4. Structure Reports, 29, 37.

IRON LANTHANUM PHOSPHORUS
Fe_4LaP_{12}

W. JEITSCHKO and D. BRAUN, 1977. Acta Cryst., B33, 3401-3406.

Cubic, Im3, a = 7.8316 Å, Z = 2, D_x = 5.08. Mo radiation, R = 0.028 for 193
reflexions, diffractometer data. Atoms are sited: La in 2(a): 0,0,0; Fe in 8(c):
1/4,1/4,1/4; P in 24(g): 0,0.3539,0.1504.

Interatomic distances (Å)

La -	8 Fe	3.39		P -	1 La	3.01	
-	12 P	3.01		-	2 Fe	2.26	
				-	2 P	2.29,	2.36
Fe -	2 La	3.39		-	4 P	2.97	
-	6 P	2.26					

   Fe_4LaP_{12}, a new structural type (Fig. 1) can be derived from the $CoAs_3$ and
WAl_{12}-type structures by filling the (somewhat distorted) icosahedral and octahedral voids with La and Fe atoms, respectively. The P atoms are coordinated by
two Fe and two P atoms forming a distorted tetrahedron augmented by a La atom
outside one face of that tetrahedron. In the polyanionic $[Fe_4P_{12}]^{3-}$ framework
the P-P bonding distances are somewhat expanded, to accommodate the large La^{3+}
cation, which in turn has shorter La-P distances than would be expected from the
La-P distances in LaP, LaP_2, and LaP_5. New compounds LnT_4P_{12} (T = Fe, Ru, Os),
isotypic with $LaFe_4P_{12}$, are given in Table I. Like $CoAs_3$ and WAl_{12}, the $LaFe_4P_{12}$
structure has fewer parameters than bonding distances; these constraints are
discussed with reference to ReO_3 and transition metal tri-pnictides; all types
occur near to Oftedal's relation y+z = 1/2, although in TP_3 (T = transition metal)
there are localized bonds, while in WAl_{12} the bonds are delocalized. The T_4P_{12}
framework is relatively rigid, compressing the large La atom but allowing extra
space for Ru and Os compared to Fe; when the lanthanide component is too small
competing phases become more stable. Bonding schemes are discussed for related
compounds.

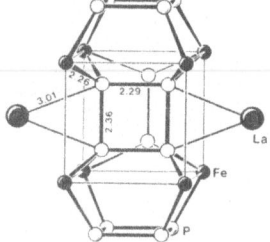

Fig. 1.   Near neighbours of the P_4 groups in $LaFe_4P_{12}$; interatomic distances
           in Å.

IRON LANTHANUM SULPHUR
$Fe_2La_2S_5$ $Fe_{1.87}La_2S_5$

F. BESREST and G. COLLIN, 1977. J. Solid State Chem., 21, 161-170.

Orthorhombic, $Cmc2_1$, Mo radiation, diffractometer data. $Fe_2La_2S_5$, a = 3.997, b =
16.485, c = 11.394 Å, D_m = 4.76, Z = 4. R = 0.037 for 797 reflexions. See also 1.
$Fe_{1.87}La_2S_5$, a = 3.9996, b = 49.508, c = 11.308 Å, Z = 12, D_x = 4.83. R = 0.068
for 922 reflexions. 27 site-sets given.

Atomic positions (all atoms in 4(a))

$Fe_2La_2S_5$	x	y	z
La(1)	0	0.3702	0.058
La(2)	0	0.3815	0.4444
Fe(1)	0	0.7034	0.2699
Fe(2)	0	0.0650	0.2133
S(1)	0	0.1920	0.1151
S(2)	0	0.2064	0.4123
S(3)	0	0.8518	0.2581
S(4)	0	0.0001	0.0335
S(5)	0	0.5475	0.3096

The structure of $Fe_2La_2S_5$ (see 1, Fig. 1) is characterized by double chains of [FeS_4] tetrahedra (mean Fe-S 2.32 Å) and [FeS_6] octahedra (mean Fe-S 2.59 Å) which share edges and apices to form the double chains; La atoms are 8-coordinated, to 6 S forming triangular prisms plus two S opposite two of the rectangular faces (mean La-S = 2.99 and 3.00 Å); Fe-Fe is 3.10 Å.

$Fe_{1.87}La_2S_5$ is a superstructure with vacancies in only two Fe sites, one tetra-hedral and the other octahedral. The latter gives distortions which result in a short Fe-Fe distance (2.82 Å).

1. Structure Reports, 38A, 112.

IRON MANGANESE SILICON
$Fe_4Mn_{77}Si_{19}$ (K-phase)

C.B. SHOEMAKER and D.P. SHOEMAKER, 1977. Acta Cryst., B33, 743-754.

Monoclinic, C2, a = 13.362, b = 11.645, c = 8.734 x 2 Å, β = 90.53°, D_m = 6.46, Z = 110 x 2 atoms/cell. There is a subcell also C2, with c_S = c/2; supercell reflexions are very weak. R = 0.062 for 1932 reflexions, diffractometer data (subcell); for supercell R = 0.073 for 2090 reflexions; 29 site-sets in the subcell.

The structure model for the substructure was derived from the aspects of the diffraction pattern that resemble the σ-FeCr and the δ-MoNi patterns. The sub-structure has rumpled layers and is entirely 'tetrahedrally close-packed' (tcp), with Mn(Fe) in the positions of coordination numbers (CN) 16, 15, and 14, and with mixtures of Mn(Fe) and Si in the positions of CN 12. The superstructure was assumed to be substitutional in regard to occupancy of CN 12 sites by Mn(Fe) or Si. Different CN 12 sites were found to be occupied by Mn(Fe) atoms, Si atoms, or mixtures of Mn(Fe) and Si. The alternation between the two subcells is such as to allow Si atoms to be everywhere coordinated only by Mn(Fe). In this the K phase resembles the ν phase, $Mn_{82}Si_{18}$, and the X phase, $Mn_{45}Co_{40}Si_{15}$, with plane-layered tcp structures, but differs from the D phase, Mn_5Si_2, with a somewhat 'non-ideal' tcp structure in which there are a few Si-Si contacts. Full details of interatomic distances and the coordinations in the K-phase and related structures are given.

IRON SILICON TUNGSTEN
$FeSiW_2 = (Fe_{0.465}Si_{0.465}W_{0.07})_7W_6$

P.I. KRIPJAKEVIC and Ja.P. JARMOLJUK, 1974. Dop. Akad. Nauk Ukr., RSR, A, 36, 460-463.

Orthorhombic, Pbam, a = 9.283, b = 7.817, c = 4.755 Å, D_m = 14.04, A = 26. R = 0.136, diffractometer data.

Atomic positions

			x	y	z	CN
W(1)	in	4(h)	0.456	0.338	1/2	15
W(2)	in	4(g)	0.133	0.047	0	14
W(3)	in	4(g)	0.411	0.158	0	16
*WFeSi	in	4(h)	0.241	0.086	1/2	12
†FeSi	in	8(i)	0.181	0.344	0.250	12
§SiFe	in	2(d)	0	1/2	1/2	12

*   WFeSi = 0.25 W, 0.25 Fe, 0.5 Si
†   FeSi   = 0.625 Fe, 0.375 Si
§   SiFe   = 0.75 Si, 0.25 Fe

[Interatomic distances (Å)]

W(1)	- 10 W	2.66 - 3.03		WFeSi	- 7 W	2.60 - 2.91
	- 4 FeSi	2.79, 2.82			- 4 FeSi	2.35, 2.41
	- 1 SiFe	2.67			- 1 SiFe	2.50
W(2)	- 10 W	2.58 - 3.09		FeSi	- 9 W	2.35 - 2.86
	- 4 FeSi	2.63, 2.65			- 2 FeSi	2.38
					- 1 SiFe	2.39
W(3)	- 8 W	2.72 - 3.09				
	- 6 FeSi	2.77 - 2.86		SiFe	- 8 W	2.50 - 2.80
	- 2 SiFe	2.80			- 4 FeSi	2.39

The structure is a Frank-Kasper tetrahedrally close-packed structure based on pentagon-triangle main layers (Fig. 1). It belongs to the A_6B_7 series including W_6Fe_7 and $Nb_6(Al_{0.25}Nb_{0.75})_7$ made up of combinations of the Zr_4Al_3 and $MgZn_2$ structure types (Fig. 2).

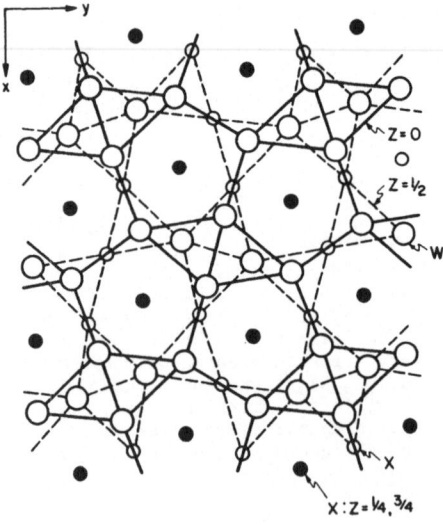

Fig. 1.   The pentagon-triangle nets in W_2FeSi.

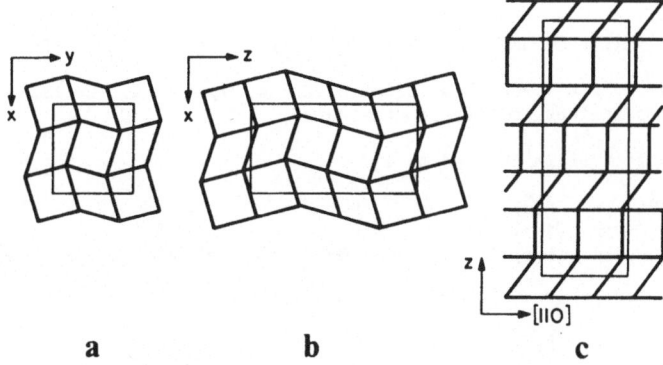

a   b   c

Fig. 2.   The related A_6B_7 structures made up of blocks of the Zr_4Al_3 structure
(squares) and $MgZn_2$ structure (parallelograms): a, $W_6(Fe,Si,W)_7$; b,
$Nb_6(Al,Ni)_7$; c, W_6Fe_7.

IRON SULPHUR   (PYRITE)
FeS_2

P. BAYLISS, 1977. Amer. Min., <u>62</u>, 1168-1172.

Triclinic, P1, a = b = c = 5.417 Å, $\alpha = \beta = \gamma = 90°$, Z = 4, D_X = 5.013. Mo radiation, R = 0.032 for 858 reflexions, diffractometer data. From Itoya mine, Yamagata,
Japan. See also <u>1</u>.

Atomic positions

	x	y	z
Fe(1)	0.0010	0.0020	0.0030
Fe(2)	0.4966	0.0001	0.5036
Fe(3)	0.5001	0.5020	0.0011
Fe(4)	-0.0006	0.5013	0.5038
S(1)	0.3857	0.3832	0.3840
S(2)	0.1149	0.6114	0.8846
S(3)	0.8854	0.1157	0.6143
S(4)	0.6153	0.8865	0.1141
S(5)	0.6151	0.6132	0.6137
S(6)	0.8854	0.3818	0.1149
S(7)	0.1147	0.8856	0.3841
S(8)	0.3857	0.1161	0.8842

Most pyrite is cubic (<u>1</u>) but some samples are weakly anisotropic; this is
associated with low formation temperature rather than chemical variation. The
present sample was pure, stoichiometric FeS_2 with a structure similar to that of
arsenian ullmannite (<u>2</u>), Fe-S 2.23-2.30, S-S 2.16 Å; S-Fe-S 85-95, Fe-S-Fe 114-117,
Fe-S-S 101-103°.

<u>1</u>. Strukturbericht, <u>1</u>, 125, 150, 215, 780; <u>2</u>, 272, 273; <u>5</u>, 52; Structure Reports,
    <u>24</u>, 231; <u>34A</u>, 97; <u>35A</u>, 75.
<u>2</u>. This volume, p. 19.

IRON SULPHUR TIN
Fe_2S_4Sn

J.C. JUMAS, E. PHILIPPOT and M. MAURIN, 1977. Acta Cryst., B33, 3850-3854.

Tetragonal, $I4_1/a$, a = 7.308, b = 10.338 Å, Z = 4, D_X = 4.32. Mo radiation, R = 0.043 for 317 reflexions, diffractometer data.

Atomic positions [origin shifted to $\bar{1}$]

			x	y	z
*Fe/Sn	in	8(d)	0	0	1/2
Fe	in	4(a)	0	1/4	0
S	in	16(f)	0.2606	-0.2503	0.0059

* 4 Fe + 4 Sn

Interatomic distances (Å)

Fe/Sn - 6 S 2.52, 2.53	S - 3 Fe/Sn 2.52, 2.53
	- 1 Fe 2.34
Fe - 4 S 2.34	- 3 S 3.49, 3.50

A small distortion of the inverse spinel structure (1) with tetrahedral interstices occupied solely by Fe^{2+} and octahedral sites by Fe/Sn.

1. Structure Reports, 15, 207; 17, 417.

IRON SULPHUR YTTERBIUM
FeS_4Yb_2

I.   A. TOMAS and M. GUITTARD, 1977. Mater. Res. Bull., 12, 1043-1046.
II. E. RIEDEL, R. KARL and R. RACKWITZ, 1977. Ibid., 12, 599-603.

Cubic, Fd3m, a = 10.69 Å, Z = 8, D_X = 5.92 for $Fe_{0.76}S_4Yb_{2.16}$. Mo radiation, R = 0.04 for 155 reflexions, diffractometer data. See 1 for earlier work and 2 for general references on spinels. The sample was quenched to avoid formation of an orthorhombic form (3). The structure is a variant of a normal spinel.

Atomic positions (I)

			x	y	z	occupancy
S	in	32(e)	0.2530	0.2530	0.2530	1
Fe	in	8(a)	1/8	1/8	1/8	0.76
Yb(1)	in	16(d)	1/2	1/2	1/2	0.96
Yb(2)	in	16(c)	0	0	0	0.12

Interatomic distances (Å)

S - 1 Fe 2.37	Yb(1) - 6 S 2.64 [2.70]
- 6 Yb 2.64, 2.70	- 6 Yb [3.78]
Fe - 4 S 2.37 [2.31]	Yb(2) - 6 S 2.70 [2.64]
- 4 Yb [2.31]	- 2 Fe [2.31]
	- 6 Yb [3.78]

1. Structure Reports, 29, 115; ref. 24.
2. Ibid., 11, 497; 13, 241; 15, 207.
3. Ibid., 32A, 132.

LANTHANUM RHENIUM SILICON
LaRe$_2$Si$_2$

V.K. PEČARSKIJ, O.I. BODAK and E.I. GLADYŠEVSKIJ, 1977. Kristallografija, 22, 630-633 [Soviet Physics - Crystallography, 22, 359-361].

Orthorhombic, Imma, a = 4.206, b = 4.116, c = 20.89 Å, Z = 4, D$_x$ = 10.42. Cu radiation, R = 0.103 (average) for 128 reflexions, photographic data.

Atomic positions (all atoms in 4(e))

	x	y	z
La	0	1/4	0.915
Re(1)	0	1/4	0.2119
Re(2)	0	1/4	0.7129
Si(1)	0	1/4	0.356
Si(2)	0	1/4	0.466

Interatomic distances (Å)

La	- 6 La	4.10 - 4.21
	- 4 Re	3.36, 3.40
	- 10 Si	3.13 - 3.26

Si(1)	- 4 La	3.19
	- 5 Re	2.51 - 3.01
	- 1 Si	2.30

Re(1)	- 2 La	3.36
	- 8 Re	2.59 - 2.94
	- 3 Si	2.54 - 3.01

Si(2)	- 6 La	3.13, 3.26
	- 3 Si	2.30, 2.50

Re(2)	- 2 La	3.40
	- 8 Re	2.59 - 2.94
	- 2 Si	2.51

The structure contains the trigonal prisms and tetragonal antiprisms common in ternary silicides. The networks of the smaller atoms (Si, Al, Ga, Re, Cr) in related structures (1-3) can be analysed as in Fig. 1.

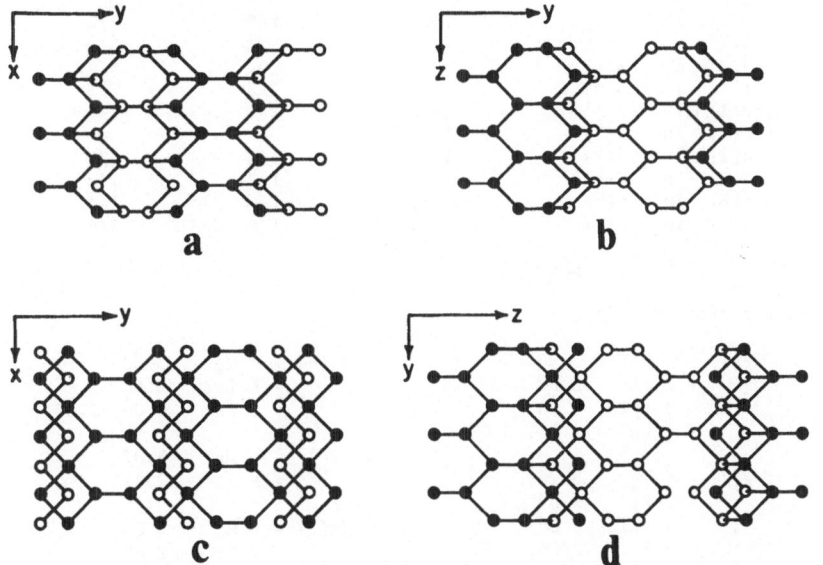

Fig. 1.   The network formed by the smaller atoms in ternary silicides of lanthanides or actinides, (a) CeAl$_2$Ga$_2$ [ThCr$_2$Si$_2$, 1], (b) CeNiSi$_2$ (2), (c) CeRe$_4$Si$_2$ (3), (d) LaRe$_2$Si$_2$.

1. Structure Reports, 30A, 129; this volume, p. 99.
2. Ibid., 34A, 63; 35A, 96.
3. This volume, p. 46.

LITHIUM MAGNESIUM ZINC
$Li_{0.11}MgZn_{1.89}$

I. P.I. KRIPJAKEVIČ and E.V. MEL'NIK, 1974. Dop. Akad. Nauk Ukr. R.S.R., A, 847-850.

Hexagonal, $P6_3/mmc$, a = 5.215, c = 59.89 Å, c/a = 11.482, D_m = 4.83, Z = 28. Cu radiation, R = 0.076, diffractometer data. 15 site-sets with occupancies are given.

The structure is a ternary Laves phase with a 14-layer structure. Mg atoms have CN 16 while the Zn and Li/Zn atoms have CN 12. Average interatomic distances are Mg-Mg 3.20, Mg-Zn 3.06, Mg-Li/Zn 3.07, Zn-Li/Zn 2.61, Zn-Zn 2.61 Å. The stacking sequence is hch hhc hhc hhh ch.

$LiMg_2Zn_3$

II. E.V. MEL'NIK and P.I. KRIPJAKEVIČ, 1974. Kristallografija, 19, 645-646
    [Soviet Physics - Crystallography, 19, 398-399].

Hexagonal, $P6_3/mmc$, a = 10.460, c = 17.050 Å, D_m = 4.03, Z = 16. Ni and Cu radiations, R = 0.119 for 190 reflexions, photographic data. Structural composition can be represented $Mg_8Li(Li_{0.5}Zn_{0.5})_6Zn_9$.

Atomic positions

			x	y	z
Mg(1)	in	12(k)	1/2	0	3/32
Mg(2)	in	12(k)	1/6	1/3	27/32
Mg(3)	in	4(f)	1/3	2/3	5/32
Mg(4)	in	4(e)	0	0	3/32
Li	in	4(f)	1/3	2/3	7/8
*LiZn(1)	in	12(j)	1/12	2/3	1/4
LiZn(2)	in	12(i)	1/4	0	0
Zn(1)	in	12(k)	1/6	1/3	1/8
Zn(2)	in	12(k)	1/4	1/2	1/2
Zn(3)	in	6(h)	1/12	1/6	1/4
Zn(4)	in	6(h)	7/12	1/6	1/4

* Li = 50%, Zn = 50% occupancy

[Interatomic distances (Å)]

Mg(1) - 4 Mg   3.20
      - 1 Li   3.07
      - 4 LiZn   3.06
      - 7 Zn   3.06, 3.07

Mg(2) - 4 Mg   3.20
      - 1 Li   3.07
      - 4 LiZn   3.06
      - 7 Zn   3.06, 3.07

Mg(3) - 4 Mg   3.20
      - 6 LiZn   3.06
      - 6 Zn   3.06, 3.07

Mg(4) - 4 Mg   3.20
      - 6 LiZn   3.06
      - 6 Zn   3.06, 3.07

Li   - 6 Mg   3.07
      - 6 Zn   2.61

LiZn(1) - 6 Mg   3.06
        - 2 LiZn 2.62
        - 4 Zn   2.61, 2.62

```
LiZn(2) - 6 Mg    3.06              Zn(3) - 6 Mg    3.06
        - 2 LiZn  2.62                    - 2 LiZn  2.62
        - 4 Zn    2.61, 2.62              - 4 Zn    2.61, 2.62

  Zn(1) - 6 Mg    3.07              Zn(4) - 6 Mg    3.06
        - 4 LiZn  2.61                    - 2 Li    2.61
        - 2 Zn    2.61                    - 2 LiZn  2.62
                                          - 2 Zn    2.62

  Zn(2) - 6 Mg    3.06
        - 1 Li    2.61
        - 2 LiZn  2.62
        - 3 Zn    2.61, 2.62
```

This is a superstructure based on the $MgNi_2$ type, with $\underline{a}$ doubled but retaining the same stacking sequence, Ca'Ba'Cb'Ab'.

LITHIUM TELLURIUM
$LiTe_3$

D.Y. VALENTINE, O.B. CAVIN and H.L. YAKEL, 1977. Acta Cryst., B$\underline{33}$, 1389-1396.

Trigonal, P$\bar{3}$c1, a = 8.7144, c = 21.35 Å, Z = 12, D_X = 5.53. A body-centred cubic subcell was given by $\underline{1}$. X-ray and neutron powder diffraction data, and X-ray single-crystal data.

   The proposed structure is based on harmonically related positional displacements of Te atoms from a reference structure that has six Te atoms at the centres of faces and edges of a cubic unit cell. The rhombohedrally centred hexagonal subcell corresponding to the diffraction symmetry has a = 8.7144 and c = 5.3363 Å, and the actual structure must be classified as a superstructure. The proposed displacements produce sections normal to $\underline{c}$ in which segments of Te-like chains can be distinguished. These sections are separated by metal-like layers that occur as the displacements become small. Li atoms are regularly distributed in channels parallel to $\underline{c}$. Li-Te distances (3 or 6-coordinated) range from 3.081 to 3.086 Å; Te-Te distances within chains from 2.86 to 3.02 Å and between chains from 3.14 to 3.32 Å; Te-Te-Te angles within chains 92 to 95° and 180°.

<u>1</u>. P.T. CUNNINGHAM, S.A. JOHNSON and E.J. CAIRNS, 1973. J. Electrochem. Soc., $\underline{120}$, 328.

MANGANESE NITROGEN
$Mn_2N_{0·86}$

I. M.N. EDDINE, E.F. BERTAUT and M. MAUNAYE, 1977. Acta Cryst., B$\underline{33}$, 2696-2698.
II. M.N. EDDINE and E.F. BERTAUT, 1977. Solid State Comm., $\underline{23}$, 147-150.

Hexagonal, $P6_322$, a = 4.8916, c = 4.5545 Å, Z = 3. Neutron diffraction data. Contrary to $\underline{1}$, $\underline{2}$, and $\underline{3}$. The magnetic structure (T < 308K) has space group C222_1, with a = 4.8552, b = 8.4088; c = 4.5327 Å at 4.2K.

Atomic positions

			x	y	z	occupancy
Mn	in	6(g)	0.33	0	0	1
N(1)	in	2(b)	0	0	1/4	0.43
N(2)	in	2(c)	1/3	2/3	1/4	0.74
N(3)	in	2(d)	1/3	2/3	3/4	0.12

[Interatomic distances (Å)]

Mn - 6 Mn 2.80	N(1) - 6 Mn 1.99
- 6 N 1.99	N(2) - 6 Mn 1.99
	N(3) - 6 Mn 1.99

The structure is essentially equivalent to that of 1, with N atoms in octa-hedral holes, but with unequal occupancies.

1. Strukturbericht, 2, 789.
2. Structure Reports, 33A, 111.
3. M. MEKATA, H. YOSHIMURA and H. TAKAKI, 1972. J. Phys. Soc. Japan, 33, 62.

MANGANESE NITROGEN SILICON
MnN_2Si

M. WINTENBERGEN, R. MARCHAND and M. MAUNAYE, 1977. Solid State Comm., 21, 733-735.

Orthorhombic, β-$NaFeO_2$($BeSiN_2$) type (1), $Pna2_1$, a = 5.248, b = 6.511, c = 5.070 Å, D_m = 4.22, Z = 4. R = 0.09, neutron diffraction data. See also 2.

Atomic positions

	x	y	z
Mn	0.072	0.628	-0.005
Si	0.071	0.130	0
N(1)	0.055	0.082	0.351
N(2)	0.102	0.655	0.410

Interatomic distances (Å)

Mn - 4 N 2.10 - 2.13	N(1) - 2 Mn 2.11, 2.13
Mean = 2.115	- 2 Si 1.71, 1.81
Si - 4 N 1.71 - 1.81	N(2) - 2 Mn 2.10, 2.12
Mean = 1.757	- 2 Si 1.73, 1.78

1. Structure Reports, 18, 422; 28, 133; 32A, 28.
2. Ibid., 38A, 97.

MANGANESE SILICON URANIUM
$Mn_3Si_5U_2$

Ja.P. JARMOLJUK, L.G. AKSEL'RUD and E.I. GLADYŠEVSKIJ, 1977. Kristallografija, 22, 627-629 [Soviet Physics - Crystallography, 22, 358-359].

Tetragonal, P4/mnc, a = 10.57, c = 5.435 Å, D_m = 8.45, Z = 4. Cu radiation, R = 0.132 for 125 reflexions, photographic data.

Atomic positions

			x	y	z
U	in	8(h)	0.0704	0.2394	0
Mn(1)	in	8(h)	0.3779	0.3401	0
Mn(2)	in	4(d)	0	1/2	1/4
Si(1)	in	8(g)	0.1839	0.6839	1/4
Si(2)	in	4(e)	0	0	0.2430*
Si(3)	in	8(h)	0.1909	0.4852	0

* also misprinted as 0.2530

Interatomic distances (Å)

U - 9 Si	2.88 - 3.12		Si(1) - 4 Si	2.50, 2.72
- 7 Mn	3.13 - 3.50		- 3 Mn	2.24, 2.75
- 4 U	3.73, 3.93		- 4 U	2.99, 3.12
Mn(1) - 6 Si	2.14 - 2.54		Si(2) - 2 Si	2.64, 2.79
- 2 Mn	3.01		- 4 Mn	2.54
- 5 U	3.13 - 3.50		- 4 U	2.95
Mn(2) - 6 Si	2.44, 2.75		Si(3) - 2 Si	2.50
- 2 Mn	2.72		- 4 Mn	2.14 - 2.50
- 4 U	3.16		- 3 U	2.88, 2.89

The structure is characterized by square-triangle nets at $z = 0$, $1/2$ and sparsely populated nets at $z = 1/4$, $3/4$, similar, for example, to those in $CuAl_2$, W_5Si_3, and Mn_5Si_3.

MARCASITE-TYPE COMPOUNDS
TX_2 (T = Os, Ru; X = As, P, Sb)

A. KJEKSHUS, T. RAKKE and A.F. ANDRESEN, 1977. Acta Chem. Scand., A31, 253-259.

Orthorhombic, FeS_2 (marcasite) type (1), Pnnm, Z = 2. Powder neutron diffraction data. See also 2. Atoms are placed: T in 2(a): 0,0,0; X in 4(g): x,y,0.

	RuP_2	OsP_2	$RuAs_2$	$OsAs_2$	$RuSb_2$	$OsSb_2$
a(Å)	5.1169	5.1012	5.4279	5.4115	5.9514	5.9411
b	5.8915	5.9022	6.1834	6.1900	6.6743	6.6873
c	2.8709	2.9183	2.9685	3.0127	3.1790	3.2109
x	0.1617	0.1634	0.1700	0.1701	0.1812	0.1848
y	0.3727	0.3723	0.3666	0.3671	0.3590	0.3596
T-X (Å) x 4	2.371	2.376	2.468	2.477	2.648	2.639
T-X (Å) x 2	2.347	2.350	2.448	2.452	2.628	2.644
X-X (Å) x 1	2.234	2.248	2.475	2.469	2.863	2.889

The x and y positional parameters are correlated in any given series from the Periodic Table.

1. Strukturbericht, 1, 216, 495; 2, 272; 5, 52; Structure Reports, 39A, 77.
2. Structure Reports, 33A, 63.

MERCURY NICKEL
HgNi

M. PUŠELJ and Z. BAN, 1977. Z. Naturf., 32B, 479.

Tetragonal, AuCu (L1$_0$) type (1), P4/mmm, a = 4.22, c = 3.14 Å, Z = 2, D$_x$ = 15.40. Cu radiation, diffractometer data. Hg in 1(a): 0,0,0 and 1(c): 1/2,1/2,0; Ni in 2(e): 1/2,0,1/2; Hg-4Hg = 2.96 and Hg-8Ni = 2.63 Å.

1. Strukturbericht, 1, 484, 505, 507.

MOLYBDENUM NICKEL PHOSPHORUS
MoNiP

I.   R. GUÉRIN and M. SERGENT, 1977. Acta Cryst., B33, 2820-2823.
II. Idem, 1977. Mater. Res. Bull., 12, 381-388.

Hexagonal, Fe$_2$P type (1), P$\bar{6}$2m, a = 5.861, c = 3.704 Å, c/a = 0.632, D$_m$ = 8.26, Z = 3. Cu radiation, R = 0.043 for 242 reflexions, diffractometer data.

Atomic positions

			x	y	z
Ni	in	3(f)	0.25034	0	0
Mo	in	3(g)	0.58647	0	1/2
P(1)	in	2(c)	1/3	2/3	0
P(2)	in	1(b)	0	0	1/2

Interatomic distances (Å)

Ni - 4 P	2.24, 2.36		Mo - 6 Ni	2.70, 2.81
- 2 Ni	2.54		- 5 P	2.42, 2.56
- 6 Mo	2.70, 2.81		- 4 Mo	3.06

   The structure of MoNiP is of the ordered Fe$_2$P (C22) type and shows tetrahedral and square-pyramidal phosphorus coordination, respectively, for the Ni and Mo atoms.

1. Strukturbericht, 2, 15, 248; Structure Reports, 23, 68; 39A, 76.

NEODYMIUM SILICON   PRASEODYMIUM SILICON
NdSi   PrSi

V.N. NGUYEN, F. TCHÉOU and J. ROSSAT-MIGNOD, 1977. Solid State Comm., 23, 821-823.

Orthorhombic, FeB type (1, 2), Pnma, Z = 4. Diffractometer data. See also 3. The magnetic structures are also reported.

	a(Å)	b(Å)	c(Å)	R	D$_x$
NdSi	8.158	3.918	5.887	0.065	6.08
PrSi	8.243	3.941	5.918	0.07	5.84

[Atomic positions*]   All atoms in 4(c): x, 1/4, z

	x		z	
	Nd	Pr	Nd	Pr
Nd or Pr	0.176	0.180	0.120	0.118
Si	0.030	0.029	0.610	0.611

* coordinates from private communication - authors.

[Interatomic distances (Å)]

		NdSi	PrSi
Nd or Pr	- 8 Nd or Pr	3.74 - 3.92	3.74 - 3.94
	- 7 Si	3.03 - 3.23	3.07 - 3.25
Si	- 7 Nd or Pr	3.03 - 3.23	3.07 - 3.25
	- 2 Si	2.40	2.42

1. D. HOHNKE and E. PARTHÉ, 1966. Acta Cryst., 20, 572.
2. Strukturbericht, 2, 7, 241; 3, 12, 619.
3. Structure Reports, 29, 127, 129.

NICKEL SCANDIUM SILICON
NiScSi$_3$

B. Ja. KOTUR, O.I. BODAK, M.G. MYS'KIV and E.I. GLADYŠEVSKIJ, 1977. Kristallo-grafija, 22, 267-270 [Soviet Physics - Crystallography, 22, 151-153].

Orthorhombic, Amm2, a = 3.815, b = 3.825, c = 20.62 Å, D_m = 4.09, Z = 4. Cu radiation, R = 0.132 for 211 reflexions, photographic data. A second refinement, based on 145 diffractometer data was judged less satisfactory.

Atomic positions

			x	y	z
Sc(1)	in	2(a)	0	0	0.1661
Sc(2)	in	2(a)	0	0	0.8282
Ni(1)	in	2(b)	1/2	0	0.3845
Ni(2)	in	2(b)	1/2	0	0.6095
Si(1)	in	2(b)	1/2	0	0.2775
Si(2)	in	2(b)	1/2	0	0.7152
Si(3)	in	2(a)	0	0	0.4406
Si(4)	in	2(a)	0	0	0.5541
Si(5)	in	2(b)	1/2	0	0.0515
Si(6)	in	2(b)	1/2	0	0.9381

Interatomic distances (Å)

Sc(1)	- 6 Sc	3.81 - 3.85		Si(2)	- 6 Sc	2.88, 3.01
	- 4 Ni	2.94			- 1 Ni	2.18
	- 10 Si	2.88 - 3.04			- 2 Si	2.30
Sc(2)	- 6 Sc	3.81 - 3.85		Si(3)	- 2 Sc	3.00
	- 4 Ni	2.94			- 2 Ni	2.23
	- 10 Si	2.90			- 5 Si	2.34, 2.70
Ni(1)	- 4 Sc	2.94		Si(4)	- 2 Sc	3.00
	- 5 Si	2.21, 2.23			- 2 Ni	2.22
Ni(2)	- 4 Sc	2.94			- 5 Si	2.34, 2.70
	- 5 Si	2.18 - 2.26		Si(5)	- 2 Ni	2.26
					- 5 Si	2.34, 2.70
Si(1)	- 6 Sc	2.87, 2.99				
	- 1 Ni	2.21		Si(6)	- 2 Ni	2.21
	- 2 Si	2.30			- 5 Si	2.34, 2.70

The structure (Fig. 1) is built from layers of the AlB$_2$ structure (1) and the CeAl$_2$Ga$_2$ [ThCr$_2$Si$_2$, 2] structure types. The tetragonal antiprisms also occur in CeRe$_4$Si$_2$ (3).

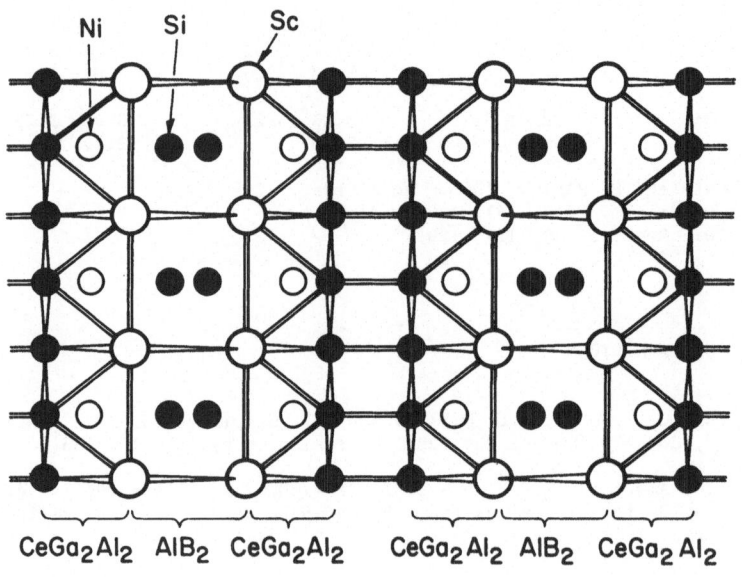

Fig. 1.   A projection onto (010) showing the trigonal prisms (AlB$_2$) and tetragonal antiprisms (CeAl$_2$Ga$_2$) as they are stacked in ScNiSi$_3$.

1. Structure Reports, 20, 51.
2. This volume, p. 99.
3. This volume, p. 46.

NICKEL SULPHUR   (HEAZLEWOODITE)
Ni$_3$S$_2$

M.E. FLEET, 1977. Amer. Min., 62, 341-345.

Rhombohedral, R32, a = 4.0821 Å, α = 89.475°, Z = 1. Fe radiation, R = 0.058 for
27 reflexions, powder data. See also 1, 2.

Atomic positions

			x	y	z
Ni	in	3(e)	1/2	0.247	-0.247
S	in	2(c)	0.255	0.255	0.255

Interatomic distances (Å)

        Ni - 4 S   2.27   S - 6 Ni   2.27
           - 4 Ni   2.51, 2.52

     The structure (1, 2) is confirmed, with Ni-Ni bonds comparable to those in
metallic Ni, and it is argued that the general valence rule is not applicable to
sulphides, contrary to 3.

1. Structure Reports, 11, 299.
2. Strukturbericht, 6, 75.
3. Structure Reports, 41A, 98.

NICKEL YTTRIUM
Ni_2Y_3

J. LE ROY, J.M. MOREAU and D. PACCARD, 1977. Acta Cryst., B33, 3406-3409.

Tetragonal, $P4_12_12$, a = 7.104, c = 36.597 Å, Z = 16, D_X = 5.52. Mo radiation, R = 0.11, diffractometer data. See also 1.

Atomic positions   (Y(1) and Y(2) in 4(a), remainder in 8(b))

	x	y	z
Y(1)	0.966	0.966	0
Y(2)	0.317	0.317	0
Y(3)	0.808	0.443	0.0156
Y(4)	0.163	0.659	0.0680
Y(5)	0.645	0.809	0.0747
Y(6)	0.012	0.163	0.0891
Y(7)	0.514	0.293	0.0933
Ni(1)	0.310	0.008	0.0484
Ni(2)	0.630	0.101	0.0257
Ni(3)	0.840	0.510	0.1063
Ni(4)	0.322	0.943	0.1167

Interatomic distances (Å)

Ni(1)	- 2 Ni	2.52, 2.54	Ni(3)	- 1 Ni	2.63
	- 7 Y	2.78 - 3.03		- 8 Y	2.79 - 3.38
Ni(2)	- 1 Ni	2.52	Ni(4)	- 3 Ni	2.54 - 3.75
	- 8 Y	2.73 - 3.60		- 7 Y	2.82 - 3.00

The structure is characterized by Ni-centred trigonal prisms of Y atoms. Four prisms are joined to form a characteristic prism grouping which has been found before in the structure of Y_8Co_5 (2). Y_3Ni_2 and Y_8Co_5 are members of a new structural series (Fig. 1) with formula $R_{2(n+1)}T_{n+2}$, where a block R_2T_2 (corresponding to a slice of the FeB type) is stacked with n blocks R_2T (corresponding to a slice of the As_2Nb or Ge_2Os type).

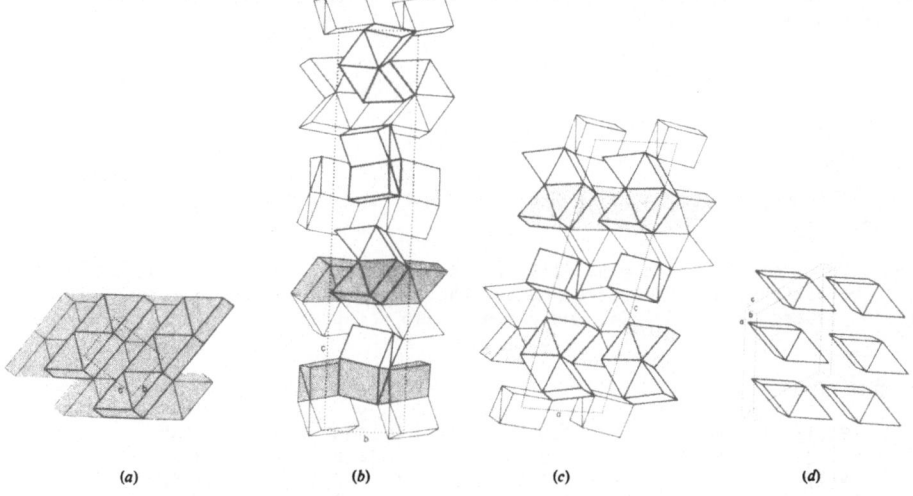

Fig. 1.   (a) FeB (n = 0), (b) Y_3Ni_2 (n = 2), (c) Y_8Co_5 (n = 3), (d) As_2Nb (n = ∞).

1. Structure Reports, 24, 198; 26, 276.
2. Ibid., 42A, 69.

NIOBIUM NITROGEN
δ-NbN

A.N. CHRISTENSEN, 1977. Acta Chem. Scand., A31, 77-78.

Cubic, NaCl type (1), Fm3m, a = 4.39 Å, Z = 4, D_x = 10.71. R = 0.037 for 16 reflexions, neutron diffraction data. Nb in 4(a): 0,0,0; N in 4(b): 1/2,1/2,1/2. See also 2. The refinement shows that δ-NNb can be stoichiometric. Nb-6N = 2.20 Å.

1. Strukturbericht, 1, 72.
2. Structure Reports, 42A, 114.

NIOBIUM SULPHUR TIN   SULPHUR TANTALUM TIN
NbS_2Sn S_2SnTa

R. EPPINGA and G.A. WIEGERS, 1977. Mater. Res. Bull., 12, 1057-1062.

Hexagonal, $P6_3/mmc$, a = 3.324, 3.307, c = 17.37, 17.44 Å, Z = 2, D_x = 5.51, 7.31 for Nb and Ta compounds, respectively. Neutron diffraction powder data, R = 0.06 for $TaSnS_2$; X-ray powder data, R = 0.146 for $NbSnS_2$. Atoms are placed Ta or Nb in 2(c): 1/3,2/3,1/4; Sn in 2(a): 0,0,0; S in 4(e): 0,0,z, z = 0.16. $TaSnS_2$ was also refined at 4.2°K.

[Interatomic distances (Å)]

		TaS_2Sn	NbS_2Sn			TaS_2Sn	NbS_2Sn
M	- 6 S	2.47	2.48	S	- 3 M	2.47	2.48
	- 6 M	3.31	3.32		- 1 Sn	2.79	2.78
					- 1 S	3.14	3.13
Sn	- 2 S	2.79	2.78		- 6 S	3.31	3.32
	- 6 Sn	3.31	3.32				

     The structure is unusual for intercalation compounds; Ta is in trigonal prismatic coordination forming TaS_2 slabs; Sn is linearly coordinated by 2 S atoms, and the Sn-Sn distance may account for the observed superconductivity as due to conduction in Sn layers.

NITROGEN SILICON
α-N_4Si_3

K. KATO, Z. INOUE, K. KIJIMA, I. KAWADA, H. TANAKA and T. YAMANE, 1975. J. Amer. Ceram. Soc., 58, 90-91.

Trigonal, P31c, a = 7.818, c = 5.591 Å, Z = 4, D_x = 3.15. Mo radiation, R = 0.041 for 1143 reflexions, diffractometer data. The oxygen content of the crystals was 0.05(3) %; traces of Fe, Mg, Ca, and Cl were also detected. See also 1; contrary to 2.

Atomic positions

			x	y	z
Si(1)	in	6(c)	0.0829	0.5135	0.6558
Si(2)	in	6(c)	0.2555	0.1682	0.4509
N(1)	in	6(c)	0.6558	0.6075	0.4320
N(2)	in	6(c)	0.3154	0.3192	0.6962
N(3)	in	2(b)	1/3	2/3	0.5926
N(4)	in	2(a)	0	0	0.4503

Confirms that α-Si₃N₄ does not contain O to any significant extent, contrary to 2 and in agreement with 3. Si(1)-N = 1.736, 1.746, 1.747, 1.759, Si(2)-N = 1.7$\overline{1}$5, 1.740, 1.744, 1.759 $\overline{\text{Å}}$.

<u>1</u>. Structure Reports, <u>21</u>, 194; <u>22</u>, 125.
<u>2</u>. Ibid., <u>40</u>A, 90, ref. 2.
<u>3</u>. Ibid., <u>34</u>A, 160; <u>40</u>A, 89.

NITROGEN TANTALUM
N₅Ta₃

I. N. TERAO, 1977. C.R. Acad. Sci. Paris, B, <u>285</u>, 17-20.

Monoclinic, Ti₃O₅ type (<u>1</u>), C2/m, a = 10.229, b = 3.875, c = 10.229 Å, β ∿ 90°, Z = 4, D$_X$ = 10.04. R = 0.38, electron diffraction data. See also <u>2</u>; [contrary to diffractometer study by <u>3</u>].

Atomic positions   (all atoms in 4(i): x,0,z)

	x	y	z
Ta(1)	0.19	0	0.25
Ta(2)	0.13	0	0.56
Ta(3)	0.13	0	0.94
N(1)	0.75	0	0.25
N(2)	0.04	0	0.12
N(3)	0.04	0	0.38
N(4)	0.31	0	0.07
N(5)	0.31	0	0.43

[Interatomic distances (Å)]

Ta(1)	- 10 Ta	3.23 - 3.88	N(2)	- 3 Ta	1.84 - 2.06
	- 6 N	2.03, 2.21		- 3 N	2.59 - 2.81
Ta(2)	- 10 Ta	2.93 - 3.89	N(3)	- 3 Ta	1.84 - 2.06
	- 6 N	1.84 - 2.30		- 3 N	2.59 - 2.81
Ta(3)	- 10 Ta	2.93 - 3.89	N(4)	- 4 Ta	2.03 - 2.27
	- 6 N	1.84 - 2.30		- 5 N	2.70 - 2.81
N(1)	- 4 Ta	2.03, 2.30	N(5)	- 4 Ta	2.03 - 2.27
	- 4 N	2.74		- 5 N	2.70 - 2.81

β-NTa₂

II. L.E. CONROY and A.N. CHRISTENSEN, 1977. J. Solid State Chem., <u>20</u>, 205-207.

Trigonal, P$\bar{3}$1m, a = 5.285, c = 4.919 Å, Z = 3, D$_X$ = 15.65, for a sample with composition N₀.₄₃Ta. R = 0.056, neutron powder diffraction data. Atoms are placed Ta in 6(k): 1/3,0,1/4; N(1) in 2(d): 1/3,2/3,1/2; N(2) in 1(a): 0,0,0. N(1) occupancy = 80%, N(2) occupancy = 97%. Contrary to <u>4</u> but equivalent to <u>2</u> by a shift of origin and change of space group.

[Interatomic distances (Å)]

Ta	- 6 Ta	3.03	N(1)	- 6 Ta	2.15
	- 3 N	2.15	N(2)	- 6 Ta	2.15

1. Structure Reports, 15, 188; 18, 459; 21, 229; 23, 335.
2. N. TERAO, 1971. Japan. J. Appl. Phys., 10, 248.
3. Structure Reports, 39A, 91.
4. Ibid., 18, 245.

NITROGEN TELLURIUM URANIUM
NTeU

R. TROJKO and Z. DESPOTOVIĆ, 1975.   Croat. Chem. Acta, 47, 121-125.

Tetragonal, PbFCl type (1), P4/nmm, a = 3.958, c = 7.630 Å, D_m = 10.24, Z = 2.   Cu
radiation, R = 0.125 for 19 reflexions, diffractometer data.   [Also given, a = 3.929,
c = 7.617 Å.]   Atoms are placed U in 2(c): z = 0.165; Te in 2(c): z = 0.624; Ni in
2(a).

Interatomic distances (Å)

U	- 5 Te	3.23, 3.50		Te	- 4 N	3.49
	- 4 N	2.35			- 4 Te	3.38
					- 4 U	3.23
N	- 4 U	2.35				
	- 4 Te	3.49				

1. Strukturbericht, 2, 45, 362; 3, 369.

NITROGEN TITANIUM
$N_{0.61}Ti$

S. NAGAKURA and T. KUSUNOKI, 1977.   J. Appl. Cryst., 10, 52-56.

Tetragonal, filled $ThSi_2$ type (1), $I4_1/amd$, a = 4.196-4.221, c = 4.296-4.305 Å
(depending on specimen preparation), A ∿ 13. Cu radiation, powder data and single-
crystal electron diffraction data.   Atoms are placed 8Ti in 8(e): 0,0,z, z = 0.268;
4N(1) in 4(a): 0,0,0; 0.88N in 4(b): 0,0,1/2.   See also 2.

[Interatomic distances (Å)]

Ti	- 2 Ti	2.79		N(1)	- 6 Ti	2.10, 2.30
	- 6 N	1.99 - 2.30		N(2)	- 6 Ti	1.99, 2.10

     In $N_{0.61}Ti$ the N atoms occupy octahedral interstices, separated as far as poss-
ible from other N atoms, while keeping the same number of N atoms along each Ti atom
row parallel to c; in the γ-Ti_2N (2) structure the N atoms occupy much narrower octahedral holes with two near Ti neighbours possibly covalently bonded, whereas $N_{0.61}Ti$
may be metallic.

1. Structure Reports, 9, 121; 26, 257.
2. Ibid., 27, 307; 34A, 115.

NITROGEN URANIUM
$β-N_3U_2$

N. MASAKI and H. TAGAWA, 1975.   J. Nucl. Mater., 58, 241-243.

Trigonal, La_2O_3 type (<u>1</u>), P$\bar{3}$m1, a = 3.700, c = 5.825 Å, Z = 1, D_x = 12.46. R = 0.058 for 24 reflexions, neutron diffractometer data; U in 2(d): 1/3,2/3,0.250; N(1) in 2(d): 1/3,2/3,0.641; N(2) in 1(a): 0,0,0. See also <u>2</u>.

Interatomic distances (Å)

U - 7 N 2.23 - 2.59	N(1) - 4 U 2.23, 2.28
- 3 U 3.61	- 3 N 2.69
	N(2) - 6 U 2.59

<u>1</u>. Strukturbericht, <u>1</u>, 242, 261, 744, 785; Structure Reports, <u>17</u>, 386; <u>30A</u>, 309.
<u>2</u>. Structure Reports, <u>11</u>, 170; <u>20</u>, 164; <u>32A</u>, 137.

PALLADIUM PHOSPHORUS
P_2Pd_{15}

I. Y. ANDERSSON, 1977. Acta Chem. Scand., A<u>31</u>, 354-358.

Rhombohedral, R$\bar{3}$, a = 7.1067, c = 17.0867 Å, Z = 3. Mo radiation, R = 0.066 for 2080 reflexions, diffractometer data. See also <u>1</u>.

Atomic positions

			x	y	z
Pd(1)	in	18(f)	0.40049	0.28437	0.96339
Pd(2)	in	18(f)	0.25807	0.23813	0.79537
Pd(3)	in	6(c)	0	0	0.07941
Pd(4)	in	3(b)	0	0	1/2
P	in	6(c)	0	0	0.28552

Interatomic distances (Å)

Pd(1) - 2 P 2.26, 2.99	Pd(3) - 13 Pd 2.64 - 3.22
- 13 Pd 2.64 - 3.35	Pd(4) - 12 Pd 2.71, 2.83
Pd(2) - 1 P 2.24	
- 13 Pd 2.64 - 3.35	P - 9 Pd 2.24 - 2.99

The structure can be described in terms of slightly distorted icosahedral building elements with six Pd(1) and six Pd(2) atoms at the corners and one Pd(4) atom at the centre. The icosahedra are arranged in nearly the same way as spheres in cubic close-packing; the 'octahedral holes' are filled with pairs of Pd(3) atoms and the 'tetrahedral holes' with phosphorus atoms (Fig. 1).

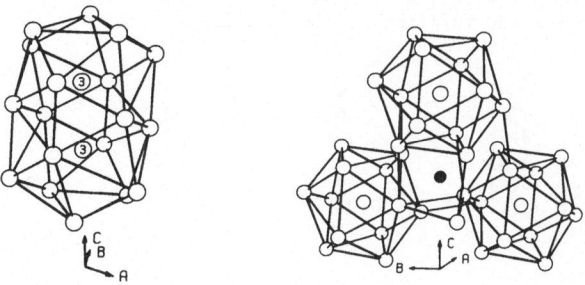

Fig. 1.   The 'octahedral hole' (left) and the 'tetrahedral hole' in $Pd_{15}P_2$ (right; for clarity only one atom of the fourth icosahedron is shown).

P_3Pd_7

II. T. MATKOVIĆ and K. SCHUBERT, 1977. J. Less-Common Metals, 55, 177-184.

Rhombohedral, R3, a = 11.976, c = 7.055 Å, Z = 6, D_x = 9.53. Cu radiation, R = 0.146 for 289 reflexions, photographic data. See also 1, 2.

Atomic positions (Pd(5) and Pd(6) in 3(a), remainder in 9(b))

	x	y	z
Pd(1)	0.093	0.260	0
Pd(2)	0.262	0.163	0.026
Pd(3)	0.209	0.241	0.389
Pd(4)	0.234	0.028	0.636
Pd(5)	0	0	0.289
Pd(6)	0	0	0.874
P(1)	0.206	0.011	0.321
P(2)	0.175	0.188	0.740

[Interatomic distances (Å)]

```
Pd(1) - 10 Pd  2.69 - 3.41        Pd(5) - 10 Pd  2.81 - 3.41
      -  3 P   2.18 - 2.51              -  3 P   2.41

Pd(2) - 10 Pd  2.69 - 3.31        Pd(6) - 10 Pd  2.87 - 3.14
      -  3 P   2.35 - 2.62              -  3 P   2.37

Pd(3) - 11 Pd  2.81 - 3.30        P(1)  -  8 Pd  2.18 - 2.78
      -  4 P   2.45 - 2.78
                                  P(2)  -  8 Pd  2.35 - 2.65
Pd(4) -  9 Pd  2.86 - 3.22
      -  4 P   2.24 - 2.65        Average P-Pd = 2.47
                                          Pd-Pd = 3.02
```

1. G. WIEHAGE, F. WEIBKE and W. BILTZ, 1936. Z. anorg. Chem., 228, 357; C.J. RAUB, W.H. ZACHARIASEN, T.H. GEBALLE and B.T. MATTHIAS, 1963. J. Phys. Chem. Solids, 24, 1093.
2. L.O. GULLMANN, 1966. J. Less-Common Metals, 11, 157.

PALLADIUM PLUTONIUM
Pd_5Pu_3

D.T. CROMER, 1976. Acta Cryst., B32, 1930-1932.

Orthorhombic, Cmcm, a = 9.201, b = 7.159, c = 9.771 Å, Z = 4, D_x = 12.98. Mo radiation, R = 0.0645 for 230 reflexions, diffractometer data.

Atomic positions

	x	y	z
Pu(1) in 4(c)	0	0.6251	1/4
Pu(2) in 8(e)	0.2018	0	0
Pd(1) in 4(c)	0	0.0254	1/4
Pd(2) in 8(f)	0	0.3147	0.4510
Pd(3) in 8(g)	0.2219	0.2863	1/4*

 *[misprinted as 0]

Interatomic distances (Å)

```
Pu(1) -  9 Pd  2.81 - 3.17        Pu(2) - 10 Pd  2.96 - 3.19
      -  8 Pu  3.78, 4.08               -  7 Pu  3.69 - 4.08
```

```
Pd(1) - 8 Pd   2.77 - 3.80        Pd(3) - 8 Pd   2.77 - 3.95
      - 5 Pu   2.87, 3.07               - 6 Pu   2.81 - 3.19


Pd(2) - 8 Pd   2.82 - 3.95
      - 6 Pu   2.95 - 3.09
```

The structure may be isostructural with Ga_5Zr_3 (1).

1. Structure Reports, 27, 203, 205.

PALLADIUM SELENIUM (PALLADSEITE)
$Pd_{17}Se_{15}$

I.   A.S. AVILOV and R.M. IMAMOV, 1976. Izv. Akad. Nauk SSSR, Neorg. Mater., 12,
     1295-1296 [Inorg. Mater., 12, 1076-1077].
II. R.J. DAVIS, A.M. CLARK and A.J. CRIDDLE, 1977. Miner. Mag., 41, 123.

I. Cubic, Pm3m, a = 10.47 Å, Z = 2, D_X = 8.66. R = 0.228 for 139 reflexions,
electron diffraction data. II. Cubic, a = 10.635 Å, Z = 2, D_X = 8.15. Confirms
structure of 1; given as $Pd_{1.1}Se$ in 1.

1. Structure Reports, 20, 167; 27, 315.

PALLADIUM TELLURIUM
Pd_3Te_2

P. MATKOVIĆ and K. SCHUBERT, 1977. J. Less-Common Metals, 52, 217-220.

Orthorhombic, Rh_3Te_2 type (1), Amam, a = 7.900, b = 12.687, c = 3.858 Å, Z = 4, D_X =
9.87. Cu radiation, R = 0.133 for 156 reflexions, photographic data.

Atomic positions

			x	y	z
Pd(1)	in	4(c)	1/4	0.5341	0
Pd(2)	in	8(f)	0.067	0.3378	0
Te(1)	in	4(a)	0	0	0
Te(2)	in	4(c)	1/4	0.7422	0

[Interatomic distances (Å)]

```
Pd(1) - 4 Pd   2.88, 2.99         Te(1) - 8 Pd   2.79, 2.87
      - 5 Te   2.64, 2.79               - 6 Te   3.82 - 3.95


Pd(2) - 5 Pd   2.88 - 3.13         Te(2) - 7 Pd   2.64, 2.70
      - 5 Te   2.70, 2.87                - 2 Te   3.82
```

1. Structure Reports, 31A, 61.

PENTLANDITES
$Co_{0.07}Fe_{3.97}Ni_{4.84}S_8$ (Frood) $Co_{5.60}Fe_{1.63}Ni_{1.87}S_8$ (Outokumpu)

V. RAJAMANI and C.T. PREWITT, 1975. Amer. Min., 60, 39-48.

Cubic, Co_9S_8 type (1), Fm3m, a = 10.04-10.16, 9.98-10.03 Å, for Frood and Outokumpu
specimens, respectively, at 24-350°C, Z = 4. Diffractometer data, R = 0.018-0.031.

Structures as previously described (2); for M(T) in 32(f), x = 0.1261-0.1265; for S(2) in 24(e), x = 0.2598-0.2632. Site occupancies as in 2.

Natural pentlandite has a smaller cell dimension than synthetic pentlandite of the same composition and some samples show non-reversible thermal expansion. The results above (Frood, M(0)-S 2.377-2.418 Å) are interpreted as due to disorder of Fe and Ni leading to an enrichment of high spin Fe^{2+} in the octahedral site relative to the natural mineral.

1. Strukturbericht, 4, 26, 137; Structure Reports, 27, 169; 41A, 53.
2. Structure Reports, 20, 125; 27, 169; 42A, 116, 117.

PHOSPHORUS SILVER SULPHUR
$Ag_4P_2S_7$

P. TOFFOLI, P. KHODADAD and N. RODIER, 1977. Acta Cryst., B33, 1492-1494.

Monoclinic, B2/b, a = 10.778, b = 16.211, c = 6.534 Å, γ = 106.8°, Z = 4, D_X = 4.36. Mo radiation, R = 0.043 for 1485 reflexions, diffractometer data.

Atomic positions   (S(4) in 4(e), remainder in 8(f))

	x	y	z
Ag(1)	0.25479	0.34128	0.3006
Ag(2)	0.03948	0.05235	0.2189
P	0.3913	0.1441	0.3250
S(1)	0.1744	0.4387	0.0603
S(2)	0.2436	0.1781	0.1832
S(3)	0.5063	0.1026	0.1331
S(4)	0	1/4	0.0035

Interatomic distances (Å)

Ag(1) - 4 S	2.52 - 2.72	average = 2.62
Ag(2) - 4 S	2.54 - 2.88	average = 2.65
P - 4 S	2.01 - 2.13	average = 2.05

The structure has pairs of PS_4 tetrahedra sharing corners to form discrete P_2S_7 groups analogous to the pyrophosphate ion $P_2O_7{}^{4-}$.

PLATINUM PLUTONIUM   PLUTONIUM RHODIUM
$Pt_{20}Pu_{31}$ $Pu_{31}Rh_{20}$

D.T. CROMER and A.C. LARSON, 1977. Acta Cryst., B33, 2620-2627.

Tetragonal, I4/mcm, a = 11.302, 11.076, c = 37.388, 36.933 Å, D_m = 15.55, 13.79 for $Pt_{20}Pu_{31}$ and $Pu_{31}Rh_{20}$, respectively, Z = 4. Mo radiation, R_W = 0.057 and 0.047 for 775 and 890 reflections, respectively, diffractometer data. 15 site-sets for each compound. [There is a misprint in the coordinates: $Pu_{31}Pt_{20}$, z of Pu(6) should be 0.1366; private communication from authors.]

The structure (Fig. 1) is closely related to that of Pu_5Ru_3 (W_5Si_3 type), being equivalent to 7 W_5Si_3 cells stacked along c with two large atoms removed, and is also related to that of Y_3Rh_2 (1) in which similar coordination polyhedra occur. The interatomic distances are all normal. Coordination numbers for the Pu atoms are 14 (x 2), 15 (x 4), and 17, and for the Pt or Rh atoms 9, 10 (x 5), and 12 (x 2). Average interatomic distances (Å) are:

		$Pu_{31}Pt_{20}$	$Pu_{31}Rh_{20}$
Pu	- Pu	3.61	3.55
Pu	- Pt/Rh	3.06	2.97
Pt/Rh	- Pt/Rh	3.15	3.11

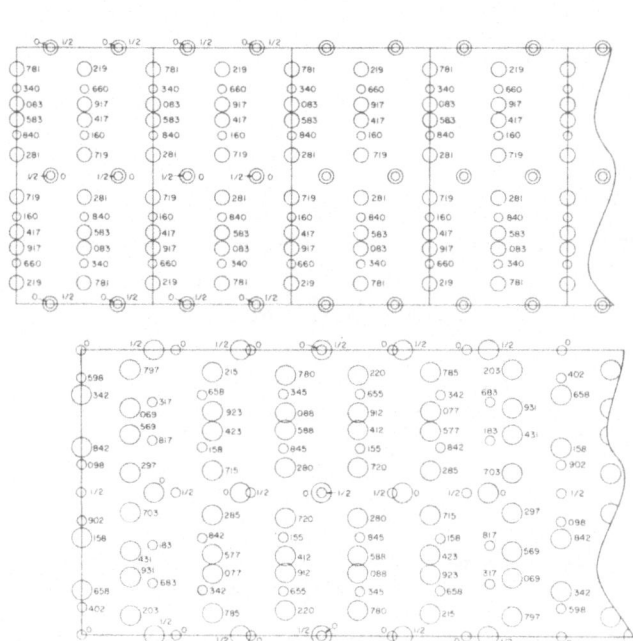

Fig. 1.   Upper part: 4 unit cells of Pu_5Ru_3 (W_5Si_3 type) projected along b; lower part: half of the $Pu_{31}Pt_{20}$ cell projected along b; a is vertical and c horizontal.

1. Structure Reports, 42A, 121.

PLATINUM SELENIUM
Pt_5Se_4

P. MATKOVIĆ and K. SCHUBERT, 1977.   J. Less-Common Metals, 55, 185-190.

Monoclinic, $P2_1/c$, a = 6.577, b = 4.610, c = 11.122 Å, β = 101.59°, Z = 2, D_x = 12.98.   Cu radiation, R = 0.151 for 422 reflexions, photographic data.   See also 1.

Atomic positions   (Pt(3) in 2(a), remainder in 4(e))

	x	y	z
Pt(1)	0.331	0.330	0.003
Pt(2)	0.223	0.663	0.174
Pt(3)	0	0	0
Se(1)	0.911	0.368	0.136
Se(2)	0.509	0.968	0.141

[Interatomic distances (Å)]

Pt(1) - 3 Pt 2.65, 2.73	Pt(2) - 2 Pt 2.65, 2.68
- 5 Se 2.41 - 3.41	- 5 Se 2.43 - 3.39

```
Pt(3) - 4 Pt  2.65, 2.68         Se(2) - 6 Pt  2.41 - 3.41
      - 6 Se  2.42 - 3.40               - 5 Se  3.13 - 4.01

Se(1) - 7 Pt  2.42 - 3.39
      - 5 Se  3.23 - 4.01
```

1. Structure Reports, 24, 210.

PLUTONIUM RHODIUM
Pu_5Rh_4

D.T. CROMER, 1977. Acta Cryst., B33, 1993-1995.

Orthorhombic, Pnma, a = 7.276, b = 14.332, c = 7.419 Å, D_m = 13.62, Z = 4. Mo radiation, R = 0.0385 for 1063 reflexions, diffractometer data. See also 1.

Atomic positions

			x	y	z
Pu(1)	in	4(c)	0.32245	1/4	0.01077
Pu(2)	in	8(d)	0.34236	0.62196	0.16361
Pu(3)	in	8(d)	0.00148	0.40796	0.17796
Rh(1)	in	4(c)	0.1898	1/4	0.3529
Rh(2)	in	4(c)	0.4602	1/4	0.6114
Rh(3)	in	8(d)	0.1845	0.5391	0.4609

Pu-Pu = 3.32-3.95, mean 3.60; Pu-Rh = 2.71 - 3.21, mean 2.96; Rh-Rh = 2.75-3.83, mean 3.22 Å. [Pu(3)-2Pu(3) = 3.792 Å is misprinted as Pu(3)-2Rh(3).]

The structure is similar to those of Sm_5Ge_4 (2) and Gd_5Si_4 (3) with slightly different coordinations.

1. Structure Reports, 42A, 99.
2. Ibid., 32A, 87.
3. Ibid., 38A, 91.

POTASSIUM SELENIUM
K_2Se_3

POTASSIUM SULPHUR
K_2S_3

P. BÖTTCHER, 1977. Z. anorg. Chem., 432, 167-172.

Orthorhombic, $Cmc2_1$, a = 7.309, 7.692, b = 9.914, 10.408, c = 7.473, 7.717 Å, D_m = 2.12, 3.31, for K_2S_3 and K_2Se_3, respectively, Z = 4. Mo radiation, diffractometer data, R = 0.019 and 0.041 for 433 and 319 reflexions.

Atomic positions   (S(2) or Se(2) in 8(b), remainder in 4(a))

K_2S_3	x	y	z	K_2Se_3	x	y	z
K(1)	0	0.1057	0.1145	K(1)	0	0.1011	0.1308
K(2)	0	0.5732	0.8371	K(2)	0	0.5804	0.8504
S(1)	0	0.7941	0.1577	Se(1)	0	0.7900	0.1810
S(2)	0.2268	0.8395	0	Se(2)	0.2417	0.8400	0

The structures contain X_3^{2-} anions (S-S = 2.083 Å, S-S-S 105.4°, and Se-Se = 2.383 Å, Se-Se-Se = 102.5°); K atoms are coordinated to 1 K (K_2S_3, 3.59 and K_2Se_3, 3.72 Å) and 8 S/Se (3.11-3.55 and 3.26-3.65 Å, respectively).

RHODIUM SAMARIUM
Rh_2Sm_3

J. LE ROY, J.M. MOREAU, D. PACCARD and E. PARTHÉ, 1977. Acta Cryst., B33, 2414-2417.

Rhombohedral, Er_3Ni_2 type (1), R$\bar{3}$, a = 8.701, c = 16.526 Å, Z = 9, D_x = 9.06. Mo radiation, R = 0.11 for 451 reflexions, diffractometer data. Isostructural compounds are given in Table I.

Atomic positions

			x	y	z
Sm(1)	in	3(b)	0	0	1/2
Sm(2)	in	6(c)	0	0	0.2962
Sm(3)	in	18(f)	0.2417	0.9861	0.0924
Rh	in	18(f)	0.5940	0.9802	0.0695

[Interatomic distances (Å)]

Sm(1)	- 8 Sm 3.37, 3.48		Sm(3)	- 11 Sm 3.48 - 4.07
	- 6 Rh 3.00			- 5 Rh 2.85 - 3.24
Sm(2)	- 10 Sm 3.37 - 4.00		Rh	- 8 Sm 2.85 - 3.24
	- 6 Rh 2.86, 3.08			- 1 Rh 2.94

The structure is composed of cubes centred by Sm atoms and double trigonal prisms centred by Rh atoms; the Er_3Ni_2 structure (Fig. 1) is closely related to that of U_3Si_2 which contains the same groupings. In the U_3Si_2 type structures 'stretched' prisms occur for borides, silicides, and germanides (c/a > 0.518) and squeezed prisms in compounds with transition elements at prism centres (c/a < 0.518). For the Er_3Ni_2 structure (1), all six cube faces are also trigonal prism faces. Thus this type will be formed only with ideal trigonal prisms having square contact faces.

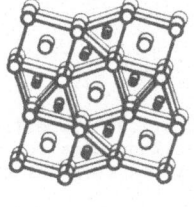 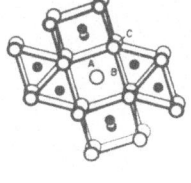

U_3Si_2 Er_3Ni_2

Fig. 1.   The U_3Si_2 and Er_3Ni_2 structures.

1. Structure Reports, 40A, 64.

SELENIUM SODIUM VANADIUM
$Na_{0.6}Se_2V$ (at 300 and 4.2°K) $NaSe_2V$ (at 300°K)

J.R. BLOEMBERGEN, R.J. HAANGE and G.A. WIEGERS, 1977. Mater. Res. Bull., 12, 1103-1110.

$NaSe_2V$, type II, $NaCrS_2$ (1), rhombohedral, R$\bar{3}$m, a = 3.735, c = 20.639 Å, at 300°K, Z = 3, D_x = 4.63. Cu radiation, diffractometer data, R = 0.044. Atoms are in sites 0,0,z, z = 0, 1/2, 0.2640 for V, Na, Se, respectively.

$Na_{0.6}Se_2V$, type I, $Na_{0.55}S_2Ti$ (2), rhombohedral, R3m, a = 3.482, 3.461, c = 22.207, 22.105 Å, at 300, 4.2°K, respectively, Z = 3, D_x = 4.76, 4.84. Cu radiation, diffractometer data, R = 0.041 and 0.039. Atoms are in sites $0,0,z$, $z = 0$, -0.1651, ±0.4014 at 300°K, 0, -0.1651, ±0.4023 at 4.2°K, for V, Na, Se, respectively. Na occupancy is 0.6; a model with 0.6 Na distributed over ±0.1651 was not significantly different (R = 0.043).

Interatomic distances (Å)

		Na_2SeV		$Na_{0.6}Se_2V$	(300°K)
Na	- 3 Se	2.95	octahedron	2.95	trigonal prism
	- 3 Se	2.95		3.00	
V	- 6 Se	2.59	octahedron	2.52	octahedron

Compounds M_xTX_2 (M = alkali metal, T = V, Ti, Zr, and X = S, Se) have two types of structure: I ($Na_{0.55}S_2Ti$), which has M in trigonal prismatic coordination by X atoms and T in octahedral coordination, and II, in which both T and M atoms have distorted octahedral coordination (1). The TX_6 octahedra are elongated in type I and compressed in type II structures.

1. Structure Reports, 37A, 135.
2. Ibid., 9, 188; 11, 437; 37A, 135; 39A, 100.

SILICON TITANIUM
Si_2Ti

W. JEITSCHKO, 1977. Acta Cryst., B33, 2347-2348.

Orthorhombic, Fddd, a = 8.2671, b = 4.8000, c = 8.5505 Å, Z = 8, D_x = 4.08. Mo radiation, R = 0.035 for 217 reflexions, diffractometer data. Atoms are placed Si in 16(e): x,1/8,1/8, x = 0.4615; Ti in 8(a): 1/8,1/8,1/8. See also 1.

Interatomic distances (Å)

Ti -	4 Ti	3.21		Si -	5 Ti	2.55 - 2.78
-	10 Si	2.55 - 2.78		-	9 Si	2.53 - 3.21

A rigid band model accounts for the electrical properties of the $TiSi_2$ and $CrSi_2$-type compounds and for the valence-electron rules for defect $TiSi_2$ (Nowotny chimney-ladder) structures, where the composition is adjusted so that no valence electrons are antibonding.

1. Strukturbericht, 7, 12, 95.

SULPHUR THALLIUM VANADIUM
S_8TlV_5

L. FOURNÈS, M. VLASSE and M. SAUX, 1977. Mater. Res. Bull. 12, 1-6.

Monoclinic, C2, a = 17.459, b = 3.354, c = 8.561 Å, β = 103.97° [given incorrectly as a = 17.465, b = 3.301, c = 8.519 Å, β = 103.94°; private communication from authors], D_m = 4.90, Z = 2. Mo radiation, R = 0.051 for 778 reflexions, diffractometer data.

Atomic positions   (Tl in 2(a), V(2) in 2(b), remainder in 4(c))

	x	y	z
Tl	0	0	0
V(1)	0.2906	0.539	0.1462
V(2)	0	0.555	1/2
V(3)	0.1507	0.044	0.4912
S(1)	0.1614	0.541	0.0013
S(2)	0.2595	0.036	0.3411
S(3)	0.0849	0.537	0.3168
S(4)	0.4261	0.521	0.3244

Interatomic distances ($\mathring{A}$)

Tl   - 10 S   3.16 - 3.35   V(3) - 6 S 2.26 - 2.59
                                       - 1 V 3.77

V(1) - 6 S   2.30 - 2.53
     - 3 V   3.05 - 3.30   S   -   S 3.23 - 3.44

V(2) - 6 S   2.34 - 2.47
     - 2 V   3.20, 3.30

The structure can be described as a framework made up of infinite layers parallel to (001) and double chains parallel to $\underline{b}$, of VS_6 octahedra sharing faces and edges. Large rectangular tunnels host the thallium atoms. The compound shows metallic behaviour. The double chains are centred by zigzag chains of V atoms (V–V = 3.05, 3.20, 3.19 $\mathring{A}$) with distances clearly less than the 3.40 $\mathring{A}$ critical distance for d-orbital overlap.

SULPHUR YTTERBIUM
ε-S_3Yb_2

I.   G.M. KUZ'MIČEVA and A.A. ELISEEV, 1977. Ž. Neorg. Khim., 22, 897-900 [Russ. J.
     Inorg. Chem., 22, 497-499].

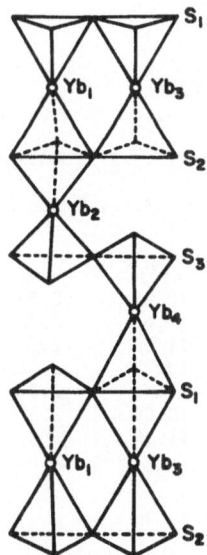

Fig. 1.   A perspective view of the YbS_6 groups in ε-Yb_2S_3.

Hexagonal, $P6_3cm$, a = 6.772, c = 18.28 Å, D_m = 6.16, Z = 6. Mo radiation, R = 0.103 for 112 reflexions, photographic data. Contrary to 1, see also 2.

Atomic positions

			x	y	z
Yb(1)	in	2(a)	0	0	0
Yb(2)	in	2(a)	0	0	0.3333
Yb(3)	in	4(b)	1/3	2/3	0
Yb(4)	in	4(b)	1/3	2/3	0.2483
S(1)	in	6(c)	0.3642	0	0.1238
S(2)	in	6(c)	0.3290	0	0.4129
S(3)	in	6(c)	0.6700	0	0.2891

The open structure (Fig. 1) has YbS_6 trigonal antiprisms (Yb-S 2.38-3.35 Å) sharing edges or faces.

$S_{1.7}Yb$

II. A.A. ELISEEV, G.M. KUZ'MIČEVA, V.V. EVDOKIMOVA and V.I. NOVOKŠONOV, 1976. Ž. Neorg. Khim., 21, 2900-2903 [Russ. J. Inorg. Chem., 21, 1600-1602].

Tetragonal, Cu_2Sb type (3), P4/nmm, a = 3.842, c = 7.747 Å, c/a = 2.02, Z = 2, D_x = 6.89. R = 0.17, diffractometer data. Yb in 2(c): 1/4,1/4,z, z = 0.24; 1.4 S(1) in 2(a): 3/4,1/4,0; S(2) in 2(c): 1/4,1/4,z, z = 0.66.

Tetragonal, $P\bar{4}b2$, a = 7.573, c = 7.764 Å, c/a = 1.025. R = 0.083, diffractometer data. Yb in 8(i): x,y,z, x = 0, y = 1/4, z = 0.27; S(1) in 4(g): x,1/2+x,0, x = 1/4; 1.2 S(2) in 2(a): 0,0,0; 1.2 S(3) in 2(c): 0,1/2,0; S(4) in 8(i): x,y,z, x = 0, y = 1/4, z = 0.66 [vacancy distribution for S(2) and S(3) not clearly specified]. Structure is a superstructure based on the Cu_2Sb type structure; Yb-S distances range from 2.67 to 3.25 Å in both structures.

1. Structure Reports, 29, 131.
2. Ibid., 41A, 133.
3. Strukturbericht, 2, 742; 3, 33, 288.

TERNARY ALKALINE-EARTH GROUP IV COMPOUNDS
AB_2X_2 (A = Ba, Ca, Sr; B = Al, Ag, Au, Cd; X = Ge, Pb, Si, Sn)

N. MAY and H. SCHÄFER, 1972. Z. Naturf., 27B, 864-865.

Tetragonal, $ThCr_2Si_2$ type (1), I4/mmm, Z = 2. Atoms are placed A in 2(a): 0,0,0; B in 4(d): 0,1/2,1/4; X in $\bar{4}$(e): 0,0,z.

AB_2X_2	a(Å)	c(Å)	c/a	D_x	z
$SrAg_2Si_2$	4.38	10.48	2.39	5.94	0.389
$SrCd_2Ge_2$	4.56	11.63	2.56	6.28	0.391
$BaAg_2Sn_2$	4.81	11.35	2.38	7.47	0.376
$CaAu_2Ge_2$	4.43	10.23	2.32	9.58	0.379
$SrAu_2Ge_2$	4.51	10.35	2.30	9.89	0.381
$BaAu_2(Au,Ge)_2$	4.65	10.56	2.27	11.94	0.375
$SrAl_2Pb_2$	4.46	11.19	2.51	8.30	0.378

Interatomic distances (Å)

	A-8B	A-8X	A-2X	B-4B	B-4X	X-1X
$SrAg_2Si_2$	3.41	3.31	4.08	3.10	2.63	2.33
$SrCd_2Ge_2$	3.69	3.46		3.22	2.81	2.54
$BaAg_2Sn_2$	3.72	3.68	4.27	3.40	2.80	2.81
$CaAu_2Ge_2$	3.38	3.37	3.89	3.13	2.58	2.48
$SrAu_2Ge_2$	3.43	3.42	3.95	3.19	2.63	2.46
$BaAu_2(Au,Ge)_2$	3.52	3.54	3.96	3.29	2.67	2.64
$SrAl_2Pb_2$	3.58	3.44	4.23	3.15	2.65	2.73

1. Structure Reports, 30A, 126; this volume, following report.

TRANSITION-METAL THORIUM SILICIDES
M_2Si_2Th (M = Cr, Mn, Ni, and Cu)

Z. BAN and M. SIKIRICA, 1965. Acta Cryst., 18, 594-599.

Tetragonal, $ThCr_2Si_2$ type, I4/mmm, Z = 2. Cu radiation, diffractometer data. Atoms
are placed Th in 2(a): 0,0,0; M in 4(d): 0,1/2,1/4; Si in 4(e): 0,0,z. Isostructural
compounds are in Table I.

ThM_2Si_2	a(Å)	c(Å)	D_X	Refl.	z(Si)
$ThCr_2Si_2$	4.043	10.577	7.53	52	0.374
$ThMn_2Si_2$	4.021	10.493	7.67*	49	0.386
$ThNi_2Si_2$	4.076	9.551	8.49	53	0.366
$ThCu_2Si_2$	4.104	9.864	8.16*	53	0.379

* D_m

Interatomic distances (Å)

		$ThCr_2Si_2$	$ThMn_2Si_2$	$ThNi_2Si_2$	$ThCu_2Si_2$
Th -	8 M	3.33	3.30	3.14	3.21
-	10 Si	3.15, 3.96	3.08, 4.05	3.15, 3.50	3.14, 3.74
-	[4 Th	4.04	4.02	4.08	4.10]
M -	4 M	2.86	2.84	2.88	2.90
-	4 Si	2.41	2.47	2.32	2.41
-	[4 Th	3.33	3.31	3.14	3.21]
[Si -	4 Th	3.15	3.08	3.15	3.14
-	4 M	2.41	2.47	2.32	2.41
-	Si	2.66	2.39	2.56	2.39]

    The structure is an ordered version of $BaAl_4$ and can be regarded as a filled-up
version of the $AgTlTe_2$ or $MoSi_2$ types.

Preliminary or duplicate accounts are not reported here; the compounds and abbreviated references are given below.

$Fe_{11}S_{12}$ (Pyrrhotite)	Acta Cryst., B31, 2759 (1975)
$Pb_7Sb_8S_{19}$ (Heteromorphite)	Neues Jb. Miner., Mh., 193 (1975)
Fe_2SnS_4	C.R. Acad. Sci. Paris, C, 284, 846 (1977)
Mn_2SnS_4	[see this volume, p. 76]

TABLE I

This Table lists substances for which structures have been assigned but not refined (usually on the basis of a powder pattern, and assumption of the atomic
parameters of the type structure).

Phase	Structure type	Reference
$Ag_4P_2Se_6$	$Ag_4P_2Se_6$	C.R. Acad. Sci. Paris, C, <u>285</u>, 559
Al_4CeCo	Al_4CoLa	Dop. Akad. Nauk Ukr.R.S.R., Ser.A, 265
Al_4CoPr	Al_4CoLa	Ibid., 265
$Al_{1.25}Ge_{0.75}Re$	$MoSi_2$	Izv. Akad. Nauk SSSR, Neorg. Mater., <u>13</u>, 926
$Al_{0.5}Mn_{0.5}Rh$	ClCs	Phys. Rev., B, <u>14</u>, 4131
$AlSe_2Tl$ (H.P.)	SeTl	Z. Naturf., <u>32B</u>, 1354
AmAs	ClNa	Radiochem. Radioanal. Lett., <u>20</u>, 371
AmBi	ClNa	Ibid., <u>20</u>, 371
AmN	ClNa	Ibid., <u>20</u>, 371
AmP (H.T.)	ClNa	Ibid., <u>20</u>, 371
AmSb	ClNa	Ibid., <u>20</u>, 371
Am_4Sb_3	P_4Th_3 (anti)	Ibid., <u>20</u>, 371
AmSi	BFe	J. Less-Common Metals, <u>56</u>, 47
As_2BaMg_2	Al_2CaSi_2	Z. Naturf., <u>32B</u>, 612
As_2CaCd_2	Al_2CaSi_2	Ibid., <u>32B</u>, 353
As_2CaZn_2	Al_2CaSi_2	Ibid., <u>32B</u>, 353
AsCm	ClNa	Radiochem. Radioanal. Lett., <u>20</u>, 371
AsCrFe	Fe_2P	Mater. Res. Bull., <u>12</u>, 381
AsDyS	AsCeS	J. Less-Common Metals, <u>53</u>, 137
AsErS	AsCeS	Ibid., <u>53</u>, 137
As_3Eu	As_3Eu	C.R. Acad. Sci. Paris, C, <u>283</u>, 479
AsGdS	AsCeS	J. Less-Common Metals, <u>53</u>, 137
AsHoS	AsCeS	Ibid., <u>53</u>, 137
AsK_3S_3	$AsNa_3S_3$	Z. anorg. Chem., <u>430</u>, 199
As_3La_4	P_4Th_3 (anti)	J. Less-Common Metals, <u>55</u>, 103
AsLaS	AsCeS	Ibid., <u>53</u>, 137
As_2Mg_2Sr	Al_2CaSi_2	Z. Naturf., <u>32B</u>, 612
AsMn	MnP	Mater. Res. Bull., <u>3</u>, 253
As_3Mo_2Ni	$As_3Mo_2Ni_{0.83}$	Rev. Chim. Minér., <u>12</u>, 335
AsNdS	AsCeS	J. Less-Common Metals, <u>53</u>, 137
AsPrS	AsCeS	Ibid., <u>53</u>, 137
AsSSm	AsCeS	Ibid., <u>53</u>, 137
AsSTb	AsCeS	Ibid., <u>53</u>, 137
AsSTm	AsCeS	Ibid., <u>53</u>, 137
AsSY	AsCeS	Ibid., <u>53</u>, 137
As_3Sr	As_3Eu	C.R. Acad. Sci. Paris, C, <u>283</u>, 479
AuCeIn	Fe_2P	Z. Metallk., <u>68</u>, 493
AuDyIn	Fe_2P	Ibid., <u>68</u>, 493
$AuDyNi_4$	Cu_4MgSn	Solid State Comm., <u>21</u>, 267
$Au_3Dy_3Sb_4$	$Au_3Sb_4Y_3$	Acta Cryst., B33, 1579
AuErIn	Fe_2P	Z. Metallk., <u>68</u>, 493
$AuErNi_4$	Cu_4MgSn	Solid State Comm., <u>21</u>, 267
$Au_3Er_3Sb_4$	$Au_3Sb_4Y_3$	Acta Cryst., B33, 1579
AuGdIn	Fe_2P	Z. Metallk., <u>68</u>, 493
$AuGdNi_4$	Cu_4MgSn	Solid State Comm., <u>21</u>, 267
$Au_3Gd_3Sb_4$	$Au_3Sb_4Y_3$	Acta Cryst., B33, 1579
AuHoIn	Fe_2P	Z. Metallk., <u>68</u>, 493
$AuHoNi_4$	Cu_4MgSn	Solid State Comm., <u>21</u>, 267
$Au_3Ho_3Sb_4$	$Au_3Sb_4Y_3$	Acta Cryst., B33, 1579
AuInLa	Fe_2P	Z. Metallk., <u>68</u>, 493
AuInNd	Fe_2P	Ibid., <u>68</u>, 493
AuInPr	Fe_2P	Ibid., <u>68</u>, 493
AuInSm	Fe_2P	Ibid., <u>68</u>, 493
AuInTb	Fe_2P	Ibid., <u>68</u>, 493

TABLE I   101

Phase	Structure type	Reference
AuInY	Fe$_2$P	Z. Metallk., 68, 493
AuInYb	Fe$_2$P	Ibid., 68, 493
Au$_3$Lu$_3$Sb$_4$	Au$_3$Sb$_4$Y$_3$	Acta Cryst., B33, 1579
Au$_3$Nd$_3$Sb$_4$	Au$_3$Sb$_4$Y$_3$	Ibid., B33, 1579
AuNi$_4$Tb	Cu$_4$MgSn	Solid State Comm., 21, 267
AuNi$_4$Tm	Cu$_4$MgSn	Ibid., 21, 267
AuNi$_4$Yb	Cu$_4$MgSn	Ibid., 21, 267
Au$_3$Sb$_4$Sm$_3$	Au$_3$Sb$_4$Y$_3$	Acta Cryst., B33, 1579
Au$_3$Sb$_4$Tb$_3$	Au$_3$Sb$_4$Y$_3$	Ibid., B33, 1579
Au$_3$Sb$_4$Tm$_3$	Au$_3$Sb$_4$Y$_3$	Ibid., B33, 1579
BCaNi$_4$	BCeCo$_4$	Z. Metallk., 68, 356
B$_6$CaNi$_{11}$	B$_6$CaNi$_{11}$	Ibid., 68, 356
B$_6$Ca$_2$Ni$_{21}$	C$_6$Cr$_{21}$W$_2$	Ibid., 68, 356
B$_4$CeRu$_4$	B$_4$LuRu$_4$	Solid State Comm., 24, 699
B$_4$CoGd	B$_4$CrY	Izv. Akad. Nauk SSSR, Neorg. Mater, 13, 923
B$_6$Co$_{12}$Gd	B$_6$Co$_{12}$Y	Ibid., 13, 923
BCo$_4$U	BCeCo$_4$	Dop. Akad. Nauk Ukr.R.S.R., Ser. A, 1029 (1974)
B$_2$Co$_3$U	B$_2$CeCo$_3$	Ibid., 1029 (1974)
B$_4$CoU	B$_4$CrY	Ibid., 652 (1975)
BCr$_2$	B$_4$Mn	Rev. Chim. Minér., 11, 547
B$_2$Cr$_2$Ir	B$_2$IrMo$_2$	Ibid., 11, 547
B$_6$Cr$_2$Nd	B$_6$CeCr$_2$	Dop. Akad. Nauk Ukr.R.S.R., Ser. A, 465 (1975)
B$_6$Cr$_2$Pr	B$_6$CeCr$_2$	Ibid., 465 (1975)
B$_6$Cr$_2$Sm	B$_6$CeCr$_2$	Ibid., 465 (1975)
B$_4$CrU	B$_4$CrY	Ibid., 652 (1975)
B$_4$DyRh$_4$	B$_4$Rh$_4$Y	Proc. Nat. Acad. Sci. U.S.A., 74, 1336
B$_4$DyRu$_4$	B$_4$LuRu$_4$	Solid State Comm., 24, 699
B$_4$ErRh$_4$	B$_4$Rh$_4$Y	Proc. Nat. Acad. Sci. U.S.A., 74, 1336
B$_4$ErRu$_4$	B$_4$LuRu$_4$	Solid State Comm., 24, 699
B$_4$EuRu$_4$	B$_4$LuRu$_4$	Ibid., 24, 699
B$_2$Fe$_3$U	B$_2$CeCo$_3$	Dop. Akad. Nauk Ukr.R.S.R., Ser. A, 1029 (1974)
B$_4$FeU	B$_4$CrY	Ibid., 652 (1975)
B$_2$Gd	AlB$_2$	J. Less-Common Metals, 56, 83
B$_{12}$Gd	B$_{12}$U	Ibid., 56, 83
B$_4$GdRh$_4$	B$_4$Rh$_4$Y	Proc. Nat. Acad. Sci. U.S.A., 74, 1336
B$_4$GdRu$_4$	B$_4$LuRu$_4$	Solid State Comm., 24, 699
B$_2$Ho	AlB$_2$	J. Less-Common Metals, 56, 83
B$_4$HoRh$_4$	B$_4$Rh$_4$Y	Proc. Nat. Acad. Sci. U.S.A., 74, 1336
B$_4$HoRu$_4$	B$_4$LuRu$_4$	Solid State Comm., 24, 699
B$_2$IrV$_2$	B$_2$IrMo$_2$	Rev. Chim. Minér., 11, 547
B$_4$LuRh$_4$	B$_4$Rh$_4$Y	Proc. Nat. Acad. Sci. U.S.A., 74, 1336
B$_2$Mo$_2$Ru	B$_2$IrMo$_2$	Rev. Chim. Minér., 11, 547
B$_4$NdRh$_4$	B$_4$Rh$_4$Y	Proc. Nat. Acad. Sci. U.S.A., 74, 1336
B$_4$NdRu$_4$	B$_4$LuRu$_4$	Solid State Comm., 24, 699
BNi$_4$U	BCeCo$_4$	Dop. Akad. Nauk Ukr.R.S.R., Ser. A, 1029 (1974)
B$_4$PrRu$_4$	B$_4$LuRu$_4$	Solid State Comm., 24, 699
B$_4$ReTh	B$_4$MoTh	Mh. Chem., 105, 1082
B$_4$Rh$_4$Sm	B$_4$Rh$_4$Y	Proc. Nat. Acad. Sci. U.S.A., 74, 1336
B$_4$Rh$_4$Tb	B$_4$Rh$_4$Y	Ibid., 74, 1336
B$_4$Rh$_4$Th	B$_4$Rh$_4$Y	Ibid., 74, 1336
B$_4$Rh$_4$Tm	B$_4$Rh$_4$Y	Ibid., 74, 1336
B$_4$Ru$_4$Sm	B$_4$LuRu$_4$	Solid State Comm., 24, 699
B$_4$Ru$_4$Tb	B$_4$LuRu$_4$	Ibid., 24, 699
B$_4$Ru$_4$Th	B$_4$LuRu$_4$	Ibid., 24, 699
B$_4$Ru$_4$Tm	B$_4$LuRu$_4$	Ibid., 24, 699
B$_4$Ru$_4$Y	B$_4$LuRu$_4$	Ibid., 24, 699

Phase	Structure type	Reference
B_4Ru_4Yb	B_4LuRu_4	Solid State Comm., $\underline{24}$, 699
B_2Sm	AlB_2	J. Less-Common Metals, $\underline{56}$, 83
$B_{12}Tb$	$B_{12}U$	Ibid., $\underline{56}$, 83
B_4ThV	B_4MoTh	Mh. Chem., $\underline{105}$, 1082
B_2Tm	AlB_2	J. Less-Common Metals, $\underline{56}$, 83
B_4UV	B_4CrY	Dop. Akad. Nauk Ukr.R.S.R., Ser. A, 652 (1975)
$BaGe_2$	$BaSi_2$	Z. Naturf., $\underline{32B}$, 1352
$BaPb$	BCr	J. Less-Common Metals, $\underline{52}$, 211
$BaPb_3$	$BaPb_3$	Ibid., $\underline{52}$, 211
Ba_2Pb	Cl_2Pb (anti)	Ibid., $\underline{52}$, 211
Ba_5Pb_3	B_3Cr_5	Ibid., $\underline{52}$, 211
$BeLiSb$	$LiSbZn$	C.R. Acad. Sci. Paris, C, $\underline{284}$, 679
$BiCm$	$ClNa$	Radiochem. Radioanal. Lett., $\underline{20}$, 371
$BiLa$	$ClNa$	J. Less-Common Metals, $\underline{52}$, 259
$BiLa_2$	La_2Sb	Ibid., $\underline{52}$, 259
Bi_2La	Bi_2La	Ibid., $\underline{52}$, 259
Bi_3La_4	P_4Th_3 (anti)	Ibid., $\underline{52}$, 259; $\underline{55}$, 103
Bi_3La_5	Mn_5Si_3	Ibid., $\underline{52}$, 259
$BiLa_4Pb_2$	P_4Th_3 (anti)	Ibid., $\underline{55}$, 103
Bi_2La_4Pb	P_4Th_3 (anti)	Ibid., $\underline{55}$, 103
Bi_2La_4Sn	P_4Th_3 (anti)	Ibid., $\underline{55}$, 103
Bi_2Mg_2Sr	Al_2CaSi_2	Z. Naturf., $\underline{32B}$, 612
$Bi_2Ni_3S_2$ (synthetic parkerite)	$Bi_2Ni_3S_2$	Neues Jb. Miner., Mh., 385 (1975)
$BiNp$	$ClNa$	Radiochem. Radioanal. Lett., $\underline{20}$, 371
$BiSmTe_3$	Bi_2STe_2	Ž. Neorg. Khim., $\underline{22}$, 1062
Bi_2Te_3	Bi_2STe_2	Ibid., $\underline{22}$, 1062
CTi	CTi	Ukr. Fiz. Ž. (Russ. Ed.), $\underline{19}$, 497
$CaCd_2P_2$	Al_2CaSi_2	Z. Naturf., $\underline{32B}$, 353
CaP_2Zn_2	Al_2CaSi_2	Ibid., $\underline{32B}$, 353
$CdCe$	$ClCs$	J. Phys. Chem., $\underline{61}$, 4666
$CdDy$	$ClCs$	Ibid., $\underline{61}$, 4666
$CdEr$	$ClCs$	Ibid., $\underline{61}$, 4666
$CdGd$	$ClCs$	Ibid., $\underline{61}$, 4666
$CdHo$	$ClCs$	Ibid., $\underline{61}$, 4666
$CdLa$	$ClCs$	Ibid., $\underline{61}$, 4666
$CdNd$	$ClCs$	Ibid., $\underline{61}$, 4666
$CdPr$	$ClCs$	Ibid., $\underline{61}$, 4666
$CdSm$	$ClCs$	Ibid., $\underline{61}$, 4666
$CdTb$	$ClCs$	Ibid., $\underline{61}$, 4666
$CeFe_4P_{12}$	Fe_4LaP_{12}	Acta Cryst., B$\underline{33}$, 3401
$CeIr_3$	Ni_3Pu	Soviet Physics - Crystallography, $\underline{19}$, 642
$CeIr_5$	$AuBe_5$	Ibid., $\underline{19}$, 642
$CeOs_4P_{12}$	Fe_4LaP_{12}	Acta Cryst., B$\underline{33}$, 3401
$CeP_{12}Ru_4$	Fe_4LaP_{12}	Ibid., B$\underline{33}$, 3401
$CePS$	$GdPS$	J. Solid State Chem., $\underline{21}$, 371
Ce_3Pt_2	Er_3Ni_2	Acta Cryst., B$\underline{33}$, 2414
Ce_3Pt_4	Pd_4Pu_3	J. Less-Common Metals, $\underline{53}$, 133
Ce_3Rh_2	Er_3Ni_2	Acta Cryst., B$\underline{33}$, 2414
CmN	$ClNa$	Radiochem. Radioanal. Lett., $\underline{20}$, 371
$CmSb$	$ClNa$	Ibid., $\underline{20}$, 371
$CoGa_3$	$CoGa_3$	Soviet Physics - Crystallography, $\underline{22}$, 188
$Co_2Ge_3Se_3$	$Co_2Ge_3S_3$	Inorg. Chem., $\underline{16}$, 2344
$Co_2Hf_3Si_3$	$Fe_2Hf_3Si_3$	Soviet Physics - Crystallography, $\underline{22}$, 726
$CoMoP$	Co_2P	Mater. Res. Bull., $\underline{12}$, 381
$CoPW$	Co_2P	Ibid., $\underline{12}$, 381
Co_2Si_2Th	Cr_2Si_2Th	Acta Cryst., $\underline{18}$, 594
$Co_2Si_3Zr_3$	$Fe_2Hf_3Si_3$	Soviet Physics - Crystallography, $\underline{22}$, 726
$CrFeP$	Co_2P	Mater. Res. Bull., $\underline{12}$, 381
$Cr_5La_6S_{16}$	$CrLaS_3$	Acta Cryst., B$\underline{33}$, 3437

TABLE I   103

Phase	Structure type	Reference
$Cr_4Nb_2Si_5$	$Cr_4Nb_2Si_5$	J. Less-Common Metals, <u>52</u>, 247
CrS_3Th	S_3ScY	Rev. Chim. Minér., <u>14</u>, 295
$CrSe_3Th$	S_3ScY	Ibid., <u>14</u>, 295
$Cr_4Si_5Ta_2$	$Cr_4Nb_2Si_5$	J. Less-Common Metals, <u>52</u>, 247
Cs_2Se	Cs_2S	Z. anorg. Chem., <u>429</u>, 118
$Cu_3FeSe_4Tl_2$ (bukov-ite)	Cr_2Se_2Th	C.R. Acad. Sci. Paris, C, <u>283</u>, 529
$Cu_3FeTe_4Tl_2$	Cr_2Se_2Th	Ibid., C, <u>283</u>, 529
$CuLa$	BFe	J. Less-Common Metals, <u>53</u>, 199
Cu_2La	AlB_2	Ibid., <u>53</u>, 199
Cu_5La	$CaCu_5$	Ibid., <u>53</u>, 199
Cu_6La (H.T.)	$CeCu_6$	Ibid., <u>53</u>, 199
Cu_7PS_6 (L.T.)	Cu_7PS_6	Z. Naturf., <u>32B</u>, 1100
Cu_7PSe_6 (L.T.)	Cu_7PS_6	Ibid., <u>32B</u>, 1100
$CuSe_2$ (krutaite)	FeS_2 (pyrite)	Bull. Soc. Fr. Minér. Crist., <u>95</u>, 475
Cu_2Se_2Tl	Cr_2Se_2Th	C.R. Acad. Sci. Paris, C, <u>283</u>, 529
Cu_2Te_2Tl	Cr_2Se_2Th	Ibid., C, <u>283</u>, 529
$DyPS$	$GdPS$	J. Solid State Chem., <u>21</u>, 371
Dy_3Pt_4	Pd_4Pu_3	J. Less-Common Metals, <u>53</u>, 133
$DyPt_2Si_2$	Cr_2Si_2Th	Ibid., <u>55</u>, 171
$ErPS$	$GdPS$	J. Solid State Chem., <u>21</u>, 371
Er_3Pt_4	Pd_4Pu_3	J. Less-Common Metals, <u>53</u>, 133
$EuFe_4P_{12}$	Fe_4LaP_{12}	Acta Cryst., B33, 3401
$EuP_{12}Ru_4$	Fe_4LaP_{12}	Ibid., B33, 3401
$EuPt_2Si_2$	Cr_2Si_2Th	J. Less-Common Metals, <u>55</u>, 171
Eu_2S_3 (α)	Eu_2S_3	Ž. Neorg. Khim., <u>22</u>, 558
$FeGeHf$	$AlNiZr$	Dop. Akad. Nauk Ukr.R.S.R., Ser. B, <u>36</u>, 1030
$FeGeZr$	$NiSiTi$	Ibid., <u>36</u>, 1030
$Fe_2Hf_3Si_3$	$Fe_2Hf_3Si_3$	Soviet Physics - Crystallography, <u>22</u>, 726
$FeMoP$	Co_2P	Mater. Res. Bull., <u>12</u>, 381
$FeNb_4P$	$CoNb_4Si$	Dop. Akad. Nauk Ukr.R.S.R., Ser. A, 262 (1977)
Fe_4NdP_{12}	Fe_4LaP_{12}	Acta Cryst., B33, 3401
$FeNp_6$	U_6Mn	J. Less-Common Metals, <u>53</u>, 147
$Fe_4P_{12}Pr$	Fe_4LaP_{12}	Acta Cryst., B33, 3401
$Fe_4P_{12}Sm$	Fe_4LaP_{12}	Ibid., B33, 3401
FeS_5Th_2	FeS_5U_2	Rev. Chim. Minér., <u>14</u>, 295
$FeSe_3Th$	FeS_3U	Ibid., <u>14</u>, 295
$FeSe_5Th_2$	FeS_5U_2	Ibid., <u>14</u>, 295
Fe_2Si_2Th	Cr_2Si_2Th	Acta Cryst., <u>18</u>, 594
$Fe_2Si_3Zr_3$	$Fe_2Hf_3Si_3$	Soviet Physics - Crystallography, <u>22</u>, 726
$Ga_{0.5}Mn_{0.5}Rh$	$ClCs$	Phys. Rev., B, <u>14</u>, 4131
Ga_3Sc	$AuCu_3$	Dop. Akad. Nauk Ukr.R.S.R., Ser.A, 166 (1977)
$GdNi$	BCr	J. Magnet. Res., <u>8</u>, 274
$GdNi_2$	Cu_2Mg	Ibid., <u>8</u>, 274
$GdNi_5$	$CaCu_5$	Ibid., <u>8</u>, 274
Gd_3Pt_4	Pd_4Pu_3	J. Less-Common Metals, <u>53</u>, 133
$GdPt_2Si_2$	Cr_2Si_2Th	Ibid., <u>55</u>, 171
GdS_3Sm	La_2S_3 (α)	Ž. Neorg. Khim., <u>22</u>, 558
$GeHfMn$	$AlNiZr$	Dop. Akad. Nauk Ukr.R.S.R., Ser. B, <u>36</u>, 1030
$GeHfMo$	$NiTiSi$	Ibid., <u>36</u>, 1030
$GeMnNb$	$AlNiZr$	Ibid., <u>36</u>, 1030
$GeMnTa$	$AlNiZr$	Ibid., <u>36</u>, 1030
$GeMnZr$	$NiSiTi$	Ibid., <u>36</u>, 1030
$GeMoZr$	$NiSiTi$	Ibid., <u>36</u>, 1030
$GeSe_2$	GeS_2 (H.T.)	J. Appl.Cryst., <u>10</u>, 202
$HfReSi$	$AlNiZr$	Dop. Akad. Nauk Ukr.R.S.R., Ser. B, <u>36</u>, 1030
$HfSiW$	$NiSiTi$	Ibid., <u>36</u>, 1030

Phase	Structure type	Reference
HoPS	GdPS	J. Solid State Chem., $\underline{21}$, 371
Ho_3Pt_4	Pd_4Pu_3	J. Less-Common Metals, $\overline{\underline{53}}$, 133
$In_{0.5}Mn_{0.5}Rh$	ClCs	Phys. Rev., B, $\underline{14}$, 4131
Ir_3La	Ni_3Pu	Soviet Physics - Crystallography, $\underline{19}$, 642
Ir_5La	$CaCu_5$	Ibid., $\underline{19}$, 642
Ir_7La_2	Ce_2Ni_7	Ibid., $\overline{\underline{19}}$, 642
K_3S_3Sb	$AsNa_3S_3$	Z. anorg. Chem., $\underline{430}$, 199
La (γ)	In	J. Less-Common Metals, $\underline{52}$, 259
$LaOs_4P_{12}$	Fe_4LaP_{12}	Acta Cryst., B$\underline{33}$, 3401
LaPS	GdPS	J. Solid State Chem., $\underline{21}$, 371
$LaP_{12}Ru_4$	Fe_4LaP_{12}	Acta Cryst., B$\underline{33}$, 3401
La_4PbSb_2	P_4Th_3 (anti)	J. Less-Common Metals, $\underline{55}$, 103
La_3Pt_2	Er_3Ni_2	Acta Cryst., B$\underline{33}$, 2414
La_3Pt_4	Pd_4Pu_3	J. Less-Common Metals, $\underline{53}$, 133
$LaPt_2Si_2$	Cr_2Si_2Th	Ibid., $\underline{55}$, 171
La_4Sb_3	P_4Th_3 (anti)	Ibid., $\overline{\underline{55}}$, 103
La_4Sb_2Sn	P_4Th_3 (anti)	Ibid., $\overline{\underline{55}}$, 103
La_3Sb_3Y	P_4Th_3 (anti)	Ibid., $\overline{\underline{55}}$, 103
Lu_3Pt_4	Pd_4Pu_3	Ibid., $\overline{\underline{53}}$, 133
$LuPt_2Si_2$	Cr_2Si_2Th	Ibid., $\overline{\underline{55}}$, 171
$Mn_4Nb_2Si_5$	$Cr_4Nb_2Si_5$	Ibid., $\overline{\underline{52}}$, 247
MnP	MnP	Mater. Res. Bull., $\underline{3}$, 253
$MnPbRh_2$	$AlCu_2Mn$	Phys. Rev., B, $\underline{14}$, $\overline{4}131$
$Mn_{0.5}RhTl_{0.5}$	ClCs	Ibid., B, $\underline{14}$, $4\overline{13}1$
$MnRh_2Sn$	$AlCu_2Mn$	Ibid., B, $\overline{14}$, 4131
$MnSe_3Th$	FeS_3U	Rev. Chim. Minér., $\underline{14}$, 295
MnSiTa	AlNiZr	Dop. Akad. Nauk Ukr.R.S.R., Ser. B, $\underline{36}$, 1030
$Mn_4Si_5Ta_2$	$Cr_4Nb_2Si_5$	J. Less-Common Metals, $\underline{52}$, 247
$Mn_4Si_5Ti_2$	$Cr_4Nb_2Si_5$	Ibid., $\underline{52}$, 247
Mo_3Re (L.T.)	OW_3	Phys. Status Solidi, $\underline{39}$, K21
NNb (δ')	AsNi	Izv. Akad. Nauk SSSR, Neorg. Mater., $\underline{12}$, 2085
$NTa_{0.83}$	ClNa	Fiz. Metal. Metalloved., $\underline{40}$, 202
NTa (ϵ)	NTa (ϵ)	J. Solid State Chem., $\underline{20}$, $\overline{2}05$
Na_3S_3Sb	$AsNa_3S_3$	Z. anorg. Chem., $\underline{430}$, $\overline{1}99$
Nb_4NiP	$CoNb_4Si$	Dop. Akad. Nauk Ukr.R.S.R., Ser. A, 262 (1977)
NbReSi	AlNiZr	Ibid., Ser.B, $\underline{36}$, 1030
$Nb_2Si_5V_4$	$Cr_4Nb_2Si_5$	J. Less-Common Metals, $\underline{52}$, 247
$Nb_4Si_5V_2$	$Cr_4Nb_2Si_5$	Ibid., $\underline{52}$, 247
Nb_3Te	OW_3	J. Phys. Chem. Solids, $\underline{35}$, 1181
$NdOs_4P_{12}$	Fe_4LaP_{12}	Acta Cryst., B$\underline{33}$, 3401
NdPS	GdPS	J. Solid State Chem., $\underline{21}$, 371
$NdP_{12}Ru_4$	Fe_4LaP_{12}	Acta Cryst., B$\underline{33}$, 3401
Nd_3Pt_2	Er_3Ni_2	Ibid., B$\underline{33}$, $24\overline{1}4$
Nd_3Pt_4	Pd_4Pu_3	J. Less-Common Metals, $\underline{53}$, 133
$NdPt_2Si_2$	Cr_2Si_2Th	Ibid., $\underline{55}$, 171
Nd_3Rh_2	Er_3Ni_2	Acta Cryst., B$\underline{33}$, 2414
NiPW	Fe_2P	Mater. Res. Bull., $\underline{12}$, 381
NiP_3W_2	$As_3Mo_2Ni_{0.83}$	Rev. Chim. Minér., $\overline{12}$, 335
NiScSi	NiSiTi	Dop. Akad. Nauk Ukr.R.S.R., Ser. A, 655 (1976)
Ni_2ScSi_2	Cr_2Si_2Th	Ibid., 655 (1976)
$Ni_4Sc_3Si_4$	$Cu_4Gd_3Ge_4$	Ibid., 655 (1976)
$Ni_{10}Sc_3Si_7$	$Ni_{10}Sc_3Si_7$	Ibid., 655 (1976)
$Ni_{16}Sc_6Si_7$	$Cu_{16}Mg_6Si_7$	Ibid., 655 (1976)
$Os_4P_{12}Pr$	Fe_4LaP_{12}	Acta Cryst., B$\underline{33}$, 3401
$P_{12}PrRu_4$	Fe_4LaP_{12}	Ibid., B$\underline{33}$, $340\overline{1}$
PPrS	GdPS	J. Solid State Chem., $\underline{21}$, 371
PSSm	GdPS	Ibid., $\underline{21}$, 371

TABLE I   105

Phase	Structure type	Reference
PSTb	GdPS	J. Solid State Chem., $\underline{21}$, 371
PSTm	GdPS	Ibid., $\underline{21}$, 371
PSY	GdPS	Ibid., $\overline{21}$, 371
$PdSe_2$	$PdSe_2$	Izv. Akad. Nauk SSSR, Neorg. Mater., $\underline{12}$, 1295
$Pd_{17}Se_{15}$ (pallad-seite)	$Pd_{17}Se_{15}$	Miner. Mag., $\underline{41}$, 123
$Pd_{20}Te_7$	$Pd_{20}Sb_7$	J. Less-Common Metals, $\underline{51}$, 35
Pd_2Y_3	Er_3Ni_2	Acta Cryst., B$\underline{33}$, 2414
PoSc (H.T.)	AsNi	Radiokhimija, $\overline{6}$, 845
Pr_3Pt_2	Er_3Ni_2	Acta Cryst., B$\overline{33}$, 2414
Pr_3Pt_4	Pd_4Pu_3	J. Less-Common Metals, $\underline{53}$, 133
$PrRe_4Si_2$	$CeRe_4Si_2$	Soviet Physics - Crystallography, $\underline{22}$, 178
Pr_3Rh_2	Er_3Ni_2	Acta Cryst., B$\underline{33}$, 2414
Pt_2Si_2Sr	Cr_2Si_2Th	J. Less-Common Metals, $\underline{55}$, 171
Pt_2Si_2Tm	Cr_2Si_2Th	Ibid., $\underline{55}$, 171
Pt_4Sm_3	Pd_4Pu_3	Ibid., $\overline{53}$, 133
Pt_4Tb_3	Pd_4Pu_3	Ibid., $\overline{53}$, 133
Pt_4Tm_3	Pd_4Pu_3	Ibid., $\overline{53}$, 133
Pt_4Yb_3	Pd_4Pu_3	Ibid., $\overline{53}$, 133
Rb_2Se	CaF_2 (anti)	Z. anorg. Chem., $\underline{429}$, 118
ReSiTa	AlNiZr	Dop. Akad. Nauk Ukr.R.S.R., Ser. B, $\underline{36}$, 1030
ReSiTi	AlNiZr	Ibid., $\underline{36}$, 1030
ReSiZr	AlNiZr	Ibid., $\overline{36}$, 1030
$Si_5Ta_2V_4$	$Cr_4Nb_2Si_5$	J. Less-Common Metals, $\underline{52}$, 247
Si_3Ti_5	Mn_5Si_3	High Temp. High Press., $\overline{6}$, 515
Sm_2Te_3 (γ)	P_4Th_3	Ž. Neorg. Khim., $\underline{22}$, 106$\overline{2}$

STRUCTURE REPORTS

SECTION II

INORGANIC COMPOUNDS

Edited by

J. Trotter

(University of British Columbia)

with the assistance of

J. M. Bree

ARRANGEMENT

To find particular inorganic compounds the subject index or formula index
should be used. The general arrangement is: elements, boron hydrides, carbonyls,
phosphorus-nitrogen and sulphur-nitrogen compounds, halides, cyanides, oxides,
double oxides (including titanates, vanadates, niobates, chromates, molybdates,
tungstates, manganates, uranates), hydroxides, borates, carbonates, nitrates,
phosphates, arsenates, sulphates, perchlorates, iodates, silicates, electron-
diffraction studies. Only complete structure analyses are described; compounds
for which only lattice parameters are determined have not been reported, and
those which have been described only in preliminary communications and for which
details will appear at a later date are tabulated.

SILICON OXYNITRIDE
Si_2N_2O

S.R. SRINIVASA, L. CARTZ, J.D. JORGENSEN, T.G. WORLTON, R.A. BEYERLEIN and M. BILLY, 1977. J. Appl. Cryst., 10, 167-171.

Neutron powder data at 0-23kbar pressure. Structure as in 1. Pressure produces a cooperative rotation of adjacent SiN_3O tetrahedra as a result of a decrease in the Si-O-Si angle, and a shortening of one Si-N bond.

1. Structure Reports, 29, 244.

NIOBIUM OXYNITRIDE   TANTALUM OXYNITRIDE
NbON   TaON

M. WEISHAUPT and J. STRÄHLE, 1977. Z. anorg. Chem., 429, 261-269.

Monoclinic, $P2_1/c$, a = 4.970, 4.968, b = 5.033, 5.037, c = 5.193, 5.185 Å, β = 100.23, 99.56°, Z = 4. Mo radiation, R = 0.084 for 366 reflexions for TaON, and 0.031 for 280 reflexions (twinned crystal) for NbON.

Atomic positions

	NbON			TaON		
	x	y	z	x	y	z
Nb,Ta	0.2911	0.0472	0.2151	0.2921	0.0440	0.2150
O	0.0636	0.3244	0.3476	0.0650	0.3331	0.3439
N	0.4402	0.7546	0.4782	0.4425	0.7529	0.4819

ZrO_2-type structure (1) as previously described for TaON (2).

1. Strukturbericht, 4, 9; Structure Reports, 23, 341; 30A, 307.
2. Structure Reports, 40A, 121.

DIBORANE
B_2H_6

D. MULLEN and E. HELLNER, 1977. Acta Cryst., B33, 3816-3822.

Refinement of the structure (1) and study of the electron-density distribution.

1. Structure Reports, 30A, 252.

HEXAKIS(AMMONIA-CYANOBORANE)SODIUM IODIDE
$[NH_3.BH_2(CN)]_6Na.I$

K.D. HARGRAVE, A.T. McPHAIL, B.F. SPIELVOGEL and P. WISIAN-NEILSON, 1977. J. Chem. Soc., Dalton, 2150-2153.

Rhombohedral, R$\bar{3}$, a = 8.506 Å, α = 82.54°, D_m = 1.34, Z = 1. Mo radiation, R = 0.063 for 1293 reflexions.

The structure (Fig. 1) contains $NH_3.BH_2(CN)$ units arranged in a regular octa-hedral manner about Na^+ and I^- ions; Na...NC = 2.487(4) Å, Na...N-C = 134°; I...HN = 2.87 Å, I...H-N = 149° (probably a hydrogen bond). B-N = 1.581, B-C = 1.579, C≡N = 1.152(8) Å.

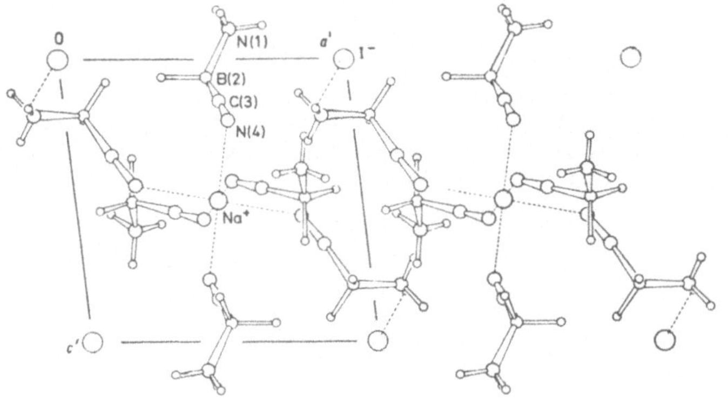

Fig. 1.   Structure of hexakis(ammonia-cyanoborane)sodium iodide.

BROMOPENTACARBONYLRHENIUM(I)
BrRe(CO)$_5$

M.C. COULDWELL and J. SIMPSON, 1977. Cryst. Struct. Comm., 6, 1-5.

Orthorhombic, Pnma, a = 11.898, b = 11.656, c = 6.189 Å, Z = 4. Mo radiation, R = 0.098 for 371 reflexions.

Isostructural with ClMn(CO)$_5$ (1). Re has octahedral coordination, Re-Br = 2.62(1), Re-C = 1.91, 1.96(3) Å.

1. Structure Reports, 37A, 170.

TRIRUTHENIUM DODECACARBONYL
Ru$_3$(CO)$_{12}$

M.R. CHURCHILL, F.J. HOLLANDER and J.P. HUTCHINSON, 1977. Inorg. Chem., 16, 2655-2659.

Monoclinic, P2$_1$/n, a = 8.117, b = 14.863, c = 14.614 Å, β = 100.67°, Z = 4. Mo radiation, R = 0.026 for 2281 reflexions.

The structure (Fig. 1) is as previously determined (1), except that axial Ru-CO bonds (1.929-1.953(5), mean 1.942 Å) are longer than equatorial Ru-CO bonds (1.908-1.934(5), mean 1.921 Å); Ru-C-O = 173 (axial), 179° (equatorial). Ru-Ru = 2.851, 2.852, 2.860(1) Å.

1. Structure Reports, 33A, 167.

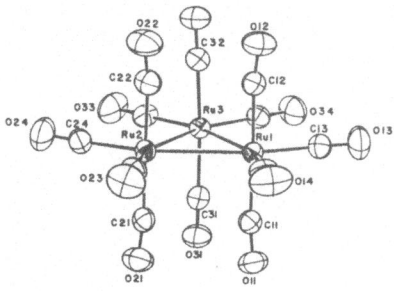

Fig. 1.   Structure of $Ru_3(CO)_{12}$.

DI-μ-HYDRIDO-DECACARBONYLTRIOSMIUM
$H_2Os_3(CO)_{10}$

M.R. CHURCHILL, F.J. HOLLANDER and J.P. HUTCHINSON, 1977.   Inorg. Chem., 16, 2697-2700.

Triclinic, PĪ, a = 8.603, b = 9.128, c = 11.893 Å, α = 91.46, β = 98.96, γ = 117.35°, Z = 2.   Mo radiation, R = 0.035 for 2049 reflexions.

The molecule (Fig. 1) has approximate C_{2v} symmetry and contains an isosceles Os_3 triangle, with one Os-Os bond bridged by two hydride ligands (not accurately located); this bond is significantly shorter than the other two (Fig. 1).   The non-bridged Os has four terminal carbonyl ligands, and the bridged Os atoms have three.

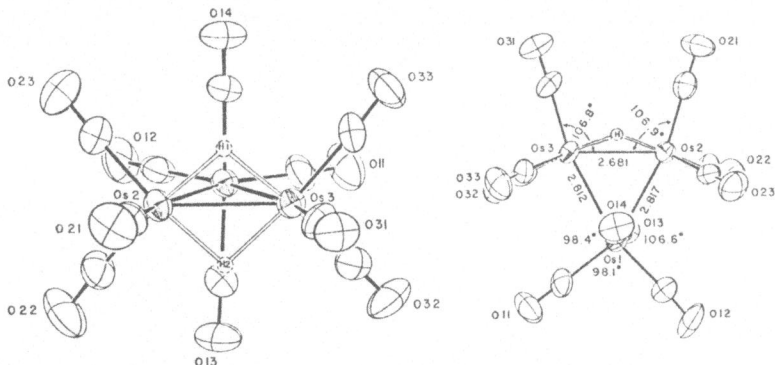

Fig. 1.   Views of $H_2Os_3(CO)_{10}$.

UNDECACARBONYLDIHYDRIDOTRIOSMIUM   DODECACARBONYLTRIOSMIUM
$H_2Os_3(CO)_{11}$ $Os_3(CO)_{12}$

M.R. CHURCHILL and B.G. DeBOER, 1977.   Inorg. Chem., 16, 878-884.

Monoclinic, $P2_1/n$, a = 8.074, 8.082, b = 14.727, 14.768, c = 14.777, 14.577 Å, β = 101.36, 100.56°, Z = 4.   Mo radiation, R = 0.037 and 0.034 for 2259 and 3040 reflexions.

The materials are isomorphous and quasi-isostructural (Fig. 1), the structure being as previously described (1), with disorder of CO groups in the hydride. The bridging hydride ligand increases one Os-Os bond length to 2.989(1) Å from an average of 2.877 Å in the dodecacarbonyl.

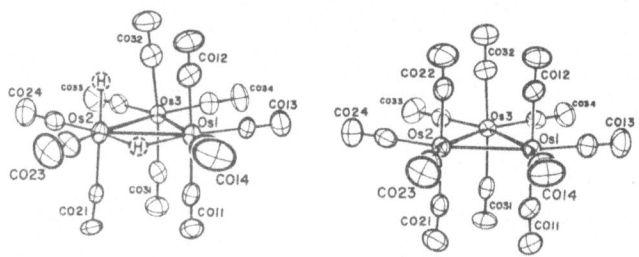

Fig. 1. $H_2Os_3(CO)_{11}$ and $Os_3(CO)_{12}$.

1. Structure Reports, 27, 651.

DODECACARBONYLDIIODOTRIOSMIUM
$Os_3(CO)_{12}I_2$

N. COOK, L. SMART and P. WOODWARD, 1977. J. Chem. Soc., Dalton, 1744-1746.

Monoclinic, $P2_1/n$, a = 10.329, b = 11.579, c = 9.334 Å, β = 100.07°, D_m = 3.3, Z = 2. Mo radiation, R = 0.062 for 2015 reflexions.

The molecule (Fig. 1) contains a linear O-C-Os-Os-Os-C-O sequence, with the I atoms occupying equatorial sites, trans to one another, on the terminal Os atoms. Terminal equatorial ligands are staggered with respect to the central carbonyl ligands, and these terminal ligands are tilted towards the central Os atom. Os-Os = 2.935(2), Os-I = 2.772(3), Os-C = 1.91-1.97(3), C-O = 1.12-1.15(3) Å.

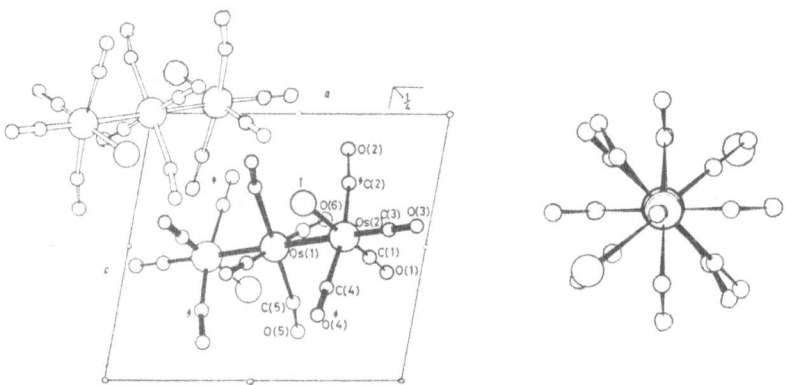

Fig. 1.   Structure of $Os_3(CO)_{12}I_2$.

HEXADECACARBONYLPENTAOSMIUM
$Os_5(CO)_{16}$

B.E. REICHERT and G.M. SHELDRICK, 1977. Acta Cryst., B33, 173-175.

Trigonal, $P3_121$, a = 9.204, c = 24.818 Å, Z = 3. Mo radiation, R = 0.033 for 809 reflexions.

The molecule, which lies on a twofold axis, contains a trigonal-bipyramidal Os_5 cluster, with four CO groups on one equatorial Os and three CO groups on the other four Os, all CO being terminal. $(CO)_4Os$-Os = 2.867, 2.889, other Os-Os = 2.738, 2.748, 2.764(3), Os-C = 1.81-1.95(4), C-O = 1.10-1.20(4) Å.

PENTADECACARBONYLHYDRIDOTRIOSMIUMRHENIUM
$HOs_3Re(CO)_{15}$

M.R. CHURCHILL and F.J. HOLLANDER, 1977. Inorg. Chem., 16, 2493-2497.

Monoclinic, $I2/m$, a = 10.557, b = 10.481, c = 10.144 Å, β = 102.14°, Z = 2. Mo radiation, R = 0.031 for 554 reflexions.

The molecule contains a planar rhombus of metal atoms, with C_{2h} crystallographic symmetry and fourfold disorder (Fig. 1); the observed pattern is consistent with two $Os(CO)_4$ groups, bridged by mutually-linked (M-M = 2.944, M-Os = 2.957 Å) $Re(CO)_4$ and $Os(CO)_3H$ groups.

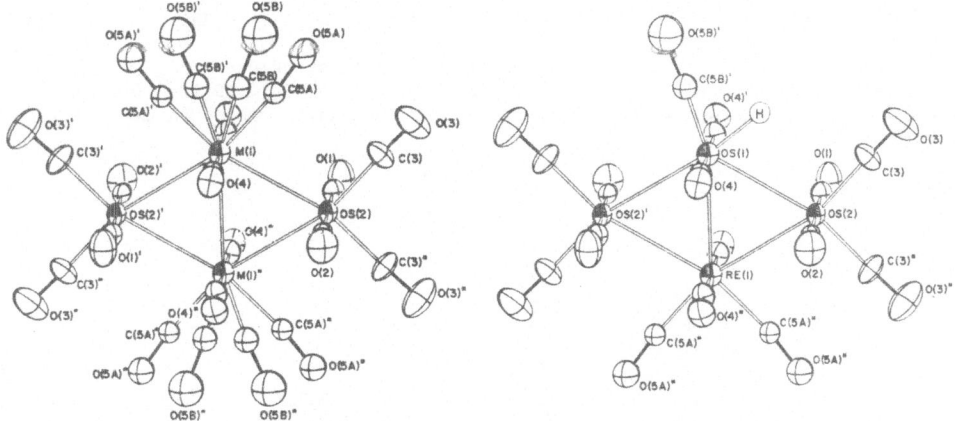

Fig. 1.   Disordered and postulated ordered structure of $HOs_3Re(CO)_{15}$; the H ligand is shown in its presumed position.

URANYL BIS(HYDROXYLAMINATE) TRIHYDRATE
$UO_2(NH_2O)_2 \cdot 3H_2O$

URANYL BIS(HYDROXYLAMINATE) TETRAHYDRATE
$UO_2(NH_2O)_2 \cdot 4H_2O$

I.   H.W.W. ADRIAN and A. van TETS, 1977. Acta Cryst., B33, 2997-3000.
II. A. van TETS and H.W.W. ADRIAN, 1977. J. Inorg. Nucl. Chem., 39, 1607-1610.

Trihydrate: orthorhombic, Pbcn, a = 5.789, 5.757, b = 11.550, 11.521, c = 11.666, 11.614 Å (at 25, -170°C, respectively), D_m = 3.27, Z = 4. Mo radiation, R = 0.039, 0.026 for 181, 194 reflexions at 25, -170°C; neutron radiation, R = 0.057, 0.065 for 291, 296 reflexions at 25, -150°C.

Tetrahydrate: triclinic, P$\bar{1}$, a = 6.295, b = 6.000, c = 5.631 Å, α = 96.05, β = 91.03, γ = 105.27°, D_m = 3.21, Z = 1. Mo radiation, R = 0.059 for 1022 reflexions.

Atomic positions

Trihydrate [-150°C, neutron data; similar values in the other refinements, except that the equivalent positions and atomic numbering scheme are different in the X-ray studies].

	x	y	z
U	1/2	1/2	1/2
O(1)	0.340	0.375	0.446
O(2)	0.662	0.384	0.643
$H_2O(3)$	0.766	0.499	0.332
N	0.828	0.379	0.554
$H_2O(4)$	0	0.313	1/4

Tetrahydrate

	x	y	z
U	1/2	1/2	1/2
O(1)	0.607	0.767	0.703
O(2)	0.796	0.372	0.607
O(3)	0.193	0.372	0.768
N	0.633	0.278	0.779
O(4)	0.109	0.911	0.789

U has pentagonal bipyramidal coordination in both structures (Fig. 1), U-N = 2.44, U-O = 1.83 (uranyl), 2.29, 2.34 (hydroxylamino), 2.48, 2.49(1) Å (water). The other water molecule is not coordinated to U, but participates in the hydrogen bonding.

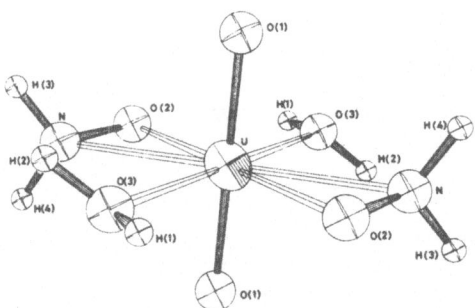

Fig. 1.   Uranium coordination in uranyl bis(hydroxylaminate) trihydrate.

SODIUM AZIDE   POTASSIUM AZIDE
β-NaN$_3$ KN$_3$

I.   E.D. STEVENS, 1977. Acta Cryst., A33, 580-584.
II.   E.D. STEVENS and H. HOPE, 1977. Ibid., A33, 723-729.
III. E.D. STEVENS, J. RYS and P. COPPENS, 1977. J. Amer. Chem. Soc., 99, 265-267.

β-NaN$_3$ (II, III)
Rhombohedral, R$\bar{3}$m, a = 3.646, c = 15.223 Å, Z = 3. Mo radiation, R = 0.02-0.03 for various refinements. z = 0.577 for N(1).

KN$_3$ (I, III)
Tetragonal, I4/mcm, a = 6.119, c = 7.102 Å, Z = 4. Mo radiation, R ∿ 0.04 for 511
reflexions (I), 0.066 for 366 high-order reflexions (III). x = 0.136 for N(1).

    Structures as previously described (1). The electron-density distribution is
studied.

1. Strukturbericht, 1, 276, 279, 288, 289; 4, 151; Structure Reports, 9, 144;
    33A, 168; 38A, 187; 42A, 156.

LEAD AZIDE
α-Pb(N$_3$)$_2$

C.S. CHOI, E. PRINCE and W.L. GARRETT, 1977. Acta Cryst., B33, 3536-3537.

Orthorhombic, Pnma, a = 6.63, b = 16.25, c = 11.31 Å (from 1), Z = 12. Neutron
radiation, R = 0.038 for 1328 reflexions.

    The structure is as previously described (2), with all four independent azide
groups being asymmetric (Fig. 1).

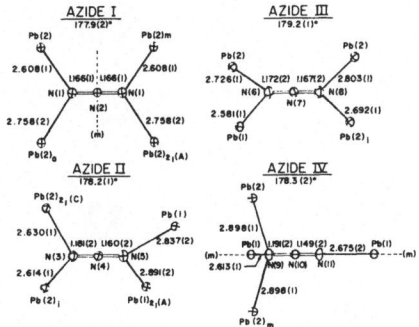

Fig. 1.   Azide groups in α-lead azide.

1. Structure Reports, 20, 358.
2. Ibid., 28, 55; 34A, 170.

POTASSIUM HEXAAMINOSTANNATE(IV)   RUBIDIUM HEXAAMINOSTANNATE(IV)
K$_2$Sn(NH$_2$)$_6$ Rb$_2$Sn(NH$_2$)$_6$

P. CHEVALIER, J. RITSMA and J. ROUXEL, 1977. Acta Cryst., B33, 1076-1079.

Potassium compound
Rhombohedral, R3̄m, a = 6.390, c = 19.601 Å, D$_m$ = 2.15, Z = 3. Mo radiation, R =
0.11 for powder data.

Rubidium compound
Trigonal, P3̄, a = 6.660, c = 5.770 Å, D$_m$ = 2.87, Z = 1. Mo radiation, R = 0.07
for 301 reflexions.

Atomic positions

$K_2Sn(NH_2)_6$		x	y	z
Sn in	3(a)	0	0	0
K	6(c)	0	0	0.1803
N	18(f)	0.1526	-0.1526	0.0643

$Rb_2Sn(NH_2)_6$				
Sn	1(a)	0	0	0
Rb	2(d)	1/3	2/3	0.7935
N	6(g)	0.8026	0.6894	0.7889

The structures contain octahedral $Sn(NH_2)_6^{2-}$ anions and K^+ or Rb^+ cations (Fig. 1). Sn-N = 2.183(2); Rb-N = 3.05-3.95 Å (12 distances).

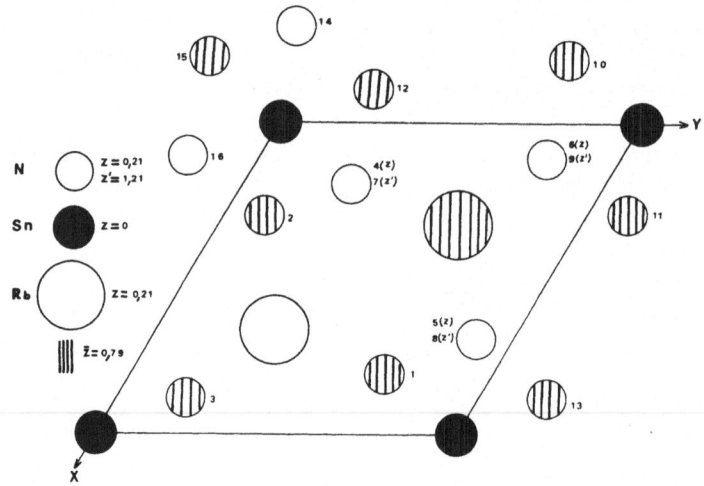

Fig. 1.   Structure of $Rb_2Sn(NH_2)_6$.

B-HEXACHLOROCYCLOTRIBORAZANE
$B_3N_3H_6Cl_6$

H. HESS, D. LUX and W. SCHWARZ, 1977. Z. Naturforsch., 32B, 982-984.

Orthorhombic, Pnma, a = 10.343, b = 12.032, c = 8.173 Å, at -150°C, D_m = 1.87, Z = 4. Mo radiation, R = 0.033 for 2335 reflexions.

The molecule, $(Cl_2BNH_2)_3$, contains a six-membered ring of alternating B and atoms, with nearly ideal chair conformation, B-N = 1.571-1.579, B-Cl = 1.836-1.854(4) Å, N-B-N = 109, B-N-B = 120°.

THIOTRITHIAZYL TRICHLOROMERCURATE(II)
$S_4N_3HgCl_3$

K. WEIDENHAMMER and M.L. ZIEGLER, 1977. Z. anorg. Chem., 434, 152-156.

Monoclinic, P2$_1$/c, a = 6.175, b = 17.373, c = 9.692 Å, β = 109.29°, Z = 4.   Mo radiation, R = 0.080 for 1474 reflexions.

The S$_4$N$_3^+$ cation contains a seven-membered ring, and the anion is a polymeric chain with Hg having trigonal bipyramidal coordination (Fig. 1); Hg-Cl = 2.34-2.46(1) Å.

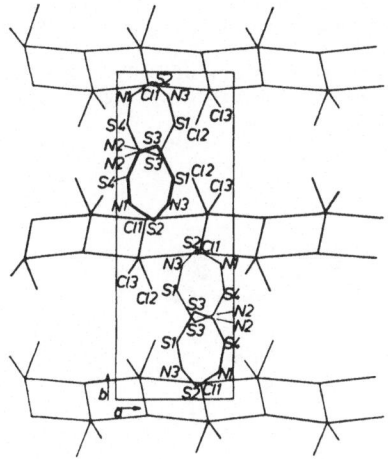

Fig. 1.   Structure of S$_4$N$_3$HgCl$_3$.

TETRASULPHURTETRAIMIDE
S$_4$N$_4$H$_4$

S.N. NABI, M.B. HURSTHOUSE and K.M.A. MALIK, 1977. Acta Cryst., B33, 2309-2310.

Orthorhombic, Pnma, a = 7.988, b = 12.244, c = 6.743 Å, D$_m$ = 1.89, Z = 4.   Mo radiation, R = 0.034 for 867 reflexions.

Atomic positions

	x	y	z
S(1)	0.0422	0.1299	0.1810
S(2)	0.1370	0.1300	-0.2420
N(1)	0.2304	1/4	-0.2816
N(2)	0.1061	1/4	0.2680
N(3)	0.1650	0.0929	-0.0071
H(1)	0.337	1/4	-0.288
H(2)	0.181	1/4	0.326
H(3)	0.238	0.059	0.014

Previous results (1) are confirmed. S-N = 1.665(1) Å, N-S-N = 109.3, S-N-S = 124.2°.

1. Structure Reports, 21, 258; 22, 224; 32A, 154.

POTASSIUM FLUORIDE DIPEROXYHYDRATE
$KF.2H_2O_2$

V.A. SARIN, V.Ja. DUDAREV, T.A. DOBRYNINA, L.E. FYKIN and V.E. ZAVODNIK, 1976.
Kristallografija, 21, 929-936 [Soviet Physics - Crystallography, 21, 531-535].

Orthorhombic, Pbca, a = 11.259, b = 6.073, c = 12.165 Å, Z = 8. Mo radiation, R =
0.031 for 275 reflexions, and neutron radiation, R = 0.049 for 200 reflexions.

The structure (Fig. 1) contains layers of H_2O_2 and F^- linked by hydrogen bonds,
and alternating layers of K^+ ions. The dihedral angles of the two independent H_2O_2
molecules are 96 and 107°.

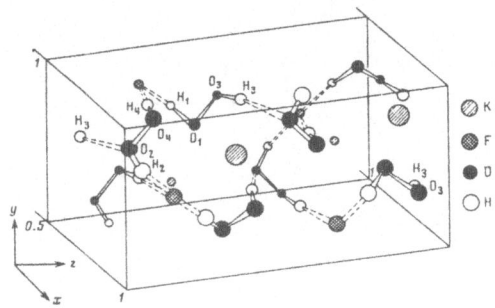

Fig. 1.   Structure of $KF.2H_2O_2$.

BARIUM FLUORIDE IODIDE   BARIUM FLUORIDE BROMIDE
BaFI   BaFBr

CALCIUM FLUORIDE CHLORIDE
CaFCl

B.W. LIEBICH and D. NICOLLIN, 1977. Acta Cryst., B33, 2790-2794.

Tetragonal, P4/nmm, a = 4.654, 4.508, 3.894, c = 7.962, 7.441, 6.818 Å, Z = 2.
Mo radiation, R = 0.077, 0.080, 0.081 for 348, 325, 114 reflexions. Ba, Ca in
2(c): z = 0.1704, 0.1911, 0.1962; F in 2(a); I, Br, Cl in 2(c): z = 0.6522,
0.6497, 0.6432 (origin at 2/m).

PbFCl-type (1) structures.

1. Strukturbericht, 2, 45, 362; 3, 369.

TRITIN(II) BROMIDE PENTAFLUORIDE   DITIN(II) CHLORIDE TRIFLUORIDE
Sn_3BrF_5 Sn_2ClF_3

J.D. DONALDSON, D.R. LAUGHLIN and D.C. PUXLEY, 1977. J. Chem. Soc., Dalton,
865-868.

Sn_3BrF_5
Monoclinic, $P2_1/n$, a = 4.27, b = 12.70, c = 12.70 Å, β = 90.0°, D_m = 4.79, Z = 4.
Mo radiation, R = 0.11 for 435 reflexions (films, densitometer intensities).

Sn_2ClF_3
Orthorhombic, $P2_12_12_1$, $a = b = c = 7.880$ Å, $D_m = 4.19$, $Z = 4$. Mo radiation, $R = 0.056$ for 740 reflexions (films, densitometer intensities).

The materials have higher-symmetry pseudo-tetragonal and pseudo-cubic cells, respectively. Both structures (Fig. 1) contain infinite Sn-F cationic networks, two-dimensional in Sn_3BrF_5 and three-dimensional in Sn_2ClF_3. Sn atoms have trigonal pyramidal coordination, Sn-F = 1.99-2.21 Å, and Br$^-$ and Cl$^-$ ions occupy holes in the cationic networks, with shortest Sn...Br and Sn...Cl = 3.29 and 3.14 Å, respectively.

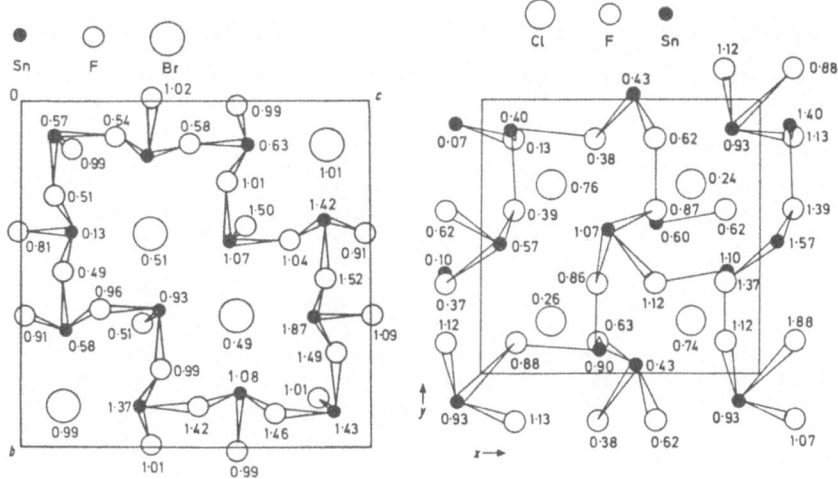

Fig. 1.   Structures of Sn_3BrF_5 and Sn_2ClF_3.

ANTIMONY TRIFLUORIDE - ANTIMONY PENTAFLUORIDE (1:1)

Sb_4F_{16} $[Sb_2F_4][SbF_6]_2$

R.J. GILLESPIE, D.R. SLIM and J.E. VEKRIS, 1977. J. Chem. Soc., Dalton, 971-974.

Monoclinic, $P2_1/c$, $a = 9.32$, $b = 12.07$, $c = 11.60$ Å, $\beta = 107.1°$, $Z = 4$. Mo radiation, $R = 0.071$ for 2048 reflexions.

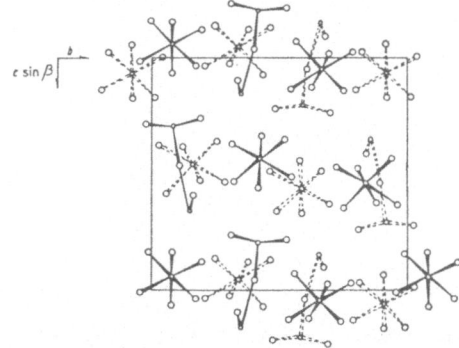

Fig. 1.   Structure of Sb_4F_{16}.

The structure (Fig. 1) contains an $[F_2Sb-F-SbF]^{2+}$ cation and two $[SbF_6]^-$ anions, with longer Sb...F contacts forming an infinite three-dimensional polymer and completing 7- and 6-coordination at the Sb atoms of the cation. In the cation, Sb-F = 1.85, 1.87, 2.01 (terminal), 1.99, 2.15 Å (bridging), F-Sb-F = 78-86, Sb-F-Sb = 148°; in the anions, Sb-F = 1.83-1.95 Å; longer Sb...F = 2.29-2.87 Å.

ANTIMONY(V) TRICHLORIDE DIFLUORIDE
$SbCl_3F_2$

J.G. BALLARD, T. BIRCHALL and D.R. SLIM, 1977. J. Chem. Soc., Dalton, 1469-1472.

Tetragonal, I4, a = 12.81, c = 7.282 Å, Z = 8. Mo radiation, R = 0.073 for 265 reflexions.

Atomic positions

	x	y	z
Sb	-0.0473	0.2214	0
Cl(1)	-0.2026	0.2979	-0.0356
Cl(2)	-0.0596	0.2624	0.0368
Cl(3)	-0.0393	0.1846	0.3077
F(1)	0.0802	0.1234	-0.0319
F(2)	-0.0410	0.2027	-0.2614

The structure contains cis-fluorine-bridged tetramers, in which Sb has distorted octahedral coordination (Fig. 1).

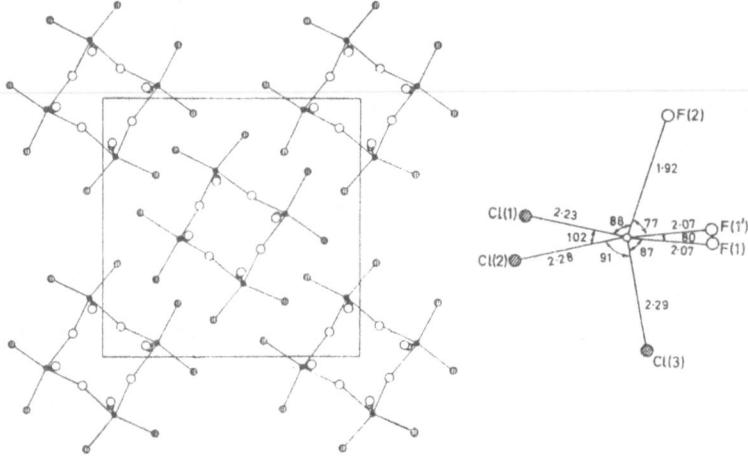

Fig. 1.   Structure of $SbCl_3F_2$.

ANTIMONY CHLORIDE FLUORIDE
$Sb_3Cl_{11}F_4$

J.G. BALLARD, T. BIRCHALL and D.R. SLIM, 1977. Canad. J. Chem., 55, 743-748.

Monoclinic, $P2_1/n$, a = 12.359, b = 16.480, c = 9.387 Å, β = 103.96°, Z = 4. Mo radiation, R = 0.106 for 1934 reflexions.

The structure contains cis fluorine-bridged trimers, with one Cl 25% substituted by F (Fig. 1). Sb atoms have distorted octahedral coordination, Sb-F =
1.87 (terminal), 2.04-2.12(1) (bridging), Sb-Cl = 2.20-2.28(1), Sb-Cl,F = 2.12(1) Å.

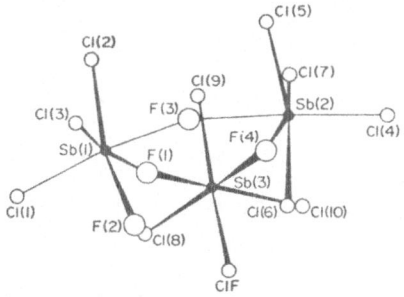

Fig. 1.   The $Sb_3Cl_{11}F_4$ molecule.

ZIRCONIUM(IV) FLUORIDE TRIHYDRATE
(DI-μ-FLUORO-HEXAFLUOROHEXAAQUODIZIRCONIUM(IV))
$ZrF_4 \cdot 3H_2O$

F. GABELA, B. KOJIĆ-PRODIĆ, M. ŠLJUKIĆ and Ž. RUŽIĆ-TOROŠ, 1977. Acta Cryst., B33,
3733-3736.

Triclinic, PĪ, a = 5.948, b = 6.964, c = 7.572 Å, α = 90.55, β = 105.06, γ =
118.72°, D_m = 2.792, Z = 2. Mo radiation, R = 0.034 for 1483 reflexions. Previous
study in 1.

The structure (Fig. 1) contains F-bridged binuclear units, $Zr_2F_8(H_2O)_6$, in
which Zr has dodecahedral 8-coordination. These units are linked by O-H...O and
O-H...F hydrogen bonds.

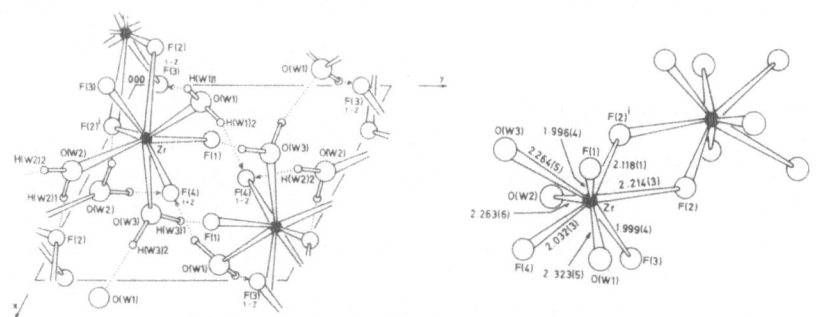

Fig. 1.   Structure of $ZrF_4 \cdot 3H_2O$ and Zr coordination.

1. T.N. WATERS, 1960. J. Inorg. Nucl. Chem., 15, 320; idem, 1964. Chem. Ind.,
   713.

DIIRON PENTAFLUORIDE DIHYDRATE
$Fe_2F_5 \cdot 2H_2O$

W. HALL, S. KIM, J. ZUBIETA, E.G. WALTON and D.B. BROWN, 1977. Inorg. Chem., 16, 1884-1887.

Orthorhombic, Imma, a = 7.489, b = 10.897, c = 6.671 Å, D_m = 2.94, Z = 4. Mo radiation, R = 0.053 for 229 reflexions.

Atomic positions

		x	y	z
$Fe^{II}(1)$ in	4(a)	0	0	0
$Fe^{III}(2)$	4(c)	1/4	1/4	1/4
F(1)	16(j)	0.2022	0.1239	0.0538
F(2)	4(e)	0	1/4	0.3368
O	8(h)	1/2	0.5640	0.1974

The structure contains columns of trans-corner-sharing Fe(III)F_6 octahedra, crosslinked by trans-Fe(II)$F_4(H_2O)_2$ octahedra (Fig. 1). Mean Fe(III)-F = 1.94, Fe(II)-F = 2.06, Fe(II)-O = 2.13 Å.

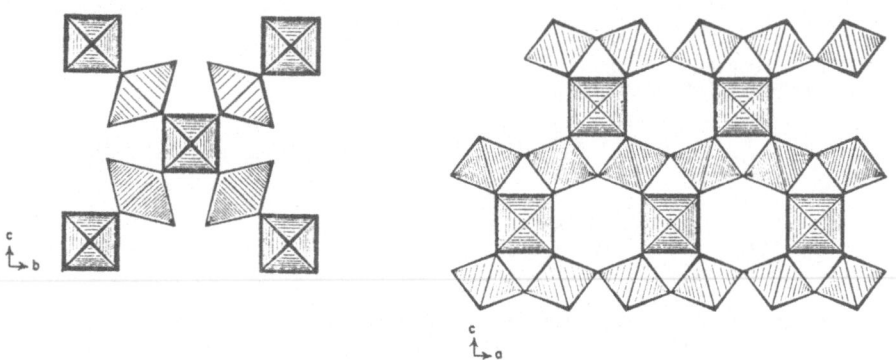

Fig. 1.   Structure of diiron pentafluoride dihydrate.

MERCURY(II) FLUORIDE DIHYDRATE
$HgF_2 \cdot 2H_2O$

B.V. BUKVECKIJ, S.A. POLIŠČUK and V.I. SIMONOV, 1976. Koordin. Khim., 2, 1208-1212.

Orthorhombic, Pbcn, a = 9.992, b = 5.805, c = 5.469 Å, D_m = 5.72, Z = 4. Mo radiation, R = 0.049 for 278 reflexions.

Atomic positions

		x	y	z
Hg in	4(c)	0	0.2317	1/4
O	8(d)	0.126	0.369	0.561
F	8(d)	0.125	0.085	0.908

The structure (Fig. 1) contains linked HgF_4O_4 polyhedra, and O-H...F hydrogen bonds.

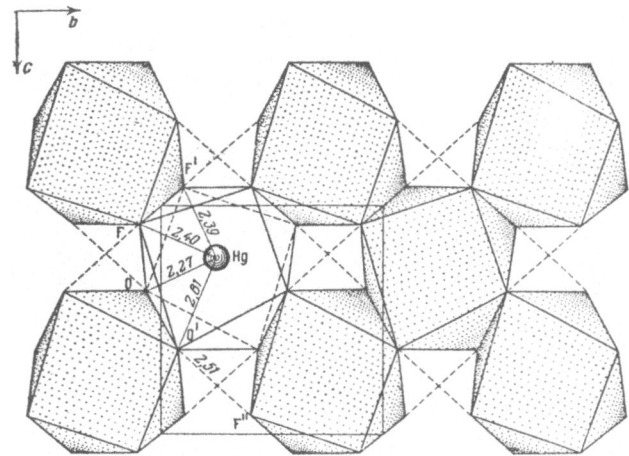

Fig. 1.   Structure of $HgF_2 \cdot 2H_2O$.

AMMONIUM COBALT(II) TETRAFLUOROBERYLLATE HEXAHYDRATE
$(NH_4)_2Co(BeF_4)_2 \cdot 6H_2O$

J. VICAT, D. TRANQUI and S. ALÉONARD, 1977. Acta Cryst., B33, 1180-1190.

Monoclinic, $P2_1/a$, a = 9.269, b = 12.541, c = 6.136 Å, β = 106.80°, Z = 2. Mo
radiation, R = 0.026 for 2268 reflexions.

    Structure as previously determined (1). Difference maps reveal bonding and
lone-pair electrons.

1. Structure Reports, 34A, 186; 41A, 153.

POTASSIUM HEPTAFLUORODIINDATE
KIn_2F_7

J.-C. CHAMPARNAUD-MESJARD and B. FRIT, 1977. Acta Cryst., B33, 3722-3726.

Monoclinic, $P2_1/m$, a = 10.753, b = 8.131, c = 6.609 Å, β = 90.71°, D_m = 4.58,
Z = 4. Mo radiation, R = 0.047 for 708 reflexions (films, densitometer intensities,
for k = 2n reflexions only).

    Chains of edge-sharing InF_7 pentagonal bipyramids are linked by corner sharing
to form a three-dimensional network, with K ions in tunnels (Fig. 1); K-F = 2.56-
3.1 Å (9- and 10-coordination).

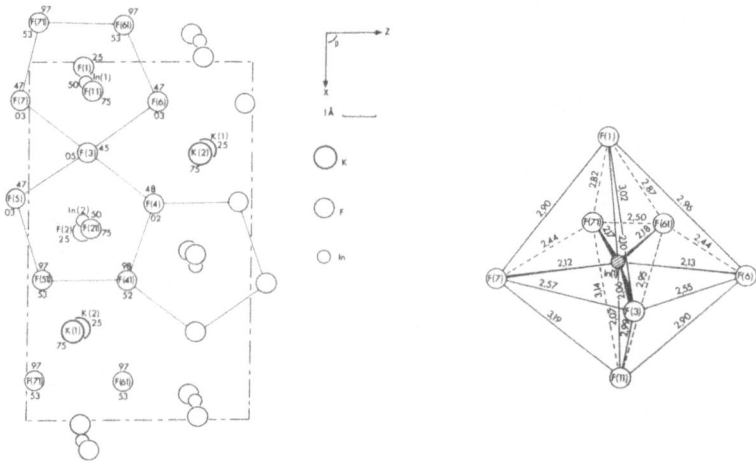

Fig. 1.   Structure of KIn₂F₇ and environment of In(1).

RUBIDIUM DECAFLUOROTRIINDATE
$RbIn_3F_{10}$

J.C. CHAMPARNAUD-MESJARD, D. MERCURIO and B. FRIT, 1977.   J. Inorg. Nucl. Chem.,
39, 947-951.

Orthorhombic, $P222_1$, a = 7.747, b = 6.655, c = 8.024 Å, D_m = 4.95, Z = 2.   Mo
radiation, R = 0.054 for 324 reflexions with ℓ = 2n (films, densitometer intensities).

Atomic positions

			x	y	z
Rb	in	2(d)	1/2	0	1/4
In(1)		2(a)	0	0	0
In(2)		2(b)	0.2685	1/2	0
In(3)		2(b)	0.7279	1/2	0
F(1)		2(b)	0	1/2	0
F(2)		2(c)	0	-0.090	1/4
F(3)		4(e)	0.492	0.319	0.028
F(4)		4(e)	0.293	0.431	0.253
F(5)		4(e)	0.189	0.193	0.059
F(6)		4(e)	0.201	0.202	0.546

    Only reflexions with ℓ = 2n have been used in the study.   Indium atoms have
octahedral and pentagonal-bipyramidal coordinations, the polyhedra sharing edges
and corners to form a three-dimensionsal $In_3F_{10}^-$ framework, with tunnels along c
which contain Rb^+ ions with (8 + 2) coordination.   In-F = 2.00-2.20(9), Rb-F =
2.85-3.16, 3.37 Å.

TIN DIFLUORIDE - ARSENIC PENTAFLUORIDE (1:1)
$SnF_2 \cdot AsF_5$ $1/3[\{(SnF)_3\}(AsF_6)_3]$

L. GOLIČ and I. LEBAN, 1977.   Acta Cryst., B33, 232-234.

Rhombohedral, R32, a = 9.123, c = 16.983 Å, Z = 9. Mo radiation, R = 0.039 for 277 reflexions.

Atomic positions

			x	y	z
Sn	in	9(e)	0.2601	0	1/2
As(1)		3(a)	0	0	0
As(2)		6(c)	0	0	0.2989
F(1)		18(f)	0.0851	0.1730	0.0592
F(2)		18(f)	0.1538	0.1512	0.2414
F(3)		18(f)	0.0039	0.1558	0.3588
F(4)		9(e)	0.1760	0.1760	1/2

The structure (Fig. 1) contains AsF_6^- anions and $(SnF)_3^{3+}$ cations, which have six-membered rings with D_{3h} symmetry, Sn-F = 2.10(1) Å, F-Sn-F = 83.0, Sn-F-Sn = 157.0°. Each Sn has six further F neighbours at 2.59-3.05 Å. The AsF_6^- ions are octahedral, As-F = 1.70-1.73(1) Å, F-As-F = 88.4-92.1°.

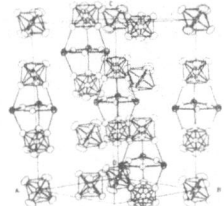

Fig. 1.   Structure of $SnF_2.AsF_5$.

SODIUM BROMOTRIFLUOROANTIMONATE(III) MONOHYDRATE
$NaSbBrF_3.H_2O$

B. DUCOURANT, B. BONNET, R. FOURCADE and G. MASCHERPA, 1977. Acta Cryst., B33, 3693-3696.

Monoclinic, $P2_1/c$, a = 8.323, b = 11.365, c = 5.765 Å, ß = 91.88°, D_m = 3.63, Z = 4. Mo radiation, R = 0.073 for 1152 reflexions.

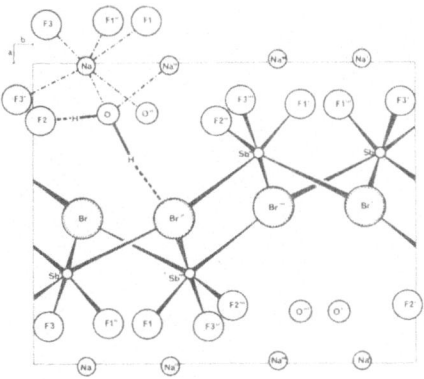

Fig. 1.   Structure of $NaSbBrF_3.H_2O$

The structure (Fig. 1) contains layers of polymeric $(SbBrF_3^-)_n$ anions, with distorted $SbBr_4F_3E$ dodecahedra linked by sharing the four Br atoms (E = lone pair directed away from the Br atoms); Sb-F = 1.94-1.98(1), Sb-Br = 3.012-3.535(5) Å. The layers are linked by Na ions with 6-coordination, Na-O = 2.39, 2.44, Na-F = 2.38-2.42 Å.

CAESIUM CHLOROTRIFLUOROANTIMONATE(III)
$CsSbClF_3$

A.A. UDOVENKO, L.M. VOLKOVA, R.L. DAVIDOVIČ and L.A. ZEMNUKHOVA, 1977.   Koordin. Khim., 3, 259-261.

Tetragonal, $I\bar{4}2m$, a = 9.94, c = 11.61 Å, D_m = 3.98, Z = 8.   R = 0.11.

The structure contains (001) layers of tetrameric $Sb_4Cl_4F_{12}^{4-}$ ions alternating with layers of Cs^+ ions. The anion contains a distorted square of four Sb atoms bridged by four chlorine atoms, with 3 F atoms and a lone pair of electrons completing distorted octahedral coordination at each Sb.

RUBIDIUM TETRAFLUOROVANDATE(III) DIHYDRATE
$RbVF_4 \cdot 2H_2O$

B.V. BUKVECKIJ, L.A. MURADJAN, R.L. DAVIDOVIČ and V.I. SIMONOV, 1976.   Koordin. Khim., 2, 1129-1134.

Monoclinic, B2/b, a = 11.921, b = 9.114, c = 6.343 Å, γ = 124.04°, Z = 4.   Mo radiation, R = 0.028 for 855 reflexions.

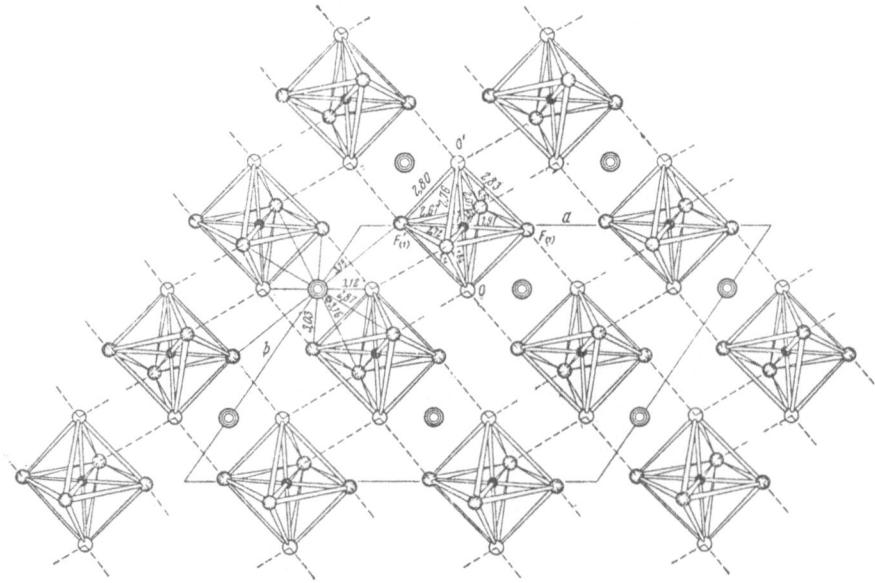

Fig. 1.   Structure of $RbVF_4 \cdot 2H_2O$.

Atomic positions

			x	y	z
Rb	in	4(e)	0	1/4	0.2224
V		4(c)	1/4	0	1/4
F(1)		8(f)	0.4115	0.0183	0.3337
F(2)		8(f)	0.2614	0.4262	0.0251
O		8(f)	0.3668	0.2584	0.1415

The structure (Fig. 1) contains trans-octahedral $VF_4(H_2O)_2^-$ anions, Rb^+ cations coordinated to 8 F and 2 O atoms, and O-H...F hydrogen bonds, 2.56 and 2.57 Å.

MANGANESE(II) PENTAFLUOROCHROMATE(III)
MnCrF$_5$

G. FÉREY, R. de PAPE, M. POULAIN, D. GRANDJEAN and A. HARDY, 1977. Acta Cryst., B33, 1409-1413.

Monoclinic, C2/c, a = 8.586, b = 6.291, c = 7.381 Å, β = 115.46°, Z = 4.   Mo radiation, R = 0.060 for 637 reflexions.

Atomic positions

			x	y	z
Mn^{2+}	in	4(e)	0	0.5333	1/4
Cr^{3+}		4(a)	0	0	0
F(1)		4(e)	0	0.9203	1/4
F(2)		8(f)	0.0155	0.7025	0.9666
F(3)		8(f)	0.7603	0.9670	0.8765

CrF_6 octahedra share   opposite corners to form chains along c (Fig. 1). Mn^{2+} ions between the chains have pentagonal bipyramidal coordination. Cr-F = 1.870-1.913(4), Mn-F = 2.018-2.434(5) Å.

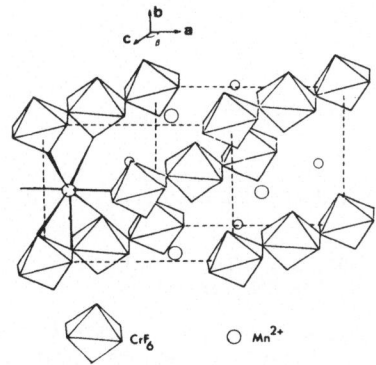

Fig. 1.   Structure of MnCrF$_5$.

SODIUM MANGANESE(II) HEXAFLUOROCHROMATE(III)
NaMnCrF$_6$

G. COURBION, C. JACOBONI and R. de PAPE, 1977. Acta Cryst., B33, 1405-1408.

Trigonal, P321, a = 8.993, c = 5.003 Å, D_m = 3.52, Z = 3. Mo radiation, R = 0.026 for 1299 reflexions.

Atomic positions

			x	y	z
Na	in	3(e)	0.3722	0	0
Mn		3(f)	0.7026	0	1/2
Cr(1)		1(a)	0	0	0
Cr(2)		2(d)	1/3	2/3	0.4971
F(1)		6(g)	0.9031	0.1033	0.7798
F(2)		6(g)	0.5390	0.4073	0.7131
F(3)		6(g)	0.2347	0.7744	0.7039

The structure is similar to that of Na_2SiF_6 (1), and contains regular octahedral CrF_6^{3-} ions, linked by Mn^{2+} and Na^+ ions which have distorted octahedral coordination (Fig. 1). Cr-F = 1.905-1.912(1), Mn-F = 2.098-2.133(1), Na-F = 2.246-2.391(2) Å.

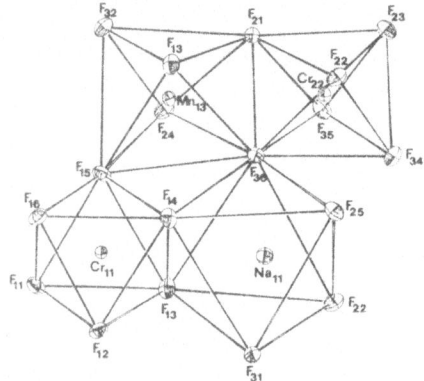

Fig. 1.   Octahedral coordination in $NaMnCrF_6$.

1. Structure Reports, 29, 264.

CAESIUM TETRAFLUOROMANGANATE(III) DIHYDRATE
$CsMnF_4 \cdot 2H_2O$

I.   P. BUKOVEC and V. KAUČIČ, 1977. J. Chem. Soc., Dalton, 945-947.
II. E. DUBLER,   L. LINOWSKY, J.-P. MATTHIEU and H.-R. OSWALD, 1977. Helv. Chim. Acta, 60, 1589-1600.

I.   Monoclinic, C2/c, a = 11.907, b = 6.597, c = 9.316 Å, β = 121.77°, D_m = 3.19, Z = 4. Mo radiation, R = 0.024 for 835 reflexions.

II. Monoclinic, C2, a = 11.891, b = 6.589, c = 10.558 Å, β = 131.46°, D_m = 3.20, Z = 4. Mo radiation, R = 0.018 for 1073 reflexions. $CsMnF_4$ is tetragonal, a = 7.936, c = 6.341 Å.

Atomic positions (I)

	x	y	z
Cs	0	0.7214	1/4
Mn	1/4	1/4	1/2
F(1)	0.0958	0.3281	0.4863
F(2)	0.2686	0.5099	0.4416
O	0.1330	0.1524	0.2331

[These two papers appear to describe the same phase, with a different choice of $\underline{c}$ axis. The positional parameters listed in II show some pseudo-symmetry relationships, but since no structure factors are listed it is not possible to make a direct comparison between the two sets of results.]

Based on I, the compound is isostructural with $RbVF_4 \cdot 2H_2O$ ($\underline{1}$). The structure contains isolated trans-octahedral $MnF_4(H_2O)_2^-$ anions and 10-coordinate Cs^+ cations; Mn-F = 1.847, Mn-O = 2.211(2), Cs-F = 3.06-3.26, Cs-O = 3.30 Å. The anions are linked by O-H...F hydrogen bonds.

$\underline{1}$. This volume, p. 126.

POTASSIUM PENTAFLUOROFERRATE(III)
K_2FeF_5

M. VLASSE, G. MATEJKA, A. TRESSAUD and B.M. WANKLYN, 1977. Acta Cryst., B$\underline{33}$, 3377-3380.

Orthorhombic, $Pn2_1a$, a = 20.39, b = 7.399, c = 12.84 Å, D_m = 3.16, Z = 16. Mo radiation, R = 0.048 for 1088 reflexions.

The structure (Fig. 1) contains infinite zigzag chains of cis-corner-sharing FeF_6 octahedra, mean Fe-F = 2.02 (bridging), 1.88 Å (terminal), Fe-F-Fe = 162-173°. K ions have 9- and 10-coordinations.

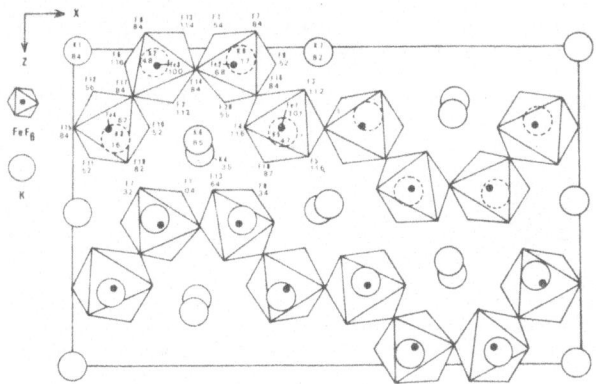

Fig. 1.   Structure of K_2FeF_5.

CAESIUM MANGANESE NICKEL FLUORIDE
Cs_2MnNiF_6

J.-M. DANCE, J. GRANNEC, A. TRESSAUD and M. PERRIN, 1977. Mater. Res. Bull., $\underline{12}$, 989-994.

Rhombohedral, $R\bar{3}m$, a = 6.209, c = 29.99 Å, Z = 6. Neutron powder data. Mn in 3(a); Ni in 3(b); (Ni,Mn), Cs(1), Cs(2) in 6(c): z = 0.408, 0.138, 0.290.

Isostructural with Cs_2NaCrF_6 ($\underline{1}$), but with a different cation distribution.

$\underline{1}$. Structure Reports, $\underline{41A}$, 163; $\underline{42A}$, 179.

BARIUM HEXAFLUOROCUPRATE(II)
Ba_2CuF_6

D. REINEN and H. WEITZEL, 1977. Z. Naturforsch., 32B, 476-478.

Orthorhombic, Cmca, a = 15.792, b = 5.915, c = 5.814 Å, Z = 4. Neutron powder data.

Atomic positions

			x	y	z
Ba	in	8(d)	0.1508	0	0
Cu		4(b)	1/2	0	0
F(1)		8(e)	1/4	0.2502	1/4
F(2)		8(f)	0	0.6729	0.2811
F(3)		8(d)	0.3832	0	0

The structure is as previously determined (1, 2). Cu has octahedral coordination, Cu-F = 1.85, 1.94, 2.33 Å (each x2).

1. Structure Reports, 32A, 168; 39A, 158.
2. C. FRIEBEL, 1974. Z. Naturforsch., 29B, 634; C. FRIEBEL, V. PROPACH and D. REINEN, 1976. Ibid., 31B, 109.

POTASSIUM TERBIUM FLUORIDE
KTb_3F_{10}

N.V. PODBEREZSKAJA, O.G. POTAPOVA, S.V. BORISOV and Ju.V. GATILOV, 1976. Ž. Strukt. Khim., 17, 948-950 [J. Struct. Chem., 17, 815-817].

Cubic, Fm3m, a = 11.611 Å, Z = 8. Mo radiation, R = 0.058 for 374 reflexions.

Atomic positions

			x	y	z
Tb	in	24(e)	0.2411	0	0
K		8(c)	1/4	1/4	1/4
F(1)		48(i)	1/2	0.333	0.333
F(2)		32(f)	0.111	0.111	0.111

Isostructural with γ-KYb_3F_{10} (1). The structure contains TbF_8 square antiprisms, which share edges to form a $Tb_6F_{32}^{14-}$ polyanion. Tb-F = 2.21, 2.37 Å. K has tetrahedral coordination (K-F = 2.79 Å), with 12 further F neighbours at 3.21 Å.

1. Following report.

POTASSIUM YTTERBIUM FLUORIDE
γ-KYb_3F_{10}

M. LABEAU, S. ALÉONARD, A. VÉDRINE, R. BOUTONNET and J.C. COUSSEINS, 1974. Mater. Res. Bull., 9, 615-624.

Cubic, Fm3m, a = 11.43 Å, Z = 8. Powder data. Yb in 24(e): x = 0.24; K in 8(c); F(1) in 48(i): x = 0.33; F(2) in 32(f): x = 0.12.

Isostructural with KTb_3F_{10} (1).

1. Preceding report.

CALCIUM CHLORIDE DIHYDRATE
$CaCl_2 \cdot 2H_2O$

A. LECLAIRE and M.M. BOREL, 1977. Acta Cryst., B33, 1608-1610.

Orthorhombic, Pbcn, a = 5.893, b = 7.469, c = 12.070 Å, D_m = 1.86, Z = 4. Mo radiation, R = 0.083 for 668 reflexions.

Atomic positions

	x	y	z
Ca	0	0.2157	1/4
Cl	-0.2725	0.4509	0.1380
O	0.2645	0.2107	0.1082

The structure (Fig. 1) contains ab layers linked by hydrogen bonds. Ca has octahedral coordination, Ca-Cl = 2.737, 2.745(3), Ca-O = 2.315(7) Å. O-H...Cl = 3.215, 3.285(8) Å.

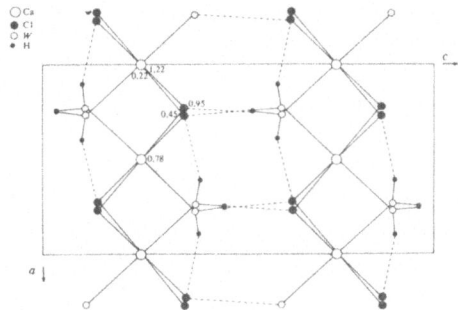

Fig. 1.   Structure of calcium chloride dihydrate.

CALCIUM CHLORIDE HEXAHYDRATE   CALCIUM BROMIDE HEXAHYDRATE
$CaCl_2 \cdot 6H_2O$ $CaBr_2 \cdot 6H_2O$

A. LECLAIRE and M.-M. BOREL, 1977. Acta Cryst., B33, 2938-2940.

Trigonal, P321, a = 7.876, 8.164, c = 3.954, 4.016 Å, D_m = 1.712 (chloride), Z = 1. Mo radiation, R = 0.068 and 0.059 for 545 and 333 reflexions. Ca in 1(a); Cl, Br in 2(d): z = 0.5749, 0.4435 (i.e. the crystals used are enantiomorphic); W(1) in 3(f): x = -0.2125, -0.2065; W(2) in 3(e): x = 0.3112, 0.3021. Previous studies in 1, 2.

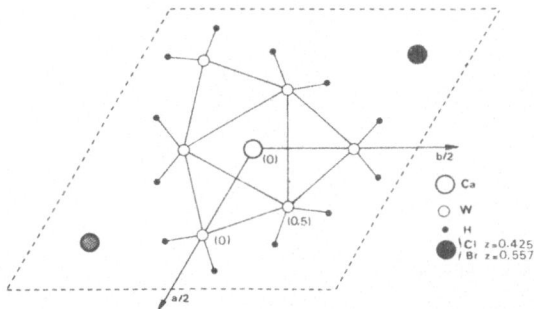

Fig. 1.   Structure of calcium chloride hexahydrate.

Isostructural with strontium chloride hexahydrate (1, 2, 3). Ca has nine water neighbours (Fig. 1), Ca-O = 2.45, 2.59 (chloride), 2.47, 2.62 Å (bromide), the polyhedra being linked by O-H...O and O-H...Cl(Br) hydrogen bonds.

1. Strukturbericht, 2, 499.
2. P.A. AGRON and W.R. BUSING, 1969. Ann. Progr. Rep. ORNL-4437, pp. 118-119.
3. Structure Reports, 8, 133.

TIN(II) CHLORIDE DIHYDRATE
$SnCl_2.2H_2O$

K. KITAHAMA and H. KIRIYAMA, 1977. Bull. Chem. Soc. Japan, 50, 3167-3176.

Monoclinic, $P2_1/c$, a = 9.320, b = 7.255, c = 8.970 Å, β = 114.91°, at 293°K, D_m = 2.710, Z = 4. Mo and neutron radiations at several temperatures in the range 88-297°K, R = 0.05-0.09 for 311-1589 reflexions.

    The structure is as previously determined (1). Hydrogen atoms are disordered above and ordered below the transition temperature of 218°K (234°K for the deuterated analogue).

1. Structure Reports, 26, 321; 39A, 160.

SCANDIUM MONOCHLORIDE
ScCl

K.R. POEPPELMEIER and J.D. CORBETT, 1977. Inorg. Chem., 16, 294-297.

Rhombohedral, $R\bar{3}m$, a = 3.473, c = 26.71 Å, Z = 6. Mo radiation, R = 0.101 for 79 reflexions.

Atomic positions

	x	y	z
Sc	0	0	0.2137
Cl	0	0	0.3914

    The structure contains sheets normal to c, and is polytypic with ZrCl (1) (Fig. 1). Sc-Sc = 3.22 (interlayer), 3.47 (intralayer), Sc-Cl = 2.59 Å.

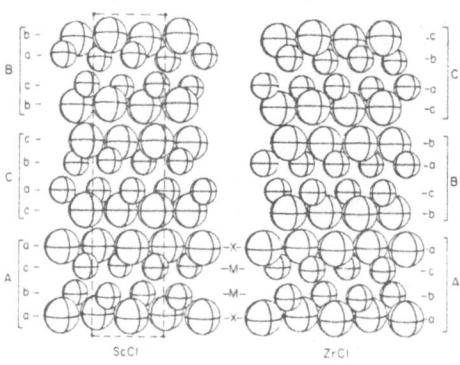

Fig. 1.   Structures of SrCl and ZrCl.

1. Structure Reports, 42A, 186.

HEPTASCANDIUM DECACHLORIDE
Sc_7Cl_{10}

K.R. POEPPELMEIER and J.D. CORBETT, 1977. Inorg. Chem., <u>16</u>, 1107-1111.

Monoclinic, C2/m, a = 18.620, b = 3.537, c = 12.250 Å, β = 91.98°, Z = 2. Mo
radiation, R = 0.059 for 705 reflexions.

The structure (Fig. 1) contains double chains of edge-sharing distorted Sc_6
octahedra along <u>b</u>; chlorine atoms cap all outward-facing metal triangles and bridge
between metal ions.

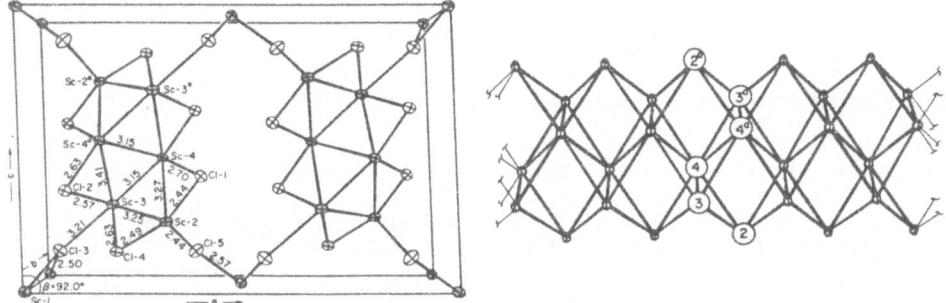

Fig. 1.   Structure of Sc_7Cl_{10} (left) and metal-chain structure (right, <u>b</u>
           horizontal).

NIOBIUM TETRACHLORIDE
$NbCl_4$

D.R. TAYLOR, J.C. CALABRESE and E.M. LARSEN, 1977. Inorg. Chem., <u>16</u>, 721-722.

Monoclinic, I2/m, a = 8.140, b = 6.823, c = 8.852 Å, β = 91.92°, Z = 4. Mo radiation, R = 0.026 for 476 reflexions.

Atomic positions

			x	y	z
Nb	in	4(g)	0	0.2220	0
Cl(1)		4(i)	0.0281	0	0.2132
Cl(2)		8(j)	-0.2789	0.2437	0.0230
Cl(3)		4(i)	0.0245	1/2	0.1873

The structure contains chains of edge-sharing $NbCl_6$ octahedra, Nb...Nb = 3.029
and 3.794(2), Nb-Cl (bridging) = 2.425, 2.523(1), Nb-Cl (terminal) = 2.291(2) Å.

fac-TRICHLOROTRIAMMINERUTHENIUM(III)
$RuCl_3(NH_3)_3$

F. BOTTOMLEY, 1977. Canad. J. Chem., <u>55</u>, 2788-2791.

Orthorhombic, $Pmn2_1$, a = 9.933, b = 6.522, c = 5.475 Å, D_m = 2.46, Z = 2. Mo radiation, R = 0.029 for 283 reflexions.

Atomic positions

	x	y	z
0.9 Ru(1)	0	0.2773	0
0.1 Ru(2)	0	0.2760	0.512
Cl(1)	0.1720	0.1397	-0.2510
Cl(2)	0	0.5931	-0.2225
N(1)	0.1469	0.3928	0.2405
N(2)	0	-0.0003	0.2038

The structure consists of chains of octahedral molecules, but is disordered, with 10% of the chains reversed in direction and displaced by c/2. Dimensions are normal.

OSMIUM(IV) CHLORIDE   (High-temperature form)
$OsCl_4$

F.A. COTTON and C.E. RICE, 1977. Inorg. Chem., 16, 1865-1867.

Orthorhombic, Cmmm, a = 7.929, b = 8.326, c = 3.560 Å, Z = 2. Mo radiation, R = 0.039 for 221 reflexions. Os in 2(a); Cl(1) in 4(g): x = 0.2852; Cl(2) in 4(j): y = 0.1894.

The structure (Fig. 1) contains infinite chains of $OsCl_6$ octahedra sharing opposite edges; Os-Cl = 2.378 (bridging), 2.261 (terminal), Os...Os = 3.560 Å.

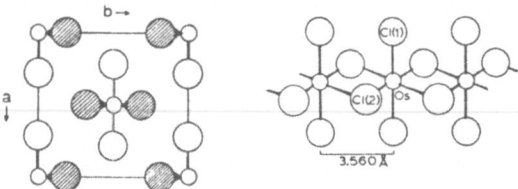

Fig. 1.   Structure of osmium(IV) chloride.

POTASSIUM MAGNESIUM CHLORIDE
$KMgCl_3$

J. BRYNESTAD, H.L. YAKEL and G.P. SMITH, 1966. J. Chem. Phys., 45, 4652-4664.

Orthorhombic, Pbnm, a = 6.954, b = 6.971, c = 9.922 Å, D_m = 2.34, Z = 4. Mo radiation, R ~ 0.15 for 41 0kℓ and 51 hk0 reflexions.

Atomic positions

			x	y	z
K	in	4(c)	-0.012	0.029	1/4
Mg		4(b)	1/2	0	0
Cl(1)		4(c)	0.045	0.491	1/4
Cl(2)		8(d)	-0.286	0.286	0.025

Orthorhombic perovskite (e.g. 1); $MgCl_6$ octahedra are regular but rotated about c. Above about 175°C the material has a cubic perovskite structure.

1. Structure Reports, 20, 273.

LITHIUM TETRACHLOROALUMINATE
LiAlCl$_4$

G. MAIRESSE, P. BARBIER, J.P. VIGNACOURT and F. BAERT, 1977. Cryst. Struct. Comm.,
<u>6</u>, 15-18.

Monoclinic, P2$_1$/c, a = 7.007, b = 6.504, c = 12.995 Å, β = 93.32°, Z = 4. Mo radiation, R = 0.029 for 1434 reflexions.

Atomic positions

	x	y	z
Al	0.7059	0.3220	0.8992
Cl(1)	0.6940	0.1833	1.0470
Cl(2)	0.8090	0.6284	0.9285
Cl(3)	0.9258	0.1816	0.8137
Cl(4)	0.4395	0.3136	0.8128
Li	0.1569	0.9831	0.3666

The structure (Fig. 1) contains distorted-tetrahedral AlCl$_4^-$ anions, linked by
octahedrally-coordinated Li$^+$ ions. Al-Cl = 2.123-2.154(1) Å, Cl-Al-Cl = 104.0-
113.2°, Li-Cl = 2.45-2.82 Å.

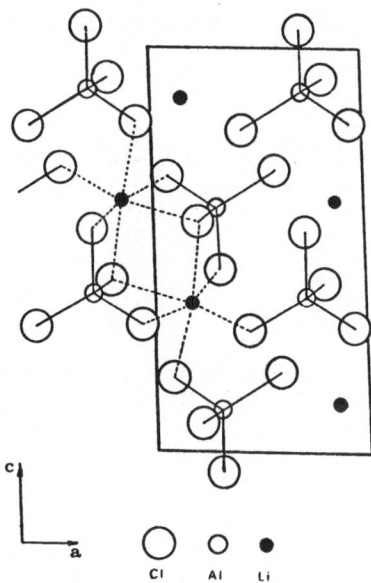

Fig. 1.   Structure of lithium tetrachloroaluminate.

SODIUM TETRACHLOROALUMINATE
NaAlCl$_4$

W. SCHEINERT and A. WEISS, 1976. Z. Naturforsch., <u>31</u>A, 1354-1369.

Orthorhombic, P2$_1$2$_1$2$_1$, a = 10.331, b = 9.905, c = 6.189 Å, D$_m$ = 2.013, Z = 4. Cu
radiation, R = 0.079 for 514 reflexions.

The structure is as previously determined (<u>1</u>).

<u>1</u>. Structure Reports, <u>15</u>, 151.

AMMONIUM DILEAD CHLORIDE
$NH_4Pb_2Cl_5$

F.G. RAS, D.J.W. IJDO and G.C. VERSCHOOR, 1977. Acta Cryst., B33, 259-260.

Monoclinic, $P2_1/c$, a = 9.018, b = 7.981, c = 12.502 Å, β = 90.09°, Z = 4. Mo
radiation, R = 0.048 for 1468 reflexions.

Ordered U_3Se_5-type structure (1), as for PbU_2Se_5 (2). Pb-Cl = 2.87-3.38(1),
NH_4-Cl = 3.27-3.57(2) Å.

1. Structure Reports, 38A, 143.
2. Ibid., 41A, 86.

THALLIUM LEAD CHLORIDE
Tl_3PbCl_5

H.-L. KELLER, 1977. Z. anorg. Chem., 432, 141-146.

Tetragonal, $P4_1$, a = 8.448, c = 14.912 Å, Z = 4. Mo radiation, R = 0.050 for 432
reflexions.

Atomic positions

	x	y	z
Tl(1)	0.9942	0.8074	0.8030
Tl(2)	0.1212	0.3198	0.8786
Tl(3)	0.6216	0.1804	0.8211
Pb	0.4940	0.6936	0.8963
Cl(1)	0.337	0.157	0.231
Cl(2)	0.687	0.008	0.354
Cl(3)	0.026	0.839	0.210
Cl(4)	0.528	0.650	0.508
Cl(5)	0.490	0.806	0.108

Tl_3PbCl_5 belongs to a new structure type, not related to those of Cs_3CoCl_5 or
$(NH_4)_3ZnCl_5$. All the cations have 7-coordination, Tl-Cl = 3.02-3.54, Pb-Cl = 2.77-
3.35 Å.

TETRACHLOROPHOSPHONIUM PENTACHLOROVANADATE(IV)
$[PCl_4][VCl_5]$

M.L. ZIEGLER, B. NUBER, K. WEIDENHAMMER and G. HOCH, 1977. Z. Naturforsch., 32B,
18-21.

Orthorhombic, Abm2, a = 6.212, b = 11.816, c = 16.307 Å, Z = 4. Mo radiation, R =
0.049 for 694 reflexions.

Atomic positions

			x	y	z
V	in	4(c)	0.6191	1/4	0.2816
P		4(a)	0	0	0.0340
Cl(1)		8(d)	0.2266	0.5857	0.2799
Cl(2)		4(c)	0.7327	3/4	0.2833
Cl(3)		4(c)	0.6176	1/4	0.4219
Cl(4)		4(c)	0.6246	1/4	0.1399
Cl(5)		8(d)	0.1789	0.4068	0.0989
Cl(6)		8(d)	0.1800	0.4052	0.4621

The structure contains tetrahedral PCl_4^+ cations and trigonal bipyramidal VCl_5^- anions. P-Cl = 1.89, 1.97(1) Å, Cl-P-Cl = 107-112°; V-Cl = 2.17, 2.19 (equatorial) and 2.29, 2.31(2) Å (axial), Cl-V-Cl = 116, 127 (equatorial), 179° (axial).

CAESIUM HEXACHLOROTUNGSTATE(V)
$CsWCl_6$

W. EICHLER and H.-J. SEIFERT, 1977. Z. anorg. Chem., <u>431</u>, 123-133.

Monoclinic, C2/c, a = 12.097, b = 6.327, c = 15.30 Å, β = 128.4°, D_m = 3.82, Z = 4. Mo radiation, R = 0.039 for 284 reflexions. Rb, K, and NH_4 compounds are isostructural.

Atomic positions

			x	y	z
W	in	4(a)	0	0	0
Cs		4(e)	0	0.3692	1/4
Cl(1)		8(f)	0.1946	0.1660	0.1587
Cl(2)		8(f)	0.1343	-0.2995	0.0403
Cl(3)		8(f)	-0.0660	-0.1327	0.1036

The structure contains nearly-regular WCl_6^- octahedra, W-Cl = 2.33 Å: Cs has 12 Cl neighbours at 3.57-3.69 Å.

RUBIDIUM TRICHLOROMANGANATE(II)
$RbMnCl_3$

J. GOODYEAR, G.A. STEIGMANN and E.M. ALI, 1977. Acta Cryst., <u>B33</u>, 256-258.

Hexagonal, $P6_3/mmc$, a = 7.16, c = 17.83 Å, D_m = 3.11, Z = 6. Mo radiation, R = 0.12 for 125 reflexions (films, densitometer and visual intensities).

Atomic positions

			x	y	z
Mn(1)	in	4(f)	1/3	2/3	0.1603
Mn(2)		2(a)	0	0	0
Rb(1)		4(f)	1/3	2/3	-0.0888
Rb(2)		2(b)	0	0	1/4
Cl(1)		12(k)	0.1616	-0.1616	0.0820
Cl(2)		6(h)	0.4928	-0.4928	1/4

The structure contains six close-packed $RbCl_3$ layers, with Mn ions between the layers coordinated octahedrally to Cl ions. Pairs of face-sharing Mn(1) octahedra share corners with Mn(2) octahedra, and Rb ions have 12-coordination. Mn-Cl = 2.48-2.55(8), Rb-Cl = 3.58-3.72(6) Å.

RUBIDIUM TETRACHLOROMANGANATE(II)
Rb_2MnCl_4

J. GOODYEAR, E.M. ALI and G.A. STEIGMANN, 1977. Acta Cryst., <u>B33</u>, 2932-2933.

Tetragonal, I4/mmm, a = 5.05, c = 16.14 Å, D_m = 2.96, Z = 2. Mo radiation, R = 0.10 for 91 reflexions (films, visual intensities). Mn in 2(a); Rb and Cl(1) in 4(e): z = 0.3551 and 0.1546; Cl(2) in 4(c).

K$_2$NiF$_4$-type (1) structure. Mn-6Cl = 2.50, 2.53(5), Rb-9Cl = 3.24, 3.44, 3.57(4) Å.

1. Structure Reports, 17, 332; 19, 323.

CAESIUM OCTACHLORODIRHENATE(III) HYDRATE
Cs$_2$Re$_2$Cl$_8$·H$_2$O

F.A. COTTON and W.T. HALL, 1977. Inorg. Chem., 16, 1867-1871.

Monoclinic, P2$_1$/c, a = 9.323, b = 13.377, c = 11.979 Å, β = 95.13°, Z = 4. Mo radiation, R = 0.037 for 1430 reflexions.

The structure is as previously reported (1), except that the Re-Re distance is longer in the Re$_2$Cl$_8$(H$_2$O)$_2$$^{2-}$ ion, 2.252(2) Å, than in the Re$_2$Cl$_8$$^{2-}$ ion, 2.237(2) Å. Re-Cl = 2.324-2.345, Re-O = 2.66 Å.

1. Structure Reports, 37A, 202; 39A, 176.

SODIUM TETRACHLOROCOBALTATE(II)
Na$_2$CoCl$_4$

SODIUM TETRACHLOROZINCATE
Na$_2$ZnCl$_4$

C.J.J. van LOON and D. VISSER, 1977. Acta Cryst., B33, 188-190.

Orthorhombic, Pnma, a = 13.713, 13.695, b = 8.073, 8.053, c = 6.428, 6.402 Å, Z = 4. Neutron powder data.

Chrysoberyl (Al$_2$BeO$_4$) structure (1), with hexagonal close-packed Cl, Co (Zn) in tetrahedral holes, Na in octahedral holes; Co-Cl = 2.20-2.42(4), Zn-Cl = 2.26-2.34(2); Na-Cl = 2.76-2.93 Å.

1. Strukturbericht, 1, 355, 400, 415; Structure Reports, 28, 141.

AMMONIUM TETRACHLOROPALLADATE(II)
(ND$_4$)$_2$PdCl$_4$

F.K. LARSEN and R.W. BERG, 1977. Acta Chem. Scand., A31, 375-378.

Tetragonal, P4/mmm, a = 7.22, 7.20, c = 4.24, 4.21 Å, at 295 and 125°K, respectively, D$_m$ = 2.1, Z = 1. Neutron diffraction data, R(F^2) = 0.077 and 0.052 for 317 and 264 reflexions at 295 and 125°K, respectively. Pd in 1(a); Cl in 4(j): x = 0.2266, 0.2266; N in 2(e); 0.5 D in 16(u): (0.0770, 0.4231, 0.3601), (0.0811, 0.4194, 0.3581) at 295 and 125°K, respectively.

The structure is as previously described for the hydrogen compound (1), with disordered D positions. The PdCl$_4$$^-$ ion is square-planar, Pd-Cl = 2.314(4) Å. The disordered ammonium ion is involved in N-D...Cl hydrogen bonding, N...Cl = 3.312 Å, N-D...Cl = 172, 173°.

1. Strukturbericht, 1, 360, 424; Structure Reports, 31A, 102.

RUBIDIUM TETRACHLOROAURATE(III)
RbAuCl$_4$

I. J. STRÄHLE and H. BÄRNIGHAUSEN, 1970. Z. Naturforsch., 25B, 1186-1187.
II. Idem, 1971. Z. Kristallogr., 134, 471-472.

Monoclinic, I2/c, a = 9.760, b = 5.902, c = 14.116 Å, β = 120.05°, Z = 4. Mo
radiation, R = 0.056 for 445 reflexions.

Atomic positions

			x	y	z
Rb	in	4(e)	0	0.4763	1/4
Au		4(a)	0	0	0
Cl(1)		8(f)	0.1161	-0.0210	0.1855
Cl(2)		8(f)	0.1633	0.2943	0.0187

The structure contains square-planar AuCl$_4^-$ ions, Au-Cl = 2.28 Å, and Rb$^+$ ions
with 10-coordination. II also gives a preliminary account of the structure of
RbAuBr$_4$ (P2$_1$/a).

SILVER(I) TETRACHLOROAURATE(III)
AgAuCl$_4$

W. WERNER and J. STRÄHLE, 1977. Z. Naturforsch., 32B, 741-744.

Monoclinic, I2/c, a = 13.223, b = 4.101, c = 13.089 Å, β = 122.72°, Z = 4. R =
0.068 for 723 reflexions.

Atomic positions

			x	y	z
Ag	in	4(e)	0	0.4915	1/4
Au		4(a)	0	0	0
Cl(1)		8(f)	0.1179	0.0272	0.2067
Cl(2)		8(f)	0.1383	0.3126	-0.0084

The structure is similar to that of RbAuCl$_4$ (1), although the unit cells differ
in shape. It contains square-planar AuCl$_4^-$ ions, Au-Cl = 2.28(1) Å, and Ag$^+$ ions
with octahedral coordination, Ag-Cl = 2.71-2.92(1) Å.

1. Preceding report.

ZIRCONIUM MONOBROMIDE
ZrBr

R.L. DAAKE and J.D. CORBETT, 1977. Inorg. Chem., 16, 2029-2033.

Rhombohedral, R3̄m, a = 3.5031, c = 28.071 Å, Z = 6. Cu radiation, powder data,
R = 0.16 for 50 lines. Atoms in 6(c): z(Zr) = 0.2092, z(Br) = 0.3917.

The structure contains the same X-Zr-Zr-X slabs as ZrCl (1), but stacked ACB
rather than ABC. SrCl (2) and HfCl are isostructural.

1. Structure Reports, 42A, 186.
2. This volume, p. 132.

POTASSIUM DECABROMODIBISMUTHATE(III) TETRAHYDRATE
$K_4Bi_2Br_{10}.4H_2O$

F. LAZARINI, 1977. Acta Cryst., B33, 1954-1956.

Orthorhombic, Pnma, a = 8.794, b = 22.737, c = 12.860 Å, D_m = 3.70, Z = 4. Cu
radiation, R = 0.079 for 1836 reflexions.

     The structure (Fig. 1) contains $Bi_2Br_{10}^{4-}$ anions (two distorted octahedra
sharing an edge), K^+ cations with 7- and 9-coordinations, and water molecules
coordinated to the cations. Bi-Br = 2.75-2.88 (terminal), 2.98 and 3.01 Å (bridging). The ammonium compound is isostructural.

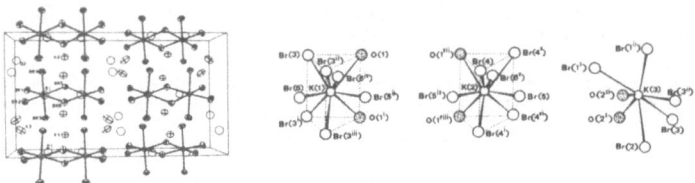

Fig. 1.   Structure and K coordination of $K_4Bi_2Br_{10}.4H_2O$.

CAESIUM ENNEABROMODIBISMUTHATE(III)
$Cs_3Bi_2Br_9$

F. LAZARINI, 1977. Acta Cryst., B33, 2961-2964.

Trigonal, P3m1, a = 7.972, c = 9.867 Å, D_m = 4.65, Z = 1. Mo radiation, R = 0.058
for 424 reflexions.

Atomic positions

			x	y	z
Bi	in	2(d)	2/3	1/3	0.1918
Cs(1)		1(a)	0	0	0
Cs(2)		2(d)	2/3	1/3	0.6669
Br(1)		3(e)	1/2	1/2	0
Br(2)		6(i)	0.3352	0.1676	0.3394

Fig. 1.   Structure of caesium enneabromodibismuthate(III).

     Cs and Br are cubic close-packed with Bi in 1/6 of the octahedral holes (Fig.
1). The $BiBr_6$ octahedra share cis-vertices with three other octahedra to form
corrugated layers. Bi-Br = 2.713 (terminal), 2.979 (shared), Cs-Br = 3.959-4.071 Å.

RUBIDIUM CADMIUM BROMIDE
$RbCdBr_3$

M. NATARAJAN IYER, R. FAGGIANI and I.D. BROWN, 1977. Acta Cryst., B33, 127-128.

Orthorhombic, Pnma, a = 9.436, b = 4.202, c = 15.607 Å, D_m = 4.68, Z = 4. Mo radiation, R = 0.050 for 497 reflexions.

Atomic positions

	x	y	z
Rb	0.4305	1/4	0.8260
Cd	0.1651	1/4	0.0564
Br(1)	0.2826	1/4	0.2089
Br(2)	0.1693	1/4	0.4950
Br(3)	0.0276	1/4	0.8987

Isostructural with NH_4CdCl_3 (1). Double columns along b of edge-sharing $CdBr_6$ octahedra are linked by nine-coordinated Rb ions. Cd-Br = 2.626-2.866(2), Rb-Br = 3.435-3.967(3) Å.

1. Strukturbericht, 6, 13, 79; 7, 19, 115.

THALLIUM LEAD IODIDE
Tl_6PbI_{10}

W. STOEGER, H. SCHULZ and A. RABENAU, 1977. Z. anorg. Chem., 432, 5-16.

Hexagonal, P6̄2c, a = 10.561, c = 13.522 Å, D_m = 6.87, Z = 2. Mo radiation, R = 0.063 for 476 reflexions.

Atomic positions

			x	y	z
	Tl(1)	in 6(g)	0.370	0	0
	Tl(2)	6(h)	0.604	0.969	1/4
0.5 Pb		4(e)	0	0	0.388
	I(1)	6(g)	0.742	0	0
	I(2)	6(h)	0.264	0.003	1/4
	I(3)	4(f)	1/3	2/3	0.399
	I(4)	4(f)	1/3	2/3	0.865

There is a sub-structure, space group $P6_3/mmc$. The structure (Fig. 1) contains a Tl_6I_6 framework with channels parallel to c, which contain Pb^{2+} and linear, nearly regular I_4^{2-} ions.

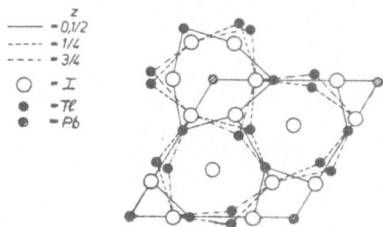

Fig. 1.   Structure of Tl_6PbI_{10}.

RUBIDIUM IODIDE TRIIODIDE HEXAIODOBISMUTHATE(III) DIHYDRATE
$Rb_5I(I_3)(BiI_6) \cdot 2H_2O$

F. LAZARINI, 1977. Acta Cryst., B33, 1957-1959.

Rhombohedral, R3̄m, a = 23.380, c = 14.959 Å, D_m = 3.9, Z = 9. Cu radiation, R = 0.104 for 878 reflexions.

Atomic positions

			x	y	z
Bi	in	9(e)	0	1/2	0
I(1)		18(h)	2y-1	0.4992	0.2050
I(2)		36(i)	0.1085	0.4619	-0.0022
I(3)		3(a)	0	0	1
I(4)		6(c)	0	0	0.3153
I(5)		9(d)	1/6	1/3	5/6
I(6)		18(h)	0.0994	2x	0.9062
Rb(1)		18(g)	0	0.3307	1/2
Rb(2)		18(h)	0.0708	2x	0.1651
0.5 Rb(3)		18(h)	0.0697	2x	0.5171
O		18(f)	0	0.2667	0

The structure contains Rb^+ cations, isolated anions (octahedral $BiI_6{}^{3-}$, linear symmetrical $I_3{}^-$, and I^-), and water molecules coordinated to the cations. Bi-I = 3.07, I-I = 2.94; Rb^+ ions have irregular coordination.

RUBIDIUM TETRAIODOCOBALTATE(II)
Rb_2CoI_4

H.J. SEIFERT and L. STÄUDEL, 1977. Z. anorg. Chem., 429, 105-117.

Monoclinic, $P2_1/m$, a = 10.383, b = 8.144, c = 7.657 Å, β = 109.8°, D_m = 4.02, Z = 2. Mo radiation, R = 0.096 for 434 reflexions.

Atomic positions

	x	y	z
Rb(1)	0.5865	1/4	0.8019
Rb(2)	0.0354	1/4	0.6963
Co	0.2144	1/4	0.3044
I(1)	-0.0498	1/4	0.1354
I(2)	0.3569	1/4	0.0813
I(3)	0.2782	-0.0040	0.5203

Sr_2GeS_4-type structure (1), containing tetrahedral $CoI_4{}^{2-}$ ions, Co-I = 2.59-2.61 Å, and Rb^+ ions, each having 6 I at 3.70-3.84 Å with two others at 4.08-4.26 Å.

Several other M_2CoX_4 compounds are isostructural, and $CsNiCl_3$-, Cs_3CoCl_5-, and β-K_2SO_4-type structures are also found in MX/CoX_2 systems.

1. Structure Reports, 35A, 65; 38A, 100.

RUBIDIUM SILVER IODIDE
Rb_2AgI_3

I.D. BROWN, H.E. HOWARD-LOCK and M. NATARAJAN, 1977. Canad. J. Chem., 55, 1511-1514.

Orthorhombic, Pnma, a = 10.258, b = 4.886, c = 20.063 Å, D_m = 4.23, Z = 4. Mo radiation, R = 0.089 for 1321 reflexions.

Atomic positions

	x	y	z
Rb(1)	0.4217	3/4	0.7113
Rb(2)	0.2522	1/4	0.4570
Ag	0.1362	1/4	0.1360
I(1)	0.1879	1/4	0.2753
I(2)	0.3803	1/4	0.0710
I(3)	0.0033	3/4	0.1018

Isostructural with K_2AgI_3 (1), as previously reported (2). The structure contains chains of corner-linked $Ag\bar{I}_4$ tetrahedra along b; Ag-I = 2.834 (terminal), 2.881 (bridging), Rb-I = 3.612-3.856 Å (7-coordination).

1. Structure Reports, 15, 169; 41A, 184.
2. Ibid., 16, 204.

RUBIDIUM CADMIUM IODIDE MONOHYDRATE
$RbCdI_3 \cdot H_2O$

M. NATARAJAN IYER, R. FAGGIANI and I.D. BROWN, 1977. Acta Cryst., B33, 129-130.

Monoclinic, Cc, a = 10.911, b = 10.030, c = 8.778 Å, β = 90.6°, D_m = 3.90, Z = 4.
Mo radiation, R = 0.061 for 1020 reflexions.

Atomic positions

	x	y	z
Rb	0.3980	0.8891	0.3267
Cd	0	0.0337	0
I(1)	0.0901	0.2868	-0.0054
I(2)	0.3407	0.4930	0.7488
I(3)	0.6733	0.3371	-0.0060
O	0.3910	0.1400	0.1544

The structure contains chains of corner-sharing CdI_4 tetrahedra along c, with Rb ions and water molecules between the chains. Cd-I = 2.723-2.823(5), Rb-$\bar{2}$ O = 2.89, 2.94, Rb-7 I = 3.75-4.13 Å.

TIN(II) OXYFLUORIDE   (TIN(II) BIS[DIFLUOROOXOSTANNATE(II)])
Sn_2OF_2 $0.5[Sn_2(Sn_2O_2F_4)]$

B. DARRIET and J. GALY, 1977. Acta Cryst., B33, 1489-1492.

Monoclinic, C2/m, a = 9.296, b = 8.076, c = 5.074 Å, β = 97.9°, D_m = 5.11, Z = 4.
Mo radiation, R = 0.036 for 381 reflexions.

Atomic positions

	x	y	z
Sn(1)	0.2167	0	0.0143
Sn(2)	0	0.2964	1/2
O	0.392	0	0.303
F	0.675	0.179	0.199

The structure (Fig. 1) contains two types of Sn(II) ion. Sn(2) has trigonal bipyramidal coordination (including an equatorial lone pair) and shares an equatorial O...O edge to form $(Sn_2O_2F_4)^{4-}$ dimers; these are connected into infinite $[(Sn_2O_2F_4)Sn_2]_n$ strings along $\underline{c}$ by Sn(1) atoms, which have tetrahedral coordination (including a lone pair). The strings are held together by weak Sn(1)...F bonds. Sn-O = 2.04, 2.11, Sn-F = 2.14, 2.39, Sn(1)...F = 2.80(1) Å.

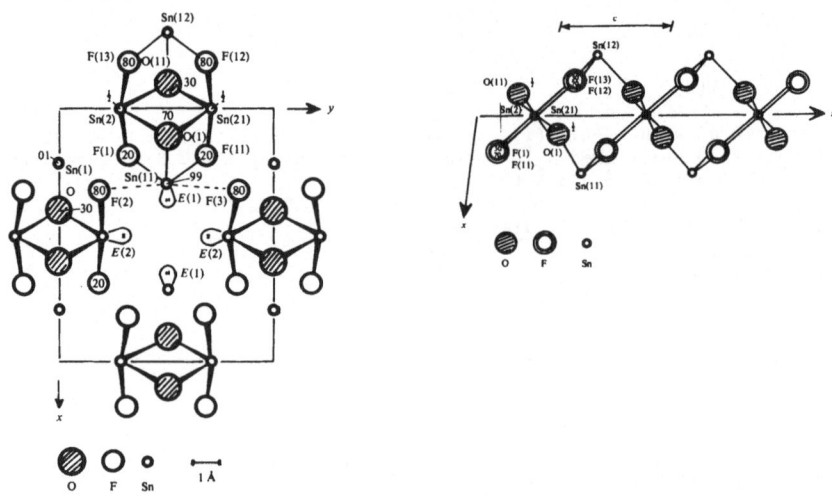

Fig. 1.   Structure of tin(II) oxyfluoride.

LEAD FLUORIDE OXIDE
Pb_2F_2O

B. AURIVILLIUS, 1976. Chem. Scripta, 10, 156-158.

Tetragonal, $P4_2/nmc$, a = 8.155, c = 5.722 Å, [D_m = 8.16 (1)], Z = 4. Mo radiation, R = 0.037 for 199 reflexions.

Atomic positions

	x	y	z
Pb	1/4	0.0253	0.0602
F	-0.0430	0.0430	1/4
O	1/4	1/4	0.367

The structure is essentially as previously described (1), and shows a strong resemblance to that of β-Bi_2O_3 ($P\bar{4}2_1c$) (2). Pb coordination is a strongly-deformed trigonal prism, Pb-O = 2.21, 2.46, Pb-F = 2.51 (x 2), 2.63 (x 2) Å, the prisms sharing edges.

1. Structure Reports, 11, 321.
2. B. AURIVILLIUS and G. MALMROS, 1972. Trans. Roy. Inst. of Technol., Stockholm, No. 291, 545.

TELLURIUM OXYFLUORIDE
(TELLURIUM(VI) TETRAKIS[OXOPENTAFLUOROTELLURATE(VI)] DIFLUORIDE)
$Te_5O_4F_{22}$ $trans-F_2Te(OTeF_5)_4$

H. PRITZKOW and K. SEPPELT, 1977.   Inorg. Chem., 16, 2685-2687.

Tetragonal, $I4_1/a$, a = 9.816, c = 20.341 Å, Z = 4.   Mo radiation, R = 0.048 for 1178
reflexions.

Atomic positions

	x	y	z
Te(1)	0	1/4	1/8
Te(2)	0.2877	0.0448	0.0963
F(1)	0	1/4	0.0341
F(2)	0.4162	0.0987	0.1543
F(3)	0.3628	0.1518	0.0344
F(4)	0.2139	0.9342	0.1579
F(5)	0.1589	0.9864	0.0385
F(6)	0.3970	0.9071	0.0700
O	0.1813	0.1909	0.1267

The structure consists of discrete molecules which have a central trans-
TeF_2O_4 octahedron linked to four TeF_5O octahedra (Fig. 1).   Mean bond lengths are
Te-F = 1.85 (central octahedron), 1.81 (outer octahedra), Te-O = 1.88(1) Å.

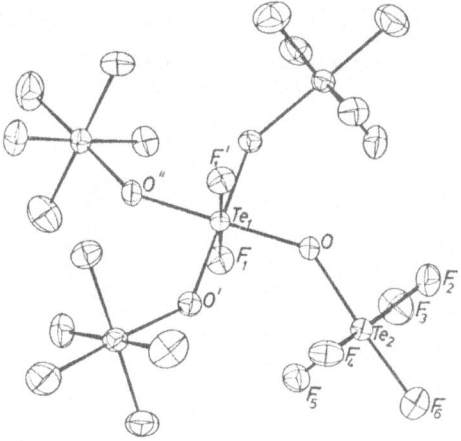

Fig. 1.   The $trans-F_2Te(OTeF_5)_4$ molecule.

μ-FLUOROSULPHATO-BIS[FLUOROXENON(II)] HEXAFLUOROARSENATE(V)
$[(FXeO)_2SOF][AsF_6]$

R.J. GILLESPIE, G.J. SCHROBILGEN and D.R. SLIM, 1977.   J. Chem. Soc., Dalton,
1003-1006.

Monoclinic, $P2_1/n$, a = 11.178, b = 8.718, c = 11.687 Å, β = 91.28°, Z = 4.   Mo
radiation, R = 0.078 for 1147 reflexions.

The structure contains discrete $[(XeF)_2SFO_3]^+$ cations and $[AsF_6]^-$ anions (Fig. 1);
Xe-F = 1.86(3), Xe-O = 2.21(2), S-O(3) = 1.39, S-O(1,2) = 1.47, S-F = 1.53(3) Å,
F-Xe-O = 178, Xe-O-S = 127°; As-F = 1.63-1.75(3) Å.

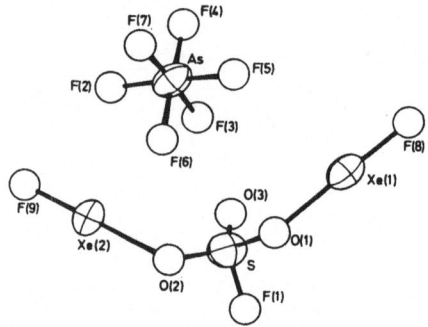

Fig. 1.   The $[(FXeO)_2SOF]^+$ and $[AsF_6]^-$ ions.

SODIUM TITANIUM ZINC OXYFLUORIDE
$Na_{0.76}(Ti,Zn)_4(O,F)_7$

M. MAYER, A. de ROY and G. PEREZ, 1976. Rev. Chim. Minér., <u>13</u>, 540-548.

Orthorhombic, Fmmm, a = 5.752, b = 11.294, c = 16.689 Å, Z = 8. Mo radiation,
R = 0.064 for 383 reflexions. The material has composition $Na_{1-x}Ti_{2+x}Zn_{2-x}O_{6+x}F_{1-x}$
(x ∿ 0.24).

Atomic positions

			x	y	z
	M(1)* in	8(f)	1/4	1/4	1/4
	M(2)	8(i)	0	0	0.1845
	M(3)	16(m)	1/2	0.1308	0.1082
0.76 Na		8(h)	0	0.1313	0
	O(1a)	16(j)	1/4	1/4	0.1292
	O(2a)	16(m)	0	0.1330	0.2668
	O(3a)	16(n)	0.2767	0	0.1051
	O(4)	8(h)	1/2	0.1656	0

Zn occupancies: M(1) = 0.33, M(2) = 0.73, M(3) = 0.35.

    The structure (Fig. 1) contains a framework of edge- and corner-sharing MO_6
octahedra, with M(2) ions in tetrahedral sites, and pairs of Na ions in cages
between two layers of octahedra.

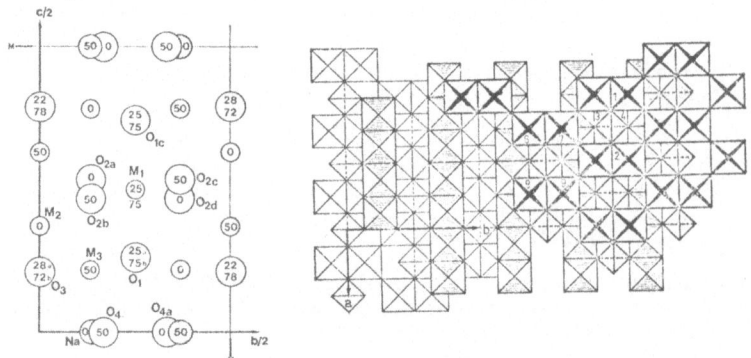

Fig. 1.   Atomic positions and idealized octahedra in $Na_{0.76}(Ti,Zn)_4(O,F)_7$.

POTASSIUM TANTALUM OXYFLUORIDE
$K_2Ta_2O_3F_6$

M. VLASSE, J.-P. CHAMINADE and M. POUCHARD, 1976. Bull. Soc. Fr. Minér. Crist., 99, 3-7.

Orthorhombic, Pnma, a = 10.212, b = 5.581, c = 14.499 Å, D_m = 4.79, Z = 4. Mo radiation, R = 0.065 for 1626 reflexions.

The structure contains double zigzag chains along b of fac-$TaO_3F_3^{4-}$ octahedra sharing the three oxygen corners; the chains are linked by the K^+ ions. Ta-F = 1.938-1.976, Ta-O 1.873-1.920(6) Å. [Independent study in 1.]

1. Structure Reports, 42A, 203.

URANIUM OXIDE TETRAFLUORIDE
α-UOF_4

J.H. LEVY, J.C. TAYLOR and P.W. WILSON, 1977. J. Inorg. Nucl. Chem., 39, 1989-1991.

Rhombohedral, R3m, a = 13.22, c = 5.72 Å, Z = 9 (1). Neutron powder data with profile fitting.

The results of a previous X-ray study (1) are confirmed.

1. Structure Reports, 41A, 193.

μ-HYDROXO-BIS[PENTAAMMINECHROMIUM(III)] CHLORIDE DIHYDRATE
$[(NH_3)_5CrOHCr(NH_3)_5]Cl_5 \cdot 2H_2O$

P. ENGEL and H.U. GÜDEL, 1977. Inorg. Chem., 16, 1589-1593.

Monoclinic, C2/c, a = 23.656, b = 7.36, c = 16.718 Å, β = 128.1°, D_m = 1.55, Z = 4. Mo radiation, R = 0.091 for 1112 reflexions.

The complex cation contains two octahedra sharing the OH corner, Cr-N = 2.081-2.091, Cr-O = 1.974 Å, Cr-O-Cr = 158°. The cations are well separated from the Cl^- ions and water molecules.

POTASSIUM PENTACHLOROOXOMOLYBDATE(V)
K_2MoOCl_5 (I)

POTASSIUM AQUOTETRABROMONITRIDOOSMATE MONOHYDRATE
$K[OsNBr_4(H_2O)] \cdot H_2O$ (II)

V.V. TKAČEV, O.N. KRASOČKA and L.O. ATOVMJAN, 1976. Ž. Strukt. Khim., 17, 940-941 [J. Struct. Chem., 17, 807-808].

I. Orthorhombic, Pnma, a = 13.430, b = 9.802, c = 6.895 Å, Z = 4. Mo radiation, R = 0.086 for 790 reflexions (film data).

II. Orthorhombic, Cmc2₁, a = 9.558, b = 13.620, c = 7.518 Å, Z = 4. Mo radiation, R = 0.106 for 474 reflexions (film data).

Atomic positions

I

			x	y	z
Mo	in	4(c)	0.0968	1/4	0.1784
K		8(d)	0.3567	0.0012	0.1782
Cl(1)		4(c)	0.0015	1/4	0.472
Cl(2)		4(c)	0.2478	1/4	0.4115
Cl(3)		4(c)	0.2213	1/4	-0.065
Cl(4)		8(d)	0.1082	0.0069	0.1989
O		4(c)	0.001	1/4	0.038

II

		x	y	z
Os	4(a)	0	0.1654	0
Br(1)	8(b)	0.1816	0.0363	0.0329
Br(2)	8(b)	0.1837	0.2883	0.0741
K	4(a)	1/2	0.1498	0.959
$H_2O(1)$	4(a)	0	0.149	0.321
$H_2O(2)$	4(a)	1/2	0.038	0.302
N	4(a)	0	0.172	-0.222

K_2MoOCl_5 is isostructural with K_2OsNCl_5 (1), and $K[OsNBr_4(H_2O)].H_2O$ is isostructural with the corresponding chloride (2). The structures contain octahedral anions; Mo-Cl = 2.37-2.59, Mo-O = 1.61(1); Os-Br = 2.486(4), Os-OH₂ = 2.42(3), Os-N = 1.67(5) Å.

1. Structure Reports, 24, 291; 33A, 227; 34A, 225.
2. Ibid., 35A, 431.

MANGANESE OXYCHLORIDE
$Mn_8O_{10}Cl_3$

G. BUISSON, 1977. Acta Cryst., B33, 1031-1034.

Tetragonal, I4/mmm, a = 9.290, c = 13.025 Å, Z = 4. Mo radiation, R = 0.105 for 461 reflexions.

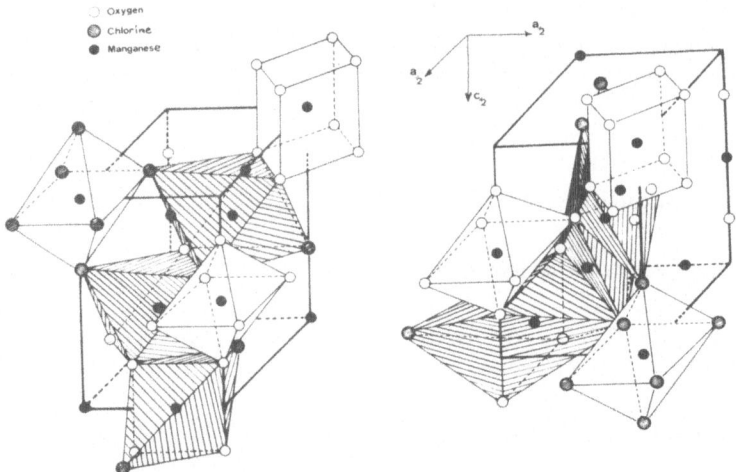

Fig. 1.   Structure of $Mn_8O_{10}Cl_3$.

The structure contains Mn coordination polyhedra, one cube and four octahedra (Fig. 1). The cube and one octahedron contain Mn^{2+}, Mn-8 O = 2.32 and Mn-6 Cl = 2.60 Å, respectively, and the other octahedra contain Mn^{3+}, Mn-O = 1.84-2.03, Mn-Cl = 2.83-2.84 Å. Above 360°C the structure becomes cubic.

AMMONIUM HEXAAMMINE-TRI-μ-HYDROXO-DICOBALT TETRACHLORIDE
$NH_4[(NH_3)_3Co(OH)_3Co(NH_3)_3]Cl_4$

G.S. MANDEL, N.S. MANDEL, R.E. MARSH and W.P. SCHAEFER, 1977. Acta Cryst., B33, 700-704.

Orthorhombic, Pnnm, a = 13.088, b = 17.276, c = 6.922 Å, D_m = 1.829, Z = 4. Co radiation, R = 0.041 for 1135 reflexions (disordered structure).

The complex cations (Fig. 1) consist of two octahedra sharing a face, and are disordered across a mirror plane, with the two orientations related by an 18° rotation about the Co...Co axis. One chloride ion is also disordered, and the other three chloride ions and the ammonium ion are possibly disordered. The disorder increases the strength and number of O-H...Cl and N-H...Cl hydrogen bonds.

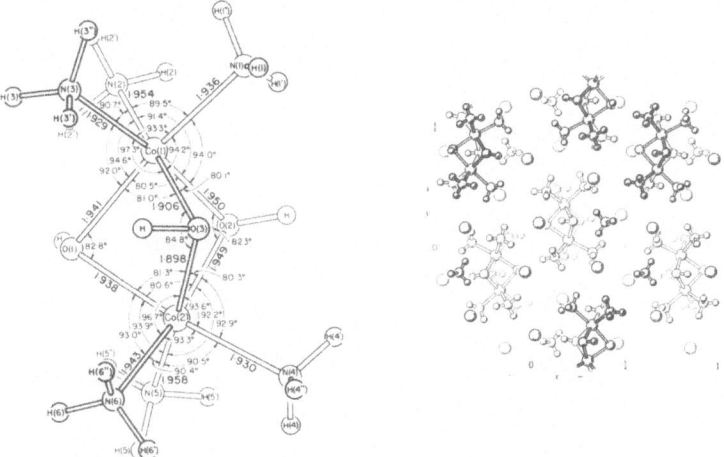

Fig. 1.   The complex cation (left, Co...Co = 2.565(1) Å) and structure (right, only one orientation shown) of ammonium hexaammine-tri-μ-hydroxo-dicobalt tetrachloride.

COPPER OXYCHLORIDE
Cu_2OCl_2

R. ARPE and H. MÜLLER-BUSCHBAUM, 1977. Z. Naturforsch., 32B, 380-382.

Orthorhombic, Fddd, a = 9.699, b = 9.603, c = 7.462 Å, Z = 8. R = 0.112 for 335 reflexions. Cu in 16(c); O in 8(a); Cl in 16(e): x = 0.176.

Cu has distorted octahedral coordination (Fig. 1), Cu-O = 1.94, Cu-Cl = 2.29, 3.13 Å (each x 2).

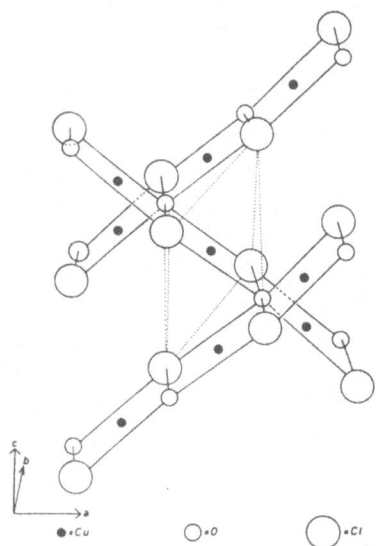

Fig. 1.   Structure of Cu_2OCl_2.

CALCIUM COPPER OXYCHLORIDE   CALCIUM COPPER OXYBROMIDE
$Ca_2CuO_2Cl_2$ $Ca_2CuO_2Br_2$

B. GRANDE and H. MÜLLER-BUSCHBAUM, 1977. Z. anorg. Chem., 429, 88-90.

Tetragonal, I4/mmm, a = 3.866, 3.875, c = 14.975, 17.264 Å, Z = 2. R = 0.091 and
0.105 for 146 and 133 reflexions. Cu in 2(a); O in 4(c); Ca in 4(e): z = 0.3957,
0.4097; Cl (Br) in 4(e): z = 0.1826, 0.1707.

        Isostructural with $Sr_2CuO_2Cl_2$ (1), K_2NiF_4-type.

1. Structure Reports, 41A, 195.

BARIUM COPPER OXYCHLORIDE
Ba_2CuO_2Cl

R. KIPKA and H. MÜLLER-BUSCHBAUM, 1977. Z. anorg. Chem., 430, 250-254.

Rhombohedral, R3̄m, a = 9.835 Å, α = 25.04° (hexagonal cell has a = 4.264, c =
28.570 Å), Z = 1. R = 0.068 for 265 reflexions. Atomic positions (hexagonal
axes): Cu in 3(b); Cl in 3(a); Ba and O in 6(c): z = 0.255 and 0.437.

        Cu has linear and Ba octahedral coordination, Cu-O = 1.80 (x 2), Ba-O =
2.57 (x 3), Ba-Cl = 3.32 (x 3) Å, the structure containing double octahedral
sheets.

BARIUM COPPER(II) OXYCHLORIDE   BARIUM COPPER(II) OXYBROMIDE
$Ba_{44}Cu_{45}O_{87}Cl_4$ $Ba_{88}Cu_{88}O_{175}Br_2$

R. KIPKA and H. MÜLLER-BUSCHBAUM, 1977. Z. Naturforsch., <u>32B</u>, 124-126.

Cubic, Im3m, a = 18.27, 18.30 Å, Z = 2, 1. R = 0.085 and 0.08 for 923 and 983 reflexions.

    The structures are similar to that of $BaCuO_2$ (<u>1</u>), but with Cl and Br in 2(a) and some Cl in 12(d).

<u>1</u>. This volume, p. 215.

CHLOROXIPHITE
$Pb_3CuO_2(OH)_2Cl_2$

J.J. FINNEY, E.J. GRAEBER, A. ROSENZWEIG and R.D. HAMILTON, 1977. Miner. Mag., <u>41</u>, 357-361.

Monoclinic, $P2_1/m$, a = 10.458, b = 5.750 [in abstract, 5.759 in text], c = 6.693 Å, β = 97.79°, D_m = 6.93, Z = 2. Mo radiation, R = 0.096 for 1192 reflexions.

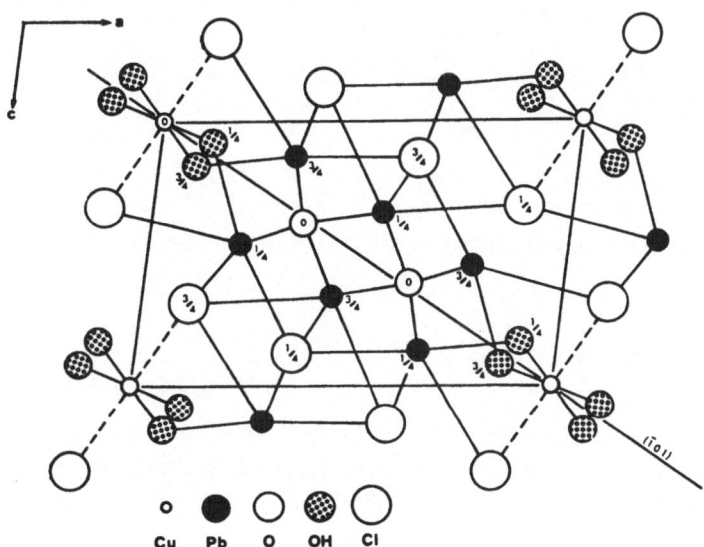

Fig. 1.   Structure of chloroxiphite.

Atomic positions

			x	y	y
Pb(1)	in	2(e)	0.2167	1/4	0.4606
Pb(2)		2(e)	0.5502	1/4	0.3432
Pb(3)		2(e)	0.6741	1/4	0.8673
Cu		2(a)	0	0	0
Cl(1)		2(e)	0.378	1/4	0.868
Cl(2)		2(e)	0.886	1/4	0.321
O(1)		4(f)	0.366	0.000	0.386
OH(1)		2(e)	0.907	1/4	0.832
OH(2)		2(e)	0.126	1/4	0.086

The structure (Fig. 1) consists of sheets parallel to ($\bar{1}01$) of composition $Pb_3Cu_2(OH)_2$, with layer sequence Pb-(O,OH,Cu)-Pb. It contains eightfold PbO_4Cl_4 and sevenfold PbO_5Cl_2 polyhedra, Pb_6 clusters, and Cu atoms in square-planar $Cu(OH)_4$ coordination.

POTASSIUM URANYL OXYCHLORIDE HYDRATE
$K_2U_4O_{11}Cl_4\cdot 7H_2O$

A. PERRIN and J.Y. LE MAROUILLE, 1977. Acta Cryst., B33, 2477-2481.

Triclinic, P$\bar{1}$, a = 12.15, b = 12.33, c = 8.026 Å, α = 110.50, β = 96.30, γ = 138.71°, D_m = 4.10, Z = 1. Mo radiation, R = 0.046 for 2940 reflexions.

The structure contains discrete tetranuclear $[(UO_2)_4O_2(OH)_2Cl_4(H_2O)_4]^{2-}$ anions (Fig. 1), held together by K^+ ions (7-coordination) and hydrogen bonds. U atoms have pentagonal bipyramidal coordination, with slightly non-linear uranyl groups (O-U-O = 176, 177°).

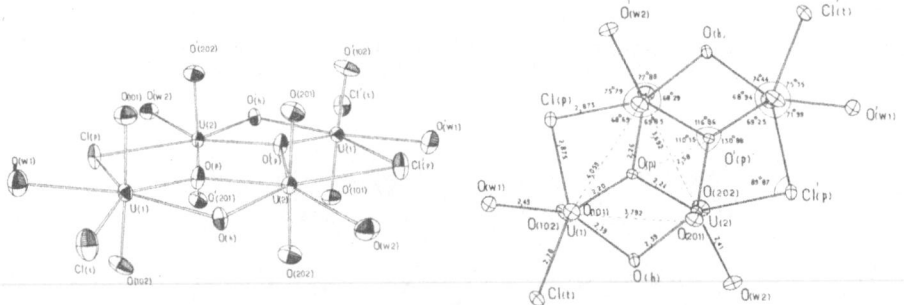

Fig. 1.   The tetranuclear $[(UO_2)_4O_2(OH)_2Cl_4(H_2O)_4]^{2-}$ anion.

RUBIDIUM URANYL OXYCHLORIDE HYDRATE
$Rb_2U_2O_5Cl_4\cdot 2H_2O$ $0.5\{Rb_4[(UO_2)_4O_2Cl_8(H_2O)_2]\cdot 2H_2O\}$

A. PERRIN, 1977. J. Inorg. Nucl. Chem., 39, 1169-1172.

Monoclinic, P2_1/c, a = 8.540, b = 8.096, c = 21.735 Å, β = 111.74°, D_m = 4.27, Z = 4. Mo radiation, R = 0.070 for 2519 reflexions.

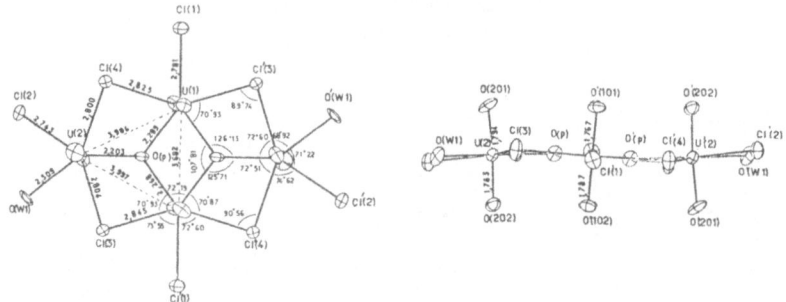

Fig. 1.   The $[(UO_2)_4O_2Cl_8(H_2O)_2]^{4-}$ anion.

The structure contains tetranuclear anions (Fig. 1), with UO_2 groups bonded by O/O and O/Cl double bridges.

TELLURIUM OXYBROMIDE
$Te_6O_{11}Br_2$

P. KHODADAD and N. RODIER, 1977. Bull. Soc. Chim. Fr., 251-253.

Orthorhombic, Ccmm, a = 6.880, b = 11.816, c = 15.823 Å, Z = 4. Mo radiation, R = 0.05 for 1134 reflexions.

Atomic positions

			x	y	z
Te(1)	in	16(h)	0.3693	0.1550	0.0857
Te(2)		8(f)	0.9284	0	0.1402
Br		8(g)	0.1899	0.3290	1/4
O(1)		8(f)	0.423	0	0.1136
O(2)		16(h)	0.164	0.3502	0.0346
O(3)		16(h)	0.080	0.1196	0.0956
O(4)		4(c)	0.062	0	1/4

The structure contains layers of atoms normal to $\underline{b}$. Te(1) has 4 O + 2 Br neighbours (distorted trigonal prism) and Te(2) has 3 $\overline{O}$ + 2 Br neighbours (distorted octahedron with a vacant apex); Te-O = 1.87-2.18(1), Te-Br = 3.110-3.500(3) Å. Br has six Te neighbours forming a distorted trigonal prism.

STRONTIUM COPPER OXYBROMIDE
$Sr_2CuO_2Br_2$

B. GRANDE and H. MÜLLER-BUSCHBAUM, 1977. Z. anorg. Chem., 433, 152-156.

Tetragonal, I4/mmm, a = 3.991, c = 17.136 Å, Z = 2. No details of data or analysis Sr in 4(e): z = 0.395; Cu in 2(a); O in 4(c); Br in 4(e): z = 0.147.

Isostructural with the chloride (1) (K_2NiF_4-type (2)). Cu-O = 2.00 (x 4), Cu-Br = 2.98 (x 2), Sr-O = 2.69 (x 4), Sr-Br = 3.06 (x 4), 3.79 Å.

1. Structure Reports, 41A, 195.
2. Ibid., 17, 332; 19, 323.

URANIUM(V) OXYBROMIDE
UO_2Br

J.-C. LEVET, M. POTEL and J.-Y. LE MAROUILLE, 1977. Acta Cryst., B33, 2542-2546.

Orthorhombic, Cmcm, a = 4.106, b = 20.200, c = 3.980 Å, D_m = 6.97, Z = 4. Mo radiation, R = 0.057 for 1580 reflexions. Atoms in 4(c): y = -0.07818, 0.18534, 0.0355, 0.4192 for U, Br, O(1), O(2).

The structure contains layers normal to $\underline{b}$ (Fig. 1), with pentagonal bipyramidal UO_5Br_2 groupings (two oxygens axial), U-Br = $\overline{2}$.939(3), U-O = 2.05-2.30(3) Å.

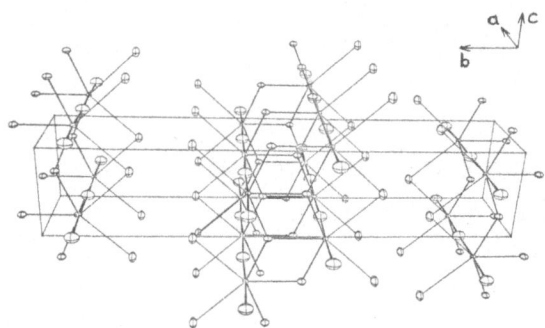

Fig. 1.   Structure of uranium(V) oxybromide.

SILVER IODIDE TETRATUNGSTATE
$Ag_{26}I_{18}W_4O_{16}$

L.Y.Y. CHAN and S. GELLER, 1977.   J. Solid State Chem., 21, 331-347.

Monoclinic, C2, a = 16.76, b = 15.52, c = 11.81 Å, β = 103.9°, D_m = 6.72, Z = 2.
Ag radiation, R = 0.092 for 1802 reflexions.

The structure contains $W_4O_{16}{}^{8-}$ ions (as in 1), and conduction passageways
which involve 90 iodide polyhedra and 56 mixed oxygen-iodide polyhedra which share
faces. About half of the Ag^+ ions are in these polyhedra and are considered to be
mobile.

1. Structure Reports, 41A, 259.

PRASEODYMIUM OXYIODIDE
PrOI

O.G. POTAPOVA, I.G. VASIL'EVA and S.V. BORISOV, 1977.   Ž. Strukt. Khim., 18,
573-577.

Tetragonal, P4/nmm, a = 4.086, c = 9.162 Å, D_m = 5.89, Z = 2.   Cu radiation, R =
0.08 for powder data.   PbFCl-type structure (1), z(Pr) = 0.131, z(I) = 0.672.
Pr-O = 2.37, Pr-I = 3.41 Å.

1. Strukturbericht, 2, 45, 362; 3, 369.

NEPTUNIUM OXYIODIDE
NpOI

D. BROWN, L. HALL, C. HURTGEN and P.T. MOSELEY, 1977.   J. Inorg. Nucl. Chem., 39,
1466-1468.

Tetragonal, P4/nmm, a = 4.051, c = 9.193 Å, Z = 2.   Cu radiation, powder data, R =
0.13 for 37 reflexions.   Np and I in 2(c): z = 0.1311 and 0.6699; O in 2(a), origin
at $\bar{4}m2$.

PbFCl-type structure (1); M-I = 3.40(2), M-O = 2.36(1) Å. The La, Er, and Tm
(2) compounds are isostructural.

1. Strukturbericht, 2, 45, 362; 3, 369.
2. Structure Reports, 26, 338.

TRITELLURIUM TRISULPHIDE HEXAFLUOROARSENATE(V)
$Te_3S_3(AsF_6)_2$ (I)

DITELLURIUM TETRASELENIDE HEXAFLUOROANTIMONATE(V)
$Te_2Se_4(SbF_6)_2$ (II)

DITELLURIUM TETRASELENIDE HEXAFLUOROARSENATE(V)
$Te_2Se_4(AsF_6)_2$ (III)

R.J. GILLESPIE, W. LUK, E. MAHARAJH and D.R. SLIM, 1977. Inorg. Chem., 16, 892-
896.

I. Monoclinic, P2$_1$/n, a = 8.421, b = 11.828, c = 15.279 Å, β = 90.92°, Z = 4.
Mo radiation, R = 0.088 for 1292 reflexions.

II and III. Orthorhombic, P2$_1$2$_1$2$_1$, a = 12.117, 12.012, b = 8.748, 8.640, c =
15.772, 15.272 Å, Z = 4. Mo radiation, R = 0.117 and 0.144 for 1465 and 838
reflexions.

The structures contain discrete ions (Figs. 1 and 2). Te-Te = 2.68-2.82,
Te-Se = 2.53-2.59, Te-S = 2.44, 2.47 Å.

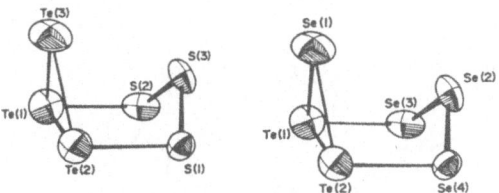

Fig. 1.   The $Te_3S_3{}^{2+}$ and $Te_2Se_4{}^{2+}$ cations.

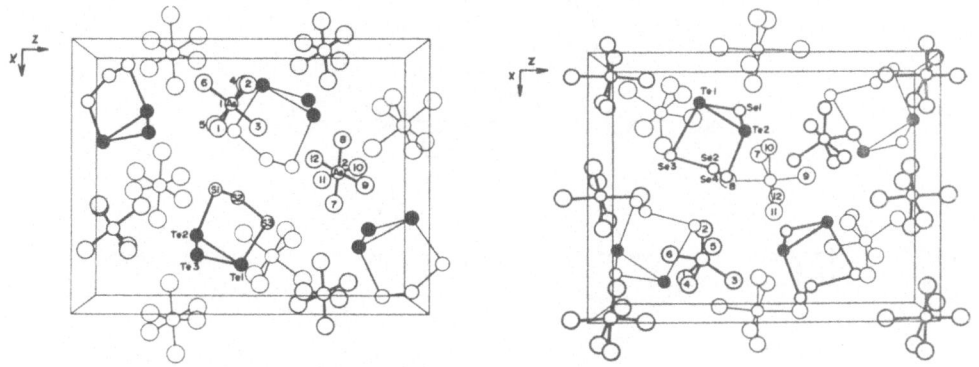

Fig. 2.   Structures of $Te_3S_3(AsF_6)_2$ and $Te_2Se_4(SbF_6)_2$.

YTTRIUM FLUOROSELENIDE
YSeF

I.   NGUYEN-HUY-DUNG and P. LARUELLE, 1977. Acta Cryst., B33, 1444-1448.
II.   Idem, 1977. Ibid., B33, 3360-3363.

10M and 8M forms
Monoclinic, $P2_1/m$, a = 9.926, 9.935, b = 31.728, 25.420, c = 4.095, 4.094 Å, γ =
96.00, 97.50°, D_m = 4.86, 4.80, Z = 20, 16. Mo radiation, R = 0.072 and 0.064 for
2139 and 672 reflexions.

    The structures are similar to those of the 2O, 6O, and 4M LnSeF polytypes
(1). The stacking sequences are: 10M, SSSTSTSTST and 8M, SSSTSTST, and there is
a partially occupied anion site.

1. Structure Reports, 39A, 199; 41A, 198, 199.

POTASSIUM LITHIUM IRON SULPHIDE CHLORIDE
$K_6LiFe_{24}S_{26}Cl$

B.S. TANI, 1977. Amer. Min., 62, 819-823.

Cubic, Pm3m, a = 10.358 Å, Z = 1. Fe radiation, R = 0.06 for 38 reflexions
(powder data).

    A model is proposed (Fig. 1) which fits the powder data. It contains clusters
of eight edge-shared FeS_4 tetrahedra, with K, Li, and Cl ions distributed among
these clusters.

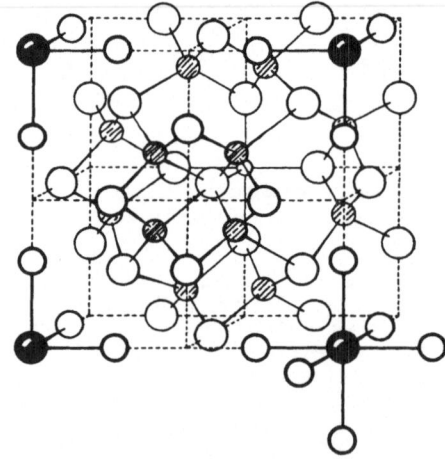

Fig. 1.   Proposed structure of $K_6LiFe_{24}S_{26}Cl$. Black circles are Cl, connected
           to K; the very small open circle is Li.

μ-SUPERSULPHIDO-BIS(PENTAAMMINERUTHENIUM) CHLORIDE DIHYDRATE
$[(NH_3)_5RuS_2Ru(NH_3)_5]Cl_4 \cdot 2H_2O$

R.C. ELDER and M. TRKULA, 1977. Inorg. Chem., 16, 1048-1051.

Orthorhombic, Pnma, a = 11.673, b = 7.216, c = 25.782 Å, D_m = 1.90, Z = 4. Mo radiation, R = 0.024 for 2048 reflexions.

The structure contains dimeric cations with a trans conformation (Fig. 1), surrounded by chloride ions and water molecules (one is disordered) which form a hydrogen bond network. The best formulation is considered to be Ru(III)-(S_2^-)-Ru(II).

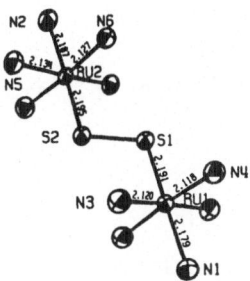

Fig. 1.   The $[(NH_3)_5RuSSRu(NH_3)_5]^{4+}$ cation.

ANTIMONY(III) IODIDE SULPHIDE
SbSI

I.   K. ITOH, H. MATSUNAGA and E. NAKAMURA, 1976. J. Phys. Soc. Japan, 41,
      1679-1680.
II.   Y. IWATA, N. KOYANO and I. SHIBUYA, 1976. Ann. Rep. Res. Reactor Inst.
      Kyoto Univ., 9, 9-18.

Orthorhombic, Pna2_1, a = 8.53, b = 10.14, c = 4.10 Å, for ferroelectric phase at 1.3°C, Z = 4. Mo radiation, R = 0.030 for 763 reflexions. II is a refinement with neutron diffraction data collected at -185, 16, and 100°C.

Atomic positions (I)

	x	y	z
Sb	0.1202	0.1234	0.2752
S	0.8450	0.0472	0.2547
I	0.5085	0.8284	0.25

The structure is as previously determined (1).

1. Structure Reports, 31A, 112; 32A, 223.

BISMUTH TELLURIUM IODIDE
BiTeI

A. TOMOKIYO, T. OKADA and S. KAWANO, 1977. Jap. J. Appl. Phys., 16, 291-298.

Trigonal, P3, a = 4.346, c = 6.835 Å, D_m = 6.9, Z = 1. Cu radiation, R = 0.17 for 152 reflexions (films, photometer intensities). Bi in 1(a): z = 0; Te in 1(c): z = 0.423; I in 1(b): z = 0.721.

The structure is similar to that of CdI_2 (if the difference between Te and I atoms is ignored). Bi-Te = 3.83, Te-I = 3.24, I-Bi = 5.53(4) Å.

NIOBIUM SELENIDE IODIDE

$NbSe_4I_{0.33}$

A. MEERSCHAUT, P. PALVADEAU and J. ROUXEL, 1977. J. Solid State Chem., 20, 21-27.

Tetragonal, P4/mnc, a = 9.489, c = 19.13 Å, D_m = 5.20, Z = 12. Mo radiation, R = 0.030 for 211 reflexions.

Atomic positions

			x	y	z
Nb(1)	in	8(f)	0	1/2	0.4200
Nb(2)		4(d)	0	1/2	1/4
I		4(e)	0	0	0.6297
Se(1)		8(h)	0.2754	0.0180	0
Se(2)		8(h)	0.3910	0.8000	0
Se(3)		16(i)	0.1265	0.6851	0.3394
Se(4)		16(i)	0.7844	0.5476	0.3283

The structure (Fig. 1) contains chains of face-sharing $NbSe_8$ antiprisms, with iodine atoms between the chains; Nb-Se = 2.58-2.73(1), Se-Se = 2.34, 2.37(2), Se-I = 3.27 Å.

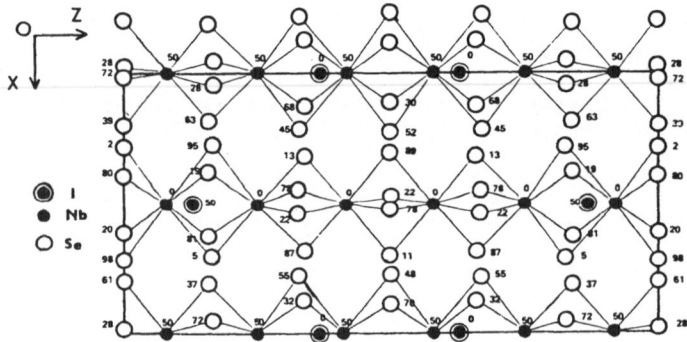

Fig. 1.   Structure of $NbSe_4I_{0.33}$.

SODIUM CYANIDE DIHYDRATE

$NaCN.2H_2O$

J.W. BATS, 1977. Acta Cryst., B33, 466-472.

Monoclinic, $P2_1/c$, a = 5.968, b = 10.508, c = 6.572 Å, β = 102.07°, at 150°K, Z = 4. Mo radiation, R = 0.043 for 2484 reflexions (conventional and high-angle refinements).

The CN group is inverted compared to a previous study (1), and is bonded to Na through N, the C being involved in O-H...C bonds of 2.97 and 2.99 Å. Na has distorted octahedral coordination to 4 O and 2 N atoms, Na-O = 2.40-2.45, Na-N = 2.44, 2.73 Å, and the water molecules form two O-H...N hydrogen bonds, 3.12 and 3.35 Å, as well as the O-H...C interactions.

Atomic positions (high-angle refinement)

	x	y	z
Na	0.0010	0.1705	0.9359
O(1)	0.7437	0.2981	0.6778
O(2)	0.1973	0.4790	0.6916
N	0.2319	0.1566	0.6194
C	0.3200	0.0676	0.7005

<u>1</u>. Structure Reports, <u>22</u>, 266.

DICYANOTRISULPHANE
NCS$_3$CN

J.W. BATS, 1977. Acta Cryst., B$\underline{33}$, 2264-2266.

Orthorhombic, Pnma, a = 10.060, b = 12.715, c = 4.240 Å, at 100°K, Z = 4. Mo
radiation, R = 0.034 for 1238 reflexions. Previous study in <u>1</u>.

Atomic positions

	x	y	z
S(1)	0.5200	1/4	0.4865
S(2)	0.4363	0.1207	0.2672
C	0.2941	0.1105	0.4815
N	0.1956	0.1012	0.6188

The structure (Fig. 1) contains NCS$_3$CN molecules, linked by S...N contacts of
3.08-3.18 Å (sum of van der Waals radii = 3.35 Å). S-S = 2.068(1), S-C = 1.700(1),
C-N = 1.155(2) Å, S-S-S = 105.3, S-S-C = 99.4, S-C-N = 177.6°.

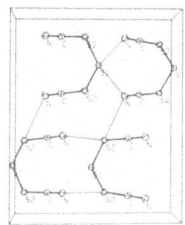

Fig. 1.   Structure of dicyanotrisulphane.

<u>1</u>. Structure Reports, <u>29</u>, 291.

POTASSIUM HEXACYANONITROSOVANADATE(I) HYDROXIDE HYDRATE
K$_4$[V(CN)$_6$NO].0·5KOH.0·5H$_2$O

M.G.B. DREW and C.F. PYGALL, 1977. Acta Cryst., B$\underline{33}$, 2838-2842.

Tetragonal, P$\bar{4}$, a = 9.035, c = 9.153 Å, D$_m$ = 1.93, Z = 2. Mo radiation, R =
0.077 for 590 reflexions.

Vanadium has sevenfold pentagonal bipyramidal coordination with the NO group
axial, but the complex anion is disordered about a twofold axis; V-N = 1.68, V-O =
2.12-2.24(3) Å. The water molecules and hydroxide ions are also disordered and
could not be distinguished, and one K position has 25% occupancy.

CAESIUM LITHIUM HEXACYANOCHROMATE(III)
$Cs_2LiCr(CN)_6$

M.R. CHOWDHURY, F.A. WEDGEWOOD, B.M. CHADWICK and H.J. WILDE, 1977. Acta Cryst.,
B33, 46-52.

Tetragonal, P4/mnc, a = 7.600, c = 10.777 Å, Z = 2. Neutron powder data with
profile analysis. Cs in 4(d); Li in 2(b); Cr in 2(a); C(1) in 8(h): (0.168, 0.201,
0); C(2) in 4(e): z = 0.206; N(1) in 8(h): (0.247, 0.322, 0); N(2) in 4(e): z =
0.314.

    The structure is closely related to that of the cubic high-temperature (above
350°K) phase (1), in comparison with which $Cr(CN)_6$ and LiN_6 octahedra are rotated
4.9 and 9.8°, respectively, about c (Fig. 1), and are tetragonally distorted.
Cr-C = 1.99, 2.22(1), C-N = 1.13(2), Li-N = 2.01, 2.35(1) Å. Cs occupies a large
hole and has high thermal motion; Cs-N = 3.55, 3.86, Cs-C = 3.75, 3.83(1) Å.

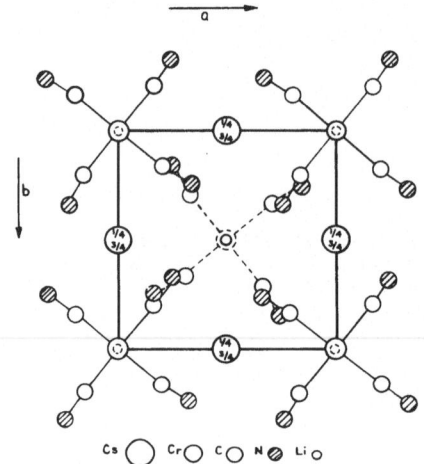

Fig. 1.   Structure of $Cs_2LiCr(CN)_6$.

1. Structure Reports, 40A, 171.

SODIUM HEXACYANOCHROMATE(II) DECAHYDRATE
$Na_4Cr(CN)_6 \cdot 10H_2O$

E. LJUNGSTRÖM, 1977. Acta Chem. Scand., A31, 104-108.

Monoclinic, $P2_1/n$, a = 9.866, b = 11.502, c = 9.152 Å, β = 98.30°, at -100°C, D_m =
1.54, Z = 2. Cu radiation, R = 0.052 for 1210 reflexions.

    Isostructural with the Fe(II) and Mn(II) compounds (1). Cr-C = 2.053(4),
C-N = 1.156(5) Å.

1. Structure Reports, 40A, 172.

HEXAAMMINECOBALT(III) HEXACYANOCHROMATE(III)
[Co(NH$_3$)$_6$][Cr(CN)$_6$]

M. IWATA, 1977. Acta Cryst., B$\underline{33}$, 59-69.

Rhombohedral, R$\bar{3}$, a = 7.372 Å, α = 97.93°, at 80°K (hexagonal cell has a = 11.122,
c = 10.864 Å), D$_m$ = 1.57 (at 291°K), Z = 1. Mo radiation, R = 0.028 for 2156
reflexions (aspherical charge refinement).

Atomic positions (hexagonal axes)

	x	y	z
Cr	0	0	0
Co	0	0	1/2
N(1)	0.2154	0.2510	0.1736
C	0.1407	0.1614	0.1097
N(2)	0.1271	0.1571	0.6060

Isostructural with [Co(NH$_3$)$_6$][Co(CN)$_6$] ($\underline{1}$). Cr-C = 2.071, C-N = 1.156(1),
Co-N = 1.979(1) Å.

$\underline{1}$. Structure Reports, $\underline{39A}$, 205.

SODIUM HEPTACYANOMOLYBDATE(II) DECAHYDRATE
Na$_5$[Mo(CN)$_7$].10H$_2$O

POTASSIUM HEPTACYANOMOLYBDATE(II) MONOHYDRATE
K$_5$[Mo(CN)$_7$].H$_2$O

M.G.B. DREW, P.C.H. MITCHELL and C.F. PYGALL, 1977. J. Chem. Soc., Dalton, 1071-
1077.

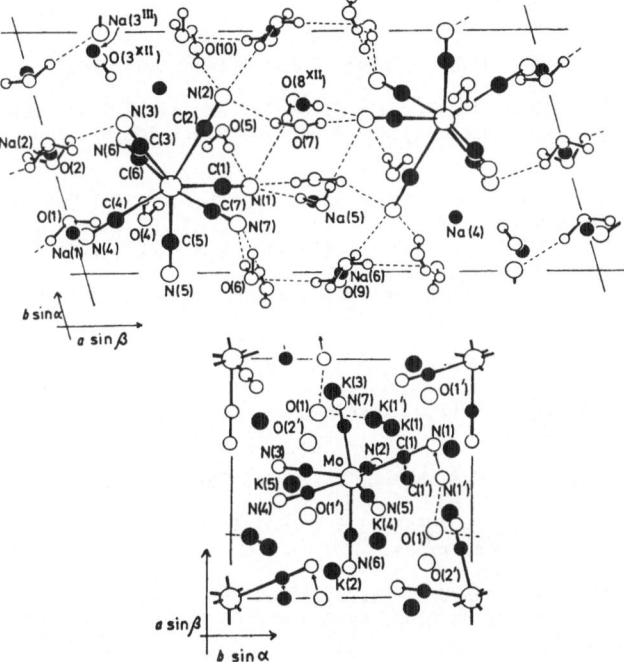

Fig. 1.   Structures of Na$_5$[Mo(CN)$_7$].10H$_2$O and K$_5$[Mo(CN)$_7$].H$_2$O.

Sodium salt
Triclinic, P$\bar{1}$, a = 16.527, b = 8.355, c = 8.324 Å, α = 105.48, β = 80.90, γ = 107.43°, D_m = 1.84, Z = 2. Mo radiation, R = 0.066 for 2790 reflexions.

Potassium salt
Triclinic, I1, a = 9.069, b = 9.218, c = 9.029 Å, α = 89.51, β = 90.17, γ = 92.47°, D_m = 2.12, Z = 2. Mo radiation, R = 0.081 for 1175 reflexions (twinned crystal).

In both salts (Fig. 1) the complex anion is best described as a pentagonal bipyramid (D_{5h} symmetry in the Na salt, but distorted to C_s in the K salt), Mo-C = 2.10-2.20 Å. Na ions have octahedral coordination and K ions have 7-9 coordination.

CAESIUM LITHIUM HEXACYANOFERRATE(III)
$Cs_2LiFe(CN)_6$

G.W. BEALL, W.O. MILLIGAN, J. KORP, I. BERNAL and R.K. McMULLAN, 1977.   Inorg. Chem., 16, 207-209.

Crystal data and structure as in 1. Neutron diffraction data, R = 0.032 for 76 reflexions, x(C) = 0.1823, x(N) = 0.2907 (in agreement with 1).

1. Structure Reports, 34A, 231; 39A, 203.

CAESIUM SODIUM HEXACYANOFERRATE(III)
$Cs_2NaFe(CN)_6$

CAESIUM POTASSIUM HEXACYANOFERRATE(III)
$Cs_2KFe(CN)_6$

S.R. FLETCHER and T.C. GIBB, 1977.   J. Chem. Soc., Dalton, 309-316.

Monoclinic, P2$_1$/n, a = 10.870, 11.140, b = 7.709, 8.131, c = 7.573, 7.660 Å, β = 89.994, 90.165°, Z = 2. Cu radiation, R = 0.060 and 0.075 for 897 and 1118 reflexions.

The structures are similar to that of cubic $Cs_2LiFe(CN)_6$ (1) but with lower symmetry; they contain octahedral Fe(CN)$_6$$^{3-}$ ions, Fe-C = 1.92-1.94(1), C-N = 1.13-1.16(2) Å. Cs-N = 3.22-3.49, Na-N = 2.46-2.49, K-N = 2.79-2.82 Å.

1. Structure Reports, 39A, 203; this volume, preceding report.

CAESIUM BARIUM HEXACYANOFERRATE(II) DIHYDRATE
$Cs_2BaFe(CN)_6 \cdot 2H_2O$

J.J. RAFALKO, B.I. SWANSON and G.W. BEALL, 1977.   J. Solid State Chem., 21, 195-201.

Monoclinic, P2$_1$/n, a = 8.799, b = 7.555, c = 11.607 Å, β = 78.56°, Z = 2. Mo radiation, R = 0.042 for 631 reflexions.

The structure contains octahedral Fe(CN)$_6$$^{4-}$ anions, Fe-C = 1.87-1.89(2) Å, 8-coordinate Ba ions (Ba-6N = 2.88-2.93, Ba-2 O = 2.93 Å), and 12-coordinate Cs ions (Cs-O = 3.05-3.70 Å).

STRONTIUM NITROPRUSSIDE TETRAHYDRATE
Sr[Fe(CN)$_5$NO].4H$_2$O

E.E. CASTELLANO, O.E. PIRO and B.E. RIVERO, 1977. Acta Cryst., B$\underline{33}$, 1725-1728.

Monoclinic, C2/m, a = 20.08, b = 7.51, c = 8.42 Å, β = 98.4°, D$_m$ = 1.96, Z = 4.
R = 0.099 for 1014 reflexions (films, visual intensities).

The structure contains distorted octahedral [Fe(CN)$_5$NO]$^{2-}$ ions which lie on
the mirror plane, linked by octahedrally-coordinated Sr^{2+} ions (Fig. 1). The
three types of water molecule are involved in hydrogen bonding. Fe-C = 1.94, Fe-N =
1.64 Å.

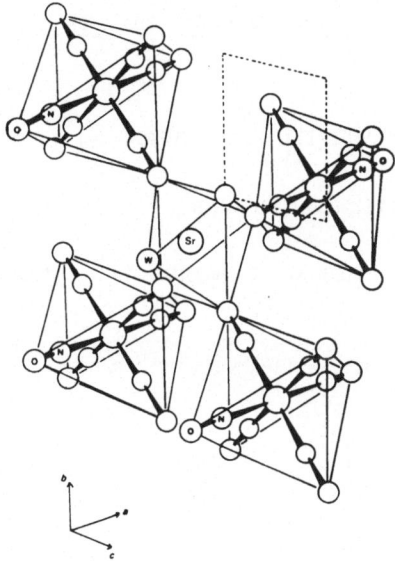

Fig. 1.   Coordination polyhedra in strontium nitroprusside tetrahydrate.

PRUSSIAN BLUE
Fe$_4$[Fe(CN)$_6$]$_3$.15H$_2$O

H.J. BUSER, D. SCHWARZENBACH, W. PETTER and A. LUDI, 1977. Inorg. Chem., $\underline{16}$, 2704-2710.

Cubic, Pm3m, a = 10.166 Å, D$_m$ = 1.75-1.81, Z = 1. Mo radiation, R = 0.032, 0.042,
0.046 for data from three crystals.

Atomic positions (3 crystals)

			Occupancies			x
			1	2	3	
Fe(III)	in	1(a)	1	1	1	
Fe(III)		3(c)	1	1	1	
Fe(II)		1(b)	0.27	0.67	0.82	
Fe(II)		3(d)	0.91	0.78	0.73	
C		6(e)	0.91	0.78	0.73	0.31
C		6(f)	0.27	0.67	0.82	0.31
C		12(h)	0.91	0.78	0.73	0.19
N		6(e)	0.91	0.78	0.73	0.20
N		6(f)	0.27	0.67	0.82	0.20

Atomic positions

			Occupancies			x
			1	2	3	
N	in	12(h)	0.91	0.78	0.73	0.30
O		6(e)	0.09	0.22	0.28	0.21
O		6(f)	0.73	0.34	0.18	0.21
O		12(h)	0.09	0.22	0.28	0.29
O		8(g)	1	1	1	0.26

The structure is similar to that of Fm3m polynuclear transition-metal cyanides (1), but the $Fe(CN)_6$ positions are only partly occupied, with partial ordering leading to the lower-symmetry Pm3m.

1. Structure Reports, 35A, 405; 38A, 229; 39A, 203; 40A, 174.

MANGANESE(II) HEXACYANOCOBALTATE(III) DODECAHYDRATE
$Mn_3[Co(CN)_6]_2 \cdot 12H_2O$

CADMIUM HEXACYANOCOBALTATE(III) DODECAHYDRATE
$Cd_3[Co(CN)_6]_2 \cdot 12H_2O$

G.W. BEALL, W.O. MILLIGAN, J. KORP and I. BERNAL, 1977. Inorg. Chem., 16, 2715-2718.

Cubic, Fm3m, a = 10.436, 10.600 Å, Z = 1.33. Mo radiation, R = 0.072, 0.043 for 160, 150 reflexions; neutron diffraction data, R = 0.088 for 72 reflexions for the Mn compound. Previous study in 1.

Atomic positions (Mn(neutron), Cd)

			x	y	z	occupancy
M	in	4(b)	0	0	1/2	1
Co		4(a)	0	0	0	2/3
C		24(e)	0	0	0.180, 0.177	2/3
N		24(e)	0	0	0.291, 0.286	2/3
O(1)		24(e)	0	0	0.292, 0.262	1/3
O(2)		4(a)	0	0	0	1/3
O(3)		32(f)	0.161, 0.217	x	x	1/6
O(4)		8(c)	1/4	1/4	1/4	1/3

The disordered structure is held together by cyanide links between Co and Mn (Cd). Eight water molecules per cell are coordinated to Mn (Cd), and the other eight are of two types, one type not hydrogen bonded to any other water molecules, and the other type hydrogen bonded to coordinated water molecules.

1. Structure Reports, 35A, 405.

SODIUM TETRACYANOPLATINATE(II) TRIHYDRATE
$Na_2Pt(CN)_4 \cdot 3H_2O$

P.L. JOHNSON, T.R. KOCH and J.M. WILLIAMS, 1977. Acta Cryst., B33, 1976-1979.

Triclinic, P1̄, a = 15.444, b = 9.082, c = 7.350 Å, α = 95.07, β = 92.73, γ = 89.04°, D_m = 2.60, Z = 4. Neutron radiation, R = 0.067 for 4067 reflexions.

The structure (Fig. 1) contains chains of eclipsed $Pt(CN)_4^{2-}$ ions, Pt...Pt = 3.65-3.75 Å. Na ions are coordinated octahedrally by N and O atoms.

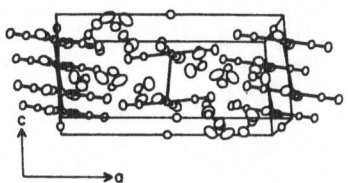

Fig. 1.   Structure of $Na_2Pt(CN)_4 \cdot 3H_2O$.

SODIUM DIBROMOTETRACYANOPLATINATE(IV) DIHYDRATE
$Na_2Pt(CN)_4Br_2 \cdot 2H_2O$

R.L. MAFFLY, P.L. JOHNSON, T.R. KOCH and J.M. WILLIAMS, 1977. Acta Cryst., B33, 558-560.

Orthorhombic, Pnma, a = 11.949, b = 15.124, c = 6.487 Å, Dm = 3.06, Z = 4.
Neutron radiation, R = 0.068 for 1447 reflexions.

     The structure (Fig. 1) contains trans-octahedral complex anions, Pt-Br = 2.480(3), Pt-C = 2.003(2), C-N = 1.156(2) Å. The Na ion has distorted trans-octa-hedral coordination, Na-4N = 2.50-2.70, Na-2 O = 2.41, 2.43 Å, and the water molecule is involved in O-H...Br and O-H...N hydrogen bonds.

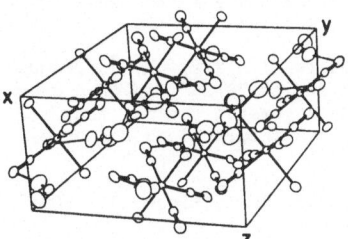

Fig. 1.   Structure of sodium dibromotetracyanoplatinate(IV) dihydrate.

POTASSIUM SODIUM TETRACYANOPLATINATE TRIHYDRATE
$KNaPt(CN)_4 \cdot 3H_2O$

P.L. JOHNSON, R.L. MUSSELMAN and J.M. WILLIAMS, 1977. Acta Cryst., B33, 3155-3159.

Monoclinic, Cc, a = 12.965, b = 13.765, c = 6.522 Å, β = 115.09°, Z = 4. Neutron radiation, R = 0.058 for 1595 reflexions.

     The structure (Fig. 1) contains chains of anions, Pt-Pt = 3.263 Å, Pt...Pt...Pt = 176°; alternate anions are staggered by 35°. Pt-C = 1.982-1.993(3), C-N = 1.153-1.583(3) Å. K and Na ions have octahedral coordination, and the water molecules participate in one O-H...O hydrogen bond, and in several (mainly bifurcated) O-H...N bonds.

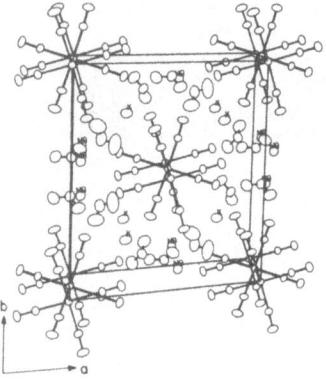

Fig. 1.   Structure of KNaPt(CN)₄.3H₂O

RUBIDIUM TETRACYANOPLATINATE(II) SESQUIHYDRATE
$Rb_2Pt(CN)_4 \cdot 1.5H_2O$

T.R. KOCH, P.L. JOHNSON and J.M. WILLIAMS, 1977. Inorg. Chem., **16**, 640-645.

Monoclinic, C2/c, a = 12.693, b = 12.809, c = 13.642 Å, β = 112.22°, D_m = 3.18, Z = 8. Neutron radiation, R = 0.073 for 2875 reflexions. Oxygen positions suggested in 1 are incorrect.

    The structure contains square-planar $Pt(CN)_4{}^{2-}$ groups stacked along c in a staggered arrangement (torsion angles 29-35°) to give a bent Pt atom chain, Pt...Pt = 3.421(2) Å. Pt(1) lies on an inversion centre, and Pt(2) on a twofold axis. The anion chains are linked by Rb^+...N interactions and by O-H...N hydrogen bonds; one water molecule is disordered.

1. L. DUPONT, 1969. Bull. Soc. R. Sci., Liège, **38**, 509; Structure Reports, **41A**, 418.

RUBIDIUM TETRACYANOPLATINATE BIFLUORIDE
$Rb_2Pt(CN)_4 \cdot (FHF)_{0.40}$

A.J. SCHULTZ, C.C. COFFEY, G.C. LEE and J.M. WILLIAMS, 1977. Inorg. Chem., **16**, 2129-2131.

Tetragonal, I4/mcm, a = 12.689, c = 5.595 Å, Z = 4. Mo radiation, R = 0.045 for 383 reflexions.

Atomic positions

			x	y	z
Pt	in	4(c)	0	0	0
C		16(k)	0.0550	0.1480	0
N		16(k)	0.0871	0.2325	0
Rb		8(h)	0.1555	0.6555	0
F(1)		8(g)	0	1/2	0.155
F(2)		4(b)	0	1/2	1/4

The structure contains columns of square-planar Pt(CN)$_4$ groups along c, Pt...
Pt = c/2 = 2.798 Å. The presumably-linear bifluoride anions are aligned parallel
to c, F...F = 2.27 Å; the sites are only partially occupied. Rb-N = 3.26-3.41,
Rb-$\bar{F}$ = 2.92-3.39 Å.

RUBIDIUM DIBROMOTETRACYANOPLATINATE(IV)
Rb$_2$[Pt(CN)$_4$Br$_2$]

G.F. NEEDHAM, P.L. JOHNSON, T.F. CORNISH and J.M. WILLIAMS, 1977. Acta Cryst., B33,
887-889.

Monoclinic, P2$_1$/c, a = 7.281, b = 9.271, c = 8.960 Å, β = 106.65°, D$_m$ = 3.5, Z = 2.
Neutron radiation, R(F^2) = 0.038 for 1322 reflexions.

The structure (Fig. 1) contains trans-octahedral anions, Pt-Br = 2.485(1),
Pt-C = 2.006(1) Å. Rb has distorted bi-capped trigonal prismatic coordination,
with 6 N in a trigonal prism at 2.998-3.303(2), and 2 Br at 3.657 and 3.678(2) Å
capping two rectangular faces.

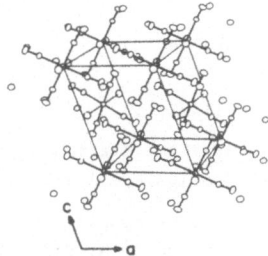

Fig. 1.   Structure of Rb$_2$[Pt(CN)$_4$Br$_2$].

CAESIUM TETRACYANOPLATINATE(II) MONOHYDRATE
Cs$_2$Pt(CN)$_4$·H$_2$O

I.   P.L. JOHNSON, T.R. KOCH and J.M. WILLIAMS, 1977. Acta Cryst., B33, 1293-1295.

II.   H.H. OTTO, H. SCHULZ, K.H. THIEMANN, H. YERSIN and G. GLIEMANN, 1977. Z.
      Naturforsch., 32B, 127-130.

Hexagonal, P6$_1$, a = 9.709, 9.687, c = 19.343, 19.336 Å, D$_m$ = 3.66, 3.67, Z = 6. I,
neutron radiation, R = 0.074 for 1348 reflexions; II, Mo radiation, R = 0.068 for
2210 reflexions.

The structure (Fig. 1) contains square-planar Pt(CN)$_4{}^{2-}$ ions stacked in a
helical fashion along c, with the planes tilted 15° from normal to c; Pt...Pt =
3.545(1) Å, Pt...Pt...$\overline{Pt}$ = 156°. Cs ions have 7- or 8- coordination, and the water
molecule forms two O-H...N hydrogen bonds.

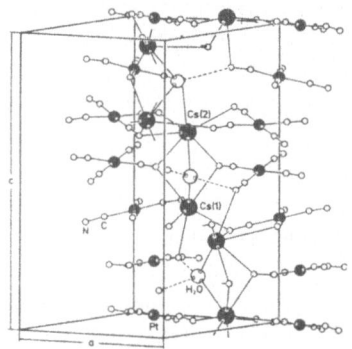

Fig. 1.   Structure of $Cs_2Pt(CN)_4.H_2O$.

BARIUM TETRACYANOPLATINATE(II) TETRAHYDRATE
$BaPt(CN)_4.4H_2O$

R.L. MAFFLY, P.L. JOHNSON and J.M. WILLIAMS, 1977. Acta Cryst., B<u>33</u>, 884-887.

Monoclinic, C2/c, a = 12.278, b = 13.882, c = 6.641 Å, β = 107.75°, D_m = 3.09, Z = 4. Neutron radiation, $R(F^2)$ = 0.052 for 1437 reflexions.

The structure contains square-planar $Pt(CN)_4{}^{2-}$ ions stacked along <u>c</u>, tilted 3° with respect to <u>c</u>, with adjacent ions staggered at 45° (Fig. 1); Pt...Pt = c/2 = 3.32 Å. Ba has $\bar{6}$ 0 and 4 N neighbours at 2.86-3.04 Å, and the water molecules are involved in hydrogen bonding.

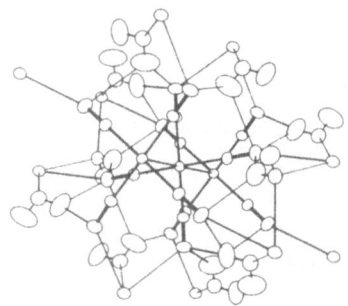

Fig. 1.   Structure of $BaPt(CN)_4.4H_2O$.

BARIUM DIBROMOTETRACYANOPLATINATE(IV) HYDRATE
$BaPt(CN)_4Br_2.4·5H_2O$

T.R. KOCH, P.L. JOHNSON, D.M. WASHECHECK, T.L. CORNISH and J.M. WILLIAMS, 1977. Acta Cryst., B<u>33</u>, 3249-3251.

Tetragonal, P4/mnc, a = 9.425, c = 17.085 Å, D_m = 2.96, Z = 4. Neutron radiation, $R(F^2)$ = 0.056 for 1217 reflexions.

Atomic positions

			x	y	z
Ba	in	4(e)	0	0	0.2988
Pt		4(c)	0	1/2	0
C		16(i)	0.1450	0.4585	0.0826
N		16(i)	0.2301	0.4381	0.1294
Br		8(h)	0.0804	0.7508	0
O(1)		16(i)	0.2082	0.1153	0.2021
0.5 O(2)		4(e)	1/2	1/2	0.0276

The structure (Fig. 1) contains trans–octahedral $Pt(CN)_4Br_2^{2-}$ anions, Ba ions with 9-coordination, and water molecules with one oxygen and all the hydrogens disordered. Pt-Br = 2.482, Pt-C = 2.003, C-N = 1.149(1) Å.

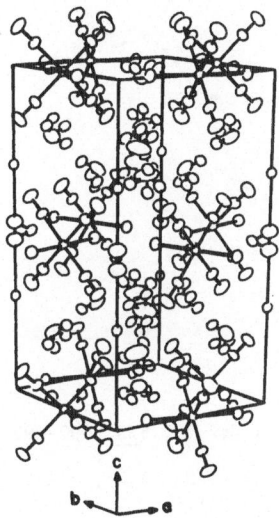

Fig. 1.   Structure of barium dibromotetracyanoplatinate(IV) hydrate.

SODIUM DICYANOCUPRATE(I) DIHYDRATE
$NaCu(CN)_2 \cdot 2H_2O$

C. KAPPENSTEIN and R.P. HUGEL, 1977. Inorg. Chem., **16**, 250-254.

Monoclinic, $P2_1/c$, a = 3.598, b = 19.655, c = 8.515 Å, β = 103.35°, D_m = 1.97, Z = 4. Cu radiation, R = 0.058 for 714 reflexions (films, densitometer intensities, twinned crystal).

The structure contains polymeric $[Cu(CN)_2^-]_\infty$ chains along <u>c</u> (Fig. 1), Cu having trigonal-planar coordination. Na has fac-octahedral coordination to 3 O and 3 N. There are two types of water molecule, one of which is hydrogen bonded to a symmetry related molecule, so that the H atoms (not located) may be disordered.

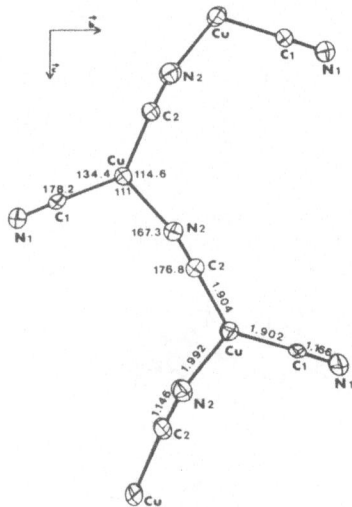

Fig. 1.   The polymeric [Cu(CN)$_2^-$]∞ chain.

SILVER DICYANAMIDE
AgN(CN)$_2$

D. BRITTON and Y.H. CHOW, 1977. Acta Cryst., B<u>33</u>, 697-699.

Trigonal, P3$_1$21, a = 3.601, c = 22.868 Å, D$_m$ > 3.24, Z = 3. Mo radiation, R = 0.058 for 247 reflexions (twinned crystal).

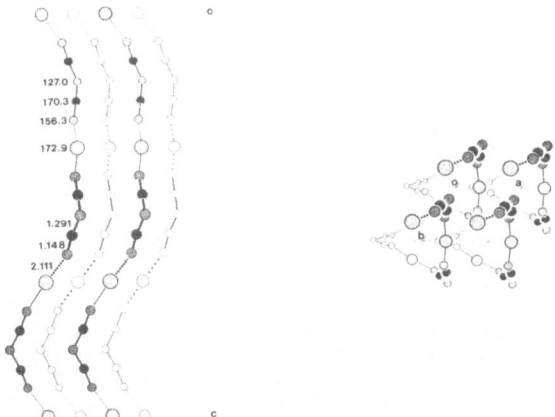

Fig. 1.   Views of the structure of silver dicyanamide perpendicular to the <u>ac</u> plane (left) and along <u>c</u> (right); σ for bond lengths ∿ 0.01 Å.

Atomic positions

			x	y	z
Ag	in	3(a)	-0.3432	-0.3432	0
N(0)		3(b)	0.7559	0.7559	1/2
N(1)		6(c)	0.5227	0.4354	0.4038
C		6(c)	0.6325	0.5591	0.4505

The structure (Fig. 1) contains infinite spiral chains -Ag-N≡C-N-C≡N-Ag- running along c and close-packed perpendicular to c.

POTASSIUM MERCURY(II) CYANATE
$KHg(NCO)_3$

G. THIELE and P. HILFRICH, 1977. Z. Naturforsch., 32B, 1239-1243.

Orthorhombic, Pnma, a = 10.152, b = 3.993, c = 17.729 Å, D_m = 3.15, Z = 4. R = 0.054 for 518 reflexions.

The structure contains K^+ and NCO^- ions, and $Hg(NCO)_2$ molecules (Hg-N = 2.07, 2.10 Å, N-Hg-N = 169°) linked by the free NCO group (Hg...N = 2.47 Å) and by one of the coordinated groups (Hg...N = 3.04 Å), so that Hg has six-coordination.

AMMONIUM THIOCYANATE
NH_4SCN

J.W. BATS and P. COPPENS, 1977. Acta Cryst., B33, 1542-1548.

Monoclinic, $P2_1/c$, a = 4.142, b = 7.063, c = 13.078 Å, β = 97.19°, at 81°K, Z = 4. Mo radiation, R = 0.020 for 1644 reflexions, neutron diffraction data, R = 0.018 for 1070 reflexions, 81°K.

Structure (Fig. 1) as previously described (1), with H atoms now located. S-C = 1.649(1), C-N = 1.176, N-H...S = 3.373 and 3.383, N-H...N = 2.948 and 2.959 Å. The deformation density in the SCN group is similar to that in NaSCN (2).

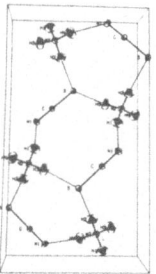

Fig. 1.   Structure of ammonium thiocyanate.

1. Structure Reports, 12, 142; 38A, 238.
2. Following report.

SODIUM THIOCYANATE
NaSCN

J.W. BATS, P. COPPENS and Å. KVICK, 1977. Acta Cryst., B33, 1534-1542.

Orthorhombic, Pmcn, a = 4.038, b = 5.602, c = 13.281 Å, at 81°K, Z = 4. Mo radiation, R = 0.034 and 0.020 for 1822 and 1499 reflexions at 150 and 81°K, respectively; neutron diffraction data, R = 0.021 for 469 reflexions at 94°K.

Atomic positions (X-ray values, 81°K)

	x	y	z
S	1/4	0.1209	0.1861
Na	1/4	-0.0596	0.3926
C	1/4	-0.1188	0.1143
N	1/4	-0.2915	0.0641

The structure is as previously described (1 [where z(S) is misprinted, and should be 1.1137]). S-C = 1.647(1), C-N = 1.178(1) Å. The deformation density around the C and N atoms suggests sp-hybridization, while the S lone-pair electrons are in a diffuse ring around S perpendicular to the C-S bond.

1. Structure Reports, 41A, 206.

ZINC THIOCYANATE
β-Zn(NCS)$_2$

L.A. ASLANOV, V.M. IONOV and K. KYNEV, 1976. Kristallografija, 21, 1198-1199 [Soviet Physics - Crystallography, 21, 693-694].

Triclinic, PĪ, a = 7.794, b = 10.803, c = 7.709 Å, α = 113.25, β = 92.79, γ = 103.40°, Z = 4. Mo radiation, R = 0.092 for 1015 reflexions.

There are two independent zinc ions bridged by thiocyanate groups to form ZnS$_4$ and ZnN$_4$ tetrahedra in layers parallel to (010). Zn-S = 2.35(1), Zn-N = 1.96(2) Å, S-Zn-S = 104-114, N-Zn-N = 107-114, Zn-S-N = 96-99, Zn-N-S = 166-175, S-C-N = 177°.

RUBIDIUM SUBOXIDE
Rb$_9$O$_2$

A. SIMON, 1977. Z. anorg. Chem., 431, 5-16.

Monoclinic, P2$_1$/m, a = 8.351, b = 14.023, c = 11.685 Å, β = 104.51°, D$_m$ = 2.009, Z = 2. Cu radiation, R = 0.081 for 1756 reflexions.

Atomic positions

			x	y	z
Rb(1)	in	4(f)	0.3886	0.1100	0.1361
Rb(2)		4(f)	0.2770	0.1097	0.5430
Rb(3)		4(f)	0.0610	0.1236	0.8159
Rb(4)		2(e)	0.9726	1/4	0.1047
Rb(5)		2(e)	0.8541	1/4	0.5035
Rb(6)		2(e)	0.4350	1/4	0.8458
O(1)		2(e)	0.2184	1/4	-0.0002
O(2)		2(e)	0.1568	1/4	0.6570

The structure contains Rb_9O_2 clusters (Fig. 1) linked by metallic bonding. The material was described previously (1) as 'Rb_3O'.

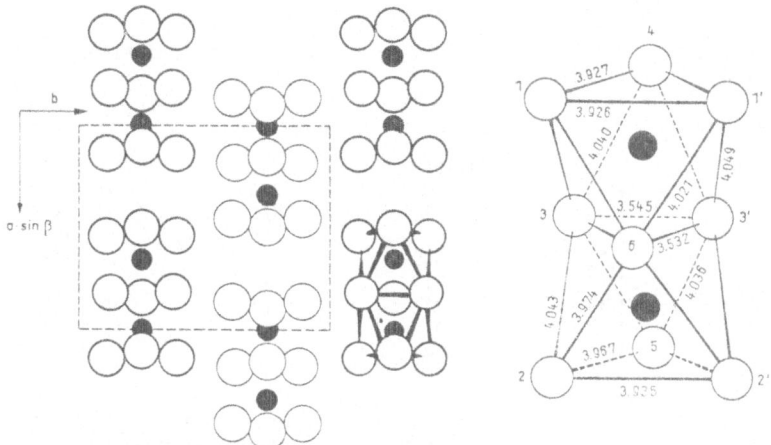

Fig. 1. Rb_9O_2 clusters; Rb-O = 2.64-2.85 Å.

1. P. TOUZAIN, 1973. Bull. Soc. Chim. Fr., 4515.

CAESIUM SUBOXIDE
$Cs_{11}O_3$

A. SIMON and E. WESTERBECK, 1977. Z. anorg. Chem., 428, 187-198.

Monoclinic, $P2_1/c$, a = 17.610, b = 9.218 Å, c =24.047 Å, β = 100.14°, D_m = 2.625, Z = 4. Mo radiation, R = 0.053 for 3031 reflexions.

The structure (Fig. 1) contains $Cs_{11}O_3$ clusters, which are quite similar to those in Cs_7O (1); Cs-Cs = 3.65-4.37, Cs-O = 2.68-2.95 Å. Between clusters Cs...Cs > 5.2 Å.

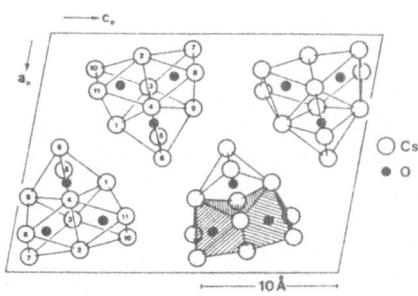

Fig. 1.   Structure of $Cs_{11}O_3$.

1. Structure Reports, 42A, 232.

ALUMINUM OXIDE HYDRATE   (ALKALI-FREE β-ALUMINA)
$11Al_2O_3 \cdot 3H_2O$

K. KATO and H. SAALFELD, 1977.   Acta Cryst., B33, 1596-1598.

Hexagonal, P6₃/mmc, a = 5.600, c = 22.702 Å, Z = 1.   Cu, Mo radiations, R = 0.067 for 1212 reflexions (film data).

Atomic positions

			x	y	z
	Al(1)	in 12(k)	0.1678	2x	-0.1060
	Al(2)	4(f)	1/3	2/3.	0.0244
	Al(3)	4(f)	1/3	2/3	0.1730
	Al(4)	2(a)	0	0	0
	O(1)	12(k)	0.1569	2x	0.0494
	O(2)	12(k)	0.5017	2x	0.1448
	O(3)	4(f)	1/3	2/3	-0.0554
	O(4)	4(e)	0	0	0.1402
1/3	O(5)	6(h)	0.3746	2x	1/4
0.131	O(6)	6(h)	0.6474	2x	1/4

    The spinel blocks characteristic of the β-alumina structure (1) are joined by an intermediate layer containing an oxygen atom, O(5), and a water molecule randomly distributed in 6(h) positions. A hydrogen atom might be bonded to O(5) to form a hydrogen bond to the nearest water molecule.

1. Strukturbericht, 5, 72; Structure Reports, 33A, 275; 37A, 226.

GALLIUM OXIDE
$β-Ga_2O_3$

S. GELLER, 1977.   J. Solid State Chem., 20, 209-210.

It is pointed out that any deviation from monoclinic symmetry (1) must be very small, since it is not detectable by a variety of physical measurements.

1. Structure Reports, 42A, 233.

ANTIMONY OXIDE
$α-Sb_2O_4$

G. THORNTON, 1977.   Acta Cryst., B33, 1271-1273.

Orthorhombic, Pna2₁, a = 5.456, b = 4.814, c = 11.787 Å, Z = 4.   Neutron powder data, with profile analysis.

Atomic positions

	x	y	z
Sb(1)	-0.032	0.036	0.009
Sb(2)	0.366	0.016	0.253
O(1)	0.340	0.177	0.096
O(2)	0.159	0.710	0.195
O(3)	0.086	0.208	0.312
O(4)	0.330	0.840	0.410

[The material seems to have the same structure as that described for cervantite (1), but some of the oxygen parameters are different.]

1. Structure Reports, 41A, 214.

TITANIUM(III) OXIDE
Ti_2O_3

C.E. RICE and W.R. ROBINSON, 1977. Acta Cryst., B33, 1342-1348.

Rhombohedral, R3̄c, a = 5.157, c = 13.61 Å, at 23°C, Z = 2. Mo radiation, R = 0.027-0.034 for 125 reflexions at eight temperatures in the range 23-595°C.

α-Corundum structure, as previously reported (1). Changes in atomic positions with temperature are consistent with a model proposed to account for the semi-conductor-metal transition.

1. Structure Reports, 28, 121; 40A, 176.

VANADIUM DIOXIDE
$V_{0.985}Al_{0.015}O_2$

M. GHEDIRA, H. VINCENT, M. MAREZIO and J.C. LAUNAY, 1977. J. Solid State Chem., 22, 423-428.

At 373°K, tetragonal, rutile structure (1), z(O) = 0.3001. R = 0.029.

At 323°K, M_2 phase (2), monoclinic, [C2/m], a = 9.060, b = 5.800, c = 4.522 Å, β = 91.85°, Z = 8. R̄ = 0.014.

At 298 and 173°K, T phase, triclinic, [C1̄], a = 9.060, 9.065, b = 5.772, 5.749, c = 4.520, 4.521 Å, α = 89.99, 90.00, β = 91.40, 91.06, γ = 89.83, 89.78°, Z = 8. R = 0.034 and 0.041.

MoKβ radiation for all data, 66, 229, 388, and 382 reflexions.

At 373°K the material has the tetragonal rutile structure. At 323°K it has the monoclinic M_2 structure (2), where half of the V atoms are paired (V-V = 2.54 and 3.26 Å) and the other half form equally spaced (2.94 Å) zigzag chains. At 298°K the structure is triclinic, with both vanadium chains containing V-V bonds, 2.55 and 2.82 Å; at 173°K these distances are 2.55 and 2.75 Å.

1. Strukturbericht, 1, 155.
2. Structure Reports, 38A, 242.

NIOBIUM DIOXIDE
NbO_2

R. PYNN, J.D. AXE and R. THOMAS, 1976. Phys. Rev., B, 13, 2965-2975.

Tetragonal, $I4_1/a$, a = 13.66, c = 5.964 Å, Z = 32. Neutron radiation, $R(F^2)$ = 0.058 for 154 reflexions.

Atomic positions

	x	y	z
Nb(1)	0.1155	0.1249	0.4746
Nb(2)	0.1356	0.1250	0.0267
O(1)	0.9866	0.1262	-0.0046
O(2)	0.9749	0.1252	0.5000
O(3)	0.2739	0.1245	0.9998
O(4)	0.2631	0.1241	0.5043

Slightly-distorted rutile structure, as previously described (1). The high-temperature (above 810°C) phase has a rutile structure (2).

1. Structure Reports, 27, 479.
2. Strukturbericht, 1, 158, 211.

TANTALUM OXIDE
TaO_3 (approximately)

V.I. KHITROVA, V.V. KLEČKOVSKAJA and Z.G. PINSKER, 1976. Kristallografija, 21, 937-942 [Soviet Physics - Crystallography, 21, 535-538].

Monoclinic, $P2_1/b$, a = 6.30, b = 12.03, c = 3.75 Å, γ = 106°55'. Electron diffraction data, R = 0.016 for 132 reflexions.

A structure is derived with Ta fully occupying position 2(a) and partially occupying (occupancy ∿ 1/8) two 4(e) positions; oxygen partially occupies 2(c) and four 4(e) positions. This structure contains layers of edge- and corner-sharing distorted TaO_6 octahedra, and would have composition Ta_5O_9 if all the sites were fully occupied.

TUNGSTEN(VI) OXIDE
WO_3

I.   B.O. LOOPSTRA and H.M. REITVELD, 1969. Acta Cryst., B25, 1420-1421.

II. E. SALJE, 1977. Ibid., B33, 574-577.

I. Monoclinic form, neutron powder data give results in agreement with 1.

II. Orthorhombic form, Pmnb, a = 7.341, b = 7.570, c = 7.754 Å, Z = 8. Perovskite-like structure, as for the monoclinic phase (1, I), but the results are not very accurate and further work is proposed.

1. Structure Reports, 24, 325; 31A, 123.

PLATINUM OXIDE
$Pt_{3.4}O_4$

B. GRANDE and H. MÜLLER-BUSCHBAUM, 1977. J. Inorg. Nucl. Chem., 39, 1084-1085.

Cubic, Pm3n, a = 5.612 Å, Z = 2. R = 0.084 for 101 reflexions.

Atomic positions

			x	y	z
0.4	Pt(1)	in 2(a)	0	0	0
	Pt(2)	6(c)	1/4	0	1/2
	O	8(e)	1/4	1/4	1/4

$Na_xPt_3O_4$-type structure (1). Pt(1)-O = 2.43 (x 8), Pt(2)-O = 1.98 (x 4) Å.

1. Structure Reports, 13, 273.

URANIUM OXIDE
α-U_3O_8

I. R.J. ACKERMAN, A.T. CHANG and C.A. SORRELL, 1977. J. Inorg. Nucl. Chem., 39, 75-85.

II. B.O. LOOPSTRA, 1977. Ibid., 39, 1713-1714.

I discusses the relationships between the structures given by 1 and 2. II suggests that the structure of 1 is incorrect.

1. Structure Reports, 22, 295, 296.
2. Ibid., 29, 312.

URANIUM TRIOXIDE
γ-UO_3

B.O. LOOPSTRA, J.C. TAYLOR and A.B. WAUGH, 1977. J. Solid State Chem., 20, 9-19.

373°K
Tetragonal, $I4_1$/amd, a = 6.901, c = 19.975 Å, Z = 16. Neutron powder data, with profile fitting.

293°K
Orthorhombic, Fddd, a = 9.787, b = 19.932, c = 9.705 Å, Z = 32. Neutron powder data, with profile fitting.

77°K
Possibly only pseudo-orthorhombic, Fddd. Further work is proposed.

Atomic positions

373°K ($I4_1$/amd)	x	y	z
U(1) in 8(e)	0	1/4	0.0618
U(2) 8(d)	0	0	1/2
O(1) 16(h)	0	0.9476	0.4073
O(2) 16(h)	0	0.0496	0.2645
O(3) 16(h)	0	0.5100	0.0660

293°K (Fddd)	x	y	z
U(1) in 16(f)	1/8	0.3115	1/8
U(2) 16(d)	1/2	1/2	1/2
O(1) 32(h)	0.0261	0.4081	0.0258
O(2) 32(h)	-0.0284	0.2652	-0.0205
O(3) 32(h)	-0.0001	-0.0663	-0.0096

The structures are essentially as previously described (1, 2), and contain chains of edge-linked $U(2)O_6$ octahedra, cross-linked by $U(1)O_8$ dodecahedra; the atomic shifts are small in the tetragonal-orthorhombic transformation (at 323°K).

1. Structure Reports, 28, 130.
2. Ibid., 37A, 354.

LITHIUM ALUMINATE
β-Li$_5$AlO$_4$

R. HOPPE and H. KÖNIG, 1977. Z. anorg. Chem., 430, 211-217.

Orthorhombic, Pmmn, a = 6.42, b = 6.30, c = 4.62 Å, Z = 2. Mo radiation, R = 0.073 for 177 reflexions.

Atomic positions

			x	y	z
Li(1)	in	8(g)	0.555	0.551	0.254
Li(2)		2(b)	1/4	3/4	0.212
Al		2(a)	1/4	1/4	0.226
O(1)		4(e)	1/4	0.016	0.430
O(2)		4(f)	0.018	1/4	0.015

The structure (Fig. 1) differs from that given previously by Stewner and Hoppe (1), and is an ordered variant of the Li$_2$O-type. Al and Li have tetrahedral coordination, Al-O = 1.76, 1.78(2), Li-O = 1.96, 2.05(2) Å.

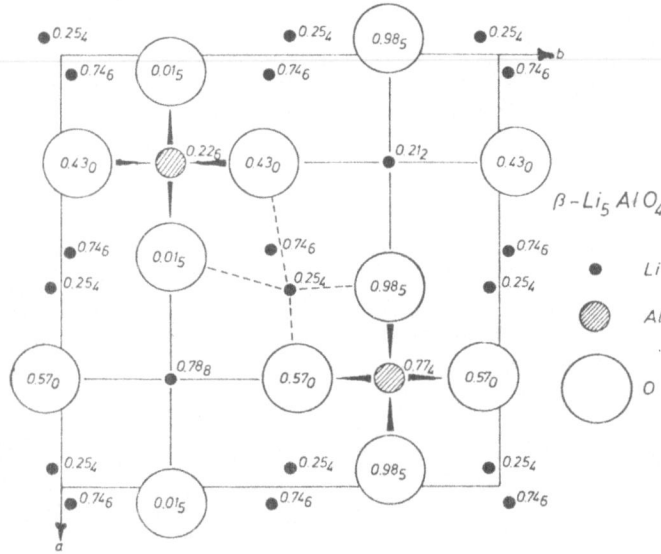

Fig. 1.   Structure of β-Li$_5$AlO$_4$.

1. Structure Reports, 37A, 227.

POTASSIUM β-ALUMINA   THALLIUM β-ALUMINA
$K_2Al_{22}O_{34}$ $Tl_2Al_{22}O_{34}$

G. COLLIN, J.P. BOILOT, A. KAHN, J. THERY and R. COMES, 1977. J. Solid State
Chem., 21, 283-292.

Hexagonal, $P6_3/mmc$, a = 5.602, 5.596, c = 22.734, 22.912 Å, Z = 1. Mo radiation,
R = 0.034 and 0.058 for 494 and 499 reflexions.

     [The structures are as previously described (1)], with distribution of K^+
and Tl^+ ions similar to those in Na β-alumina (2).

1. Structure Reports, 42A, 239.
2. Ibid., 37A, 226.

CALCIUM IRON ALUMINUM OXIDE
$CaFe_6Al_6O_{19}$

M. HARDER and H. MÜLLER-BUSCHBAUM, 1977. Z. Naturforsch., 32B, 833-834.

Hexagonal, $P6_3/mmc$, a = 5.628, c = 22.084 Å, Z = 2. R = 0.09 for 231 reflexions.

Atomic positions

		x	y	z
2 Ca	2(d)	1/3	2/3	3/4
1 Al + 1 Fe in 2(a)	0	0	0
1 Al + 1 Fe	2(b)	0	0	1/4
1 Al + 3 Fe	4(f)	1/3	2/3	0.0291
1 Al + 3 Fe	4(f)	1/3	2/3	0.1904
8 Al + 4 Fe	12(k)	0.1690	0.3380	0.8911
O(1)	4(e)	0*	0*	0.149
O(2)	4(f)	1/3	2/3	0.944
O(3)	6(h)	0.181	0.362	1/4
O(4)	12(k)	0.150	0.300	0.052
O(5)	12(k)	0.504	0.008	0.148

* [Not 1/3, 2/3, z]

     Magnetoplumbite structure (1).

1. Strukturbericht, 6, 74.

HOLMIUM ALUMINATE
$HoAlO_3$

J. HAMMANN and M. OCIO, 1977. Acta Cryst., A33, 975-978.

Orthorhombic, Pbnm, a = 5.182, b = 5.324, c = 7.37 Å, Z = 4. Neutron powder data.

Atomic positions

		x	y	z
Ho in 4(c)	-0.007	0.059	1/4
Al	4(b)	1/2	0	0
O(1)	4(c)	0.094	0.493	1/4
O(2)	8(d)	-0.297	0.290	0.038

     Deformed perovskite.

STRONTIUM GADOLINIUM GALLATE
Sr_2GdGaO_5

NGUYEN-TRUT-DINH, J. FAVA and G. LE FLEM, 1977. Z. anorg. Chem., **433**, 275-283.

Tetragonal, I4/mcm, a = 6.781, c = 11.18 Å, Z = 4. R = 0.05 for 28 reflexions (powder data).

Atomic positions

			x	y	z
Sr	in	4(a)	0	0	1/4
(Sr,Gd)		8(h)	0.181	0.681	0
Ga		4(b)	0	1/2	1/4
O(1)		4(c)	0	0	0
O(2)		16(ℓ)	0.152	0.652	0.652

Cs_3CoCl_5-type structure (1).

1. Strukturbericht, **3**, 134, 498; Structure Reports, **29**, 277.

MAGNESIUM GERMANATE (SPINEL PHASE)
Mg_2GeO_4

R.B. von DREELE, A. NAVROTSKY and A.L. BOWMAN, 1977. Acta Cryst., B**33**, 2287-2288.

Cubic, Fd3m, a = 8.250 Å. Neutron powder data, R = 0.021 for 15 reflexions. Material synthesized from the olivine phase by heat and pressure. Normal spinel, x(O) = 0.3758.

BARIUM ALUMINUM GERMANATE
$BaAl_2Ge_2O_8$

M. CALLERI and G. GAZZONI, 1977. Acta Cryst., B**33**, 3275-3282.

Monoclinic, I2/c, a = 8.799, b = 13.371, c = 14.727 Å, β = 114.93°, Z = 8. Mo radiation, R = 0.036 for 1726 reflexions.

Celsian structure (1, 2) with highly ordered arrangement of cations. Mean bond distances are Ba-O = 2.853, Al-O = 1.744, Ge-O = 1.735 Å.   ·

1. Structure Reports, **24**, 491.
2. Ibid., **42A**, 245.

BARIUM IRON(II) DIGERMANATE
$Ba_2FeGe_2O_7$

Ju.A. MALINOVSKIJ, E.A. POBEDIMSKAJA and N.V. BELOV, 1976. Kristallografija, **21**, 1195-1197 [Soviet Physics - Crystallography, **21**, 691-692].

Tetragonal, $P\bar{4}2_1m$, a = 8.475, c = 5.496 Å, Z = 2. Mo radiation, R = 0.059 for 980 reflexions.

Atomic positions

			x	y	z
Ba	in	4(e)	0.3336	1/2-x	-0.0077
Fe		2(b)	0	0	1/2
Ge		4(e)	0.1381	1/2-x	0.5352
O(1)		2(c)	0	1/2	0.662
O(2)		4(e)	0.136	1/2-x	0.221
O(3)		8(f)	0.079	0.188	0.686

Melilite-type structure ($\underline{1}$), containing layers of BaO_8 polyhedra alternating along $\underline{c}$ with $FeGe_2O_7$ layers which consist of FeO_4 tetrahedra and tetrahedral Ge_2O_7 groups. Ba-O = 2.69-3.00, Fe-O = 2.01, Ge-O = 1.73-1.79 Å, Ge-O-Ge = 135°.

1. Strukturbericht, $\underline{2}$, 541; Structure Reports, $\underline{17}$, 574.

CALCIUM COPPER GERMANATE
$CaCu_3Ge_4O_{12}$

Y. OZAKI, M. GHEDIRA, J. CHENAVAS, J.C. JOUBERT and M. MAREZIO, 1977. Acta Cryst., B$\underline{33}$, 3615-3617.

Cubic, Im3, a = 7.202 Å, Z = 2. Ag radiation, R = 0.011 for 94 reflexions. Ca in 2(a); Cu in 6(b); Ge in 8(c); O in 24(g): (0.3012, 0.1859, 0).

Cubic perovskite-like structure similar to that of $CaCu_3Mn_4O_{12}$ ($\underline{1}$, $\underline{2}$). Interatomic distances are Ca-O = 2.549 (icosahedral), Cu-O = 1.960, 2.677, 3.134 (three sets of four oxygen neighbours), Ge-O = 1.895 Å (octahedral).

1. Structure Reports, $\underline{39}$A, 247; $\underline{42}$A, 284.
2. Ibid., $\underline{41}$A, 260.

SODIUM TITANIUM ZINC GERMANATE
$Na_2TiZn_2(GeO_4)O_3$

P.A. SANDOMIPSKIJ, T.L. EVSTIGNEEVA, M.A. SIMONOV and N.V. BELOV, 1976. Ž. Strukt. Khim., $\underline{17}$, 1080-1083 [J. Struct. Chem., $\underline{17}$, 918-920].

Orthorhombic, Pbca, a = 20.453, b = 11.844, c = 5.221 Å, D_m = 4.8, Z = 8. Mo radiation, R = 0.044 for 1100 reflexions.

Isostructural with $Na_2TiZn_2(SiO_4)O_3$ ($\underline{1}$). Ge and Zn have tetrahedral coordination, Ti octahedral, and Na ions have 4 near oxygen neighbours, with 2 or 3 others at slightly greater distances.

1. Structure Reports, $\underline{42}$A, 432.

CADMIUM HEPTAGERMANATE
$Cd_2Ge_7O_{16}$

E. PLATTNER and H. VÖLLENKLE, 1977. Mh. Chem., $\underline{108}$, 443-449.

Tetragonal, P$\bar{4}$b2, a = 11.31, c = 4.63 Å, D_m = 5.60, Z = 2. Cu radiation, R = 0.063 for 230 reflexions (films, photometer intensities).

Atomic positions

			x	y	z
Cd	in	4(g)	0.1624	0.6624	0
Ge(1)		2(d)	0	1/2	1/2
Ge(2)		4(g)	0.3645	0.8645	0
Ge(3)		8(i)	0.0707	0.1856	0.5100
O(1)		8(i)	0.003	0.383	0.744
O(2)		8(i)	0.034	0.277	0.288
O(3)		8(i)	0.174	0.229	0.769
O(4)		8(i)	0.138	0.062	0.343

The structure (Fig. 1) contains GeO_4 tetrahedra and GeO_6 octahedra linked into a three-dimensional framework, with Cd ions (6-coordination) in cavities. Mean distances are Ge-O = 1.74 (tetrahedra), 1.89 (octahedra), Cd-O = 2.36 Å.

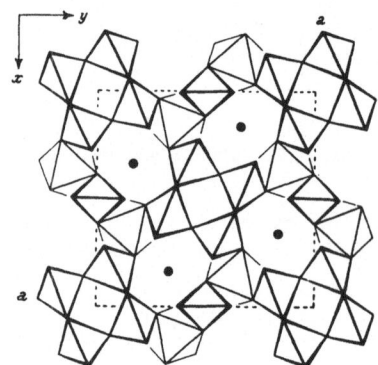

Fig. 1.   Structure of $Cd_2Ge_7O_{16}$.

SODIUM LANTHANUM HYDROXYGERMANATE
$Na_2LaGeO_4(OH)$

D.Ju. PUŠČAROVSKIJ, E.A. POBEDIMSKAJA, O.V. KUDRJAVCEVA and B. GETTAŠ, 1976.
Kristallografija, 21, 1126-1128 [Soviet Physics - Crystallography, 21, 651-652].

Orthorhombic, Pnma, a = 9.574, b = 7.550, c = 7.004 Å, Z = 4. Mo radiation, R = 0.048 for 1809 reflexions.

Atomic positions

	x	y	z
La	0.0434	1/4	0.1135
Ge	0.0920	3/4	0.3227
Na	0.3298	0.5237	0.1453
O(1)	0.0611	0.5680	0.1717
O(2)	0.2677	3/4	0.3844
O(3)	0.4789	3/4	-0.0194
OH	0.2838	1/4	0.3000

The structure contains GeO_4 tetrahedra, LaO_8 polyhedra, and irregular Na coordination polyhedra (5 oxygen neighbours, with 2 others at greater distances) Ge-O = 1.74-1.76, La-O = 2.42-2.65, Na-O = 2.37-2.60, 2.90, 2.91 Å.

LITHIUM STANNATE
Li_8SnO_6

R. HOPPE and R.M. BRAUN, 1977. Z. anorg. Chem., **433**, 181-188; **437**, 304.

Rhombohedral, R$\bar{3}$, a = 5.99 Å, α = 54.3° (hexagonal cell has a = 5.46, c = 15.28 Å), Z = 1. Mo radiation, R = 0.034 for 197 reflexions. Previous study in **1**.

Atomic positions (hexagonal axes)

			x	y	y
Li(1)	in	18(f)	0.378	0.357	0.1214
Li(2)		6(c)	0	0	0.3426
Sn		3(a)	0	0	0
O		18(f)	0.303	0.988	0.0809

Sn and Li(2) have octahedral and Li(1) tetrahedral coordination. Sn-O = 2.09, Li(2)-O = 2.14, 2.38, Li(1)-O = 1.88-2.00 Å.

1. Structure Reports, **33**A, 288; **34**A, 263.

EUROPIUM ARSENIC OXIDE
Eu_4As_2O

Y. WANG, L.D. CALVERT, E.J. GABE and J.B. TAYLOR, 1977. Acta Cryst., B**33**, 3122-3125.

Tetragonal, I4/mmm, a = 4.7924, c = 16.1933 Å, Z = 2. Mo radiation, R = 0.050 for 529 reflexions. Eu(1) in 4(c); Eu(2) in 4(e): z = 0.3261; As in 4(e): z = 0.1356; O in 2(b).

The structure is a filled version of the La_2Sb type, and is closely related to the K_2NiF_4 and GeTeU types. Atom coordinations are shown in Fig. 1.

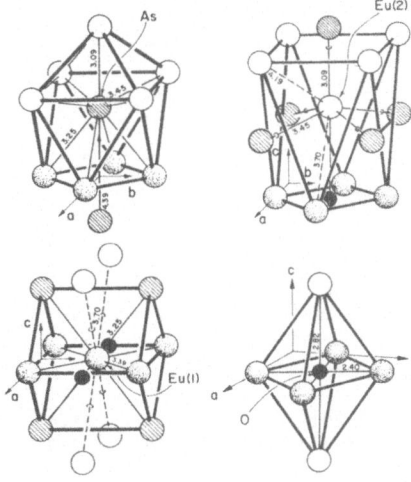

Fig. 1.   Atom coordinations in europium arsenic oxide.

SODIUM TETRAANTIMONATE(III)
Na$_2$Sb$_4$O$_7$

P. MARAINE and G. PÉREZ, 1977. Acta Cryst., B33, 1158-1163.

Monoclinic, C2/c, a = 11.03, b = 16.92, c = 9.64 Å, β = 149.4°, D$_m$ = 4.58, Z = 4.
Mo radiation, R = 0.067 for 2162 reflexions.

Atomic positions

	x	y	z
Sb(1)	0.1947	0.6810	0.246
Sb(2)	0.2874	0.5091	0.179
O(1)	0.964	0.7636	0.926
O(2)	0.113	0.6179	0.005
O(3)	0.417	0.5644	0.471
O(4)	1/2	0.5599	1/4
Na	0.2435	0.8411	0.080

The structure (Fig. 1) contains (Sb$_4$O$_7$$^{2-}$) chains parallel to [Ī01], linked
to form tunnels into which Sb(III) lone pairs project; each Sb has 3 O at 1.95-
2.15 Å, and 1 O at 2.20, 2.36 Å. Na has 5 O at 2.34-2.52 Å.

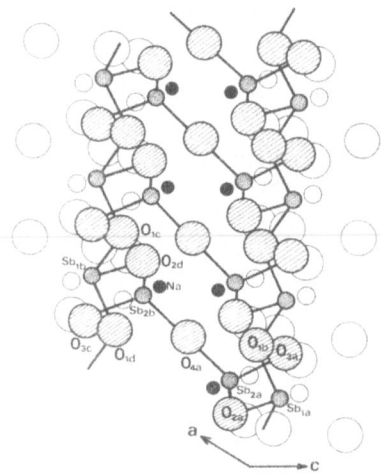

Fig. 1.   Structure of Na$_2$Sb$_4$O$_7$.

SODIUM PENTAANTIMONATE(III)
NaSb$_5$O$_8$

J.P. SAUVAGE, P. MARAINE and G. PEREZ, 1976. Rev. Chim. Minér., 13, 556-563.

Triclinic, PĪ, a = 9.638, b = 7.291, c = 7.172 Å, α = 85.44, β = 105.60, γ =
106.11°, D$_m$ = 5.36, Z = 2. Mo radiation, R = 0.08 for 3293 reflexions.

The structure (Fig. 1) contains (Sb$_5$O$_8$$^-$)$_n$ layers parallel to (010). Sb(1),
Sb(4), and Sb(5) have tetrahedral coordination including the lone electron-pair
(Sb-O = 1.94-2.05 Å), and Sb(2) and Sb(3) have trigonal bipyramidal coordination
including an equatorial lone pair (Sb-O = 1.95-2.27 Å). The layers are linked by
the longer Sb-O bonds and by the 7-coordinate Na ion.

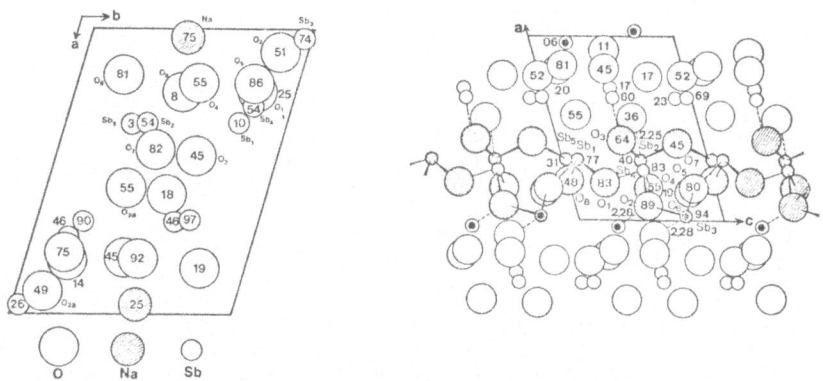

Fig. 1.   Atomic positions and $(Sb_5O_8^-)_n$ layers in $NaSb_5O_8$.

POTASSIUM TRIANTIMONATE(III)
KSb_3O_5

J.P. SAUVAGE, P. MARAINE and G. PEREZ, 1976. Rev. Chim. Minér., 13, 549-555.

Monoclinic, $P2_1/c$, a = 7.505, b = 13.946, c = 7.860 Å, β = 126.43°, D_m = 4.84, Z = 4. Mo radiation, R = 0.070 for 3472 reflexions.

     The structure (Fig. 1) contains $(Sb_3O_5^-)_n$ layers; Sb(2) and Sb(3) have tetrahedral coordination including the lone electron-pair (Sb-O = 1.95-2.01 Å), and Sb(1) has trigonal bipyramidal coordination including an equatorial lone pair (Sb-O = 1.98-2.24 Å, O(ax)-Sb-O(ax) = 157°). K is coordinated to seven oxygen atoms all in the same layer, and the layers are linked by the longest Sb-O bond (2.24 Å).

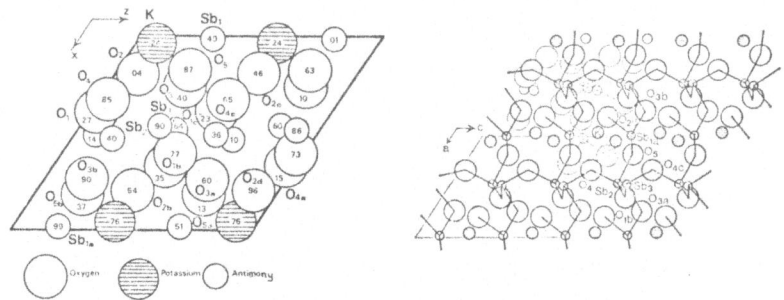

Fig. 1.   Atomic positions and $(Sb_3O_5^-)_n$ layers in KSb_3O_5.

BARIUM NICKEL ANTIMONATE
$Ba_3NiSb_2O_9$

P. KÖHL and D. REINEN, 1977. Z. anorg. Chem., 433, 81-93.

Hexagonal, P6₃mc, a = 5.837, c = 14.392 Å, D_m = 6.64, Z = 2. Mo radiation, R = 0.030 for 1309 reflexions.

Atomic positions

			x	y	z
Ba(1)	in	2(a)	0	0	0.2465
Ba(2)		2(b)	1/3	2/3	0.0963
Ba(3)		2(b)	1/3	2/3	0.3998
Ni		2(b)	1/3	2/3	0.6552
Sb(1)		2(a)	0	0	0.4982
Sb(2)		2(b)	1/3	2/3	0.8413
O(1)		6(c)	0.160	-0.160	0.5794
O(2)		6(c)	0.486	-0.486	0.7474
O(3)		6(c)	0.165	-0.165	0.9211

The material is a hexagonal elpasolite. The structure (Fig. 1) can be described in terms of close-packed BaO_3 layers in the sequence cchcch. It contains groups of two face-sharing octahedra (Ni and Sb(2)) linked by corner sharing with an Sb(1)O₆ octahedron. Ni-O = 2.04, 2.06, Sb-O = 2.00-2.05, Ba-O = 2.84-3.05 Å (12-coordination). $Ba_3CuSb_2O_9$ is isostructural, and its detailed structure is to be described in another publication.

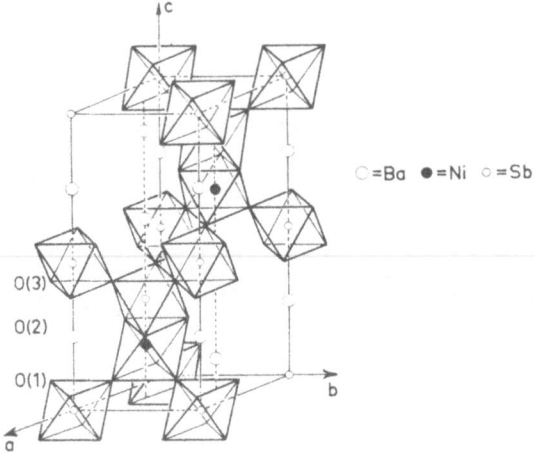

C=Ba ●=Ni ○=Sb

Fig. 1.   Structure of $Ba_3NiSb_2O_9$.

BARIUM BISMUTHATE
$BaBiO_3$

R. ARPE and H. MÜLLER-BUSCHBAUM, 1977. Z. anorg. Chem., 434, 73-77.

Tetragonal, P4̄2m, a = 4.364, c = 4.518 Å, D_m = 7.60, Z = 1. R = 0.074 for 108 reflexions. 0·5 Ba in 2(g): z = 0.0477; Bi in 1(b); 0.5 O(1) in 4(m): z = 0.655; 0·5 O(2) in 2(h): z = 0.961.

The derived structure contains a statistical distribution of Ba^{2+} and O^{2-} ions, and there are two possible 4+2 coordination polyhedra for Bi (Bi^{3+} or Bi^{5+}); Bi-O = 2.29 (x 4), 2.08 or 2.44 (x 2) Å. Ba has eight oxygen neighbours at 2.81 and 3.08 Å.

ALUMINUM BISMUTH OXIDE
$Al_4Bi_2O_9$

R. ARPE and H. MÜLLER-BUSCHBAUM, 1977. J. Inorg. Nucl. Chem., 39, 233-235.

Orthorhombic, Pbam, a = 7.722, b = 8.112, c = 5.703 Å, D_m = 6.12, Z = 2. R = 0.047 for 674 reflexions.

Atomic positions

			x	y	z
Al(1)	in	4(e)	0	0	0.2606
Al(2)		4(h)	0.3526	0.1596	1/2
Bi		4(g)	0.1711	0.3321	0
O(1)		4(g)	0.143	0.071	0
O(2)		4(h)	0.139	0.087	1/2
O(3)		8(i)	0.372	0.294	0.251
O(4)		2(d)	1/2	0	1/2

Al(1) has octahedral coordination (Al-O = 1.87-1.94 Å) and Al(2) has tetrahedral coordination (Al-O = 1.72-1.79 Å), two tetrahedra sharing a corner to form an Al_2O_7 group (Fig. 1). Bi has irregular coordination to 6 oxygen atoms at 2.13-2.91 Å.

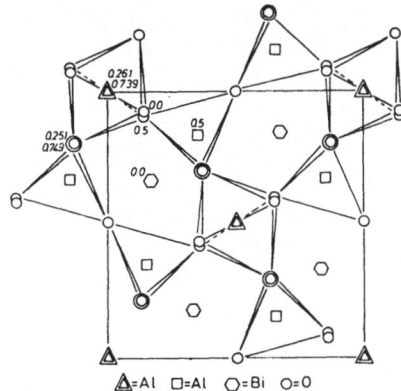

△=Al □=Al ○=Bi ⊙=O

Fig. 1.   Structure of $Al_4Bi_2O_9$.

COPPER BISMUTH OXIDE
$CuBi_2O_4$

J.-C. BOIVIN, J. TREHOUX and D. THOMAS, 1976. Bull. Soc. Fr. Minér. Crist., 99, 193-196.

Tetragonal, P4/ncc, a = 8.510, c = 5.814 Å, D_m = 8.56, Z = 4. Cu radiation, R = 0.054 for 68 reflexions.

Atomic positions (origin at $\bar{4}$)

			x	y	z
Cu	in	4(c)	0	1/2	0.080
Bi		8(f)	0.1685	0.1685	1/4
O		16(g)	0.295	0.114	-0.097

The structure contains $(BiO_2)_n$ chains linked by Cu^{2+} ions with square-planar coordination, Cu-O = 2.04 Å. The Bi^{3+} lone-pair exhibits stereochemical activity; Bi-O = 2.12-2.73 Å. [Compare the different results in 1.]

<u>1</u>. Structure Reports, <u>42</u>A, 251.

ZIRCONIUM YTTRIUM OXIDE
$Zr_3Y_4O_{12}$

H.G. SCOTT, 1977. Acta Cryst., B<u>33</u>, 281-282.

Rhombohedral, $R\overline{3}$, a = 9.735, c = 9.109 Å, Z = 3. Cu radiation, $R_W(I)$ = 0.032
for 88 reflexions (powder data).

Atomic positions

	x	y	z
M(1)	0	0	0
M(2)	0.4109	0.2905	-0.0174
O(1)	0.1544	0.4522	0.2734
O(2)	0.4599	0.2965	0.2281

The material has the δ-phase structure (<u>1</u>), which is derived from the defect-fluorite type by ordering anion vacancies. M(1) has octahedral coordination, M(1)-O = 2.09 Å (hence M(1) is probably Zr) and M(2) has seven oxygen neighbours at seven of the eight corners of a distorted cube, M(2)-O = 2.18-2.53 Å (M(2) is probably (Zr + 2 Y)/3).

<u>1</u>. Structure Reports, <u>33</u>A, 291.

BARIUM ANTIMONY TITANIUM OXIDE   STRONTIUM NIOBIUM TITANIUM OXIDE
$Ba_3Sb_4Ti_4O_{21}$ $Sr_3Nb_4Ti_4O_{21}$

STRONTIUM TANTALUM TITANIUM OXIDE
$Sr_3Ta_4Ti_4O_{21}$

C. SAUREL, D. GROULT and B. RAVEAU, 1977. Mater. Res. Bull., <u>12</u>, 629-635.

Hexagonal, $P6_3/mcm$, a = 8.966, 9.0, 9.0, c = 11.857, 11.5, 11.5 Å, Z = 2. Powder
data.

Atomic positions (for Ba compound, similar values for Sr compounds)

			x	y	z
Ba	in	6(g)	0.604	0	1/4
M(1)		12(k)	0.232	0	0.100
M(2)		4(d)	1/3	2/3	0
O(1)		6(g)	0.238	0	1/4
O(2)		12(k)	0.825	0	0.086
O(3)		24(ℓ)	0.181	0.492	0.110

M(1) = (Sb,Ti), (Ti,Nb), (Ti,Ta)
M(2) = Sb, (Ti,Nb), (Ti,Ta)

Isostructural with $A_3(Nb_{8-x}M_x)O_{21}$ compounds (<u>1</u>).

<u>1</u>. Structure Reports, <u>42</u>A, 261.

TITANIUM VANADIUM OXIDE   TITANIUM ALUMINUM OXIDE
$(Ti,V)_2O_3$ $(Ti,Al)_2O_3$

TITANIUM SCANDIUM OXIDE
$(Ti,Sc)_2O_3$

I.   C.E. RICE and W.R. ROBINSON, 1977.   J. Solid State Chem., 21, 145-154.
II. Idem, 1977.   Ibid., 21, 155-160.

Rhombohedral, R3c, a ∿ 5, c ∿ 14 Å, for various solid solutions, Z = 6.   Mo radiation, R = 0.02-0.03.

    α-Al_2O_3 structure (1).   V, Al, and Sc substitution in Ti_2O_3 causes an increase
in the metal-metal distance along c (2.578 Å in pure Ti_2O_3).

1. Strukturbericht, 1, 240.

ARMALCOLITE
$(Fe,Mg)Ti_2O_5$

B.A. WECHSLER, 1977.   Amer. Min., 62, 913-920.

Orthorhombic, Bbmm, a = 9.77-9.90, b = 10.02-10.18, c = 3.74-3.76 Å, at 24-1100°C,
Z = 4.   Mo radiation, R = 0.03-0.08 for 300-429 reflexions.

    Pseudobrookite structure as previously described (1).   M-O distances suggest
disordered cation distribution at room temperature, order at 400°C, with disorder
reappearing at 1100°C.

1. Structure Reports, 41A, 232.

COPPER TITANATE
Cu_3TiO_4

K.-J. RANGE and F. KETTERL, 1977.   Z. Naturforsch., 32B, 1356-1357.

α-Form: rhombohedral, R3m, a = 3.040, c = 17.193 Å, Z = 1.5.   Mo radiation, R =
0.084.   3 Cu in 3(a); 1·5 Cu + 1·5 Ti in 3(b); 6 O in 6(c): z = 0.1075.

β-Form: hexagonal, $P6_3/mmc$, a = 3.040, c = 11.459 Å, Z = 1.   Mo radiation, R =
0.074.   2 Cu in 2(c); 1 Cu + 1 Ti in 2(a); 4 O in 4(f): z = 0.0898.

    The structures are very similar, with oxygen layers AABBCC in α, and AABB in
β. Cu^+ has linear coordination, Cu-O = 1.84 Å, and (Cu^{2+},Ti^{4+}) has distorted
octahedral coordination, M-O = 2.03 Å.   Independent study in 1.

1. Structure Reports, 40A, 188.

VANADATE PYROXENES
KVO_3 NH_4VO_3 $RbVO_3$ $CsVO_3$

F.C. HAWTHORNE and C. CALVO, 1977. J. Solid State Chem., 22, 157-170.

Orthorhombic, Pbcm, a = 5.176, 4.909, 5.261, 5.393, b = 10.794, 11.786, 11.425,
12.249, c = 5.680, 5.830, 5.715, 5.786 Å, Z = 4. Mo radiation, R = 0.056, 0.045,
0.049, 0.030 for 475, 569, 460, 487 reflexions.

The structures are as previously described (1).

1. Structure Reports, 13, 306; 18, 442, 443; 24, 442.

CAESIUM VANADATES
$Cs_2V_5O_{13}$ CsV_2O_5

I.   K. WALTERSSON and B. FORSLUND, 1977. Acta Cryst., B33, 784-789.
II. Idem, 1977. Ibid., B33, 789-793.

$Cs_2V_5O_{13}$
Tetragonal, I4mm, a = 7.762, c = 11.74 Å, D_m = 3.39, Z = 2. Mo radiation, R =
0.031 for 267 reflexions.

CsV_2O_5
Monoclinic, $P2_1/c$, a = 7.021, b = 9.898, c = 7.783 Å, β = 90.65°, D_m = 3.87, Z =
4. Mo radiation, R = 0.032 for 1247 reflexions.

Atomic positions

$Cs_2V_5O_{13}$		x	y	z
Cs	in 4(b)	0	1/2	1/4
V(1)	8(c)	0.2239	0.2239	0.0029
V(2)	2(a)	0	0	0.6493
O(1)	8(c)	0.3164	0.3164	0.1164
O(2)	8(c)	0.3059	0.3059	0.8902
O(3)	8(d)	0.2715	0	-0.0104
O(4)	8(a)	0	0	0.7844
CsV_2O_5				
Cs	in 4(e)	0.6168	0.1255	0.2208
V(1)	4(e)	0.8825	0.4774	0.1628
V(2)	4(e)	0.1212	0.2950	0.4868
O(1)	4(e)	0.1138	0.1279	0.1838
O(2)	4(e)	0.3324	0.3206	0.4130
O(3)	4(e)	0.6706	0.4258	0.1170
O(4)	4(e)	0.9499	0.3649	0.3586
O(5)	4(e)	0.0678	0.3851	0.0215

The $Cs_2V_5O_{13}$ structure (Fig. 1) contains corner-sharing VO_4 tetrahedra and
VO_5 square pyramids which form $(V_5O_{13}{}^{2-})_n$ layers, held together by Cs^+ ions;
alternatively, it can be considered as $V_4O_{12}{}^{4-}$ tetramers, VO^{2+}, and Cs^+ ions.
Cs has 12 O neighbours at 3.24-3.54 Å.

The CsV_2O_5 structure is essentially as previously described (1), but with
shifts in oxygen parameters which lead to a slightly different description of the
V(IV) coordination. The structure contains $V(V)O_4$ tetrahedra and $V(IV)O_5$ square
pyramids (Fig. 2). Cs has 9 O at 3.10-3.54, with two additional O at 3.68 and
3.81 Å.

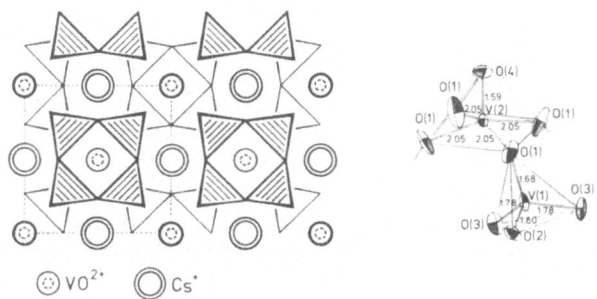

Fig. 1.   Structure and vanadium coordination in $Cs_2V_5O_{13}$.

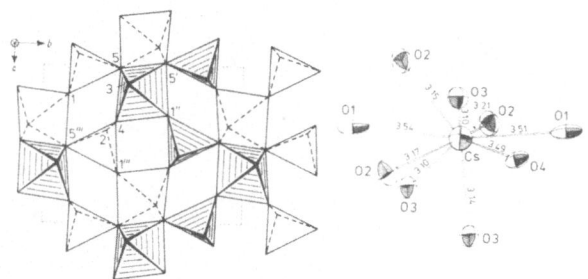

Fig. 2.   Vanadium coordination in CsV_2O_5.

CAESIUM VANADIUM BRONZES
$Cs_{0.35}V_3O_7$ $Cs_{0.3}V_2O_5$

III.   K. WALTERSSON and B. FORSLUND, 1977. Acta Cryst., B33, 775-779.
IV.   Idem, 1977. Ibid., B33, 780-784.

$Cs_{0.35}V_3O_7$
Hexagonal, a = 29.64, c = 18.03 Å; sub-cell, $P6_3/m$, a' = a/3 = 9.880, c' = c/5 =
3.605 Å, Z = 2. Mo radiation, R = 0.047 for 100 reflexions.

$Cs_{0.3}V_2O_5$
Hexagonal, sub-cell, $P6_3/m$, a' = 14.360, c'= 3.611 Å, D_m = 3.45, Z = 6 (the true
cell has c = 7c'). Mo radiation, R = 0.040 for 487 reflexions.

Atomic positions (sub-cells)

$Cs_{0.35}V_3O_7$			x	y	z
0.37 Cs(1)	in	2(b)	0	0	0
0.32 Cs(2)		2(a)	0	0	1/4
6 V		6(h)	0.1219	-0.3814	1/4
6 O(1)		6(h)	0.0963	-0.4516	3/4
6 O(2)		6(h)	0.2162	0.3511	1/4
2 O(3)		2(c)	1/3	2/3	1/4

$Cs_{0.3}V_2O_5$			x	y	z
1.24 Cs(1)	in	2(c)	1/3	2/3	1/4
0.65 Cs(2)		4(e)	0	0	0.1318
6 V(1)		6(h)	0.3525	0.2626	1/4
6 V(2)		6(h)	0.4259	0.0870	1/4
6 O(1)		6(h)	0.0639	0.3725	1/4
6 O(2)		6(h)	0.3173	0.3800	1/4
6 O(3)		6(h)	0.4538	0.2240	1/4
6 O(4)		6(h)	0.2450	0.1515	1/4
6 O(5)		6(h)	0.5422	0.0986	1/4

Both structures (Fig. 3) contain a three-dimensional linking of zigzag strings of edge-sharing VO_5 square pyramids, forming tunnels along c which accommodate the Cs ions. Partial ordering of Cs results in the larger cells.

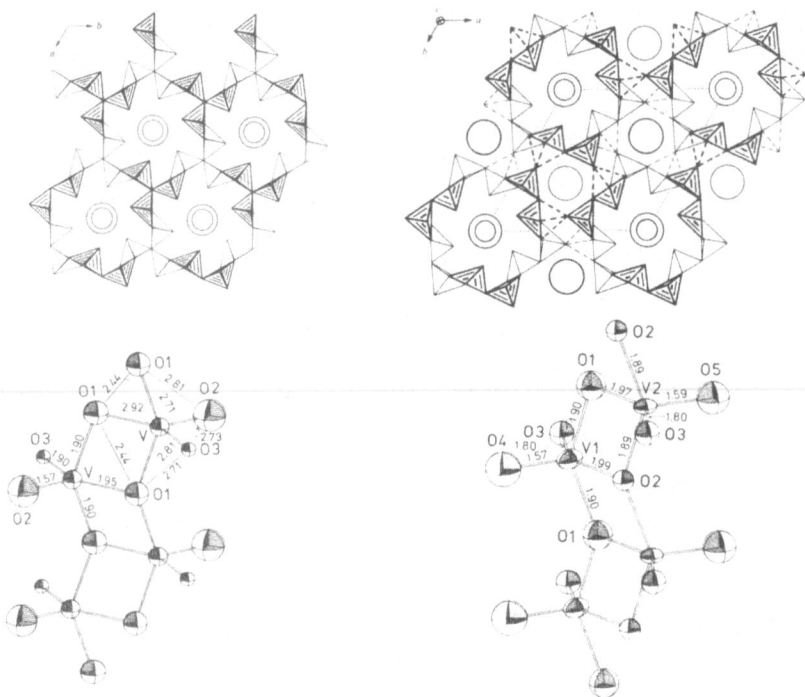

Fig. 3.   Structures and VO_5 square pyramids of $Cs_{0.35}V_3O_7$ (left), and $Cs_{0.3}V_2O_5$ (right).

1.   Structure Reports, 37A, 235.

POTASSIUM CADMIUM VANADATE
$KCd_4(VO_4)_3$

E. HOLT, S. DRAI, R. OLAZCUAGA and M. VLASSE, 1977. Acta Cryst., B33, 95-98.

Monoclinic, Cc, a = 12.89, b = 13.37, c = 7.092 Å, β = 114.84°, D_m = 4.95, Z = 4.
Mo radiation, R = 0.061 for 1130 reflexions.

The structure (Fig. 1) is of distorted scheelite type. It contains a three-dimensional $[Cd_2(VO_4)_3^{5-}]_n$ framework which consists of two edge-sharing octahedra (Cd(1) and Cd(4)) joined by corners to VO_4 tetrahedra. Voids in the framework accommodate Cd(3) (octahedral coordination), and Cd(2) and K (trigonal prismatic coordinations). V-O = 1.58-1.82, Cd-O = 2.15-2.85, K-O = 2.59-2.95 Å.

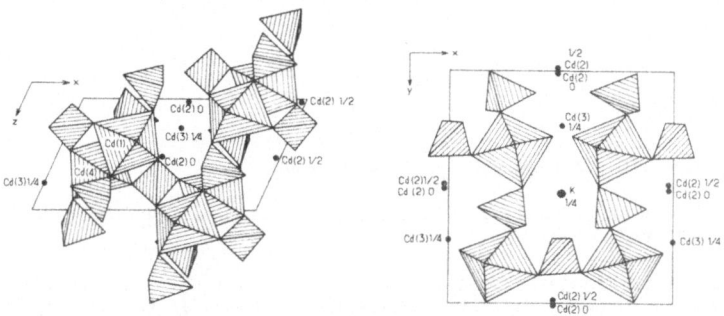

Fig. 1.   Structure of potassium cadmium vanadate.

RUBIDIUM NIOBATE
$Rb_4Nb_{11}O_{30}$

G.D. FALLON and B.M. GATEHOUSE, 1977.   J. Solid State Chem., 22, 405-409.

Rhombohedral, R3̄m, a = 7.527, c = 43.17 Å, D_m = 4.40, Z = 3. Mo radiation, R = 0.072 for 675 reflexions.

Atomic positions

			x	y	z
9 Nb(1)	in	9(e)	1/2	0	0
6 Nb(2)		6(c)	0	0	0.4001
18 Nb(3)		18(h)	0.1657	-0.1657	0.1379
6 Rb(1)		6(c)	0	0	0.2313
4 Rb(2)		18(h)	0.0488	-0.0488	-0.0473
2 Rb(3)		6(c)	0	0	0.0452
18 O(1)		18(h)	0.2080	-0.2080	-0.0084
18 O(2)		18(h)	0.4567	-0.4567	0.0440
18 O(3)		18(h)	0.2103	-0.2103	0.0975
18 O(4)		18(h)	0.4753	-0.4753	0.1473
18 O(5)		18(h)	-0.1165	0.1165	0.1359

The structure (Fig. 1) contains layers of corner-sharing groups of six edge-shared octahedra, separated by pyrochlore-like layers of octahedra; Nb-O = 1.84-2.11 Å. Rb ions are in tunnels and have 9-coordination, Rb-O = 2.98-3.79 Å.

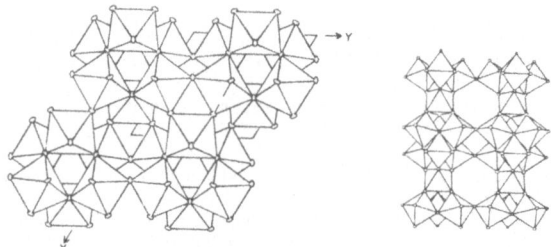

Fig. 1.   Structure of $Rb_4Nb_{11}O_{30}$.

THALLIUM NIOBATES
$TlNb_3O_9$

I. M. GASPERIN, 1977. Acta Cryst., B$\underline{33}$, 2306-2308.

Orthorhombic, C222$_1$, a = 7.551, b = 13.005, c = 7.734 Å, Z = 4. Mo radiation,
R = 0.069 for 535 reflexions.

The material is pseudo-hexagonal, and the structure (Fig. 1) is similar to
that of a hexagonal bronze, but with an excess of cations at the usually empty
trigonal sites and probably a deficiency of anions. The formula suggested is
$TlNb_{3.175}(O_{8.25}F_{0.375})$.

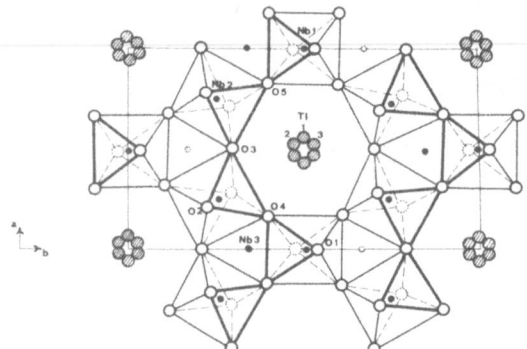

Fig. 1.   Structure of $TlNb_3O_9$.

$Tl_8Nb_{22}O_{59}$

II. M. GASPERIN, 1977. Acta Cryst., B$\underline{33}$, 398-402.

Rhombohedral, R$\bar{3}$m, a = 7.51, c = 43.29 Å, Z = 1.5. Mo radiation, R = 0.07 for
700 reflexions.

The structure (Fig. 2) contains Nb coordination octahedra linked in 15 levels
along $\underline{c}$, two edge-sharing levels alternating with three corner-sharing levels. Tl
ions are in large cavities with 6 O neighbours; six of the twelve Tl are statistically distributed over three sites. Nb-O = 1.83-2.13, Tl-O = 2.89, 3.01 Å.

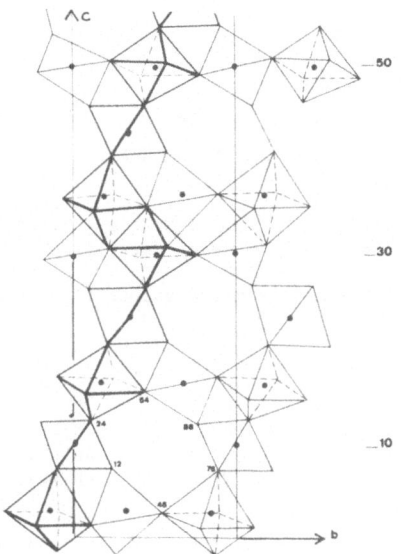

Fig. 2.   Part of the structure of thallium niobate.

POTASSIUM GERMANIUM NIOBIUM OXIDE
$K_{10}Nb_{22}Ge_4O_{68}$

J. CHOISNET, M. HERVIEU, D. GROULT and B. RAVEAU, 1977. Mater. Res. Bull., 12, 621-627.

Hexagonal, P$\bar{6}$2m, a = 9.112, c = 20.01 Å, D_m = 4.43, Z = 1. Powder data.

A structural model is derived by comparison with $K_6Nb_6X_4O_{26}$ compounds (1). It contains an intergrowth of Nb_8O_{21} and $Nb_6Ge_4O_{26}$ layers; Nb has octahedral and Ge tetrahedral coordination.

1. Structure Reports, 42A, 396.

LEAD METANIOBATE
$PbNb_2O_6$

Ph. LABBÉ, M. FREY, B. RAVEAU and J.C. MONIER, 1977. Acta Cryst., B33, 2201-2212.

Orthorhombic, Bb2$_1$m, a = 35.293, b = 17.943, c = 7.747 Å, Z = 40 (subcell, Cm2m, a' = a/2, b' = b, c' = c/2). Mo radiation, R = 0.062 for 763 reflexions (0.057 for 353 reflexions for average subcell structure).

The subcell structure is as previously described (1). The true structure involves ordering of Pb in the pentagonal tunnels and statistical occupation of all the tetragonal tunnels. With respect to the oxygen framework all the metal atoms are displaced along the ferroelectric b axis; average Nb shift is 0.18 Å and maximum Pb shift 0.61 Å.

1. Structure Reports, 39A, 220.

TITANOBIOBATES AND TITANOTANTALATES
$Rb_2Nb_6TiO_{18}$ $Rb_2Ta_6TiO_{18}$ $Cs_2Nb_6TiO_{18}$ $Cs_2Ta_6TiO_{18}$

G. DESGARDIN, C. ROBERT, D. GROULT and B. RAVEAU, 1977. J. Solid State Chem., 22, 101-111.

Trigonal, $P\bar{3}m1$, a = 7.529, 7.512, 7.533, 7.513, c = 8.194, 8.231, 8.189, 8.227 Å, D_m = 4.38, 6.62, 4.78, 6.93, Z = 1. Powder data, R(I) = 0.06-0.08 for 64 lines.

Atomic positions (for $Rb_2Nb_6TiO_{18}$; similar values for the others)

			x	y	z
Rb	in	2(d)	1/3	2/3	0.665
Nb,Ti		1(b)	0	0	1/2
Nb,Ti		6(i)	0.1696	-0.1696	0.1468
O(1)		6(i)	0.447	-0.447	0.155
O(2)		6(i)	0.854	-0.854	0.128
O(3)		6(i)	0.124	-0.124	0.358

The structure contains a basic M_6O_{24} unit of 3 x 2 octahedra sharing edges and corners, M-O = 1.8-2.2 Å, and is related to the hexagonal tungsten bronze and pyrochlore structures.

BISMUTH CHROMIUM NIOBIUM OXIDE
$Bi_{1.34}CrNbO_6$

Y. TORII and K. HASEGAWA, 1977. Bull. Chem. Soc. Japan, 50, 2638-2642.

Cubic, Fd3m, a = 10.455 Å, Z = 8. Cu radiation, powder data, R = 0.056 for 23 lines. 0.67 Bi in 16(d); Cr/Nb in 16(c); O in 48(f): x = 0.330.

Pyrochlore structure (1) assumed, with vacancies in the 16(d) and 8(b) sites.

1. Strukturbericht, 2, 58.

CALCIUM THORIUM NIOBATE
$CaThNb_2O_8$

G. FONTENEAU, H. L'HELGOUALCH and J. LUCAS, 1977. Mater. Res. Bull., 12, 25-34.

Monoclinic, I2/c, a = 5.446, b = 11.207, c = 5.137 Å, β = 94.70°, D_m = 6.18, Z = 2. Cu radiation, $R(F^2)$ = 0.066 for 73 reflexions (powder data).

Atomic positions

			x	y	z
Nb	in	4(e)	0	0.1036	1/4
Ca,Th		4(e)	0	0.6285	1/4
O(1)		8(f)	0.240	0.034	0.048
O(2)		8(f)	0.159	0.210	0.503

The structure is similar to that of (Y,Yb)NbO$_4$ (1) (which does not deviate significantly from I2/c). The SrTh, CdTh, and CaU compounds are isostructural, and all four materials transform to tetragonal (fergusonite) phases at high temperatures (535-850°C).

1. Structure Reports, 23, 389.

LITHIUM TANTALATE
M-LiTa$_3$O$_8$

A. SANTORO, R.S. ROTH and D. MINOR, 1977. Acta Cryst., B33, 3945-3947.

Monoclinic, C2/c, a = 9.410, b = 11.521, c = 5.051 Å, β = 91.11°, Z = 4. Neutron powder data with profile analysis.

Structure as determined by X-ray methods (1), with confirmation of Li position at (0,0.8453,1/4).

1. Structure Reports, 42A, 267.

SODIUM TANTALATE
NaTaO$_3$

M. AHTEE and L. UNONIUS, 1977. Acta Cryst., A33, 150-154.

Orthorhombic, Pcmn, a = 5.513, b = 7.751, c = 5.494 Å, Z = 4. Cu radiation, powder data for 15 lines.

Atomic positions

	x	y	z
Ta	0	0	1/2
Na	0	1/4	-0.02
O(1)	0.054	1/4	1/2
O(2)	0.275	-0.027	0.275

The structure is essentially as in 1, but in the higher-symmetry space group.

1. Structure Reports, 21, 317.

TANTALATES
K$_3$(Ta$_7$Ti)O$_{21}$ K$_3$(Ta$_{7.5}$Fe$_{0.5}$)O$_{21}$ Ba$_3$(Ta$_4$Ti$_4$)O$_{21}$ Ba$_3$(Ta$_6$Fe$_2$)O$_{21}$

D. GROULT, J.M. CHAILLEUX, B. RAVEAU and A. DESCHANVRES, 1977. Rev. Chim. Minér., 14, 1-10.

Hexagonal, P6$_3$/mcm, a ∿ 9, c ∿ 12 Å, Z = 2. Powder data. Isostructural with K$_3$Nb$_8$O$_{21}$ (1), with Ta preferentially occupying the 4(d) sites.

1. Structure Reports, 42A, 261.

CALCIUM CHROMATE MONOHYDRATE
$CaCrO_4 \cdot H_2O$

O. BARS, J.Y. LE MAROUILLE and D. GRANDJEAN, 1977. Acta Cryst., B33, 3751-3755.

Orthorhombic, Pbca, a = 8.014, b = 8.147, c = 12.761 Å, D_m = 2.793, Z = 8. R = 0.048 for 2326 reflexions.

The structure (Fig. 1) contains CrO_4 tetrahedra, $CaO_6(H_2O)_2$ polyhedra, and hydrogen bonds. Cr-O = 1.623-1.660(2), Ca-O = 2.364-2.684(2), O-H...O = 2.781, 2.866 Å.

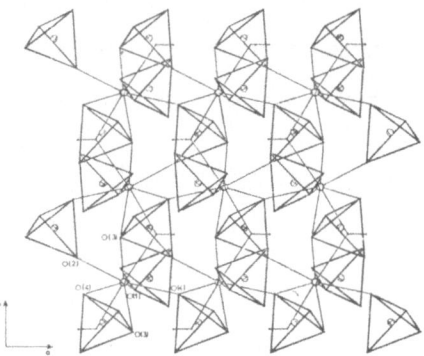

Fig. 1.   Structure of $CaCrO_4 \cdot H_2O$.

ALUMINUM CHROMATE DICHROMATE TETRAHYDRATE
$\alpha\text{-}Al_2(CrO_4)_2Cr_2O_7 \cdot 4H_2O$

Y. CUDENNEC, 1977. J. Inorg. Nucl. Chem., 39, 1711-1713.

Monoclinic, P2/n, a = 5.329, b = 10.48, c = 13.30 Å, β = 90.00°, Z = 2. Mo radiation, R = 0.082 for 483 reflexions (films, densitometer intensities).

Isostructural with the iron compound (1).

1. This volume, p. 199.

BARIUM PHOSPHOCHROMATE MONOHYDRATE   BARIUM PHOSPHOCHROMATE TRIHYDRATE
$BaHCr_2PO_{10} \cdot H_2O$ $BaHCr_2PO_{10} \cdot 3H_2O$

M.T. AVERBUCH-POUCHOT, A. DURIF and J.C. GUITEL, 1977. Acta Cryst., B33, 1431-1435.

Monohydrate
Triclinic, P$\bar{1}$, a = 9.333, b = 7.779, c = 7.526 Å, α = 106.28, β = 105.37, γ = 94.14°, Z = 2. Ag radiation, R = 0.041 for 1643 reflexions.

Trihydrate
Triclinic, P$\bar{1}$, a = 10.189, b = 8.207, c = 7.749 Å, α = 108.80, β = 107.14, γ = 89.04°, Z = 2. Ag radiation, R = 0.059 for 2340 reflexions.

The structures (Figs. 1 and 2) contain $Cr_2PO_{10}^{3-}$ ions, Cr-O = 1.59-1.63 (terminal), 1.83 (bridging), P-O = 1.48-1.58 (terminal), 1.54, 1.55(1) Å (bridging), angles 106-115°. Ba has 9-coordination in the monohydrate and 10-coordination in the trihydrate. In the trihydrate, one water molecule, O(W3), is not coordinated to Ba.

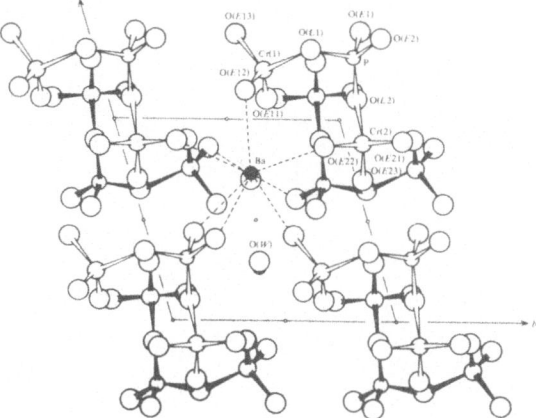

Fig. 1.   Structure of barium phosphochromate monohydrate projected on the bc plane.

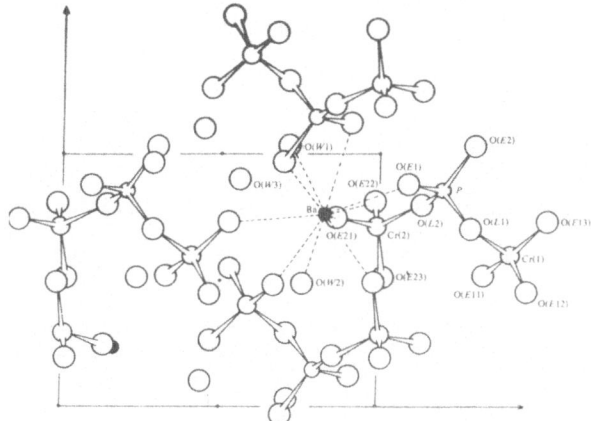

Fig. 2.   Structure of barium phosphochromate trihydrate projected on the ab plane.

IRON(III) CHROMATE DICHROMATE TETRAHYDRATE
α-$Fe_2(CrO_4)_2Cr_2O_7 \cdot 4H_2O$

Y. GÉRAULT and A. BONNIN, 1976. Bull. Soc. Fr. Minér. Crist., 99, 197-202.

Monoclinic, P2/n, a = 5.440, b = 10.684, c = 13.482 Å, β = 90.3°, D_m = 2.69, Z = 2. Mo radiation, R = 0.089 for 1080 reflexions. A β-form is orthorhombic.

The structure (Fig. 1) contains CrO_4^{2-} and $Cr_2O_7^{2-}$ ions which link $FeO_4(H_2O)_2$ octahedra into layers parallel to (010); the layers are linked by hydrogen bonds. $Fe-OH_2 = 2.04$, $Fe-O = 1.91-1.97$, $Cr-O = 1.57-1.73(1)$ Å.

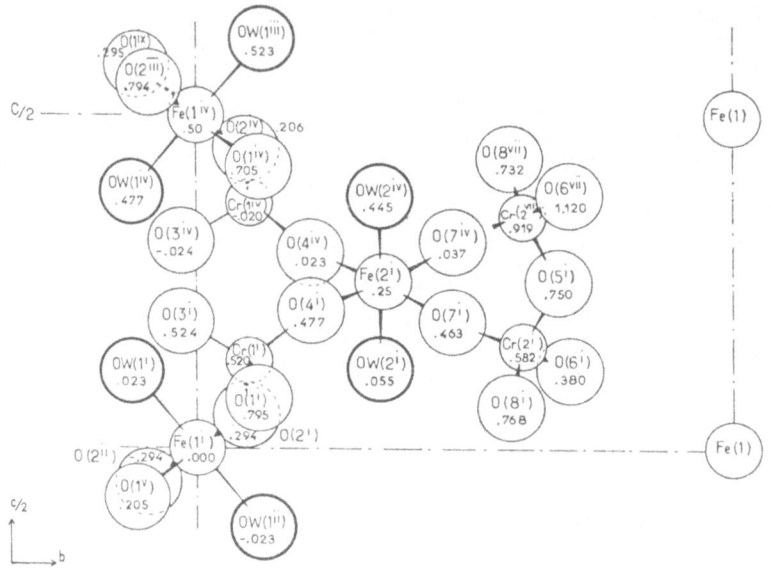

Fig. 1.   Structure of $\alpha-Fe_2(CrO_4)Cr_2O_7.4H_2O$.

AMMONIUM IRON(III) CHROMATE
$\alpha-NH_4Fe(CrO_4)_2$

P. GRAVEREAU, A. HARDY and A. BONNIN, 1977. Acta Cryst., B33, 1362-1367.

Monoclinic, $P2_1$, a = 8.192, b = 15.136, c = 5.520 Å, β = 90.59°, D_m = 2.98, Z = 4. Mo radiation, R = 0.048 for 864 reflexions.

The structure (Fig. 1) contains layers of FeO_6 octahedra and CrO_4 octahedra, linked by 12-coordinate NH_4^+ ions. Fe-O = 1.94-2.06, Cr-O = 1.54-1.73, N-O = 2.79-3.53(6) Å.

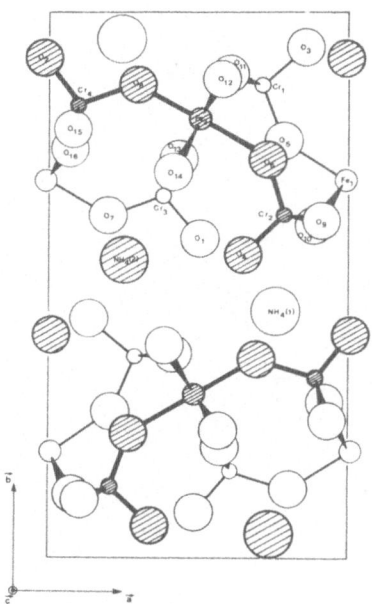

Fig. 1.   Structure of α-NH$_4$Fe(CrO$_4$)$_2$.

POTASSIUM IRON(III) CHROMATE
KFe(CrO$_4$)$_2$

P. GRAVEREAU and A. HARDY, 1976. Bull. Soc. Fr. Minér. Crist., <u>99</u>, 206-210.

Monoclinic, C2/m, a = 8.64, b = 5.50, c = 7.64 Å, β = 94.7°, Z = 2. Mo radiation,
R = 0.073 for 152 reflexions.

Atomic positions

			x	y	z
Fe	in	2(c)	1/2	1/2	1/2
Cr		4(i)	0.6238	0	0.2923
K		2(a)	0	0	0
O(1)		4(i)	0.7725	0	0.4468
O(2)		4(i)	0.6832	0	0.1062
O(3)		8(j)	0.5194	0.7444	0.3169

      Isostructural with KCr$_3$O$_8$ (<u>1</u>) and the mineral yavapaiite, KFe(SO$_4$)$_2$ (<u>2</u>).
FeO$_6$ octahedra and CrO$_4$ tetrahedra are linked into layers held together by K ions
(Fig. 1).   Fe-O = 1.97, Cr-O = 1.56-1.70, K-O = 2.75-2.93(3) Å.

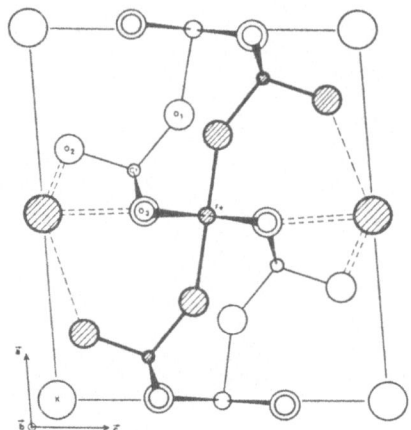

Fig. 1.   Structure of KFe(CrO₄)₂.

1. Structure Reports, 22, 330.
2. Ibid., 37A, 308; 38A, 336.

LANTHANUM CHROMITE   LANTHANUM STRONTIUM CHROMITE
LaCrO₃   (La,Sr)CrO₃

I.   C.P. KHATTAK and D.E. COX, 1977. Mater. Res. Bull., 12, 463-471.
II.   Ibid., 1977. J. Appl. Cryst., 10, 405-411.

LaCrO₃
Orthorhombic, Pbnm (1), a = 5.520, b = 5.483, c = 7.765 Å, Z = 4. Neutron powder
data.

(La,Sr)CrO₃
Rhombohedral, R3̄c, a = 5.50, c = 13.31 Å, Z = 2. X-ray and neutron powder data.

Atomic positions

	LaCrO₃		x	y	z
La	in	4(c)	-0.0046	0.0196	1/4
Cr		4(b)	1/2	0	0
O(1)		4(c)	0.0676	0.4935	1/4
O(2)		8(d)	0.2265	0.2265	0.5338
	La₀.₇₅Sr₀.₂₅CrO₃				
La,Sr		2(a)	1/4	1/4	1/4
Cr		2(b)	0	0	0
O		6(e)	0.2031	0.2969	-1/4

1. Structure Reports, 21, 308.

CERIUM(IV) CHROMATE DIHYDRATE
$Ce(CrO_4)_2 \cdot 2H_2O$

O. LINDGREN, 1977. Acta Chem. Scand., A31, 167-170.

Monoclinic, $P2_1/m$, a = 6.587, b = 10.672, c = 5.670 Å, β = 92.59°, D_m = 3.37, Z = 2. Cu radiation, R = 0.084 for 584 reflexions (films, visual intensities).

Atomic positions

	x	y	z
Ce	0.0439	1/4	0.2206
Cr	0.2570	0.0433	0.7454
O(1)	0.8401	0.9014	0.0092
O(2)	0.7749	0.3866	0.2649
O(3)	0.5081	0.9255	0.2545
O(4)	0.8570	0.8891	0.4833
O(5)	0.4124	1/4	0.2672
O(6)	0.1761	3/4	0.1665

The structure (Fig. 1) contains CrO_4^{2-} ions and Ce(IV) ions with eightfold bicapped trigonal prismatic coordination to six chromate oxygens and two water molecules, Cr-O = 1.58 - 1.69(2), Ce-O = 2.23-2.58 Å, and probable hydrogen bonds, O-H...O = 2.79 and 2.91 Å.

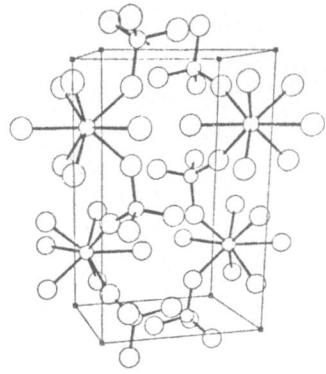

Fig. 1.   Structure of $Ce(CrO_4)_2 \cdot 2H_2O$.

TERBIUM CHROMATE(V)
$TbCrO_4$

G. BUISSON, F. TCHÉOU, F. SAYETAT and K. SCHEUNEMANN, 1976. Solid State Comm., 18, 871-875.

Tetragonal, $I4_1/amd$, cell parameters not given, Z = 4. Neutron powder data. Tb in 4(a); Cr in 4(b); O in 16(h): (0, 0.181, 0.327). Zircon structure (1), with transition to an orthorhombic phase below 48°K.

1. Strukturbericht, 1, 345.

POTASSIUM MOLYBDENUM BRONZE   (RED)
$K_{0.33}MoO_3$

F.A. SCHRÖDER and W. SCHUCKMANN, 1977.   Z. Naturforsch., 32B, 365-368.

Monoclinic, C2/m, a = 14.299, b = 7.737, c = 6.394 Å, β = 92.62°, Z = 12.   Mo
radiation, R = 0.023 for 859 reflexions.

    The structure (Fig. 1) is as previously described (1); it contains distorted
MoO_6 octahedra, Mo-O = 1.68-2.39 Å, and K^+ ions with 8 oxygen neighbours at
2.65-3.02 Å.

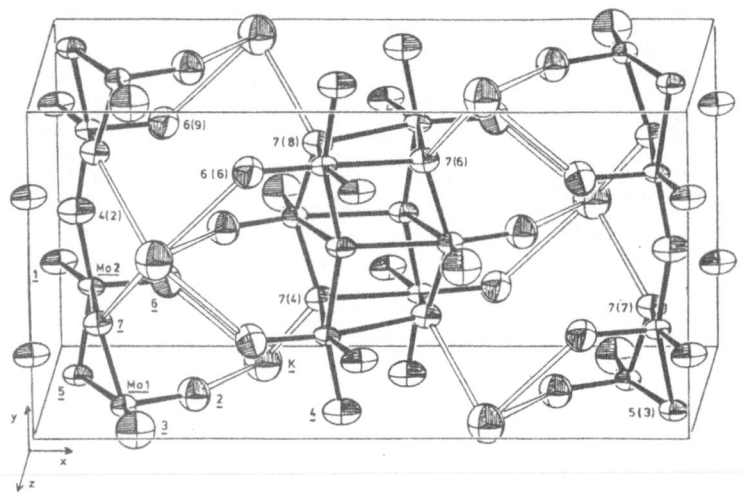

Fig. 1.   Structure of $K_{0.33}MoO_3$.

1. Structure Reports, 30A, 341.

MAGNESIUM MOLYBDATE PENTAHYDRATE
$MgMoO_4 \cdot 5H_2O$

O. BARS, J.Y. LE MAROUILLE and D. GRANDJEAN, 1977.   Acta Cryst., B33, 1155-1157.

Triclinic, P$\bar{1}$, a = 6.529, b = 10.706, c = 6.341 Å, α = 76.44, β = 109.03, γ =
90.31°, D_m = 2.29, Z = 2.   R = 0.065 for 3041 reflexions.

    $Mg(H_2O)_4O_2$ octahedra share corners with MoO_4 tetrahedra to form chains along
[110] (Fig. 1); the chains are linked by hydrogen bonds involving the fifth water
molecule. Mo-O = 1.747-1.768(4), Mg-O = 2.024-2.097(5) Å. The material is not
isostructural with $MgSO_4 \cdot 5H_2O$ (1), in spite of the similarity between the unit
cells.

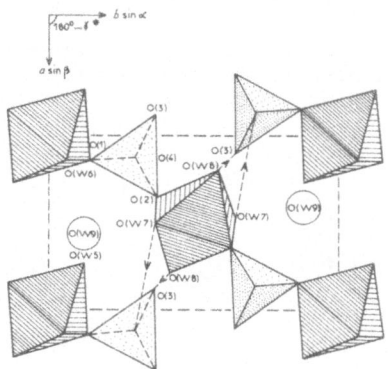

Fig. 1.   Structure of MgMoO₄.5H₂O.

<u>1</u>. Structure Reports, <u>38</u>A, 331.

MAGNESIUM DIMOLYBDATE
MgMo₂O₇

K. STADNICKA, J. HABER and R. KOZLOWSKI, 1977. Acta Cryst., B<u>33</u>, 3859-3862.

Monoclinic, P2₁/c, a = 8.111, b = 5.700, c = 15.002 Å, β = 115.26°, Dₘ = 3.22, Z = 4. Mo radiation, R = 0.095 for 2558 reflexions.

    The $Mo_2O_7^{2-}$ anion consists of two corner-sharing MoO_4 tetrahedra, which are twisted 12° from the ideal eclipsed configuration; Mo-O = 1.707-1.781 (terminal), 1.857, 1.881(4) Å (bridging), O-Mo-O = 106.0-115.5, Mo-O-Mo = 160.7°. Anion sheets parallel to (100) are linked by distorted MgO_6 octahedra which share one edge to form Mg_2O_{10} units (Fig. 1); Mg-O = 2.031-2.119(4) Å.

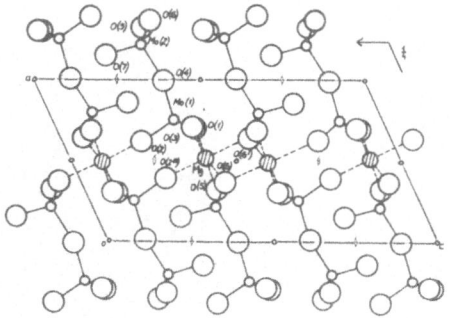

Fig. 1.   Structure of MgMo₂O₇.

SODIUM DODECAMOLYBDOGERMANATE OCTAHYDRATE
Na₄Mo₁₂GeO₄₀.8H₂O

R. STRANDBERG, 1977. Acta Cryst., B<u>33</u>, 3090-3096.

Triclinic, P$\bar{1}$, a = 14.421, b = 13.187, c = 11.596 Å, α = 114.31, β = 103.88, γ = 76.45°, Z = 2. Mo radiation, R = 0.029 for 9988 reflexions.

The anion (Fig. 1) has the well-known Keggin structure (α-isomer), mean Mo-O = 1.69, 1.82, 2.05, 2.29 Å, depending on coordination number, Ge-O = 1.73 Å. The anions are linked by Na ions and by hydrogen bonds from water molecules.

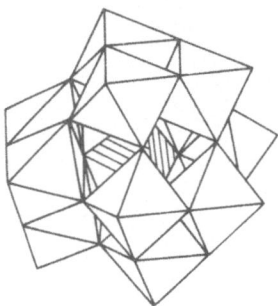

Fig. 1.   The $Mo_{12}GeO_{40}{}^{4-}$ anion.

SODIUM PENTAMOLYBDODIPHOSPHATE TETRADECAHYDRATE
$Na_6Mo_5P_2O_{23} \cdot 14H_2O$

B. HEDMAN, 1977. Acta Cryst., B**33**, 3083-3090.

Orthorhombic, P$2_12_12_1$, a = 15.830, b = 19.876, c = 10.683 Å, D_m = 2.56, Z = 4. Mo radiation, R = 0.024 for 6171 reflexions.

The anion contains a ring of five MoO_6 octahedra with one PO_4 tetrahedron attached to each side (Fig. 1); Mo-O = 1.70-1.73, 1.90-1.95, 2.17-2.39 Å, depending on coordination number, P-O = 1.50-1.57 Å. Anions are linked by Na ions (5- and 6-coordination) and hydrogen bonds from water molecules.

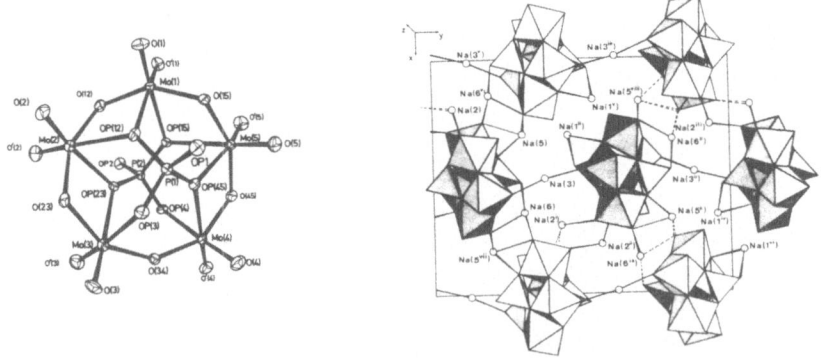

Fig. 1.   The $Mo_5P_2O_{23}{}^{6-}$ anion and structure of $Na_6Mo_5P_2O_{23} \cdot 14H_2O$.

LITHIUM SCANDIUM MOLYBDATE   LITHIUM YTTRIUM MOLYBDATE
$LiScMo_3O_8$ $LiYMo_3O_8$

I.   J. DeBENEDITTIS and L. KATZ, 1965. Inorg. Chem., $\underline{4}$, 1836-1839.
II.   W.H. McCARROLL, 1977. Ibid., $\underline{16}$, 3351-3353.

Trigonal, P3m1, a = 5.724, 5.781, c = 4.943, 5.153 Å, D_m = 5.51, 5.72, Z = 1.
Powder data.

Atomic positions

			x	y	z
Li	in	1(c)	2/3	1/3	0.714
Y or Sc		1(b)	1/3	2/3	0.720
Mo		3(d)	0.8534	-0.8534	0.250
O(1)		1(a)	0	0	0.495
O(2)		1(c)	2/3	1/3	-0.005
O(3)		3(d)	0.498	-0.498	0.464
O(4)		3(d)	0.167	-0.167	0.062

    Hexagonal close-packed oxygens with Li and Sc in tetrahedral and octahedral
holes, respectively, and trigonal clusters of three MoO_6 octahedra. Compounds
$LiMMo_3O_8$ (M = In, Sm, Gd, Yb, Lu) are isostructural. $Zn_2Mo_3O_8$ has a related
structure with different oxygen packing ($\underline{1}$).

1. Structure Reports, $\underline{21}$, 333; $\underline{31A}$, 155.

POTASSIUM NONAMOLYBDOMANGANATE(IV) HEXAHYDRATE
$K_6MnMo_9O_{32}.6H_2O$

T.J.R. WEAKLEY, 1977. J. Less-Common Metals, $\underline{54}$, 289-296.

Rhombohedral, R3, a = 15.59, c = 12.44 Å, D_m = 2.34, Z = 3. Cu radiation, R =
0.069 for 1087 reflexions.

    The $MnMo_9O_{32}^{6-}$ anion is similar to that found in the ammonium salt ($\underline{1}$), but
has only C_3 rather than D_3 symmetry.

1. Structure Reports, $\underline{18}$, 534; $\underline{41A}$, 246.

LITHIUM NICKEL MOLYBDATE
$Li_2Ni_2Mo_3O_{12}$

M. OZIMA, S. SATO and T. ZOLTAI , 1977. Acta Cryst., B$\underline{33}$, 2175-2181.

Orthorhombic, Pmcn, a = 10.423, b = 17.525, c = 5.074 Å, D_m = 4.36, Z = 4. Mo
radiation, R = 0.053 for 2717 reflexions.

    The structure is very similar to those of $NaCo_{2.31}Mo_3O_{12}$ ($\underline{1}$) and $Li_3FeMo_3O_{12}$
and $Li_2Fe_2Mo_3O_{12}$ ($\underline{2}$). It contains close-packed oxygen sheets, with Li in trigonal-
prismatic voids, Li/Na disordered in distorted and regular trigonal-antiprismatic
voids, and Mo in elongated tetrahedral voids. Mean interatomic distances are
Mo-O = 1.774, Li/Ni-O = 2.097 Å.

1. Structure Reports, $\underline{29}$, 329.
2. Ibid., $\underline{35A}$, 250.

SODIUM ZINC HYDROXIDE MOLYBDATE   (DISODIUM DI-μ-HYDROXO-DIZINC MOLYBDATE)
$NaZn(OH)MoO_4$

A. CLEARFIELD, R. GOPAL and C.H. SALDARRIAGA-MOLINA, 1977. Inorg. Chem., 16,
628-631.

Orthorhombic, $Pna2_1$, a = 7.850, b = 9.292, c = 6.148 Å, $D_m \sim 4$, Z = 4. Mo radiation, R = 0.038 for 843 reflexions.

    The structure (Fig. 1) contains chains of edge-sharing ZnO_6 octahedra, linked
by molybdate tetrahedra to form sheets, which are further connected by molybdate
groups and Na^+ ions (7-coordination). Zn-O = 1.88-2.33, Mo-O = 1.67-1.79, Na-O =
2.36-2.80(1) Å. The H atom is probably bonded to O(4).

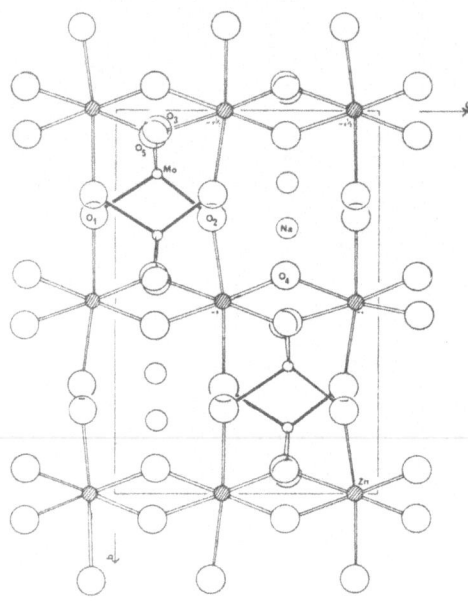

Fig. 1.   Structure of $NaZn(OH)MoO_4$.

LITHIUM TUNGSTATE(VI)   (High-pressure)
Li_2WO_4

I.   K.-A. WILHELMI, K. WALTERSSON AND P. LÖFGREN, 1977. Cryst. Struct. Comm.,
      6, 219-223.
II.   K. WALTERSSON, P.-E. WERNER and K.-A. WILHELMI, 1977. Ibid., 6, 225-230.
III. Idem, 1977. Ibid., 6, 231-235.

Preliminary reports on three high-pressure forms [results are in fact given in
full, the implication being that duplicate accounts will be published later].

POTASSIUM HEXATUNGSTATE   POTASSIUM OCTATUNGSTATE
$K_2W_6O_{19}$ $K_2W_8O_{25}$

A. KLUG, 1977. Mater. Res. Bull., 12, 837-845.

Hexatungstate, orthorhombic, Pmmm, a = 7.305, b = 2 x 12.705, c = 7.631 Å. Powder data.

Octatungstate, orthorhombic, Cmmm substructure, a = 7.302, b = 5 x 12.713, c = 7.627 Å. R = 0.086 for 2200 reflexions, for a substructure with b' = b/5, c' = c/2.

Substructure of the hexagonal tungsten bronze type (1).

1. Structure Reports, 17, 401.

RUBIDIUM POLYTUNGSTATE
$Rb_{22}W_{32}O_{107}$

K. OKADA, F. MARUMO and S. IWAI, 1977. Acta Cryst., B33, 3345-3349.

Tetragonal, I$\bar{4}$, a = 15.966, c = 10.099 Å, D_m = 6.13, Z = 1. Mo radiation, R = 0.078 for 2259 reflexions.

The structure (Fig. 1) contains four WO_6 octahedra which share corners to form a W_4O_{18} group; these groups are further linked by corner-sharing to form a three-dimensional $(W_{32}O_{107}^{22-})_n$ anion. One-eighth of one oxygen site is vacant to satisfy the stoichiometry. W-O = 1.66-2.25 Å. Rb ions have 12, 14, 15, 16, and 18 coordinations, Rb-O = 2.94-3.73 Å.

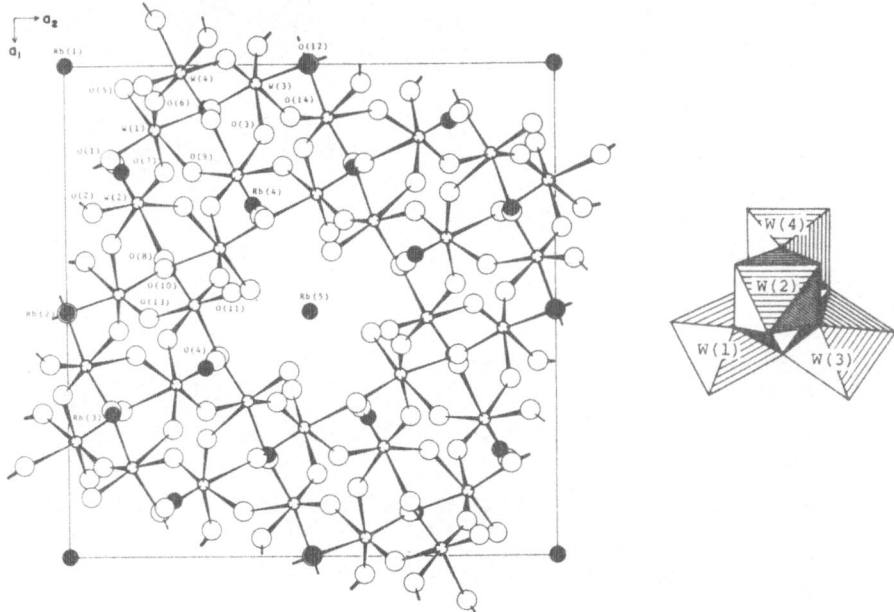

Fig. 1.   Structure of $Rb_{22}W_{32}O_{107}$ (left) and idealized W_4O_{18} group (right).

RASPITE
PbWO$_4$

T. FUJITA, I. KAWADA and K. KATO, 1977. Acta Cryst., B33, 162-164.

Monoclinic, P2$_1$/a, a = 13.555, b = 4.976, c = 5.561 Å, β = 107.63°, Z = 4. Mo radiation, R = 0.080 for 2144 reflexions.

Atomic positions

	x	y	z
Pb	0.1496	0.1941	0.1667
W	0.0771	0.7494	0.6119
O(1)	0.0163	0.0515	0.7290
O(2)	0.0595	0.4346	0.3882
O(3)	0.1510	0.6148	0.9000
O(4)	0.1903	0.8829	0.5386

The structure (Fig. 1) contains chains along b of edge-sharing WO$_6$ octahedra, the chains being linked by Pb atoms which have irregular 7-coordination. W-O = 1.70-2.17(2), Pb-O = 2.31-2.85(2) Å.

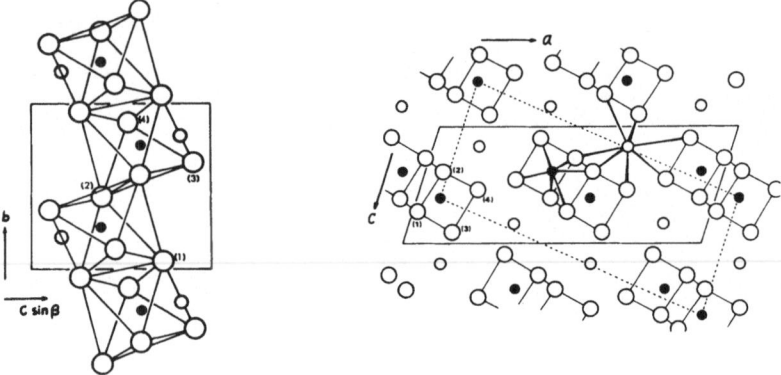

Fig. 1.   Structure of raspite.

DODECATUNGSTOPHOSPHORIC ACID HEXAHYDRATE
H$_3$PW$_{12}$O$_{40}$·6H$_2$O   (H$_5$O$_2^+$)$_3$(PW$_{12}$O$_{40}^{3-}$)

G.M. BROWN, M.-R. NOE-SPIRLET, W.R. BUSING and H.A. LEVY, 1977. Acta Cryst., B33, 1038-1046.

Cubic, Pn3m, a = 12.506 Å, D$_m$ = 5.60, Z = 2. Mo radiation, R = 0.059 for 528 reflexions, and neutron radiation, R = 0.035 for 541 reflexions.

Atomic positions (neutron data, origin at $\bar{3}$m)

			x	y	z
P	in	2(c)	3/4	3/4	3/4
W		24(k)	0.7582	0.9569	0.9569
O(1)		8(e)	0.8227	0.8227	0.8227
O(2)		24(k)	0.6568	0.8432	0.9932
O(3)		24(k)	0.8716	0.8716	1.0245
O(4)		24(k)	0.7328	1.0544	1.0544
0.5 O(W)		24(h)	3/4	1.1525	1¼
0.5 H(W)		48(ℓ)	0.7453	1.1126	1.1829
H(A)		6(d)	3/4	1¼	1¼

Keggin's 'pentahydrate' (<u>1</u>) is a hexahydrate. The structure of the $PW_{12}O_{40}{}^{3-}$ anion (Fig. 1) is essentially as previously described (<u>1</u>, <u>2</u>). The six water molecules are paired in nearly-planar, twofold-disordered $\bar{H}_5O_2{}^+$ ions with linear symmetric O...H...O bonds (O...O = 2.414 Å); an alternative description involves pyramidal geometry at O and further disorder.

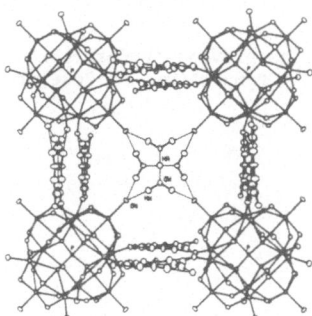

Fig. 1.   Structure of dodecatungstophosphoric acid hexahydrate.

<u>1</u>. Strukturbericht, <u>3</u>, 113, 463.
<u>2</u>. Structure Reports, <u>41A</u>, 255.

WOLFRAMITE AND NIOBITE STRUCTURES
$MnWO_4$ $CoWO_4$ $NiWO_4$ $MnNb_2O_6$ $FeNb_2O_6$ $CoNb_2O_6$ $MnTa_2O_6$

H. WEITZEL, 1976. Z. Kristallogr., <u>144</u>, 238-258.

MWO_4 (M = Mn, Co, Ni)
Monoclinic, P2/c, a = 4.824, 4.670, 4.599, b = 5.750, 5.687, 5.661, c = 4.990, 4.952, 4.907 Å, β = 91.18, 90.00, 90.03°, Z = 2. Neutron powder data.

MNb_2O_6 (M = Mn, Fe, Co), $MnTa_2O_6$
Orthorhombic, Pbcn, a ∿ 14, b ∿ 5.7, c ∿ 5.1 Å, Z = 4. Neutron powder data.

Atomic positions

MnWO₄ (similar values for M = Co, Ni)

	x	y	z
Mn	1/2	0.6866	1/4
W	0	0.1853	1/4
O(1)	0.2132	0.1026	0.9394
O(2)	0.2524	0.3707	0.3918

MnNb₂O₆ (similar values for the other compounds)

	x	y	z
Mn	0	0.1903	1/4
Nb	0.1616	0.3206	0.7478
O(1)	0.0973	0.3976	0.4363
O(2)	0.0846	0.1178	0.9014
O(3)	0.2555	0.1248	0.5819

The MWO_4 compounds have the wolframite structure (e.g. <u>1</u>), and the MNb_2O_6 compounds and $MnTa_2O_6$ have the columbite [niobite] structure (<u>2</u>).

<u>1</u>. Structure Reports, <u>21</u>, 305; <u>31A</u>, 158; <u>32A</u>, 315; <u>33A</u>, 346, 347; <u>34A</u>, 285.
<u>2</u>. Strukturbericht, <u>2</u>, 55, 337; Structure Reports, <u>37A</u>, 243.

IRON(III) TUNGSTATE
Fe_2WO_6

H. PINTO, M. MELAMUD and H. SHAKED, 1977. Acta Cryst., A33, 663-667.

Orthorhombic, Pbcn, a = 4.576, b = 16.766, c = 4.967 Å (from 1), Z = 4. Neutron powder data at 300-4.2°K.

Tri-α-PbO$_2$ type structure, as previously described (1). Appearance of systematically absent reflexions as the temperature is lowered is interpreted in terms of the magnetic structure.

1. Structure Reports, 40A, 207.

RUBIDIUM DYSPROSIUM TUNGSTATE
$RbDy(WO_4)_2$

E.N. IPATOVA, R.F. KLEVCOVA and L.P. SOLOV'EVA, 1976. Kristallografija, 21, 1121-1126 [Soviet Physics - Crystallography, 21, 648-651].

Monoclinic, $P2_1/c$, a = 8.90, b = 10.25, c = 7.93 Å, β = 99°, Z = 4. Mo radiation, R = 0.13 for 1700 reflexions (films, visual intensities).

The structure contains isolated $W_4O_{16}{}^{8-}$ groups of edge-sharing distorted octahedra, W-O = 1.68-2.59 Å, linked by wide bands along c of 8-coordinate Rb and Dy polyhedra, Rb-O = 2.76-3.17, Dy-O = 2.32-2.75 Å.

COPPER(II) MANGANESE(IV) OXIDE
$Cu_2Mn_3O_8$

A. RIOU and A. LECERF, 1977. Acta Cryst., B33, 1896-1900.

Monoclinic, C2/m, a = 9.695, b = 5.635, c = 4.912 Å, β = 103.31°, D_m = 5.23, Z = 2. Mo radiation, R = 0.096 for 602 reflexions (films, densitometer intensities).

Atomic positions

			x	y	z
Mn(1)	in	2(c)	0	0	1/2
Mn(2)		4(g)	0	0.2589	0
Cu		4(i)	0.2803	0	0.4307
O(1)		8(j)	0.1188	0.2272	0.3665
O(2)		4(i)	0.3964	0	0.1113
O(3)		4(i)	0.1021	0	0.8634

The structure (Fig. 1) contains $(Mn_3O_8{}^{4-})_n$ sheets connected by Cu^{2+} ions; similar sheets are found in $Co_2Mn_3O_8$ (1). Mn^{4+} ions have octahedral coordination, Mn-O = 1.84-1.95 Å, and Cu^{2+} has square-pyramidal coordination, Cu-O = 1.98, 2.00 (basal), 2.10(1) Å (apical).

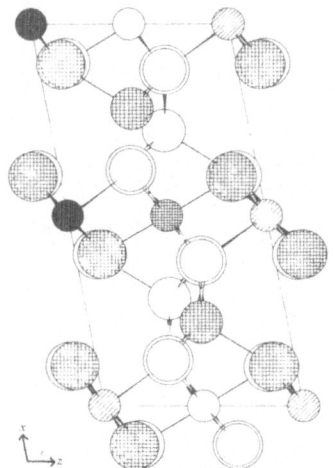

Fig. 1.   Structure of $Cu_2Mn_3O_8$; small circles Mn, medium circles Cu, large
            circles O.

<u>1</u>. Structure Reports, <u>41</u>A, 260.

YTTERBIUM PERRHENATE TETRAHYDRATE
$Yb(ReO_4)_3 \cdot 4H_2O$

E.D. BAKHAREVA, M.B. VARFOLOMEEV, V.P. MAŠONKIN and V.V. ILJUKHIN, 1976.   Koordin.
Khim., <u>2</u>, 1135-1141.

Triclinic, P$\bar{1}$,   a = 7.169, b = 8.696, c = 11.884 Å, α = 74.47, β = 96.54, γ =
102.32°, Z = 2. Mo radiation, R = 0.039 for 1100 reflexions.

     The structure (Fig. 1) contains ReO_4 tetrahedra (Re-O = 1.61-1.76 Å) and YbO_8
polyhedra (Yb-O = 2.28-2.40 Å), linked to form $[Yb_2(ReO_4)_6 \cdot (H_2O)_6]_\infty$ units, and a
system of O-H...O hydrogen bonds.

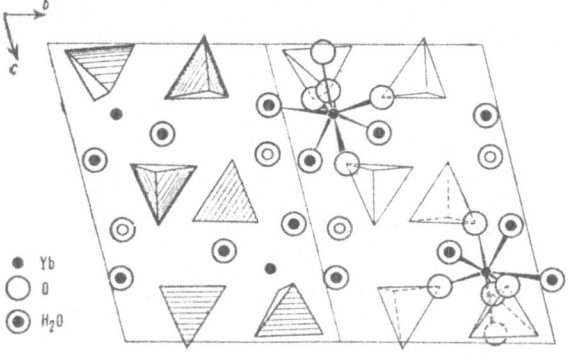

Fig. 1.   Structure of ytterbium perrhenate tetrahydrate.

LANTHANUM RUTHENIUM OXIDE   LANTHANUM OSMIUM OXIDE
$La_4Ru_6O_{19}$ $La_4Os_6O_{19}$

F. ABRAHAM, J. TRÉHOUX and D. THOMAS, 1977. Mater. Res. Bull., 12, 43-51.

Cubic, I23, a = 8.981, 9.050 Å, D_m = 6.71 (Ru), Z = 2. Mo radiation, R = 0.022
and 0.024 for 356 and 419 reflexions. La in 8(c): x = 0.1628 (0.1602); Ru (Os) in
12(e): x = 0.3615 (0.3619); O(1) in 12(d): x = 0.3331 (0.3264); O(2) in 2(a);
O(3) in 24(f): x = 0.3404 (0.3389), y = 0.2858 (0.2849), z = 0.9712 (0.9746).

    Isostructural with $La_4Re_6O_{19}$ (1). Ru-Ru = 2.488(1), Os-Os = 2.499(1) Å.

1. Structure Reports, 33A, 353.

BARIUM COBALTATE(IV)
$BaCoO_3$

H. TAGUCHI, Y. TAKEDA, F. KANAMARU, M. SHIMADA and M. KOIZUMI, 1977. Acta Cryst.,
B33, 1298-1299.

Hexagonal, $P6_3/mmc$, a = 5.645, c = 4.752 Å, D_m = 6.1, Z = 2. Mo radiation, R =
0.060 for 226 reflexions. Ba in 2(d); Co in 2(a); O in 6(h): x = 0.1462.

    Isostructural with $BaNiO_3$ (1). Co-O = 1.874 Å.

1. Structure Reports, 42A, 287.

LANTHANUM NICKELATE
La_2NiO_4

B. GRANDE and H. MÜLLER-BUSCHBAUM, 1977. Z. anorg. Chem., 433, 152-156.

Tetragonal, I4/mmm, a = 3.876, c = 12.683 Å, Z = 2. No details of data or analysis.
La in 4(e): z = 0.363; Ni in 2(a); O(1) in 4(c); O(2) in 4(e): z = 0.180.

    K_2NiF_4-type structure (1). Ni-O = 1.94 (x 4), 2.28 (x 2), La-O = 2.61,
2.79 (each x 4), 2.32 Å.

1. Structure Reports, 17, 332; 19, 323.

BARIUM PLATINUM OXIDE
$Ba_3Pt_2O_7$

P.S. HARADEM, B.L. CHAMBERLAND, L. KATZ and A. GLEIZES, 1977. J. Solid State
Chem., 21, 217-223.

Hexagonal, $P\bar{6}2c$, a = 10.108, c = 8.638 Å, D_m = 7.99, Z = 4. Mo radiation, R = 0.080
for 401 reflexions.

Atomic positions

				x	y	z
	Ba(1)	in	6(h)	0.3337	0.0192	3/4
	Ba(2)		6(g)	0.6646	0	0
	Pt(1)		4(f)	1/3	2/3	0.0907
0.25	Pt(2)		4(f)	1/3	2/3	0.6984
0.17	Pt(3)		6(h)	0.4025	0.7378	3/4
0.33	Pt(4)		4(e)	0	0	0.1554
0.18	Pt(5)		4(e)	0	0	0.0877
0.12	Pt(6)		2(a)	0	0	0
	O(1)		12(i)	0.824	0.336	0.055
	O(2)		6(h)	0.168	0.643	1/4
	O(3)		6(h)	0.140	0.170	1/4
0.74	O(4)		6(g)	0.156	0	0

The structure contains Ba-O layers with a double-hexagonal four-layer stacking sequence (ABAB). Pt is mainly in face-sharing octahedra, but is also distributed over some sites with square-planar and some with trigonal-prismatic coordination.

BARIUM OXOCUPRATE(II)
$BaCuO_2$

R. KIPKA and H. MÜLLER-BUSCHBAUM, 1977. Z. Naturforsch., 32B, 121-123.

Cubic, Im3m, a = 18.27 Å, Z = 90. R = 0.06 for 725 reflexions.

Atomic positions

				x	y	z
	Ba(1)	in	48(j)		0.151	0.310
	Ba(2)		24(h)	0.364		
	Ba(3)		16(f)	0.177		
	Ba(4)		2(a)			
	Cu(1)		48(i)	0.150		
	Cu(2)		24(h)	0.125		
	Cu(3)		12(e)	0.206		
1/2	Cu(4)		12(e)	0.430		
	O(1)		48(k)	0.072		0.186
	O(2)		48(k)	0.144		0.343
	O(3)		48(k)	0.267		0.085
	O(4)		12(d)			
	O(5)		12(e)	0.338		
1/4	O(6)		48(j)		0.112	0.440

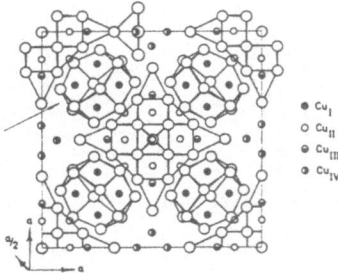

Fig. 1.   Copper and oxygen positions in $BaCuO_2$.

The structure (Fig. 1) contains a complicated polyhedral framework. Cu ions have square-planar and square-pyramidal coordinations, Cu-O = 1.80-2.41 Å, and Ba ions have 8-, 8-, 9-, and 24-coordinations.

LANTHANUM OXOCUPRATE   GADOLINIUM OXOCUPRATE
La_2CuO_4 Gd_2CuO_4

B. GRANDE, H. MÜLLER-BUSCHBAUM and M. SCHWEIZER, 1977. Z. anorg. Chem., **428**, 120-124.

La_2CuO_4
Orthorhombic, Abma, a = 5.406, b = 5.370, c = 13.15 Å, Z = 4. R = 0.087 for 461 reflexions.

Gd_2CuO_4
Tetragonal, [I4/mmm], a = 3.800, c = 11.80 Å, Z = 4. R = 0.070. Gd in 4(e): z = 0.350; Cu in 2(a); O in 4(c) and 4(d).

Atomic positions

La_2CuO_4		x	y	z
La	in 8(f)	0.007	0	0.362
Cu	4(a)	0	0	0
O(1)	8(e)	1/4	1/4	0.007
O(2)	8(f)	0.969	0	0.187

Gd_2CuO_4 (transformed for comparison with La_2CuO_4)			
Gd	0	0	0.350
Cu	0	0	0
O(1)	1/4	1/4	0
O(2a)	1/4	1/4	1/4

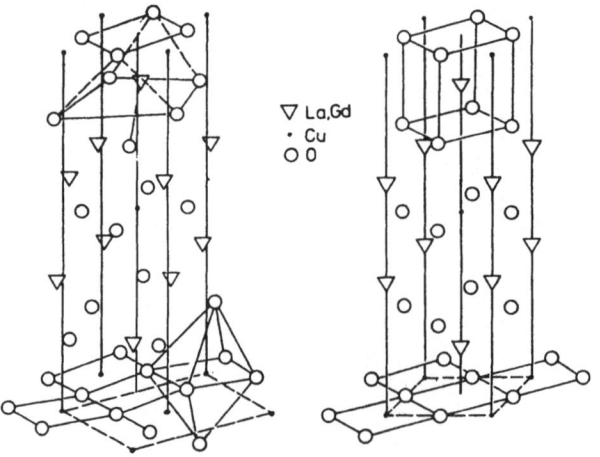

▽ La,Gd
• Cu
○ O

Fig. 1.   Structures of La_2CuO_4 (left) and Gd_2CuO_4 (right).

Gd_2CuO_4 is isostructural with Nd_2CuO_4 (**1**), and La_2CuO_4 has a slightly-distorted K_2NiF_4-type (**2**) structure (Fig. 1). Cu has square-planar coordination in Gd_2CuO_4 (Cu-O = 1.95 (x 4), with next nearest O at 3.55 Å), and distorted octahedral coordination in La_2CuO_4 (Cu-O = 1.91 (x 4), 2.46 (x 2) Å). Gd has eight- and La nine-coordination (Fig. 1).

1. Structure Reports, 41A, 269.
2. Ibid., 17, 332; 19, 323.

HOLMIUM OXOCUPRATE
$Ho_2Cu_2O_5$

H.-R. FREUND and H. MÜLLER-BUSCHBAUM, 1977. Z. Naturforsch., 32B, 609-611.

Orthorhombic, P2₁nb, a = 12.478, b = 10.813, c = 3.495 Å, Z = 4. R = 0.107 for 1079 reflexions.

The structure (Fig. 1) contains Cu ions with very distorted tetrahedral coordinations (fifth oxygens at 2.74 and 2.91 Å), and octahedrally-coordinated Ho ions, Ho-O = 2.10-2.43 Å.

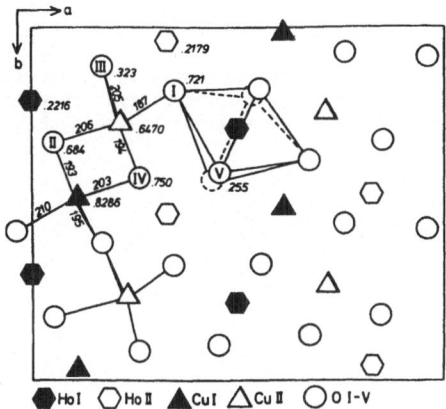

Fig. 1.   Structure of $Ho_2Cu_2O_5$ (Cu-O in Å x 10^2).

LUTETIUM OXOCUPRATE
$Lu_2Cu_2O_5$

SCANDIUM OXOCUPRATE
$Sc_2Cu_2O_5$

H.-R. FREUND and H. MÜLLER-BUSCHBAUM, 1977. Z. Naturforsch., 32B, 1123-1124.

Orthorhombic, P2₁nb, a = 12.350, 12.041, b = 10.698, 10.452, c = 3.405, 3.220 Å, Z = 4. R = 0.084 and 0.032 for 367 and 724 reflexions.

Isostructural with $Ho_2Cu_2O_5$ (1).

1. Preceding report.

CAESIUM PRASEODYMATE
Cs_2PrO_3

RUBIDIUM TERBATE
Rb_2TbO_3

RUBIDIUM CERATE
Rb_2CeO_3

H. BRUNN and R. HOPPE, 1977. Z. anorg. Chem., 433, 189-199.

Cs_2PrO_3, Rb_2TbO_3 (low-temperature)
Orthorhombic, $Cmc2_1$, a = 11.47, 10.91, b = 7.72, 7.39, c = 6.43, 6.10 Å, D_m = 5.18, 5.05, Z = 4. Mo radiation, R = 0.096 for 254 reflexions, for Cs_2PrO_3; Cu radiation, powder data for Rb_2TbO_3.

Rb_2CeO_3 (high-temperature)
Rhombohedral, $R\bar{3}m$, a = 3.84, c = 18.47 Å, D_m = 4.94, Z = 2. Cu radiation, powder data.

Atomic positions
Cs_2PrO_3 (low-temperature Rb_2TbO_3)

			x	y	z
Cs (Rb)	in	8(b)	0.3344 (0.336)	0.1458 (0.141)	1/4
Pr (Tb)		4(a)	0	0.0888 (0.094)	1/4
O(1)		8(b)	0.1195 (0.133)	0	0
O(2)		4(a)	0	0.346 (0.360)	0.167 (0.15)

Rb_2CeO_3 (high-temperature)

Rb in 3(a); 1/3 Rb + 2/3 Ce in 3(b); O in 6(c): z = 0.23.

Cs_2PrO_3 and low-temperature Rb_2TbO_3 are isostructural with Cs_2PbO_3 ([1]); Cs_2CeO_3 and Cs_2TbO_3 also belong to this structure type (powder data). High-temperature Rb_2CeO_3 and Rb_2TbO_3 are isostructural with K_2TbO_3 ([2]) (α-$NaFeO_2$-type ([3])).

[1]. Structure Reports, 38A, 251.
[2]. R. HOPPE, 1969. Angew. Chem., 76, 691.
[3]. Strukturbericht, 3, 75, 392.

TERBIUM IRON GARNET
$Tb_3Fe_5O_{12}$

H. FUESS, G. BASSI, M. BONNET and A. DELAPALME, 1976. Solid State Comm., 18, 557-562.

Cubic, $Ia3d$, a = 12.436 Å, Z = 8. X-ray data, $R(F^2)$ = 0.039 for 348 reflexions, and neutron powder data. Tb in 24(c); Fe in 16(a) and 24(d); O in 96(h): (-0.0279, 0.0559, 0.1499). Garnet structure ([1]).

[1]. Strukturbericht, 1, 363.

CALCIUM HOLMIUM OXIDES   CALCIUM THULIUM OXIDE
$Ca_{0.07}Ho_{1.86}O_{2.86}$, $CaHoO_{2.5}$ $Ca_{0.5}Tm_{1.5}O_{2.75}$

W. MUSCHICK and H. MÜLLER-BUSCHBAUM, 1977. Z. Naturforsch., 32B, 495-498.

$Ca_{0.07}Ho_{1.86}O_{2.86}$
Monoclinic, C2/m, a = 13.901, b = 3.502, c = 8.606 Å, β = 100.4°, Z = 6. R = 0.093 for 738 reflexions.

$CaHoO_{2.5}$, $Ca_{0.5}Tm_{1.5}O_{2.75}$
Monoclinic, $P2_1/m$, a = 6.566, 6.502, b = 3.567, 3.524, c = 5.894, 5.845 Å, β = 92.3, 92.3°, Z = 2. R = 0.099 and 0.12 for 220 and 254 reflexions.

Atomic positions

$Ca_{0.07}Ho_{1.86}O_{2.86}$	x	y	z
Ho/Ca(1)	0.6347	0	0.4895
Ho/Ca(2)	0.6897	0	0.1375
Ho/Ca(3)	0.9667	0	0.1866
O(1)	1/2	0	0
O(2)	0.126	0	0.284
O(3)	0.825	0	0.030
O(4)	0.796	0	0.374
O(5)	0.473	0	0.345

$CaHoO_{2.5}$			
Ca/Ho(1)	0.4310	1/4	0.2451
Ca/Ho(2)	0.1193	1/4	0.7722
O(1)	0.686	1/4	0.013
O(2)	0.089	1/4	0.209
O(3)	0.416	1/4	0.608

$Ca_{0.5}Tm_{1.5}O_{2.75}$			
Tm	0.4256	1/4	0.2403
Ca/Tm	0.1283	1/4	0.7704
O(1)	0.618	1/4	0.973
O(2)	0.071	1/4	0.182
O(3)	0.424	1/4	0.593

$Ca_{0.07}Ho_{1.86}O_{2.86}$ has a B-type rare-earth oxide structure ([1]); M-O = 2.17-3.10 Å. The other two compounds represent a new structure type. In all three compounds the oxygen deficiency is distributed over all the oxygen sites.

[1]. Structure Reports, 21, 225.

BOEHMITE
γ-AlOOH

L. FARKAS, P. GADÓ and P.-E. WERNER, 1977. Mater. Res. Bull., 12, 1213-1219.

Orthorhombic, Cmcm, a = 3.695, 3.701, b = 12.232, 12.234, c = 2.868, 2.874 Å, for synthetic and natural material, Z = 4. Powder data. y = -0.3208 (-0.3165), 0.2943 (0.2891), 0.0834 (0.0775) for Al, O, OH in synthetic (natural) specimens.

   Structure as previously described ([1]).

[1]. Structure Reports, 10, 99; 20, 284.

GALLIUM OXIDE HYDROXIDE
α-GaOOH

M.F. PYE, J.J. BIRTILL and P.G. DICKENS, 1977. Acta Cryst., B33, 3224-3226.

Orthorhombic, Pbnm, a = 4.516, b = 9.779, c = 2.966 Å (25°C), Z = 4. Neutron powder data with profile analysis for GaOOD at 25°C and 4.2°K.

Atomic positions (atoms in 4(c), z = 1/4)

	25°C		4.2°K	
	x	y	x	y
Ga	0.0512	-0.1447	0.0515	-0.1443
O(1)	0.7020	0.1953	0.7006	0.1955
O(2)	0.1945	0.0546	0.1943	0.0551
D	0.4035	0.0869	0.4031	0.0883

Isostructural with diaspore (<u>1</u>). The hydrogen bond is asymmetric and slightly bent (Fig. 1), O-D = 0.995(5), D...O = 1.715 Å, O...O-D = 12.5° (25°C). Ga-O = 1.920-2.054(5) Å.

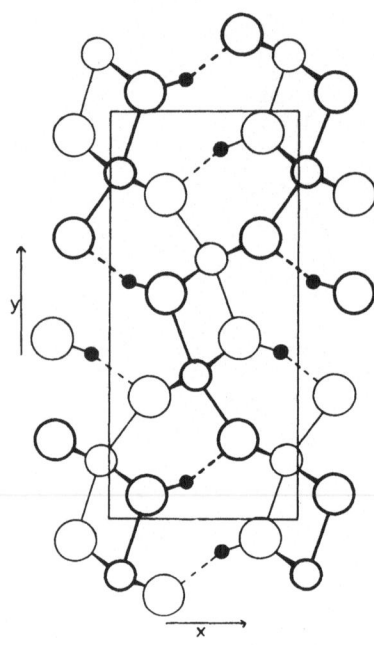

Fig. 1.   Structure of GaOOD.

<u>1</u>. Strukturbericht, <u>2</u>, 46, 350; <u>3</u>, 64, 371; Structure Reports, <u>9</u>, 150; <u>10</u>, 98; <u>22</u>, 276.

CALCIUM HEXAHYDROXOPLUMBATE(IV)   CADMIUM HEXAHYDROXOPLUMBATE(IV)
CaPb(OH)$_6$ CdPb(OH)$_6$

C. LÉVY-CLÉMENT and Y. BILLIET, 1976. Bull. Soc. Fr. Minér. Crist., <u>99</u>, 361-372.

Cubic, Pn3, a = 8.252, 8.150 Å, D_m = 4.16, 5.04, Z = 4.   Cu radiation, powder data. Ca, Cd in 4(b); Pb in 4(c); O in 24(h): (0.083,-0.078,0.263), (0.102,-0.053,0.262) for Ca, Cd compounds (origin at $\bar{3}$).

The structure is a derivative of that of ReO_3, related to those of other AB(OH)$_6$ compounds.

POTASSIUM BISMUTH OXIDE HYDROXIDE
$(K_{1.14}Bi_{0.37})Bi_2(O_5OH)OH$

J. TREHOUX, F. ABRAHAM and D. THOMAS, 1977. J. Solid State Chem., 21, 203-209.

Cubic, Fd3m, a = 10.965 Å, D_m = 6.58, Z = 8. Cu radiation, R = 0.04 for 28 reflexions (powder data). K,Bi(III) in 16(d); Bi(III),Bi(V) in 16(c); O,OH in 48(f): x = 0.324; OH in 32(e): x = 0.410. Disordered pyrochlore (1) structure.

1. Strukturbericht, 2, 58.

CHROMIUM(III) OXIDE HYDROXIDE
α-CrOOH
α-CrOOD

I.   A.N. CHRISTENSEN, P. HANSEN and M.S. LEHMANN, 1977. J. Solid State Chem., 21, 325-329.

CrOOH
Rhombohedral, R3̄m, a = 2.979, c = 13.37 Å, Z = 3. Neutron powder data. Cr in 3(a); O in 6(c): z = 0.4076; 0.5 H in 6(c): z = 0.4940 (at 4.2°K, similar values at 300°K).

CrOOD
Rhombohedral, R3m, a = 2.985, c = 13.48 Å, Z = 3. Neutron powder data. Atoms in 3(a): z = 0, 0.4091, 0.6000, 0.4868 for Cr, O(1), O(2), D.

The structure is as previously determined (1), except that an ordered arrangement of D atoms in R3m is preferred for CrOOD. O-H = 1.07, O...O = 2.47; O-D = 1.05, O...O = 2.57 Å.

β-CrOOD

II. A.N. CHRISTENSEN and P. HANSEN, 1976. Acta Chem. Scand., A30, 835-836.

Orthorhombic, $P2_1nm$, a = 4.873, b = 4.332, c = 2.963 Å, Z = 2 [from 2]. Magnetic cell with doubled b and c axes. Neutron powder data.

Structure as previously described (2), with $P2_1nm$ now being used. The magnetic structure is described.

1. Structure Reports, 21, 337; 28, 223.
2. Ibid., 42A, 297.

VANDENBRANDEITE
$CuUO_2(OH)_4$ $\qquad\qquad\qquad\qquad\qquad\qquad\qquad$ $CuUO_4 \cdot 2H_2O$

A. ROSENZWEIG and R.R. RYAN, 1977. Cryst. Struct. Comm., 6, 53-56.

Triclinic, P1̄, a = 7.855, b = 5.449, c = 6.089 Å, α = 91.44, β = 101.90, γ = 89.2°, Z = 2. Mo radiation, R = 0.037 for 898 reflexions.

Atomic positions

	x	y	z
U	0.2840	0.1477	0.3422
Cu	0.0921	0.3509	0.8548
O(1)	0.394	0.406	0.269

	x	y	z
O(2)	0.836	0.113	0.604
O(3)	0.219	0.049	0.945
O(4)	0.010	0.316	0.138
O(5)	0.517	0.899	0.304
O(6)	0.186	0.402	0.599

The structure contains layers of interconnected Cu_2O_6 and U_2O_{12} dimeric units parallel to (110). Cu has square-pyramidal and U pentagonal-bipyramidal coordination; Cu-O = 1.88-1.98 (basal), 2.59 (apical), U-O = 2.31-2.44 (equatorial), 1.77, 1.78 Å (axial, uranyl). Layers are joined by Cu...O interactions and O-H...O hydrogen bonds.

LANTHANON HYDROXIDES
$Ln(OH)_3$

I.   G.W. BEALL and W.O. MILLIGAN, 1977. J. Inorg. Nucl. Chem., 39, 65-70.

Hexagonal, $P6_3/m$, a = 6.5-6.2, c = 3.8-3.5 Å, for Ln = La, Nd, Sm, Gd, Tb, Dy, Ho, Er, Y, Z = 2. Mo radiation, R = 0.02-0.03 for 324-720 reflexions. Ln in 2(c): (1/3,2/3,1/4); O in 6(h): (x,y,1/4), x = 0.39, y = 0.31; H in 6(h): x ∿ 0.28, y ∿ 0.15.

UCl₃-type structure (1), as previously reported (e.g. 2). M-O decreases from La to Er.

TERBIUM HYDROXIDE
$Tb(OH)_3$

II. G.W. BEALL, W.O. MILLIGAN, J. KORP and I. BERNAL, 1977. Acta Cryst., B33, 3134-3136.

Hexagonal, $P6_3/m$, a = 6.308, c = 3.600 Å, Z = 2. Neutron radiation, R = 0.074 for 83 reflexions. Structure as in I, x(O) = 0.3946, y(O) = 0.3117; x(H) = 0.2754, y(H) = 0.1426. The results agree with those in 3, if the lattice constants of the present study are used. Tb-O = 2.443, 2.455(4), O-H = 0.95(1) Å.

1. Structure Reports, 11, 278; 40A, 149.
2. Ibid., 32A, 246; 42A, 299, 300.
3. G.H. LANDER and Z.O. BRUN, 1973. Acta Cryst., A29, 684.

POTASSIUM NIOBIUM SULPHIDE HYDRATE   POTASSIUM TANTALUM SULPHIDE HYDRATE
$K_{0.5}NbS_2 \cdot 0.4H_2O$ $K_{0.33}TaS_2 \cdot 0.6H_2O$

H.A. GRAF, A. LERF and R. SCHÖLLHORN, 1977. J. Less-Common Metals, 55, 213-220.

Hexagonal, $P\bar{6}2c$, a = 3.35, 3.336, c = 17.94, 18.18 Å, D_m = 3.35, 4.98, Z = 2. Mo radiation, R = 0.102 and 0.094 for 786 and 229 reflexions. S in 4(e): z = 0.1635 and 0.1648; Nb and Ta in 2(d); K and H_2O not located.

The structures contain S-M-S sandwich layers similar to those in unhydrated phases, Nb and Ta having trigonal prismatic coordination. Stacking disorder and high thermal mobility prevented location of interlayer K and O atoms.

LITHIUM BORACITE
$Li_4B_7O_{12}Cl$

W. JEITSCHKO, T.A. BITHER and P.E. BIERSTEDT, 1977. Acta Cryst., B33, 2767-2775.

γ-Form (above 348°K)
Cubic, F$\bar{4}$3c, a = 12.167 Å, Z = 8. Mo radiation, R = 0.025 for 179 reflexions
(at 353°K).

β-Form (310-348°K)
Cubic, P$\bar{4}$3n, a = 12.161 Å, Z = 8. Mo radiation, R = 0.025 for 152 F$\bar{4}$3c sub-cell
reflexions (at 328°K).

α-Form (below 310°K)
Rhombohedral, R3, a = 12.141, α = 90.083° (hexagonal cell has a = 17.182, c =
20.998 Å), Z = 8. Mo radiation, R = 0.122 for 187 F$\bar{4}$3c sub-cell reflexions
(at 298°K).

Atomic positions (γ-form)

		x	y	z
0.32 Li(1) in	32(e)	0.8643	x	x
0.94 Li(2)	24(c)	0	1/4	1/4
B(1)	24(d)	1/4	0	0
B(2)	32(e)	0.1008	x	x
O	96(h)	0.0226	0.0977	0.1821
Cl	8(b)	1/4	1/4	1/4

Boracite-type ($Mg_3B_7O_{13}Cl$) structure (1), with additional Li in 32(e), and
the absence of the set of oxygens in 8(a). Li ions are located in channels and
have high mobility. The principal difference between the γ-form and the sub-cells
of the β- and α-forms is in Li occupancies of the 32(e) and 24(c) positions (β =
0.28 : 0.97, α = 0.25 : 1.00). [Compare 2.]

1. Strukturbericht, 2, 407; 15, 282; 39A, 264.
2. Structure Reports, 39A, 265.

SODIUM BORATE
Na_3BO_3

H. KÖNIG and R. HOPPE, 1977. Z. anorg. Chem., 434, 225-232.

Monoclinic, P2₁/c, a = 5.687, b = 7.530, c = 9.993 Å, β = 127.15°, Z = 4. Mo
radiation, R = 0.076 for 722 reflexions.

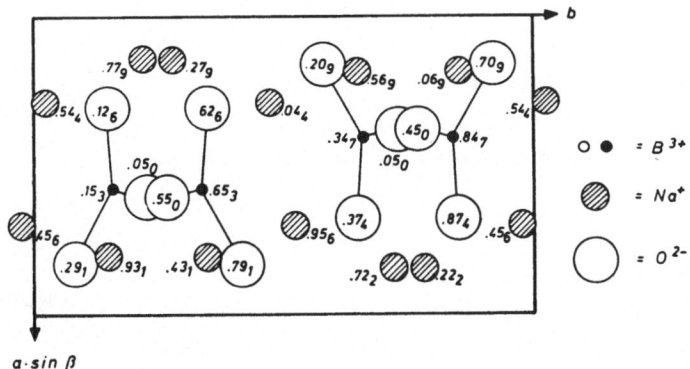

Fig. 1.   Structure of Na_3BO_3.

Atomic positions

	x	y	z
Na(1)	0.1509	0.2187	0.7785
Na(2)	0.2956	0.4745	0.0437
Na(3)	0.1909	0.8512	0.0693
B	0.5844	0.1588	0.1533
O(1)	0.3162	0.1483	0.1263
O(2)	0.6117	0.2697	0.0499
O(3)	0.8334	0.0804	0.2914

The structure (Fig. 1) contains planar triangular BO_3^{3-} anions, B-O = 1.38-1.41 Å, and Na^+ ions with 5, 5, and 4 close oxygen neighbours, Na-O = 2.26-2.57 Å. α-Li_3BO_3 ([1]) is not isostructural.

[1]. Structure Reports, **37**A, 271.

SODIUM PENTABORATE MONOHYDRATE
$Na_3[B_5O_8(OH)_2].H_2O$

S. MENCHETTI and C. SABELLI, 1977. Acta Cryst., B**33**, 3730-3733.

Orthorhombic, Pbca, a = 8.804, b = 18.371, c = 10.924 Å, Z = 8.' Cu radiation, R = 0.073 for 627 reflexions.

The structure (Fig. 1) contains two-dimensional pentaborate polyanions, $[B_5O_8(OH)_2^{3-}]_n$, which contain three tetrahedra and two triangles, linked to form two approximately perpendicular six-membered rings; B-O = 1.43-1.53 (tetrahedra), 1.35-1.40(1) Å (triangles). The polyanion sheets are linked by Na ions (6-, 6-, and 5-coordinations) and hydrogen bonds; Na-O = 2.33-2.65, O-H...O = 2.68-2.82 Å.

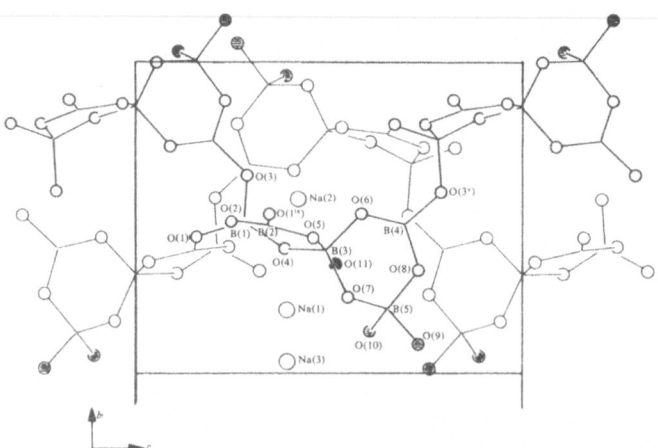

Fig. 1.   Part of the B-O sheet in sodium pentaborate monohydrate.

POTASSIUM METABORATE HYDRATE
$K_3[B_3O_4(OH)_4].2H_2O$ $3[KBO_2.4/3H_2O]$

J.R. CLARK and C.L. CHRIST, 1977. Acta Cryst., B**33**, 3272-3273.

The hydrogen bonding scheme in the structure (1) suggests the possibility of a new class of water molecule with four cations oriented in a directed way about the lone-pair orbitals. Unusual hydrogen bonds may also be present in kaliborite (2) and preobrazhenskite (3).

1. Structure Reports, 41A, 281.
2. Ibid., 31A, 172.
3. I.M. RUMANOVA, Z.P. RAZMANOVA and N.V. BELOV, 1972. Sov. Phys. Dokl. 16, 518.

HUNGCHAOITE

MgB$_4$O$_5$(OH)$_4$.7H$_2$O   MgO.2B$_2$O$_3$.9H$_2$O

CHE'NG WAN and S. GHOSE, 1977. Amer. Min., 62, 1135-1143.

Triclinic, PĪ, a = 8.807, b = 10.657, c = 7.897 Å, α = 103.39, β = 108.53, γ = 97.18°, D$_m$ = 1.706, Z = 2. Mo radiation, R = 0.049 for 4860 reflexions.

The structure (Fig. 1) contains a molecular complex, [Mg(H$_2$O)$_5$B$_4$O$_5$(OH)$_4$], and two water molecules, hydrogen bonded into a three-dimensional framework. Mean distances are Mg-O = 2.08, B-O = 1.48 (tetrahedral), 1.37 (triangular), O-H...O = 2.78 Å.

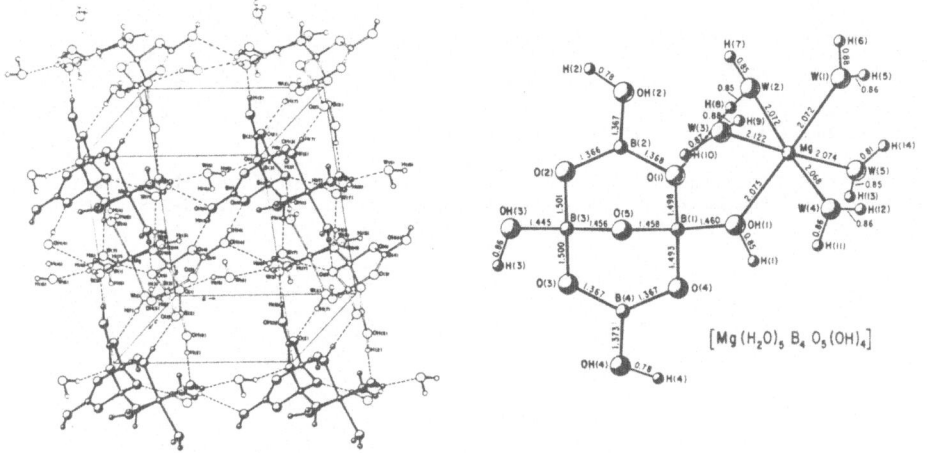

Fig. 1.   Structure of hungchaoite.

ARISTARAINITE

Na$_2$Mg[B$_6$O$_8$(OH)$_4$]$_2$.4H$_2$O

S. GHOSE and CHE'NG WAN, 1977. Amer. Min., 62, 979-989.

Monoclinic, P2$_1$/a, a = 18.886, b = 7.521, c = 7.815 Å, β = 97.72°, D$_m$ = 2.027, Z = 2. Mo radiation, R = 0.036 for 1941 reflexions.

The structure contains hexaborate polyanions, B$_6$O$_9$(OH)$_4^{4-}$, of three triangles and three tetrahedra sharing corners (Fig. 1); these units share an oxygen to form [B$_6$O$_8$(OH)$_4^{2-}$]$_n$ chains along b. Mg (octahedral coordination) and Na (distorted square-pyramidal coordination) link the chains into sheets parallel to (001), which are cross-linked by hydrogen bonds. Mean interatomic distances are: B-O = 1.47

(tetrahedral), 1.36 (triangle), Mg-O = 2.09, Na-O = 2.51, O-H...O = 2.78 Å.

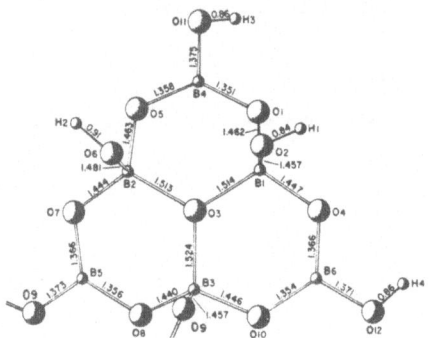

Fig. 1.   The hexaborate polyanion in aristarainite.

CALCIUM BORATE
CaB_4O_7

N.V. ZAJAKINA and A.A. BROVKIN, 1977. Kristallografija, 22, 275-280 [Soviet
Physics - Crystallography, 22, 156-159].

Monoclinic, $P2_1/n$, a = 12.34, b = 9.95, c = 7.85 Å, β = 92.1°, D_m = 2.67, Z = 8.
Mo radiation, R = 0.13 for 1874 reflexions (film data).

The structure (Fig. 1) contains a $B_8O_{14}^{4-}$ polyanion of four tetrahedra and
four triangles, and Ca ions with 7- and 8-coordination. B-O = 1.40-1.55 (tetrahedra), 1.32-1.43 (triangles), Ca-O = 2.28-2.89 Å.

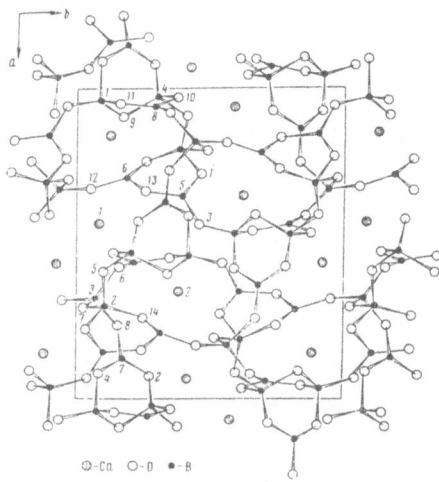

Fig. 1.   Structure of CaB_4O_7.

PENTAHYDROBORITE
$Ca[B_2O(OH)_6].2H_2O$ $CaB_2O_4.5H_2O$

E.V. KAZANSKAJA, T.N. ČEMODINA, Ju.K. EGOROV-TISMENKO, M.A. SIMONOV and N.V. BELOV, 1977. Kristallografija, 22, 66-68 [Soviet Physics - Crystallography, 22, 35-36].

Triclinic, P$\bar{1}$, a = 7.845, b = 6.525, c = 8.124 Å, α = 111.62, β = 111.19, γ = 73.44°, D_m = 2.0, Z = 2. Mo radiation, R = 0.065 for 1120 reflexions.

The structure contains $B_2O(OH)_6^{2-}$ ions (two tetrahedra sharing the O corner) and pairs of 7-coordinate Ca polyhedra (Fig. 1), joined by hydrogen bonds, directly and via the water molecules. B-O = 1.45-1.51(1) Å, B-O-B = 128°, Ca-O = 2.38-2.58 Å. Previous study in 1.

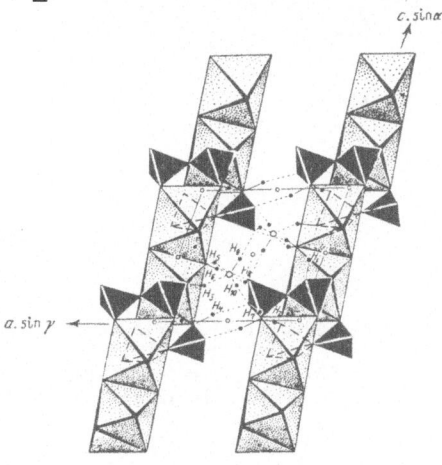

Fig. 1.   Structure of pentahydroborite.

1. Structure Reports, 39A, 363.

SOLONGOITE
$Ca_2[B_3O_4(OH)_4]Cl$

N.A. JAMNOVA, M.A. SIMONOV and N.V. BELOV, 1977. Kristallografija, 22, 624-626 [Soviet Physics - Crystallography, 22, 356-357].

Monoclinic, P2_1/b, a = 7.975, b = 12.571, c = 7.237 Å, γ = 86.14°, D_m = 2.514, Z = 4. Mo radiation, R = 0.053 for 3212 reflexions. Previous study in 1.

The structure (Fig. 1) contains a $B_3O_4(OH)_4^{3-}$ ion of two tetrahedra and one triangle, CaO_6Cl_2 and CaO_8 polyhedra, and O-H...O and O-H...Cl hydrogen bonds.

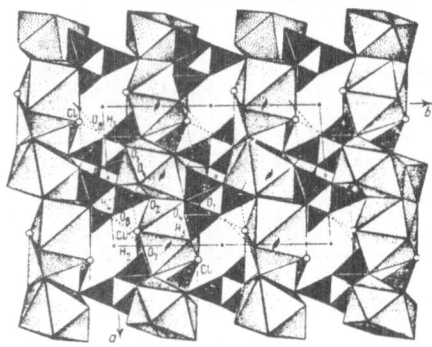

Fig. 1.   Structure of solongoite.

<u>1</u>. Structure Reports, <u>41</u>A, 409.

ALUMINUM ORTHOBORATE
$AlBO_3$

A. VEGAS, F.H. CANO and S. GARCÍA-BLANCO, 1977. Acta Cryst., B<u>33</u>, 3607-3609.

Rhombohedral, R$\bar{3}$c, a = 4.4638, c = 13.745 Å, Z = 6. Mo radiation, R = 0.015
for 278 reflexions. Al in 6(b); B in 6(a); O in 18(e): x = 0.6909.

     Calcite structure (<u>1</u>). B-O = 1.380, Al-O = 1.923 Å.

<u>1</u>. Strukturbericht, <u>1</u>, 292; Structure Reports, <u>21</u>, 354; <u>30</u>A, 408; <u>42</u>A, 422.

HARKERITE
$Ca_{24}Mg_8[AlSi_4(O,OH)_{16}]_2(BO_3)_8(CO_3)_8(H_2O,Cl)$

G. GIUSEPPETTI, F. MAZZI and C. TADINI, 1977. Amer. Min., <u>62</u>, 263-272.

Rhombohedral, R$\bar{3}$m, a = 18.131 Å, α = 33.46°, Z = 1. Mo radiation, R = 0.07 for
1175 reflexions.

     Apparent cubic symmetry previously reported (<u>1</u>) results from twinning. The
structure can be compared with that of sakhaite (<u>2</u>), both showing marked Fd3m
pseudo-symmetry, a = 14.7 Å. The structure (Fig. 1) contains incomplete cubic
close packing of Ca and O, with Mg in octahedral sites, Si and Al in tetrahedral
sites, and B or C in triangular holes, but with some structural disorder.

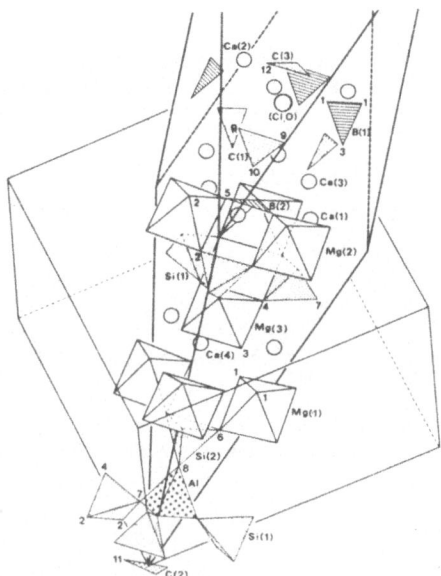

Fig. 1.   Structure of harkerite.

1. Structure Reports, 42A, 310.
2. Ibid., 41A, 409.

POTASSIUM BORONIOBATE
$K_3Nb_3B_2O_{12}$

J. CHOISNET, D. GROULT, B. RAVEAU and M. GASPERIN, 1977. Acta Cryst., B33, 1841-1845.

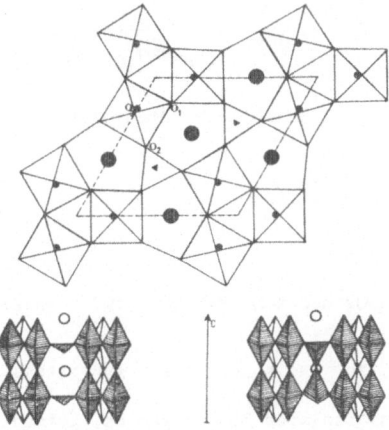

Fig. 1.   Structure of the sub-cell of $K_3Nb_3B_2O_{12}$, and comparison with $K_6Nb_6Si_4O_{26}$.

Trigonal, P31m, a = 34.01, c = 3.97 Å, Z = 16. Mo radiation, R = 0.08 for 1500 reflexions.

There is a sub-cell with a' = a/4 and structure very similar to that of $K_6Nb_6Si_4O_{26}$ (1), with BO_3 triangles replacing a Si_2O_7 group (Fig. 1). The 16 sub-cells are nearly identical [coordinates are given for all the atoms in the true cell, but it is not clear how these were derived].

1. Structure Reports, 35A, 269; 42A, 396.

NICKEL HEXABORATE OCTAHYDRATE
Ni[$B_6O_7(OH)_6$].$8H_2O$

E.Ja. SILIN' and A.F. IEVIN'Š, 1977. Kristallografija, 22, 505-509 [Soviet Physics - Crystallography, 22, 288-291].

Triclinic, PĪ, a = 9.09, b = 8.64, c = 9.99 Å, α = 106°50', β = 106°40', γ = 81°40', D_m = 1.96, Z = 2. Ni radiation, R = 0.123 for 1325 reflexions (films, visual intensities).

The structure (Fig. 1) contains neutral Ni(H_2O)$_4$[$B_6O_7(OH)_6$] complexes with an intra-complex hydrogen bond, and an additional water molecule. The borate anion consists of three tetrahedra and three triangles, and Ni has octahedral coordination. B-O = 1.39-1.54 (tetrahedra), 1.31-1.42 (triangles), Ni-O = 2.01-2.11 Å.

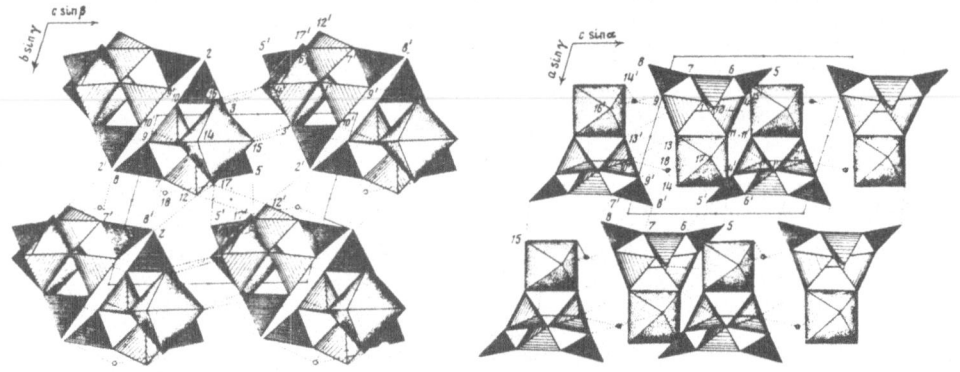

Fig. 1.   Structure of nickel hexaborate octahydrate.

LANTHANUM BORATE
$LaBO_3$

G.K. ABDULLAEV, G.G. DŽAFAROV and K.S. MAMEDOV, 1976. Azerbajdž. Khim. Ž., 117-120.

Orthorhombic, Pmcn, a = 5.130, b = 8.300, c = 5.880 Å, Z = 4. R = 0.11.

The structure contains isolated BO_3 triangles, joined by LaO_9 polyhedra which share corners and edges to form columns.

NEODYMIUM COBALT METABORATE
$NdCo(BO_2)_5$

G.K. ABDULLAEV, 1976. Ž. Strukt. Khim., 17, 1128-1131 [J. Struct. Chem., 17, 961-963].

Monoclinic, $P2_1/n$, a = 8.65, b = 7.61, c = 9.48 Å, β = 92.5°, Z = 4. Cu radiation, R = 0.128 for 600 reflexions (film data).

Isostructural with the Sm and La compounds (1). The structure contains infinite $B_5O_{10}^{5-}$ polyanions, with three BO_4 tetrahedra and two BO_3 triangles. Nd has 10-coordination and Co distorted octahedral coordination.

1. Structure Reports, 40A, 226; 41A, 290.

LITHIUM NEODYMIUM BORATE
$Li_3Nd_2(BO_3)_3$

G.K. ABDULLAEV and K.S. MAMEDOV, 1977. Kristallografija, 22, 271-274 [Soviet Physics - Crystallography, 22, 154-156].

Monoclinic, $P2_1/n$, a = 8.814, b = 14.143, c = 5.776 Å, β = 103.75°, Z = 4. Mo radiation, R = 0.08 for 1920 reflexions.

The structure (Fig. 1) contains BO_3 triangles linked by Li ions (tetrahedral coordination) and Nd ions (9-coordination). B-O = 1.34-1.41, Li-O = 1.84-2.19, Nd-O = 2.39-2.82 Å.

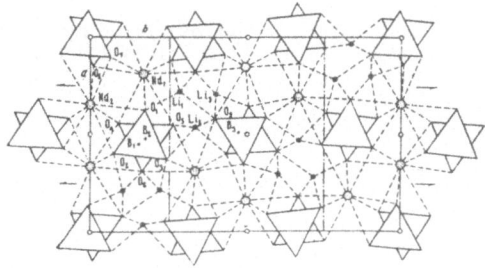

Fig. 1.   Structure of $Li_3Nd_2(BO_3)_3$.

LITHIUM PRASEODYMIUM BORATE
$Li_3Pr_2(BO_3)_3$

G.K. ABDULLAEV, K.S. MAMEDOV, I.R. AMIRASLANOV and A.I. MAGERRAMOV, 1977. Ž. Strukt. Khim., 18, 410-413 [J. Struct. Chem., 18, 331-333].

Monoclinic, $P2_1/n$, a = 8.816, b = 14.127, c = 5.812 Å, β = 103.72°, Z = 4. Cu radiation, R = 0.092 for 882 reflexions.

The structure (Fig. 1) contains triangular BO_3^{3-} ions, Li^+ ions with tetrahedral coordination, and Pr^{3+} ions with 9-coordination. B-O = 1.35-1.41, Li-O = 1.87-2.18, Pr-O = 2.41-2.86 Å. [The Nd compound (1) is isostructural.]

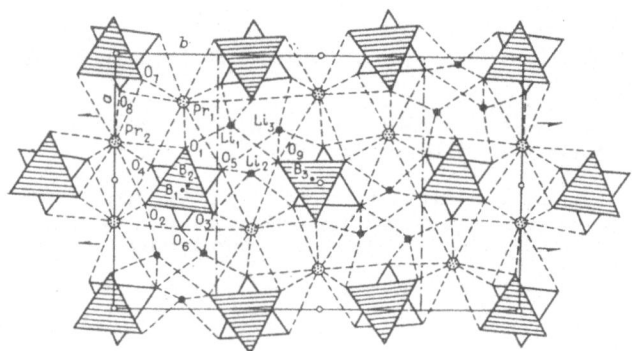

Fig. 1.   Structure of $Li_3Pr_2(BO_3)_3$.

<u>1</u>. Preceding report.

LITHIUM YTTERBIUM BORATE
$Li_6Yb(BO_3)_3$

G.K. ABDULLAEV and K.S. MAMEDOV, 1977. Kristallografija, <u>22</u>, 389-392 [Soviet
Physics - Crystallography, <u>22</u>, 220-222].

Monoclinic, $P2_1/b$, a = 7.115, b = 6.578, c = 16.324 Å, γ = 105.06°, Z = 4.   Mo
radiation, R = 0.068 for 2187 reflexions.

       The structure (Fig. 1) contains BO_3 triangles linked by Li ions (4- and 5-
coordination) and Yb ions (8-coordination).   B-O = 1.34-1.42, Li-O = 1.83-2.42,
Yb-O = 2.32-2.51 Å.

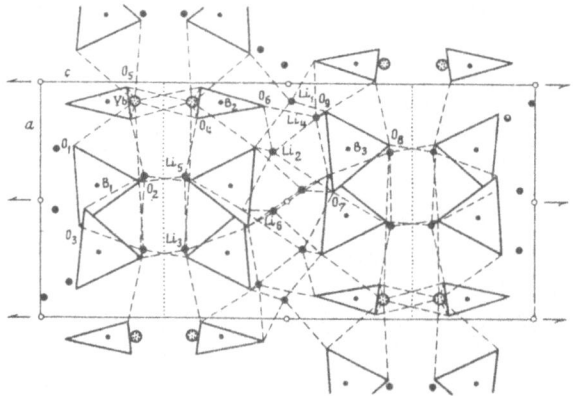

Fig. 1.   Structure of $Li_6Yb(BO_3)_3$.

SODIUM PERCARBONATE
$Na_2CO_3 \cdot 1 \cdot 5H_2O_2$

J.M. ADAMS and R.G. PRITCHARD, 1977. Acta Cryst., B<u>33</u>, 3650-3653.

Orthorhombic, Aba2, a = 9.224, b = 15.805, c = 6.747 Å, D_m = 2.05, Z = 8.   R = 0.114 (films, densitometer intensities).

The H_2O_2 molecules (one disordered) and CO_3^{2-} ions lie in planes about 3.5 Å apart, with Na^+ ions midway between these sheets (Fig. 1).   C-O = 1.28-1.32(2), O-O = 1.49, 1.51(7), Na-O = 2.25-2.75(5) Å.

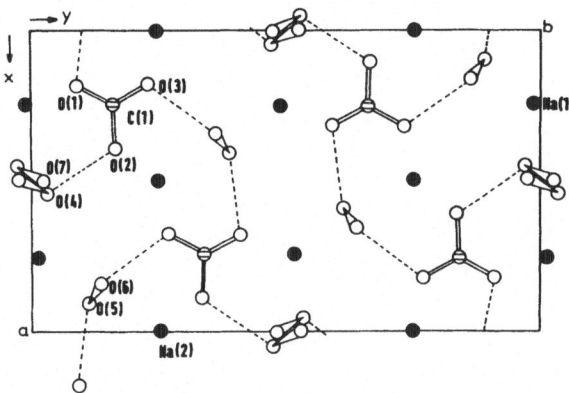

Fig. 1.   Structure of sodium percarbonate; broken lines are hydrogen bonds.

ARTINITE
$Mg_2CO_3(OH)_2 \cdot 3H_2O$

M. AKAO and S. IWAI, 1977. Acta Cryst., B33, 3951-3953.

Monoclinic, C2/m, a = 16.560, b = 3.153, c = 6.231 Å, β = 99.10°, D_m = 2.03, Z = 2. Mo radiation, R = 0.049 for 640 reflexions.

Atomic positions

	x	y	z
Mg	0.3159	0	0.1520
O(H)	0.2388	1/2	0.1708
O(W)	0.3999	1/2	0.0899
O(1)	0.3670	0	0.4685
* O(2)	0.4516	0.3582	0.7036
* C	0.4233	0	0.6203
H(1)	0.222	1/2	0.275
H(2)	0.443	1/2	0.174
H(3)	0.408	1/2	-0.039
* H(4)	0.395	0.195	0.528

* Occupancy = 0.5

The y-coordinates of previous studies (1) are incorrect. MgO_6 octahedra share edges to form chains along b, which are cross-linked by hydrogen bonds (Fig. 1). The carbonate groups and O(1) water molecules alternate statistically along b (oscillation photographs exhibit weak diffuse layer-lines corresponding to a doubled b-axis)

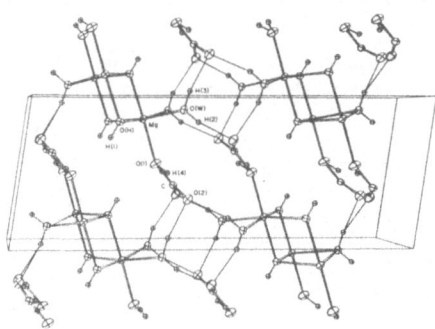

Fig. 1.   Structure of artinite.

<u>1</u>. Structure Reports, <u>16</u>, 328; <u>30</u>A, 408.

HYDROMAGNESITE
$Mg_5(CO_3)_4(OH)_2 \cdot 4H_2O$

M. AKAO and S. IWAI, 1977. Acta Cryst., B<u>33</u>, 1273-1275.

Monoclinic, $P2_1/c$, a = 10.105, b = 8.954, c = 8.378 Å, β = 114.44°, D_m = 2.25, Z = 2. Mo radiation, R = 0.052 for 2598 reflexions.

     The structure is as previously described (<u>1</u>, <u>2</u>), and hydrogen atoms have now been located (Fig. 1).

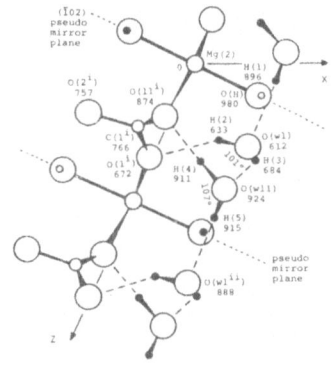

Fig. 1.   Hydrogen bonding in hydromagnesite.

<u>1</u>. Structure Reports, <u>40</u>A, 227.
<u>2</u>. G.W. STEPHAN, 1974. Doctoral Thesis, Amsterdam.

NYEREREITE
$(Na,K)_2Ca(CO_3)_2$

D. McKIE and E.J. FRANKIS, 1977. Z. Kristallogr., <u>145</u>, 73-95.

Low nyerereite
Orthorhombic, $Cmc2_1$, a = 5.044, b = 8.809, c = 12.743 Å, D_m = 2.54, Z = 4. Cu
radiation, 858 reflexions, structure not fully determined.

High nyerereite
Hexagonal, $P6_3mc$, a = 5.05, c = 12.85 Å, Z = 2. No intensity data.

A partial average structure for low nyerereite is given, but there are
additional non-Bragg reflexions whose significance is to be discussed in a subsequent
publication. From this average structure, a structure is proposed for high
nyerereite.

DOLOMITE   ANKERITE   MAGNESIAN CALCITE
$CaMg(CO_3)_2$ $Ca(Mg,Fe)(CO_3)_2$ $(Ca,Mg)CO_3$

I.   P.L. ALTHOFF, 1977. Amer. Min., 62, 772-783.
II. A. BERAN and J. ZEMANN, 1977. Tschermaks Min. Petr. Mitt., 24, 279-286.

Dolomite (I)
Rhombohedral, $R\bar{3}$, a = 4.8033, c = 15.984 Å, Z = 3. Mo radiation, R = 0.045 for
178 reflexions. Ca in 3(a); Mg in 3(b); C in 6(c): z = -0.2423; O in 18(f):
(0.2829, 0.0350, -0.2440). Ca-O = 2.378(1), Mg-O = 2.081(1), C-O = 1.284(2) Å.

Dolomite, ankerite (II)
Rhombohedral, $R\bar{3}$, a = 4.812, 4.830, c = 16.020, 16.167 Å, Z = 3. Mo radiation,
R = 0.029 and 0.030 for 394 and 401 reflexions. Ca in 3(a); Mg/Fe in 3(b); C in
6(c): z = 0.2429, 0.2442; O in 18(f): (0.2485, -0.0343, 0.2439), (0.2506, -0.0283,
0.2449). C-O = 1.286, 1.284(1), Ca-O = 2.381, 2.371, M-O = 2.087, 2.126(1) Å.

Magnesian calcite, $Ca_{0.9}Mg_{0.1}CO_3$ (I)
Rhombohedral, $R\bar{3}c$, a = 4.941, c = 16.854 Å, Z = 6. Mo radiation, R = 0.025 for
84 reflexions. Ca,Mg in 6(b); C in 6(a); O in 18(e): x = 0.2583. (Ca,Mg)-O =
2.331(1), C-O = 1.276(3) Å.

Dolomite and calcite structures as previously described (1, 2).

1. Strukturbericht, 1, 303, 317, 324; Structure Reports, 23, 419.
2. Strukturbericht, 1, 292, 317, 318; Structure Reports, 21, 354; 30A, 408.

DAWSONITE
$NaAl(OH)_2CO_3$

E. CORAZZA, C. SABELLI and S. VANNUCCI, 1977. Neues Jb. Miner., Mh., 381-397.

Orthorhombic, Imma, a = 6.759, b = 5.585, c = 10.425 Å, D_m = 2.436, Z = 4. Mo
radiation, R = 0.058 for 406 reflexions.

Atomic positions

	x	y	z
Al	0	0	1/2
Na	1/4	3/4	1/4
C	0	1/4	0.2526
O(1)	0	1/4	0.1326
O(2)	0	0.0474	0.3156
O(3)	0.1811	1/4	0.5250
H	0.275	1/4	0.478

The structure is as previously described (1). It contains edge-shared $AlO_2(OH)_4$ octahedra along b and edge-shared $NaO_4(OH)_2$ octahedra along a, linked by the CO_3 group and by hydrogen bonds.

1. Structure Reports, 32A, 419.

LANTHANITE
$(La,Ce)_2(CO_3)_3 \cdot 8H_2O$

A. DAL NEGRO, G. ROSSI and V. TAZZOLI, 1977. Amer. Min., 62, 142-146.

Orthorhombic, Pbnb, a = 9.504, b = 16.943, c = 8.937 Å, Z = 4. Mo radiation, R = 0.025 for 1022 reflexions.

The structure (Fig. 1) contains ac layers of LnO_{10} polyhedra and CO_3^{2-} groups; the layers are connected to each other only by hydrogen bonds. Ln-O = 2.50-2.76, C-O = 1.23-1.31, O-H...O = 2.64-2.84 Å. [The structure has been described previously for synthetic material (1).]

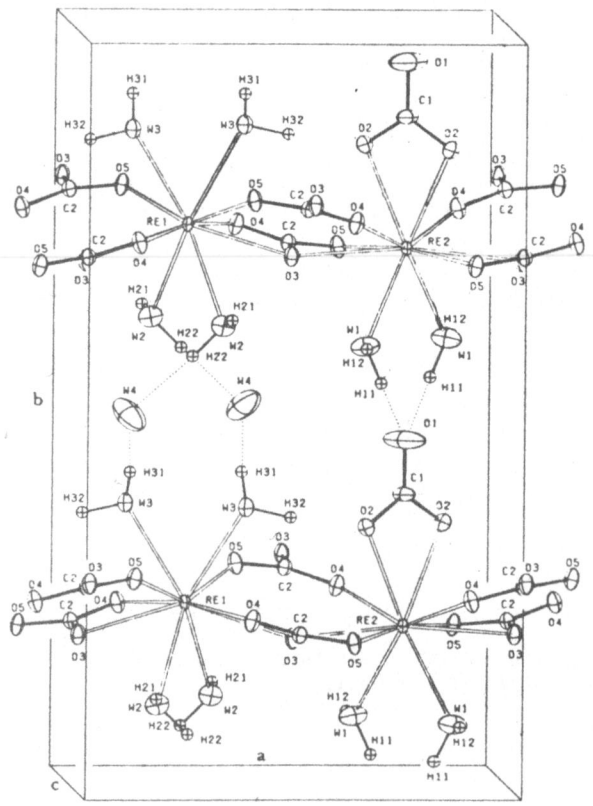

Fig. 1.   Structure of lanthanite.

1. Structure Reports, 33A, 433.

POTASSIUM PENTANITROCUPRATE(II)
$K_3Cu(NO_2)_5$

K.A. KLANDERMAN, W.C. HAMILTON and I. BERNAL, 1977. Inorg. Chim. Acta, <u>23</u>, 117-129.

Orthorhombic, Pbnm, a = 21.70, b = 18.94, c = 10.88 Å, D_m = 2.46, Z = 16. Mo radiation, R = 0.056 for 1225 reflexions.

There are four independent anions in the unit cell, which are nearly identical in pairs (Fig. 1); all modes of coordination of NO_2^- are exhibited. The eight independent K^+ ions have irregular coordination.

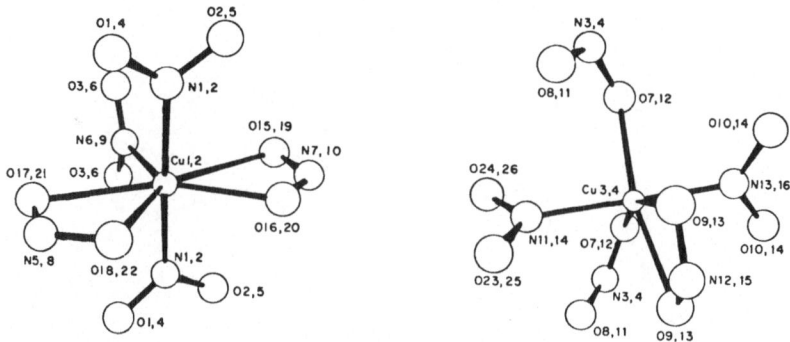

Fig. 1.   The $Cu(NO_2)_5^{3-}$ anions.

POTASSIUM LEAD(II) HEXANITROCUPRATE(II)   (Orthorhombic)
$K_2PbCu(NO_2)_6$

M.D. JOESTEN, S. TAKAGI and P.G. LENHERT, 1977. Inorg. Chem., <u>16</u>, 2680-2685.

Orthorhombic, Fmmm, a = 10.741, b = 10.734, c = 10.538 Å (at 276°K), Z = 4. Mo radiation, R = 0.062 for 1291 reflexions (276°K), 0.081 for 1240 reflexions (193°K).

The structure is similar to that of other orthorhombic hexanitrocuprates (<u>1</u>). Cu coordination is observed to be tetragonally-compressed octahedral, Cu-O = 2.16 (4 bonds), 2.06 Å (2 bonds), but examination of the thermal parameters suggests that this results from a dynamic average of two tetragonally-elongated octahedra.

<u>1</u>. Structure Reports, <u>40</u>A, 230; <u>41</u>A, 299; <u>42</u>A, 319.

LITHIUM NITRATE TRIHYDRATE
$LiNO_3 \cdot 3H_2O$

K. HERMANSSON, J.O. THOMAS and I. OLOVSSON, 1977. Acta Cryst., B<u>33</u>, 2857-2861.

Orthorhombic, Cmcm, a = 6.802, b = 12.713, c = 5.999 Å, D_m = 1.55, Z = 4. Mo radiation, R = 0.040 for 367 reflexions.

The structure (Fig. 1) contains planes of nitrate and lithium ions and one water molecule, adjacent planes being linked through the second water molecule. Li has octahedral coordination (Fig. 1) and the two water molecules have different environments (Fig. 2).

Atomic positions

	x	y	z
Li	0	0	0
N	0	0.2175	1/4
O(1)	0	0.1703	0.0700
O(2)	0	0.3166	1/4
H₂O(3)	0.2915	0.4783	1/4
H(1)	0.233	0.427	1/4
H(2)	0.220	0.524	1/4
H₂O(4)	0	0.6376	1/4
H(3)	0	0.673	0.144

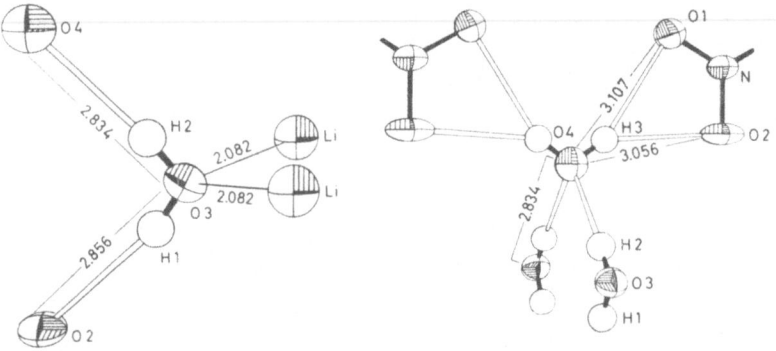

Fig. 1.   Structure of and lithium coordination in lithium nitrate trihydrate.

Fig. 2.   Water invironments in lithium nitrate trihydrate.

BERYLLIUM NITRATE TETRAHYDRATE
Be(NO₃)₂·4H₂O

V. DIVJAKOVIĆ, A. EDENHARTER, W. NOWACKI and B. RIBÁR, 1976. Z. Kristallogr., 144, 314-322.

Tetragonal, I4̄2d, a = 11.636, c = 6.252 Å, Dₘ = 1.59, Z = 4.   Cu radiation, R = 0.062 for 258 reflexions.

Atomic positions

			x	y	z
Be	in	4(b)	0	0	1/2
N		8(d)	0.1946	1/4	1/8
O(1)		8(d)	0.3034	1/4	1/8
O(2)		16(e)	0.1440	0.1574	0.1012
O(3)		16(e)	0.3907	0.0349	0.1045
H(1)		16(e)	0.37	0.11	0.04
H(2)		16(e)	0.04	0.10	0.30

The structure (Fig. 1) contains tetrahedral $Be(H_2O)_4{}^{2+}$ cations, hydrogen bonded to $NO_3{}^-$ anions. Be-O = 1.616(5), N-O = 1.24, 1.27(1), O-H...O = 2.66, 2.70 Å, O-Be-O = 108.5, 111.5, O-N-O = 118.4, 123.2, H-O-H = 134, O-H...O = 139, 147°.

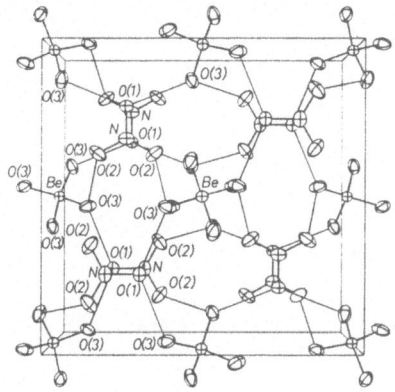

Fig. 1.   Structure of $Be(NO_3)_2.4H_2O$.

CALCIUM NITRATE TETRAHYDRATE
α-$Ca(NO_3)_2.4H_2O$

A. LECLAIRE and J.-C. MONIER, 1977. Acta Cryst., B33, 1861-1866.

Monoclinic, $P2_1/c$, a = 6.279, b = 9.155, c = 14.900 Å, β = 106.22°, Z = 4. Mo radiation, R = 0.028 for 3232 reflexions.

The structure (Fig. 1) is as previously described (1, 2). One hydrogen atom of water molecule W(2) is disordered.

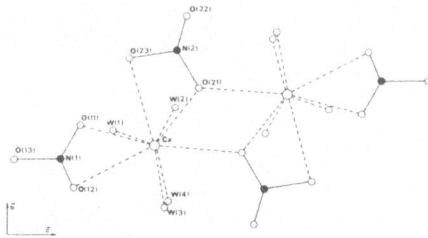

Fig. 1.   Structure of α-$Ca(NO_3)_2.4H_2O$.

1. A. LECLAIRE and J.-C. MONIER, 1970. C.R. Acad. Sci. Paris, C, 271, 1555; Structure Reports, 40A, 317.

<u>2</u>. Structure Reports, <u>39</u>A, 275; <u>41</u>A, 438.

MANGANESE(II) NITRATE MONOHYDRATE
$Mn(NO_3)_2.H_2O$

N. MILINSKI, B. RIBÁR, Ž. ĆULUM and S. DJURIĆ, 1977. Acta Cryst., B<u>33</u>, 1678-1682.

Orthorhombic, $C222_1$, a = 6.115, b = 13.244, c = 13.087 Å, Z = 8. Mo radiation, R = 0.054 for 727 reflexions.

The structure contains 8- and 6-coordinated Mn ions (Fig. 1) linked into layers which are joined by hydrogen bonds. Mean Mn-O = 2.34 (8-coordinate), 2.20 Å (6-coordinate).

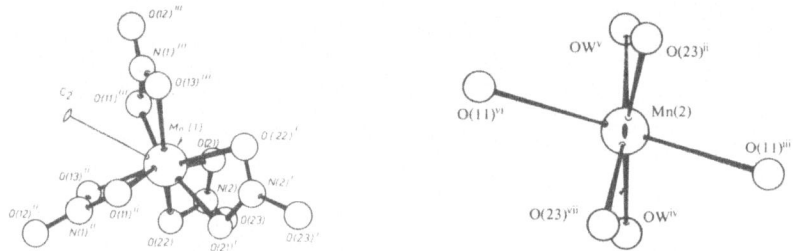

Fig. 1.   Mn coordinations in manganese(II) nitrate monohydrate

MANGANESE(II) NITRATE HEXAHYDRATE
$Mn(NO_3)_2.6H_2O$

B. RIBÁR, D. PETROVIĆ, S. DJURIĆ and I. KRSTANOVIĆ, 1976.   Z. Kristallogr., <u>144</u>, 334-340.

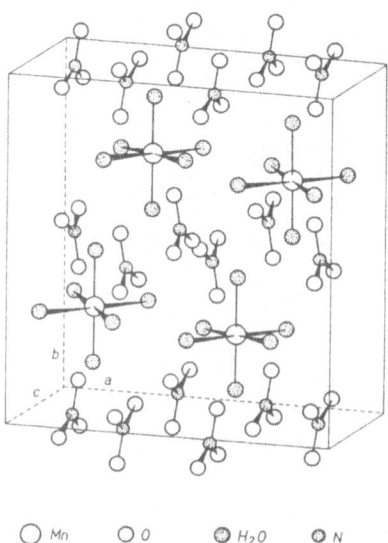

○ Mn   ○ O   ◉ H₂O   ◉ N

Fig. 1.   Structure of $Mn(NO_3)_2.6H_2O$.

Orthorhombic, Pnma, a = 12.484, b = 12.983, c = 6.336 Å, D_m = 1.81, Z = 4. Mo radiation, R = 0.058 for 844 reflexions.

Isostructural with the Zn salt (1). The structure (Fig. 1) contains octahedral $Mn(H_2O)_6^{2+}$ cations, hydrogen bonded to NO_3^- anions. Mn-O = 2.16-2.21(1), N-O = 1.24-1.25, O-H ...O = 2.78-2.99 Å.

1. Structure Reports, 32A, 352.

IRON(III) NITRATE NONAHYDRATE   (HEXAAQUOIRON(III) NITRATE TRIHYDRATE)
$Fe(NO_3)_3 \cdot 9H_2O$ $[Fe(H_2O)_6](NO_3)_3 \cdot 3H_2O$

N.J. HAIR and J.K. BEATTIE, 1977. Inorg. Chem., 16, 245-250.

Monoclinic, $P2_1/c$, a = 13.989, b = 9.701, c = 11.029 Å, β = 95.52°, D_m = 1.81, Z = 4. Mo radiation, R = 0.042 for 2209 reflexions.

The structure (Fig. 1) contains two independent $Fe(H_2O)_6^{3+}$ octahedra, each lying on a centre of symmetry, joined by a complex hydrogen-bonded network involving the nitrate anions and lattice water molecules. Mean Fe-O = 1.986(7) Å.

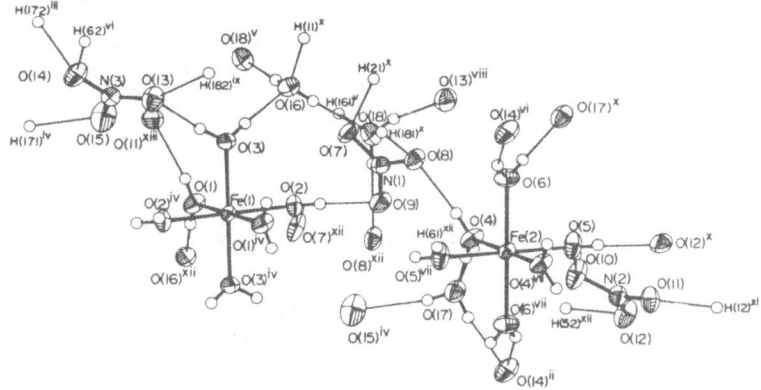

Fig. 1.   Structure of $[Fe(H_2O)_6](NO_3)_3 \cdot 3H_2O$.

cis-DINITRATODIAMMINEPLATINUM(II)
$Pt(NH_3)_2(NO_3)_2$

B. LIPPERT, C.J.L. LOCK, B. ROSENBERG and M. ZVAGULIS, 1977. Inorg. Chem., 16, 1525-1529.

Orthorhombic, Pbca, a = 9.760, b = 10.087, c = 13.495 Å, D_m = 3.7, Z = 8. Mo radiation, R = 0.054 for 1319 reflexions.

The molecule has cis-square-planar coordination at Pt, with each nitrate group coordinated through one oxygen atom; both nitrate groups are below, with their planes approximately perpendicular to the coordination plane. Pt-O = 2.01, Pt-N = 2.00(1) Å, Pt-O-N = 120°. The molecules are linked by N-H...O hydrogen bonds.

DI-μ-HYDROXO-BIS[DIAMMINEPLATINUM(II)] NITRATE
$[(NH_3)_2Pt(OH)_2Pt(NH_3)_2](NO_3)_2$

R. FAGGIANI, B. LIPPERT, C.J.L. LOCK and B. ROSENBERG, 1977. J. Amer. Chem. Soc.,
99, 777-781.

Triclinic, PĪ, a = 6.763, b = 7.890, c = 7.256 Å, α = 92.3, β = 133.1, γ = 91.0°,
D_m = 3.9, Z = 1. Mo radiation, R = 0.051 for 1229 reflexions.

The cation is a centrosymmetric, approximately planar hydroxo-bridged dimer,
each Pt having square-planar coordination, Pt-O = 2.03, Pt-N = 2.02(1) Å. The
cations and the nitrate anions are linked by hydrogen bonds.

CYCLO-TRI-μ-HYDROXO-TRIS[cis-DIAMMINEPLATINUM(II)] NITRATE
$[Pt_3(NH_3)_6(OH)_3](NO_3)_3$

R. FAGGIANI, B. LIPPERT, C.J.L. LOCK and B. ROSENBERG, 1977. Inorg. Chem., 16,
1192-1196.

Triclinic, PĪ, a = 9.683, b = 12.513, c = 8.832 Å, α = 120.90, β = 98.67, γ =
98.32°, D_m = 3.8, Z = 2. Mo radiation, R = 0.061 for 2486 reflexions.

The cation (Fig. 1) is a hydroxo-bridged trimer, with Pt atoms having square
planar coordination; O(3) is in the plane of the three Pt atoms with O(1) and O(2)
about 1 Å above and below the plane. Pt-O = 2.01, Pt-N = 2.02(3) Å. The structure
contains layers of cations hydrogen bonded to the nitrate anions.

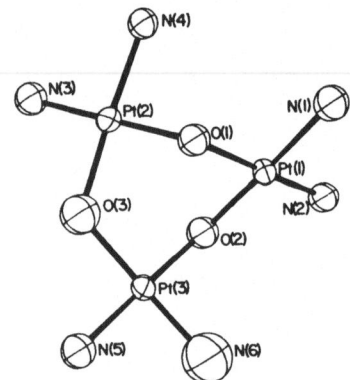

Fig. 1.   The $Pt_3(NH_3)_6(OH)_3^{3+}$ cation.

LIKASITE
$Cu_3P_2H_3(NO_3)(OH)_2 \cdot H_2O$

J.-P. DECLERCQ, G. GERMAIN and P. PIRET, 1977. Acta Cryst., B33, 1422-1427.

Orthorhombic, Pc2_1n, a = 5.828, b = 6.769, c = 21.690 Å, D_m = 2.885, Z = 4. Cu
radiation, R = 0.127 for 488 reflexions.

The structure contains infinite $Cu_3P_2H_3(OH)_2$ layers, linked by NO_3^- ions
and H_2O molecules (Fig. 1). Cu atoms have 4- and 5-coordination, Cu-P = 1.94-
2.01, Cu-O = 1.85-2.87 Å, and P atoms have tetrahedral coordination.

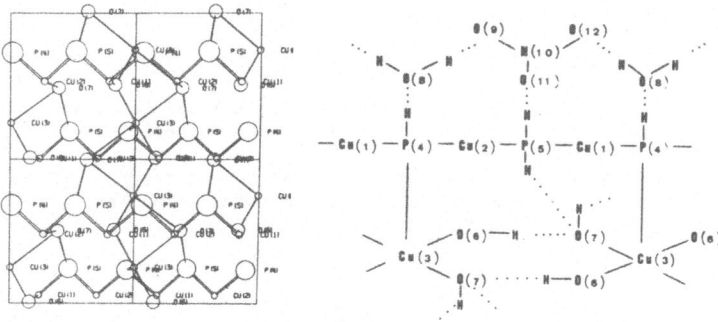

Fig. 1.   Structure of and hydrogen bonding scheme in likasite.

SILVER BROMIDE NITRATE
Ag_2BrNO_3

K. PERSSON and B. HOLMBERG, 1977. Acta Cryst., B<u>33</u>, 3768-3772.

Orthorhombic, Pnma, a = 6.846, b = 5.132, c = 12.823 Å, D_m = 5.2, Z = 4. Mo radiation, R = 0.056 for 965 reflexions.

Atomic positions

	x	y	z
Ag(1)	0.2504	1/4	0.0335
Ag(2)	0.1451	3/4	0.2564
Br	0.3829	3/4	0.0896
N	0.3527	1/4	0.3508
O(1)	0.2127	1/4	0.2914
O(2)	0.4258	0.4596	0.3807

The structure (Fig. 1) contains Ag_5Br trigonal bipyramids which are linked by edges and corners, creating cavities in which the NO_3 groups are located. Ag-Br = 2.665-2.959(1), Ag-O = 2.646-3.317(4), N-O = 1.224-1.247(5) Å.

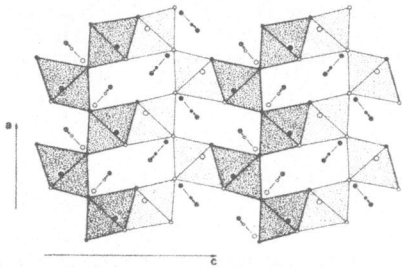

Fig. 1.   Structure of Ag_2BrNO_3.

LANTHANUM MAGNESIUM NITRATE HYDRATE
$La_2Mg_3(NO_3)_{12}.24H_2O$

M.R. ANDERSON, G.T. JENKIN and J.W. WHITE, 1977. Acta Cryst., B33, 3933-3936.

Rhombohedral, R$\bar{3}$, a = 13.172 Å, α = 49.29° (hexagonal cell has a = 10.989, c = 34.63 Å), Z = 1. Neutron radiation, R = 0.065 for 1662 reflexions.

    Isostructural with the Ce compound (1). The structure (Fig. 1) contains $Mg(H_2O)_6{}^{2+}$ and $La(NO_3)_6{}^{3-}$ ions and uncoordinated water molecules, linked by hydrogen bonds. Mg-O = 2.05-2.06(1), La-O = 2.64-2.70(1), O-H...O = 2.77-3.10 Å.

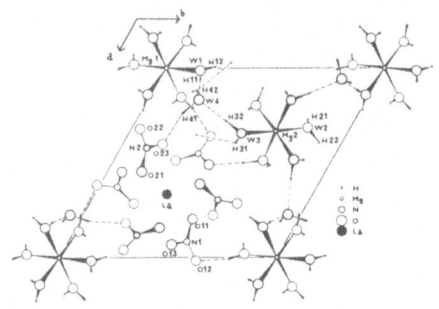

Fig. 1.   Structure of $La_2Mg_3(NO_3)_{12}.24H_2O$.

1. Structure Reports, 27, 571; 28, 181.

ERBIUM HYDROXIDE NITRATE
$Er_4O_2(OH)_8.HNO_3$

H.A. WOLCOTT, W.O. MILLIGAN and G.W. BEALL, 1977. J. Inorg. Nucl. Chem., 39, 59-63.

Monoclinic, P2$_1$, a = 9.338, b = 16.369, c = 3.608 Å, β = 101.16°, Z = 2. Mo radiation, R = 0.030 for 1187 reflexions.

    One Er has 9-coordination to OH groups (mean Er-O = 2.44(4) Å), and the other three Er have 7-coordination to 5 OH (mean Er-O = 2.35 Å) and 2 O (mean Er-O = 2.24 Å). The HNO$_3$ molecule randomly occupies two orientations in a hole in the structure.

BERYLLIUM POLYPHOSPHATE   (Form II)
$Be(PO_3)_2$

M.T. AVERBUCH-POUCHOT, A. DURIF and I. TORDJMAN, 1977. Acta Cryst., B33, 3462-3464.

Monoclinic, P2$_1$/n, a = 6.959, b = 12.853, c = 4.839 Å, β = 106.79°, Z = 4. Mo radiation, R = 0.034 for 1834 reflexions.

The structure (Fig. 1) contains $(PO_3)_\infty$ chains along <u>a</u> with a period of four tetrahedra, linked into a three-dimensional network by BeO_4 tetrahedra. P-O = 1.576-1.598 (bridging), 1.479-1.485 Å (terminal), P-O-P = 129.5, 139.0°, Be-O = 1.604-1.622 Å, Be-O-P = 139.3-149.3°.

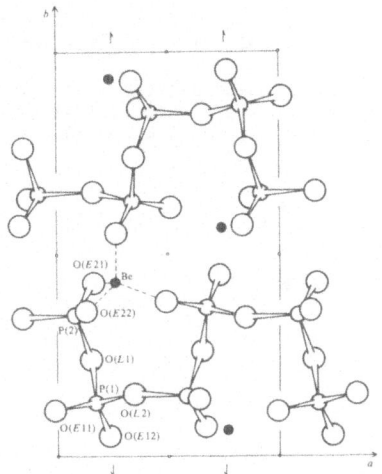

Fig. 1.   Structure of $Be(PO_3)_2$.

AMMONIUM PHOSPHOBERYLLATE
$NH_4P_3Be_2O_{10}$

M.T. AVERBUCH-POUCHOT, A. DURIF, J. COING-BOYAT and J.C. GUITEL, 1977. Acta Cryst., B<u>33</u>, 203-205.

Monoclinic, C2/c, a = 12.202, b = 8.645, c = 8.949 Å, β = 117.41°, Z = 4. Mo radiation, R = 0.02 for 1505 reflexions.

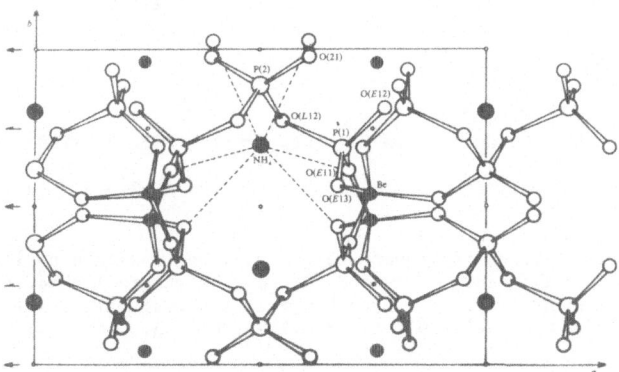

Fig. 1.   Structure of ammonium phosphoberyllate.

The structure (Fig. 1) contains a three-dimensional $(P_3Be_2O_{10})n$ polyanion of BeO_4 and PO_4 tetrahedra; P-O(E) = 1.496-1.511, P-O(L) = 1.584-1.631, Be-O = 1.608-1.662(1) Å. NH_4 has six oxygen neighbours at 2.935-3.099 Å.

LITHIUM LEAD(II) POLYPHOSPHATE
LiPb(PO₃)₃

J.C. GUITEL and M. BRUNEL-LAÜGT, 1977. Acta Cryst., B33, 2713-2716.

Triclinic, P1̄, a = 7.245, b = 7.409, c = 6.795 Å, α = 100.76, β = 97.96, γ = 83.74°, Z = 2. Ag radiation, R = 0.032 for 3546 reflexions.

The structure contains (PO₃)ₙ chains along c with a repeat period of three tetrahedra (Fig. 1); P-O = 1.47-1.50 (terminal), 1.57-1.63 Å (bridging), P-O-P = 125-138°. LiO₄ tetrahedra and PbO₈ polyhedra share edges; Li-O = 1.97-2.15, Pb-O = 2.49-2.76 Å.

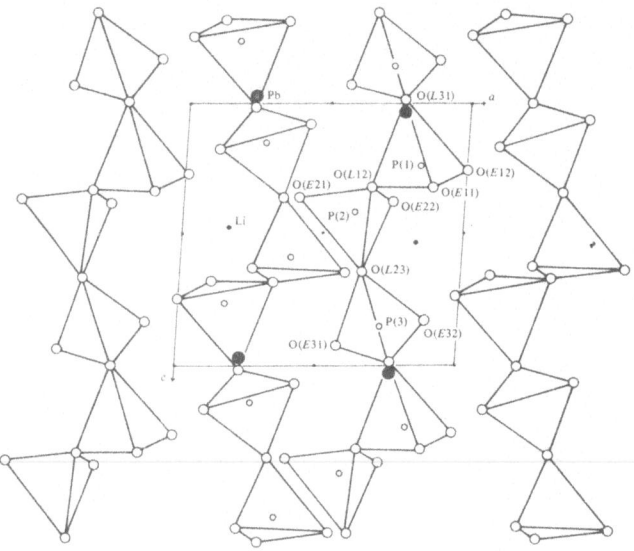

Fig. 1.   Structure of lithium lead(II) tripolyphosphate.

POTASSIUM LEAD(II) POLYPHOSPHATE
K₂Pb(PO₃)₄

M. BRUNEL-LAÜGT and J.-C. GUITEL, 1977. Acta Cryst., B33, 937-939.

Orthorhombic, Pbca, a = 15.510, b = 15.550, c = 9.249 Å, Z = 8. Ag radiation, R = 0.049 for 1420 reflexions.

The structure contains polyphosphate chains along a with a period of eight PO₄ tetrahedra (Fig. 1). PbO₇, KO₈, and KO₁₁ polyhedra are linked in a three-dimensional arrangement (Fig. 1). Mean P-O = 1.61 (bridging), 1.48(1) Å (terminal), P-O-P = 129-133°; Pb-O = 2.45-2.97, K-O = 2.73-3.36(1) Å.

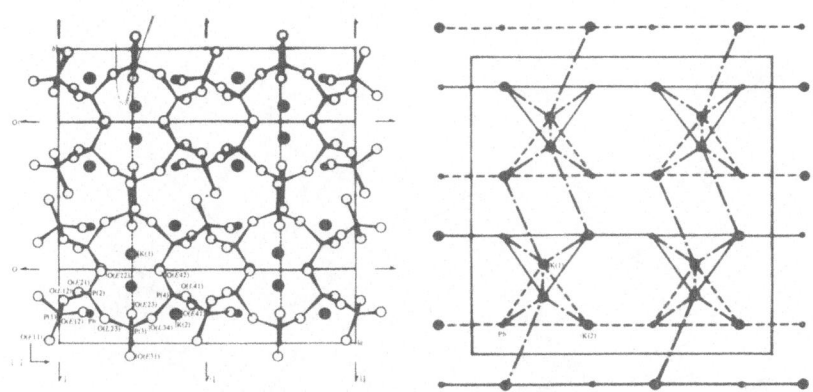

Fig. 1.   Structure and cation linkages in $K_2Pb(PO_3)_4$.

VANADIUM(III) METAPHOSPHATE
$V(PO_3)_3$

N. MIDDLEMISS, F. HAWTHORNE and C. CALVO, 1977.   Canad. J. Chem., 55, 1673-1679.

Monoclinic, Ic, a = 10.615, b = 19.095, c = 9.432 Å, β = 97.94°, D_m = 3.0, Z =
12. Mo radiation, R = 0.091 for 2467 reflexions (R = 0.134 for reflexions with
k ≠ 3n).

    Isostructural with $Al(PO_3)_3$ (1); there is a similar subcell with b' = b/3.

1. Structure Reports, 42A, 332.

DISODIUM HYDROGEN PHOSPHATE DIHYDRATE
$Na_2HPO_4 \cdot 2H_2O$

M. CATTI, G. FERRARIS and M. FRANCHINI-ANGELA, 1977.   Acta Cryst., B33, 3449-3452.

Orthorhombic, Pbca, a = 16.872, b = 10.359, c = 6.599 Å, Z = 8.   Mo radiation, R =
0.036 for 1268 reflexions.

    The structure (Fig. 1) contains PO_3H tetrahedra joined along c by hydrogen
bonding, directly and via one of the water molecules. Na ions have 5+1 and 6+1
coordinations, Na-O = 2.32-2.51, 3.15 and 2.33-2.57, 3.18 Å, respectively.

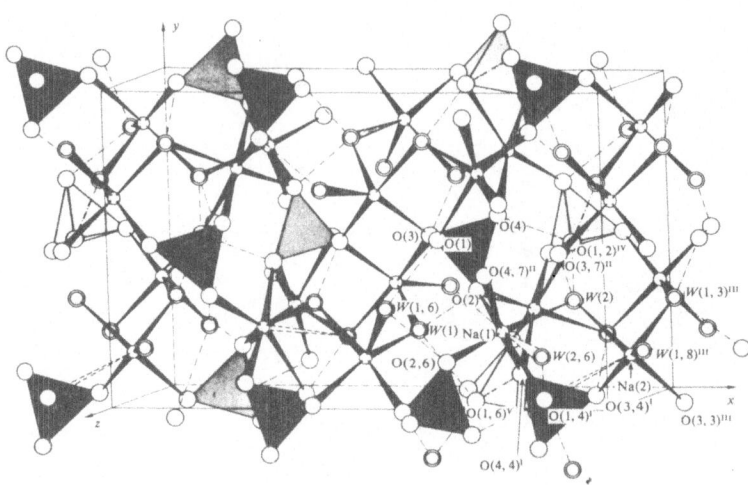

Fig. 1.   Structure of $Na_2HPO_4 \cdot 2H_2O$.

ROSCHERITE   (Triclinic)
$CaMn_2(Fe,Mn,Al)Be_2(PO_4)_3(OH)_2 \cdot 3H_2O$

L. FANFANI, P.F. ZANAZZI and A.R. ZANZARI, 1977. Tschermaks Min. Petr. Mitt., 24, 169-178.

Triclinic, $C\bar{1}$, a = 15.921, b = 11.965, c = 6.741 Å, α = 91°4', β = 94°21', γ = 89°59.5', Z = 4. Mo radiation, R = 0.060 for 2380 reflexions, material from Foote Mine, California.

     The atomic arrangement is similar to that in monoclinic roscherite (1), with the lower symmetry resulting from segregation of trivalent cations into only half of the sites of a monoclinic point position: 0,0,0 and 1/2,1/2,0 are empty, and 0,0,1/2 and 1/2,1/2,1/2 have two-thirds occupancy.

1. Structure Reports, 41A, 314.

CALCIUM PHOSPHATE
$\alpha-Ca_3(PO_4)_2$

M. MATHEW, L.W. SCHROEDER, B. DICKENS and W.E. BROWN, 1977. Acta Cryst., B33, 1325-1333.

Monoclinic, $P2_1/a$, a = 12.887, b = 27.280, c = 15.219 Å, β = 126.20°, D_m = 2.81, Z = 24. Mo radiation, R = 0.051 for 7002 reflexions.

     There is a subcell with b' = b/3. The structure contains columns of cations, and columns of cations and anions with compensating cation vacancies, and is closely related to the structure of glaserite, $K_3Na(SO_4)_2$ (Fig. 1). Ca ions exhibit a wide range of coordination numbers and geometries.

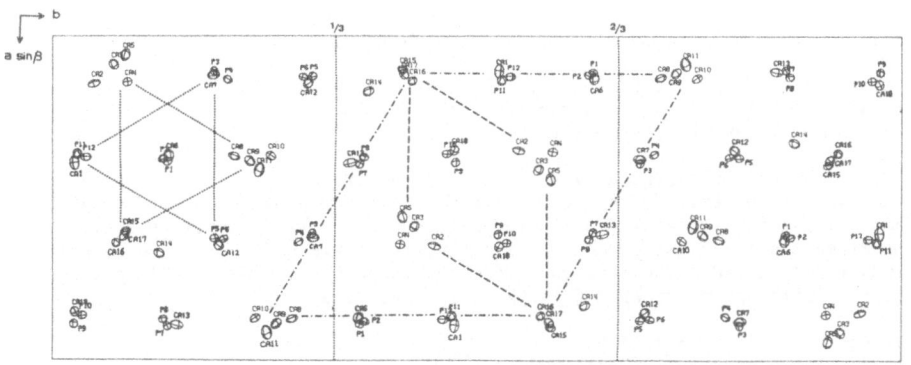

Fig. 1.   Ca and P positions in α-Ca₃(PO₄)₂. Dashed lines outline a glaserite
             unit cell, and dash-dot lines a hydroxyapatite cell.

CALCIUM PHOSPHATE   CALCIUM MAGNESIUM PHOSPHATE
β-Ca$_3$(PO$_4$)$_2$ β-(Ca,Mg)$_3$(PO$_4$)$_2$

L.W. SCHROEDER, B. DICKENS and W.E. BROWN, 1977.   J. Solid State Chem., $\underline{22}$, 253-262.

Ca$_{3-x}$Mg$_x$(PO$_4$)$_2$, rhombohedral, R3c, a = 10.439, 10.401, 10.337, c = 37.375, 37.316,
37.068 Å, for x = 0, 0.11, 0.29, respectively, Z = 21.   Mo radiation, R = 0.02-
0.04 for about 2500 reflexions.

        The structure (Fig. 1) is as previously described $(\underline{1})$.

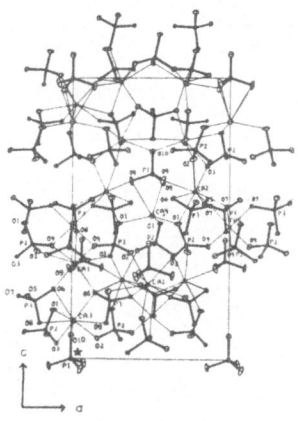

Fig. 1.   Structure of Ca$_{3-x}$Mg$_x$(PO$_4$)$_2$.

$\underline{1}$. Structure Reports, $\underline{40A}$, 240.

MONETITE
CaHPO$_4$

M. CATTI, G. FERRARIS and A. FILHOL, 1977. Acta Cryst., B33, 1223-1229.

Triclinic, P1 or P$\bar{1}$, a = 6.910, b = 6.627, c = 6.998 Å, α = 96.34, β = 103.82, γ = 88.33° (from 1), Z = 4. Neutron radiation, R = 0.0286 and 0.0283 for 1749 reflexions for P1 and P$\bar{1}$, respectively.

The structure is essentially as previously described (1, 2). In the P$\bar{1}$ model a proton lies on a centre of symmetry forming a symmetric hydrogen bond (2.459 Å), and another is statistically distributed between two centrosymmetric positions; in the P1 models the first proton is slightly off centre and the second is either ordered or disordered. The existence of a low-temperature ordered (P1) phase and high-temperature disordered (P$\bar{1}$) phase is suggested.

1. Structure Reports, 38A, 308.
2. Ibid., 19, 432; 26, 458; 37A, 293.

BARIUM DIHYDROGENPHOSPHATE
Ba(H$_2$PO$_4$)$_2$

J.D. GILBERT, P.G. LENHERT and L.K. WILSON, 1977. Acta Cryst., B33, 3533-3535.

Orthorhombic, Pccn, a = 7.796, b = 10.257, c = 8.565 Å, D$_m$ = 3.21, Z = 4. Mo radiation, R = 0.033 for 2846 reflexions.

Atomic positions

	x	y	z
Ba	3/4	1/4	0.4051
P	0.0719	0.0756	0.2103
O(1)	0.2353	0.0711	0.1190
O(2)	0.0322	-0.0721	0.2478
O(3)	-0.0808	0.1412	0.1411
O(4)	0.0941	0.1451	0.3731

The results agree with those in 1. The structure (Fig. 1) contains a corrugated network of hydrogen-bonded phosphate groups parallel to (010), with Ba-O bonds between phosphate layers. P-OH = 1.578, P-O = 1.493(2), Ba-O = 2.659-2.903 Å.

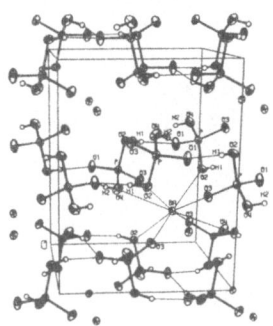

Fig. 1.   Structure of Ba(H$_2$PO$_4$)$_2$ (a horizontal, b vertical).

1. Structure Reports, 42A, 339.

VARISCITE
AlPO$_4$.2H$_2$O

R. KNIEP, D. MOOTZ and A. VEGAS, 1977. Acta Cryst., B$\underline{33}$, 263-265.

Orthorhombic, Pbca, a = 9.822, b = 8.561, c = 9.630 Å, Z = 8. Mo radiation,
R = 0.031 for 960 reflexions.

Isostructural with InPO$_4$.2H$_2$O ($\underline{1}$). AlO$_4$(OH$_2$)$_2$ cis-octahedra and PO$_4$ tetrahedra share corners to form a three-dimensional structure, which shows some
resemblance (Fig. 1) to that of metavariscite ($\underline{2}$). Al-O = 1.856-1.899, Al-OH$_2$ =
1.909, 1.963, P-O = 1.528-1.540(2) Å. Three of the four H atoms participate in
hydrogen bonds, O-H...O = 2.59-2.67 Å.

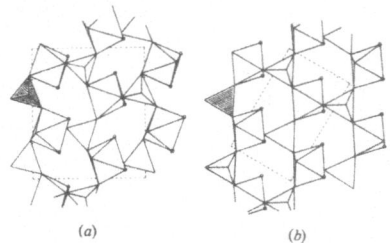

(a)   (b)

Fig. 1.   Crystal structures of (a) variscite along $\underline{c}$, and (b) metavariscite
            along $\underline{b}$.

$\underline{1}$. Structure Reports, $\underline{26}$, 464.
$\underline{2}$. Ibid., $\underline{31A}$, 185; $\underline{39A}$, 285.

MINYULITE
KAl$_2$(PO$_4$)$_2$F.4H$_2$O

A.R. KAMPF, 1977. Amer. Min., $\underline{62}$, 256-262.

Orthorhombic, Pba2, a = 9.337, b = 9.740, c = 5.522 Å, D$_m$ = 2.46, Z = 2. Mo
radiation, R = 0.022 for 618 reflexions.

The structure (Fig. 1) contains [Al$_2$F(H$_2$O)$_4$(PO$_4$)$_2$O$_2$] dimeric clusters linked
to form sheets parallel to (001). Cavities in the sheets are occupied by K$^+$
ions. One tetrahedral vertex is unshared and accepts four hydrogen bonds from
water molecules coordinated to Al^{3+}, which provide linkage between the sheets.

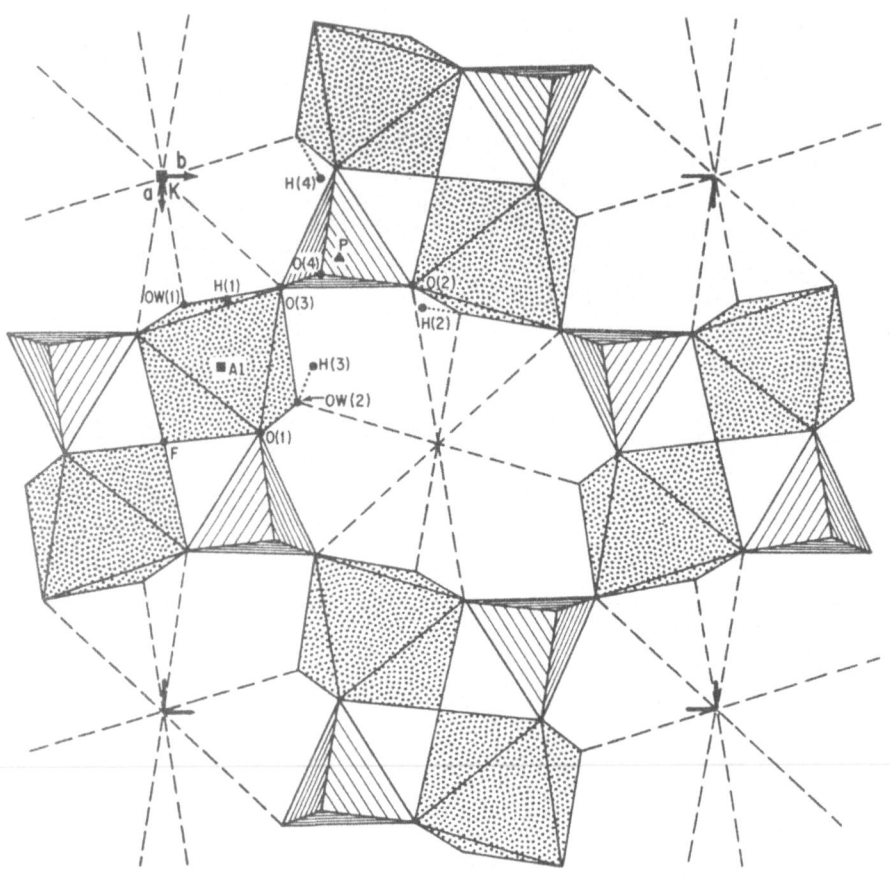

Fig. 1.   Structure of minyulite.

OVERITE   SEGELERITE
MgCaAl(OH)(PO$_4$)$_2$·4H$_2$O   MgCaFe(OH)(PO$_4$)$_2$·4H$_2$O

P.B. MOORE and T. ARAKI, 1977. Amer. Min., 62, 692-702.

Orthorhombic, Pbca, a = 14.723, 14.826, b = 18.746, 18.751, c = 7.107, 7.307 Å,
D$_m$ = 2.53, 2.67, Z = 16. Mo radiation, R = 0.090 and 0.236 for 1915 and 2387
reflexions.

    The structures of overite, segelerite (Fig. 1), and jahnsite (1) are based on
dense slabs of composition [A$_2$B$_2$(OH)$_2$(PO$_4$)$_4$]$^{4-}$ whose maximal symmetry is Bmam.
The slabs are linked by MgO$_2$(H$_2$O)$_4$ octahedra (trans octahedra in overite and
segelerite, and alternately cis and trans in jahnsite). Mean distances in overite
are Mg-O = 2.08, Ca-O = 2.40, Al-O = 1.91, P-O = 1.54 Å. There is a herringbone
pattern of hydrogen bonds parallel to (010).

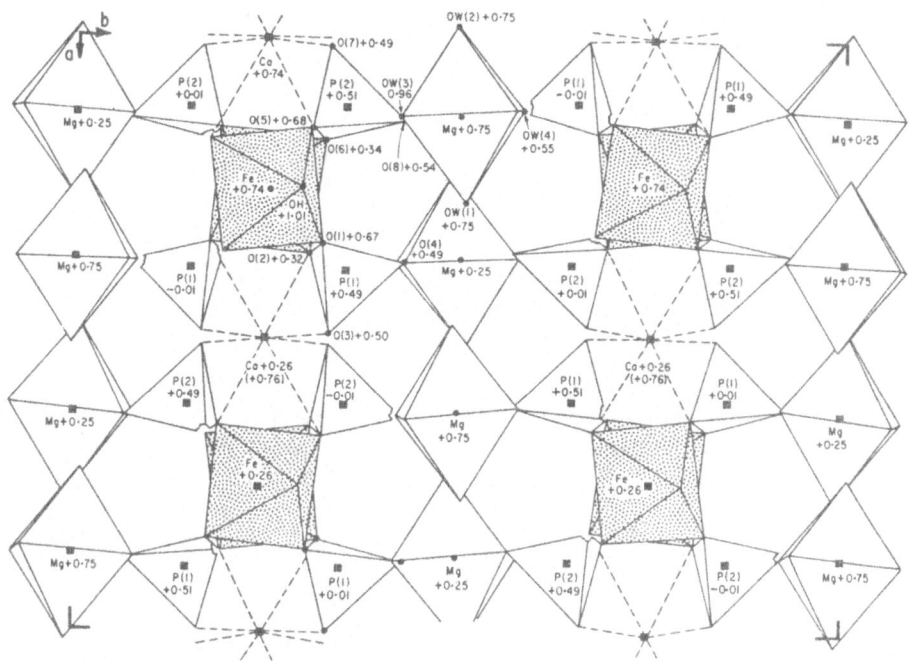

Fig. 1.   Structure of segelerite.

<u>1</u>. Structure Reports, <u>40</u>A, 248.

SAMUELSONITE

$Ba_{0.5}Ca_9(Mn,Fe,Na)_4Al_2(OH)_2(PO_4)_{10}$

P.B. MOORE and T. ARAKI, 1977. Amer. Min., <u>62</u>, 229-245.

Monoclinic, C2/m, a = 18.495, b = 6.805, c = 14.000 Å, β = 112.75°, D_m = 3.353, Z = 2. Mo radiation, R = 0.084 for 2571 reflexions.

The structure (Fig. 1) contains regions which are isomorphous with apatite (<u>1</u>) and octacalcium phosphate (<u>2</u>). The underlying unit is a column of composition $[\overline{C}a_4(PO_4)_{12}]$, linked to form double columns of composition $[Ca_4(PO_4)_{10}]$, which are joined by the other coordination polyhedra.

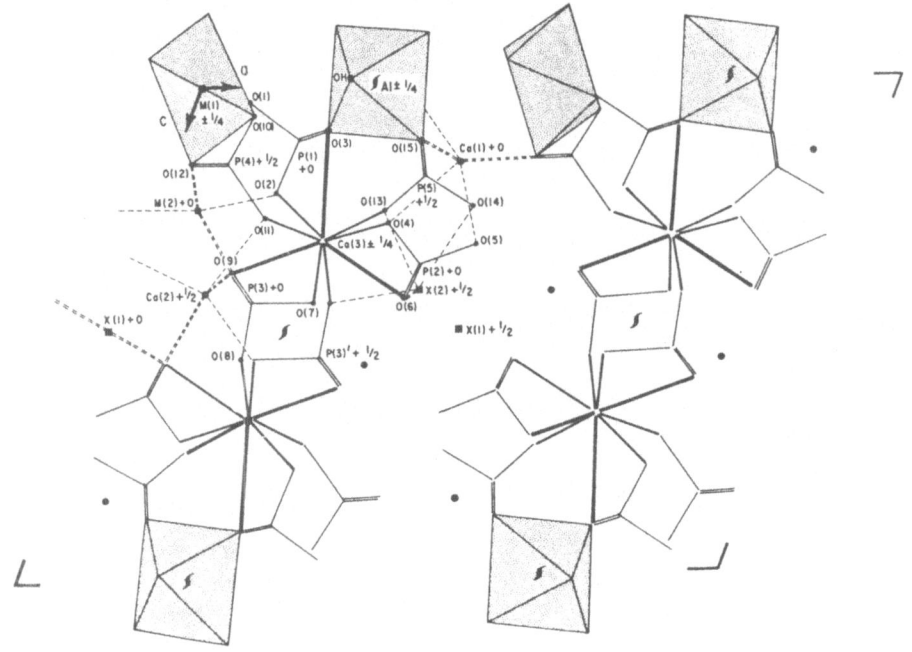

Fig. 1.   Structure of samuelsonite.

<u>1</u>. Strukturbericht, <u>2</u>, 99.
<u>2</u>. Structure Reports, <u>27</u>, 575.

TIN(II) PHOSPHATE
$Sn_3(PO_4)_2$

M. MATHEW, L.W. SCHROEDER and T.H. JORDAN, 1977. Acta Cryst., B<u>33</u>, 1812-1816.

Monoclinic, $P2_1/c$, a = 11.092, b = 4.830, c = 16.401 Å, β = 94.28°, Z = 4. Mo
radiation, R = 0.047 for 1813 reflexions.

    The structure (Fig. 1) contains alternating layers of Sn and PO_4 ions parallel
to (010), with open channels along <u>b</u>. The three independent Sn ions have trigonal
pyramidal coordination; two of the pyramids share an edge across a centre of
symmetry, resulting in O-Sn-O angles of 69°. Sn-O = 2.08-2.32 Å, with some longer
Sn...O contacts of 2.86-3.05 Å; P-O = 1.51-1.54 Å.

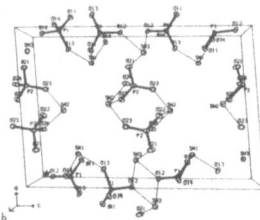

Fig. 1.   Structure of tin(II) phosphate.

ZIRCONIUM HYDROGEN PHOSPHATE MONOHYDRATE
$Zr(HPO_4)_2 \cdot H_2O$

J.M. TROUP and A. CLEARFIELD, 1977. Inorg. Chem., 16, 3311-3314.

Monoclinic, $P2_1/n$, a = 9.060, b = 5.297, c = 15.414 Å, β = 101.71°, Z = 4. Mo
radiation, R = 0.027 for 2005 reflexions.

The structure is essentially as previously described (1). The Zr-O octahedron is regular, Zr-O = 2.064(5) Å. The water molecule accepts hydrogen bonds
from two P-OH groups and acts as a donor to one P-OH group; the second H atom of
the water molecule is not involved in hydrogen bonding.

1. Structure Reports, 34A, 328.

MARICITE
$NaFePO_4$

Y. LE PAGE and G. DONNAY, 1977. Canad. Miner., 15, 518-521.

Orthorhombic, Pmnb, a = 6.861, b = 8.987, c = 5.045 Å, D_m = 3.66, Z = 4. Mo
radiation, R = 0.030 for 488 reflexions.

Atomic positions

	x	y	z
Na	1/4	0.8508	0.5305
Fe	0	0	0
P	1/4	0.1760	0.4640
O(1)	1/4	0.1164	0.7521
O(2)	1/4	0.3492	0.4557
O(3)	0.0692	0.1213	0.3174

The structure is like that of high-temperature $CoSO_4$ and $CuSO_4$ (1), with
additional Na.

1. Structure Reports, 22, 452, 454; 26, 438; 30A, 370.

MELONJOSEPHITE
$CaFe_2(OH)(PO_4)_2$

A.R. KAMPF and P.B. MOORE, 1977. Amer. Min., 62, 60-66.

Orthorhombic, Pbam, a = 9.542, b = 10.834, c = 6.374 Å, Z = 4. Mo radiation,
R = 0.043 for 728 reflexions.

The structure (Fig. 1) contains chains of edge-sharing FeO_6 octahedra along c,
linked into $Fe_2(OH)O_7$ (010) sheets by sharing corners with dimers of edge-sharing
octahedra. The PO_4 tetrahedra link within and between sheets, with a CaO_7 polyhedron in a pocket between the sheets. Fe^{2+}/Fe^{3+} are distributed over the two
independent Fe sites. Mean distances are Fe-O = 2.07, P-O = 1.54, Ca-O = 2.47 Å.

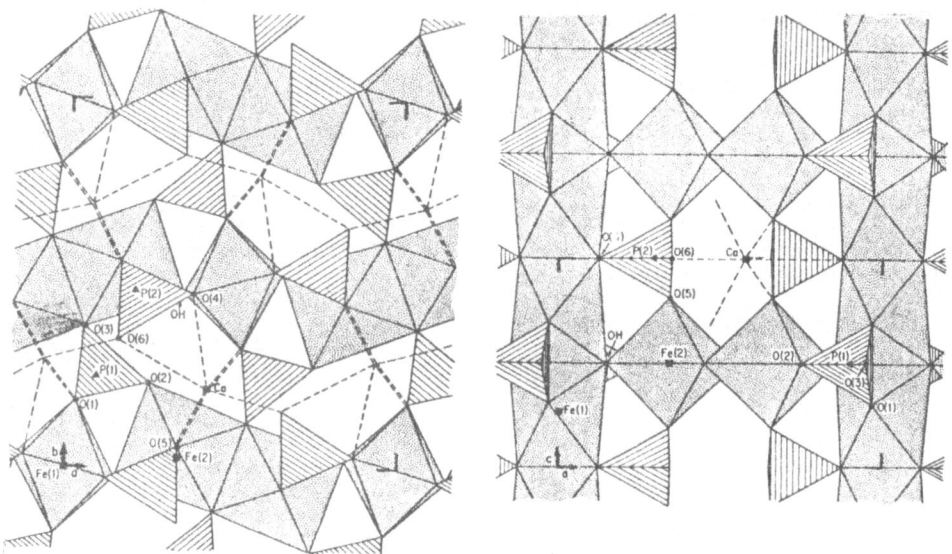

Fig. 1.   Structure of melonjosephite viewed along c (left) and b (right).

MITRIDATITE

$Ca_2Fe_3O_2(PO_4)_3 \cdot 3H_2O$

P.B. MOORE and T. ARAKI, 1977. Inorg. Chem., **16**, 1096-1106.

Monoclinic, Aa, a = 17.553, b = 19.354, c = 11.248 Å, β = 95.84°, D_m = 3.24, Z = 12. Mo radiation, R = 0.068 for 5547 reflexions.

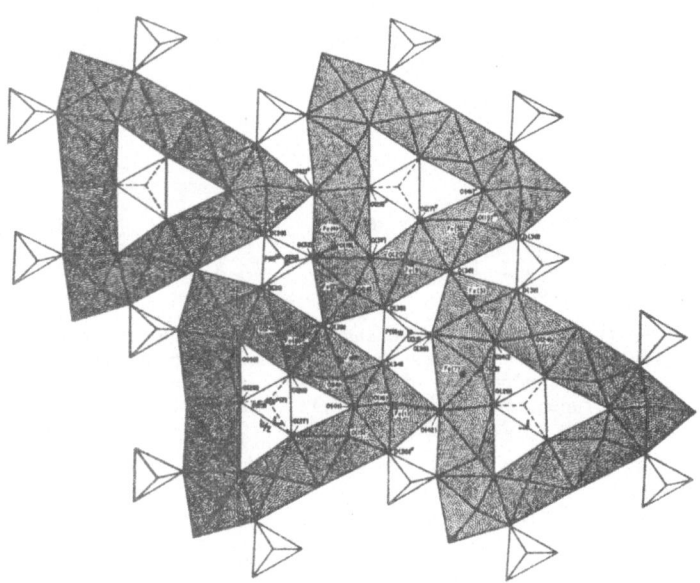

Fig. 1.   The $Fe_9O_6(PO_4)_9^{12-}$ sheets in mitridatite.

The structure contains compact $Fe_9O_6(PO_4)_9{}^{12-}$ sheets parallel to (100), consisting of edge-sharing FeO_6 octahedra and PO_4 tetrahedra (Fig. 1). These sheets are interleaved by $CaO_5(H_2O)_2$ polyhedra and hydrogen-bonded water molecules.

COBALT(II) PHOSPHATE HYDROXIDE   MANGANESE(II) PHOSPHATE HYDROXIDE
$Co_5(PO_4)_2(OH)_4$ $Mn_5(PO_4)_2(OH)_4$

F.A. RUSZALA, J.B. ANDERSON and E. KOSTINER, 1977.   Inorg. Chem., 16, 2417-2422.

Orthorhombic, $P2_12_12_1$, a = 8.903, 9.110, b = 17.397, 18.032, c = 5.5154, 5.6923 Å for Co, Mn, respectively, Z = 4. Mo radiation, R = 0.059 and 0.033 for 2176 and 2132 reflexions.

Isostructural with arsenoclasite (1). However all five independent Co (Mn) atoms can be described as octahedrally coordinated (the description in 1 was based on close-packed oxygen atoms, with 4-, 5-, and 6-coordinations for the metal ions). The structure (Fig. 1) contains a ladder-like chain of edge-shared octahedra and a chain of edge- and corner-shared octahedra coupled by phosphate tetrahedra. Co-O = 1.98-2.41, Mn-O = 2.08-2.59, P-O = 1.52-1.56 Å.

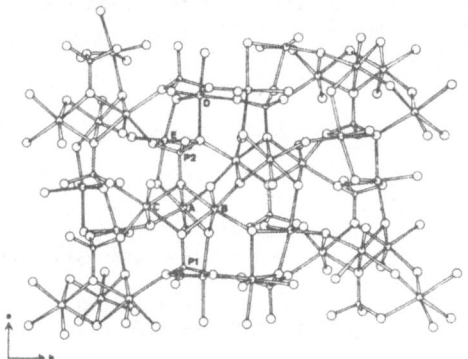

Fig. 1.   Structure of $M_5(PO_4)_2(OH)_4$, M = Co, Mn.

1. Structure Reports, 37A, 303.

COPPER(II) PHOSPHATE
$Cu_3(PO_4)_2$

G.L. SHOEMAKER, J.B. ANDERSON and E. KOSTINER, 1977.   Acta Cryst., B33, 2969-2972.

Triclinic, $P\bar{1}$, a = 4.854, b = 5.286, c = 6.182 Å, α = 72.35, β = 86.99, γ = 68.54°, Z = 1. Mo radiation, R = 0.038 for 1240 reflexions.

Atomic positions

	x	y	z
Cu(1)	0	0	0
Cu(2)	0.2776	0.2258	0.3157
P	0.3586	0.3534	0.7785
O(1)	-0.1536	0.3443	0.3389
O(2)	0.3324	0.6515	0.1695
O(3)	0.2303	0.2274	0.0049
O(4)	0.3786	0.1498	0.6334

Isostructural with the mineral stranskiite, $Zn_2Cu(AsO_4)_2$ (1). Cu(1) has elongated octahedral coordination (Fig. 1), Cu-O = 1.92, 1.98, $\overline{2}$.95 Å (each x 2), and Cu(2) has five oxygen neighbours (four at 1.93-2.02, the fifth at 2.27 Å; next nearest at 3.06 Å). P-O = 1.51-1.57 Å, O-P-O = 104.6-111.5°.

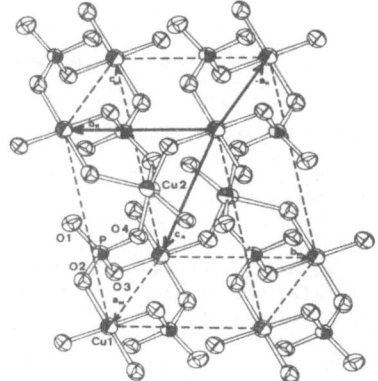

Fig. 1.   Structure of copper(II) phosphate (dotted lines indicate the reduced cell, full lines a morphological cell).

1. Structure Reports, 32A, 389; 41A, 425.

COPPER OXIDE PHOSPHATE
$Cu_5O_2(PO_4)_2$

M. BRUNEL-LAÜGT and J.-C. GUITEL, 1977. Acta Cryst., B33, 3465-3468.

Triclinic, PĪ, a = 7.603, b = 5.304, c = 5.200 Å, α = 111.66, β = 90.19, γ = 82.56°, Z = 1. Mo radiation, R = 0.040 for 1087 reflexions.

The structure (Fig. 1) contains PO_4 tetrahedra, Cu ions with tetrahedrally-elongated octahedral and five-coordinations, and an oxide ion with a tetrahedral environment of Cu ions. P-O = 1.503-1.579, Cu-O = 1.89-3.01 Å.

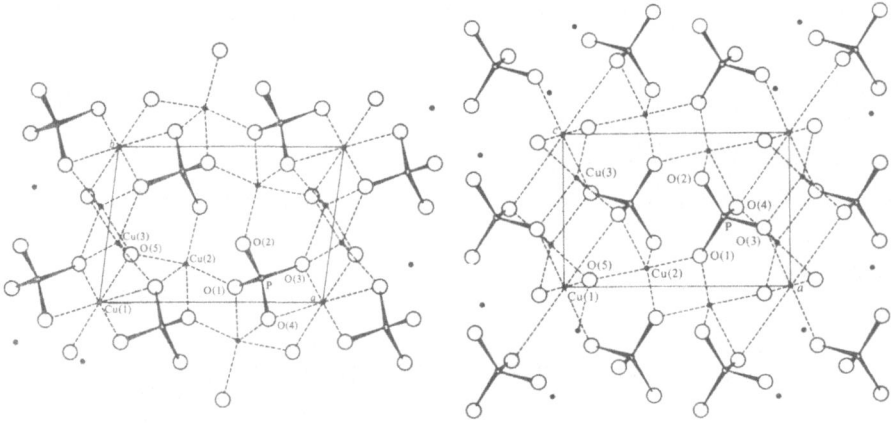

Fig. 1.   Views of the structure of $Cu_5O_2(PO_4)_2$.

COPPER(II) PHOSPHATE HYDROXIDE
$Cu_5(PO_4)_2(OH)_4$

J.B. ANDERSON, G.L. SHOEMAKER, E. KOSTINER and F.A. RUSZALA, 1977. Amer. Min., 62, 115-121.

Monoclinic, $P2_1/a$, a = 9.186, b = 10.684, c = 4.461 Å, β = 92.31°, Z = 2. Mo radiation, R = 0.056 for 1275 reflexions.

The material is a synthetic polymorph of pseudomalachite (1). The structure contains ab sheets of two CuO_6 distorted octahedra and one CuO_5 square pyramid which share edges; the sheets are linked by distorted PO_4 tetrahedra. In the octahedra Cu-O = 1.92-1.98 (4 bonds), 2.38-2.72 Å (2 bonds), and in the square pyramids Cu-O = 1.95-1.98 (4 bonds), 2.30 Å (sixth oxygen at 2.91 Å); P-O = 1.52-1.57 Å. The structure of pseudomalachite (1) differs in the polyhedral linkages within the sheets.

1. Structure Reports, 28, 192; this volume, following report.

PSEUDOMALACHITE
$Cu_5(PO_4)_2(OH)_4$

G.L. SHOEMAKER, J.B. ANDERSON and E. KOSTINER, 1977. Amer. Min., 62, 1042-1048.

Monoclinic, $P2_1/c$, a = 4.4728, b = 5.7469, c = 17.032 Å, β = 91.043°, Z = 2. Mo radiation, R = 0.040 for 1411 reflexions.

The structure is as previously determined (1). It contains sheets of edge-sharing CuO_6 polyhedra in the bc plane, linked in the a-direction by distorted PO_4 tetrahedra (Fig. 1); Cu-O = 1.88-2.01 (4 bonds), 2.35-2.76 (2 bonds), P-O = 1.51-1.59 Å. Hydrogen bonds are present.

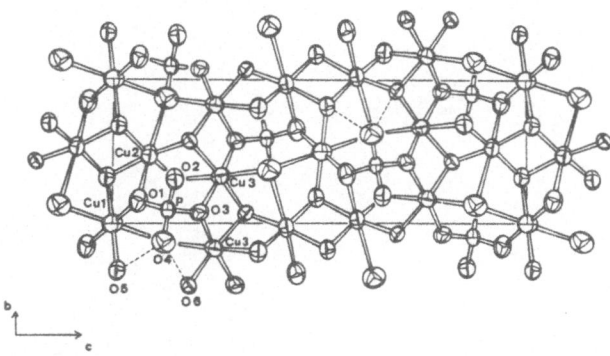

Fig. 1.   Structure of pseudomalachite.

1. Structure Reports, 28, 192.

ZINC MAGNESIUM PHOSPHATE
$Zn_2Mg(PO_4)_2$

A.G. NORD, 1977. Mater. Res. Bull., 12, 563-568.

Monoclinic, $P2_1/n$, a = 7.569, b = 8.355, c = 5.059 Å, β = 94.95°, Z = 2.   Cu
radiation, powder data with profile fitting.

Atomic positions

	x	y	z
$Zn_{0.88}Mg_{0.12}$	0.616	0.138	0.086
$Zn_{0.24}Mg_{0.76}$	0	0	1/2
P	0.198	0.191	0.036
O(1)	0.053	0.135	0.831
O(2)	0.125	0.205	0.320
O(3)	0.250	0.360	0.949
O(4)	0.358	0.084	0.055

γ-$Zn_3(PO_4)_2$-type structure (1), with Mg preferentially in the six-coordinate
site.

1. Structure Reports, 28, 189.

SCHOONERITE
$ZnMnFe_3(OH)_2(PO_4)_3 \cdot 9H_2O$

I.   P.B. MOORE and A.R. KAMPF, 1977.   Amer. Min., 62, 246-249.
II. A.R. KAMPF, 1977.   Ibid., 62, 250-255.

Orthorhombic, Pmab, a = 11.119, b = 25.546, c = 6.437 Å, Dm = 2.9, Z = 4.   Mo
radiation, R = 0.085 for 897 reflexions.

The structure (Fig. 1) contains edge-sharing Fe^{2+}-O octahedral chains joined
by sharing corners with Fe^{3+}-O octahedra and PO_4 tetrahedra to form a sheet
parallel to (010). To this sheet are linked additional PO_4 tetrahedra, $Mn^{2+}O_6$
octahedra, and ZnO_5 polyhedra. The resultant slabs are joined in the b direction
by hydrogen bonds, which involve interlayer water molecules.

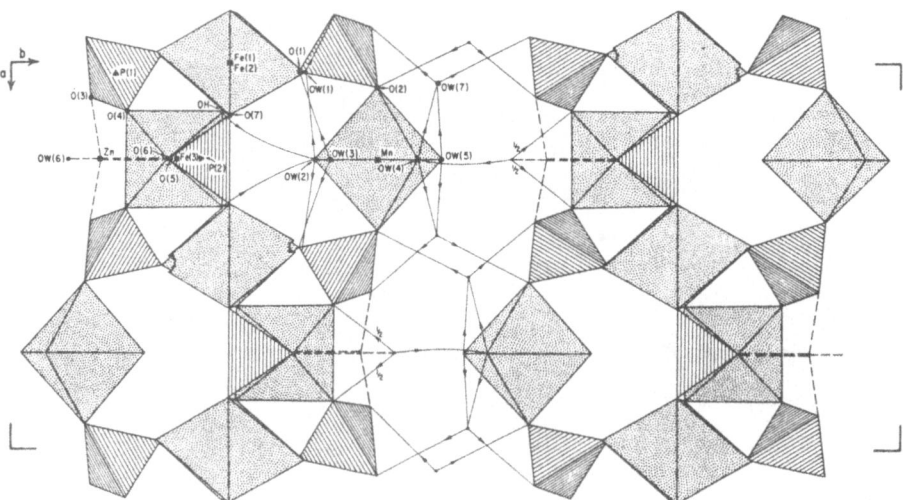

Fig. 1.   Structure of schoonerite.

PHOSPHOPHYLLITE
$Zn_2Fe(PO_4)_2 \cdot 4H_2O$

R.J. HILL, 1977. Amer. Min., 62, 812-817.

Monoclinic, $P2_1/c$, a = 10.378, b = 5.084, c = 10.553 Å, β = 121.14°, Z = 2. Mo
radiation, R = 0.032 for 1999 reflexions.

The structure (Fig. 1) is essentially as previously described (1). It contains $[Zn_2P_2O_7]$ tetrahedral sheets identical to those in hopeite, interleaved
with $[FeO \cdot 4H_2O]$ octahedral sheets similar to those in parahopeite. P-O = 1.508-
1.572, Zn-O = 1.894-1.996, Fe-O = 2.119-2.141 Å.

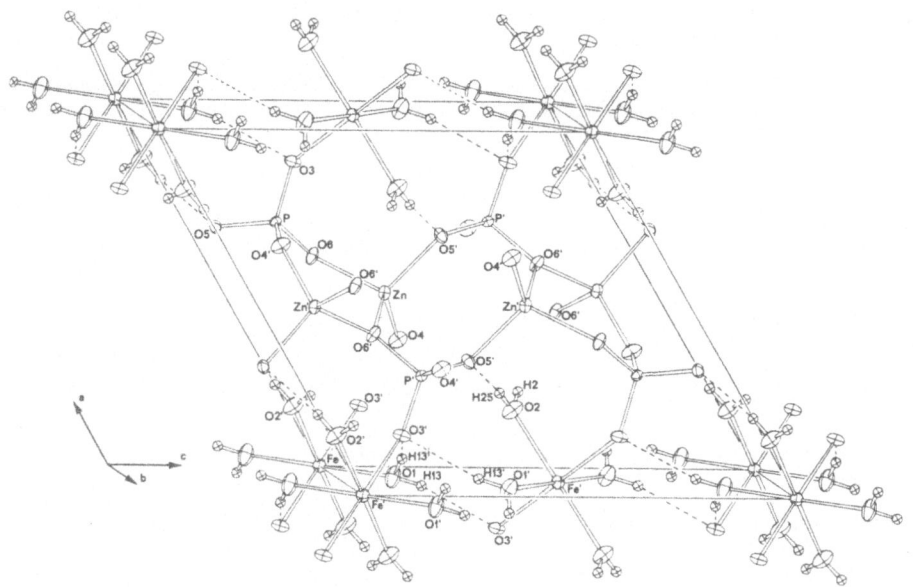

Fig. 1.   Structure of phosphophyllite.

1. Structure Reports, 26, 463.

CADMIUM APATITES
$Cd_5(PO_4)_3Cl$ $Cd_5(PO_4)_3Br$ $Cd_5(AsO_4)_3Br$ $Cd_5(VO_4)_3Br$ $Cd_5(VO_4)_3I$

I.   K. SUDARSANAN, R.A. YOUNG and A.J.C. WILSON, 1977. Acta Cryst., B33,
      3136-3142.
II.   A.J.C. WILSON, K. SUDARSANAN and R.A. YOUNG, 1977. Acta Cryst., B33,
      3142-3154.

Hexagonal, $P6_3/m$, a = 9.633, 9.733, 10.100, 10.173, 10.307, c = 6.484, 6.468,
6.519, 6.532, 6.496 Å, Z = 2. Mo radiation, R = 0.024-0.058 for 1234-1770
reflexions.

Apatite structures, as previously described for cadmium chlorapatite (1),
except that the halogen ions are not in unique positions, but are distributed
along the hexad axis and are present in less than stoichiometric amounts. The
deficiency appears to be due to steric hindrance and can be correlated with the
amount by which the ionic diameters of the halogens exceed c/2. The formulae

should be written $Cd_{5-x}(MO_4)_3X_{1-2x}$ (x as large at 0.13, for the iodide).

1. Structure Reports, 39A, 288.

AMMONIUM CADMIUM PHOSPHATE MONOHYDRATE   (PHASE B)
$NH_4CdPO_4 \cdot H_2O$

Ju.A. IVANOV, M.A. SIMONOV and N.V. BELOV, 1977. Kristallografija, 22, 171-173
[Soviet Physics - Crystallography, 22, 97-98].

Orthorhombic, Pnma, a = 17.090, b = 5.902, c = 5.133 Å, Z = 4. Mo radiation,
R = 0.035 for 1190 reflexions.

     The structure is as previously described (1), except that a water molecule is
now considered to be present rather than OH groups. Compare phase A (2), which is
considered to contain OH groups.

1. Structure Reports, 41A, 326.
2. Ibid., 42A, 350.

LANTHANUM PHOSPHATE   FLUORAPATITE   QUARTZ
$LaPO_4$ $Ca_5(PO_4)_3F$ SiO_2

R.A. YOUNG, P.E. MACKIE and R.B. von DREELE, 1977. J. Appl. Cryst., 10, 262-
269.

X-ray powder data with profile fitting; results agree with previous studies
(1, 2, 3).

1. Structure Reports, 38A, 318.
2. Ibid., 38A, 308.
3. Ibid., 27, 674; 28, 119; 30A, 420.

CALCIUM AMMONIUM HYDROGENPYROPHOSPHATE
$CaNH_4HP_2O_7$ $CaNH_4[O_3P-O-PO_2(OH)]$

M MATHEW and L.W. SCHROEDER, 1977. Acta Cryst., B33, 3025-3028.

Monoclinic, $P2_1/n$, a = 10.523, b = 17.672, c = 7.266 Å, β = 90.47°, Z = 8. Mo
radiation, R = 0.039 for 2895 reflexions.

     The structure (Fig. 1) contains calcium, ammonium, and hydrogenpyrophosphate
ions, held together by hydrogen bonds and Ca...O interactions. Two anions are
linked by a short, probably symmetric hydrogen bond (O...O = 2.473 Å), with
further hydrogen bonds to other anions and to ammonium ions. Calcium ions have
sevenfold pentagonal bipyramidal coordination. P-OH = 1.560, 1.576, P-O = 1.486-
1.528 (terminal), 1.604-1.623 Å (bridging), P-O-P = 125, 139°.

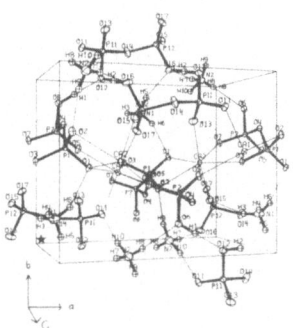

Fig. 1.   Structure of calcium ammonium hydrogenpyrophosphate.

AMMONIUM SODIUM CADMIUM PYROPHOSPHATE TRIHYDRATE
$NH_4NaCdP_2O_7 \cdot 3H_2O$

M.T. AVERBUCH-POUCHOT and J.C. GUITEL, 1977. Acta Cryst., B<u>33</u>, 3460-3462.

Monoclinic, Cc, a = 10.211, b = 16.56, c = 5.632 Å, β = 103.73°, Z = 4. Mo
radiation, R = 0.049 for 1032 reflexions.

     The structure (Fig. 1) contains $P_2O_7^{4-}$ anions linked by CdO_6, NaO_6, and
$(NH_4)O_6$ octahedra. P-O = 1.62 (bridging), 1.51 (terminal), Cd-O = 2.26-2.36,
Na-O = 2.31-2.47, NH_4-O = 2.16-3.09 Å. The Ca salt is isostructural.

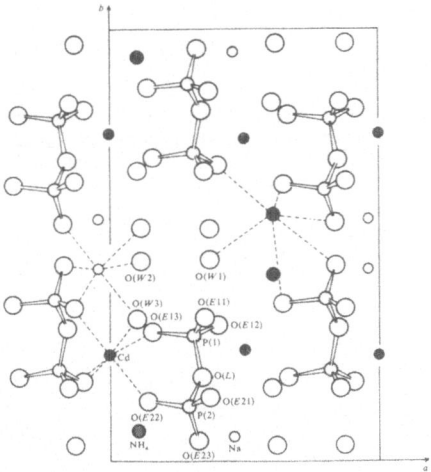

Fig. 1.   Structure of $NH_4NaCdP_2O_7 \cdot 3H_2O$.

CAESIUM TRIMETAPHOSPHATE MONOHYDRATE
$Cs_3P_3O_9 \cdot H_2O$

I. TORDJMAN, R. MASSE and J.-C. GUITEL, 1977. Acta Cryst., B33, 585-586.

Triclinic, P$\bar{1}$, a = 10.610, b = 7.966, c = 8.172 Å, α = 96.64, β = 68.84, γ = 95.42°, Z = 2. Ag radiation, R = 0.05 for 1120 reflexions.

The structure (Fig. 1) contains a cyclic trimetaphosphate anion and three octahedrally-coordinated Cs ions. P-O = 1.45-1.49 (terminal), 1.58-1.66(2) Å (ring), P-O-P = 123-131°, Cs-O = 3.01-3.36 Å.

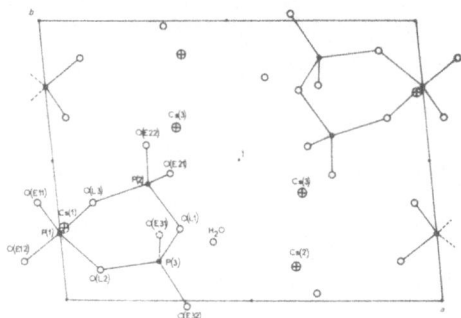

Fig. 1.   Structure of caesium trimetaphosphate monohydrate.

AMMONIUM HYDROGEN ALUMINUM TRIPOLYPHOSPHATE
$NH_4HAlP_3O_{10}$

M.T. AVERBUCH-POUCHOT, A. DURIF and J.C. GUITEL, 1977. Acta Cryst., B33, 1436-1438.

Monoclinic, P2/a, a = 11.643, b = 4.918, c = 8.705 Å, β = 119.27°, Z = 2. Ag radiation, R = 0.066 for 469 reflexions.

The structure (Fig. 1) contains a two-dimensional framework of P_3O_{10} groups linked by AlO_6 octahedra in the ab plane; these layers are joined by ammonium ions which have octahedral coordination. P-O = 1.48-1.52 (terminal), 1.58, 1.61 (bridging), Al-O = 1.87-1.90, N-O = 2.92-3.12(1) Å. The K salt is isostructural.

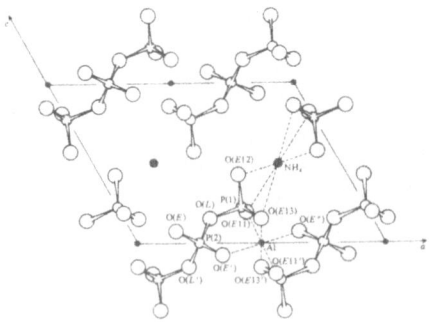

Fig. 1.   Structure of $NH_4HAlP_3O_{10}$.

IRON(III) DIHYDROGEN TRIPOLYPHOSPHATE MONOHYDRATE
$FeH_2P_3O_{10} \cdot H_2O$

M.T. AVERBUCH and J.C. GUITEL, 1977. Acta Cryst., B$\underline{33}$, 1613-1615.

Monoclinic, C2/c, a = 12.076, b = 8.443, c = 9.352 Å, β = 112.10°, Z = 4. Mo radiation, R = 0.043 for 1282 reflexions.

The structure (Fig. 1) contains $P_3O_{10}^{5-}$ ions linked by FeO_6 octahedra (hydrogen atoms were not located). P-O = 1.49-1.52 (terminal), 1.58-1.61 (bridging), Fe-O = 1.97-2.01 Å.

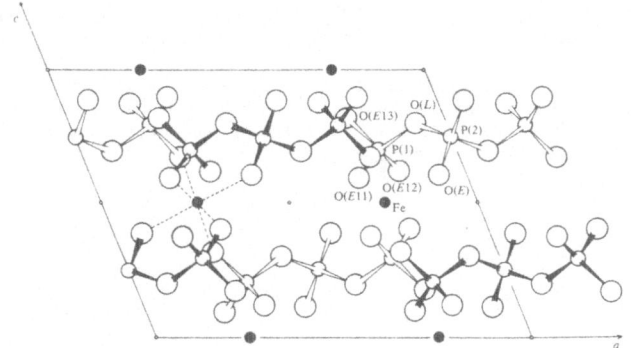

Fig. 1.   Structure of $FeH_2P_3O_{10} \cdot H_2O$.

BARIUM SILVER TRIMETAPHOSPHATE TETRAHYDRATE
$BaAgP_3O_9 \cdot 4H_2O$

D. SEETHANEN, A. DURIF and J.C. GUITEL, 1977. Acta Cryst., B$\underline{33}$, 2716-2719.

Monoclinic, C2/c, a = 21.35, b = 7.163, c = 18.35 Å, β = 121.72°, Z = 8. Ag radiation, R = 0.035 for 2400 reflexions.

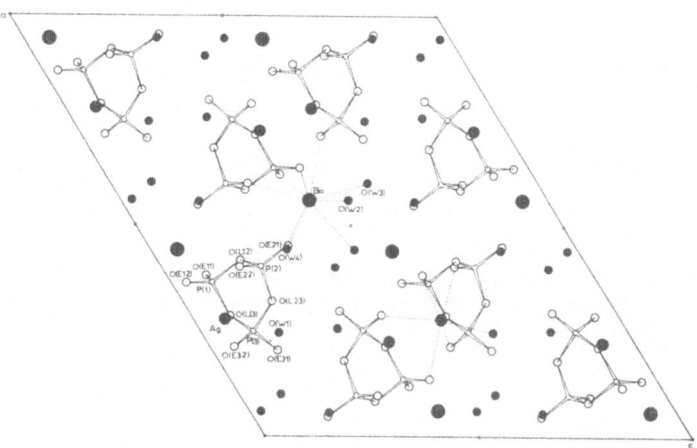

Fig. 1.   Structure of barium silver trimetaphosphate tetrahydrate.

The structure (Fig. 1) contains cyclic $P_3O_9^{3-}$ anions linked by the cations and the water molecules. P-O = 1.47-1.49 (terminal), 1.59-1.61 (bridging), Ba-O = 2.71-2.99 (9-coordination), Ag-O = 2.37-2.71 Å (6-coordination), P-O-P = 125-132°.

SODIUM ZINC TRIPOLYPHOSPHATE NONAHYDRATE
$NaZn_2P_3O_{10}\cdot9H_2O$

M.T. AVERBUCH-POUCHOT and J.C. GUITEL, 1977. Acta Cryst., B33, 1427-1431.

Triclinic, P1̄, a = 10.454, b = 10.675, c = 8.629 Å, α = 101.14, β = 109.85, γ = 99.03°, Z = 2. Ag radiation, R = 0.041 for 2862 reflexions.

The structure (Fig. 1) is similar to that of $Ag_{0.62}H_{0.38}Zn_2P_3O_{10}\cdot9H_2O$ (1). It contains $P_3O_{10}^{5-}$ anions, two octahedrally-coordinated Zn^{2+} ions, one tetrahedrally-coordinated Zn^{2+} ion, and a Na^+ ion with octahedral coordination. P-O = 1.48-1.62 (terminal), 1.59-1.63 (bridging), Zn-O = 2.04-2.16 (octahedral), 1.93-1.94 (tetrahedral), Na-O = 2.35-2.51 Å.

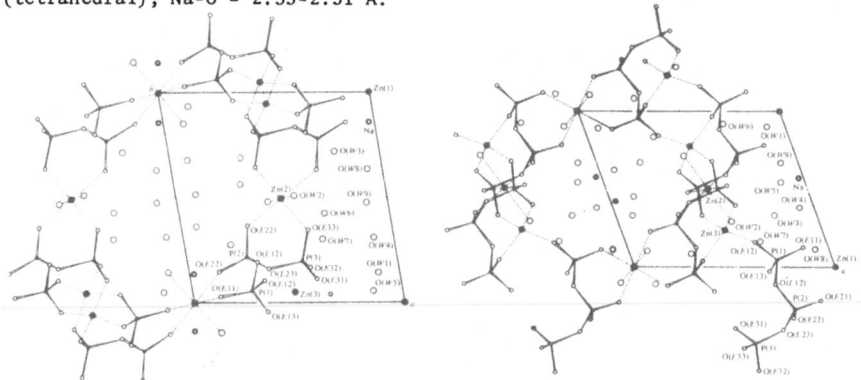

Fig. 1.   Structure of $NaZn_2P_3O_{10}\cdot9H_2O$ projected on the ab (left) and ac (right) planes.

1. Structure Reports, 42A, 359.

CAESIUM CADMIUM TRIMETAPHOSPHATE
$CsCdP_3O_9$

M.T. AVERBUCH-POUCHOT and A. DURIF, 1977. Acta Cryst., B33, 3114-3116.

Orthorhombic, Pmcn, a = 7.508, b = 12.684, c = 9.530 Å, Z = 4. Mo radiation, R = 0.032 for 917 reflexions.

The structure contains cyclic anions (Fig. 1), Cd ions with octahedral coordination, and Cs ions in large cavities with eleven oxygen neighbours. P-O = 1.47-1.49 (terminal), 1.60-1.61 Å (bridging), P-O-P = 126, 135°, Cd-O = 2.23-2.31, Cs-O = 3.11-3.57 Å.

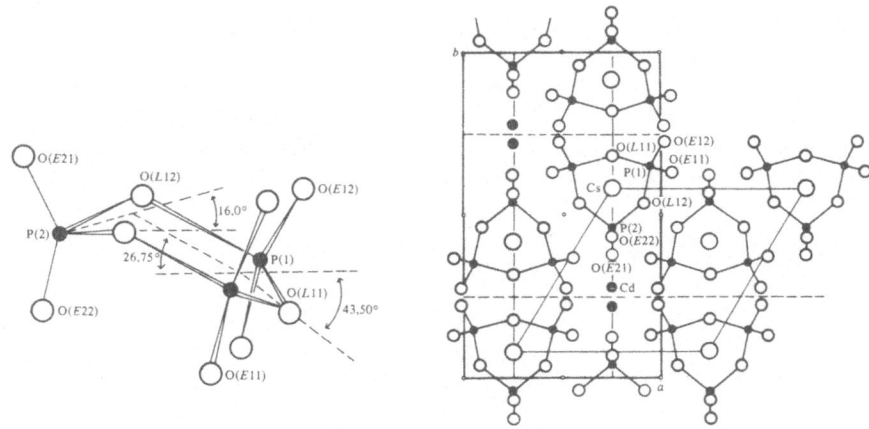

Fig. 1.   The $P_3O_9^{3-}$ ion and structure of caesium cadmium trimetaphosphate
(a pseudo-hexagonal benitoite-type cell is outlined).

TETRAMETAPHOSPHIMIC ACID DIHYDRATE
$(NHPO_2H)_4 \cdot 2H_2O$

R. ATTIG and D. MOOTZ, 1977. Acta Cryst., B33, 605-607.

Orthorhombic, $P2_12_12$, a = 13.995, b = 8.334, c = 5.064 Å, D_m = 1.95, Z = 2.   Mo
radiation, R = 0.030 for 1026 reflexions.

The structure is as previously described (1).   The hydrogen atoms have now
been located (Fig. 1) and indicate the presence of hydroxonium ions and formally
symmetrical hydrogen bonds (O...H...N = 2.46 Å) between anions.

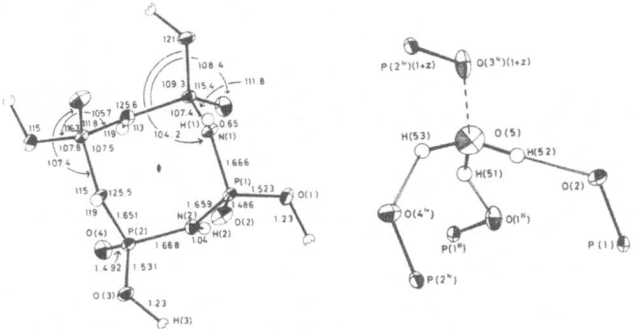

Fig. 1.   The anion and hydroxonium ion in tetrametaphosphimic acid dihydrate.

1. Structure Reports, 30A, 391.

RUBIDIUM NEODYMIUM TETRAMETAPHOSPHATE
RbNdP$_4$O$_{12}$

H. KOIZUMI and J. NAKANO, 1977.   Acta Cryst., B33, 2680-2684.

Monoclinic, C2/c, a = 7.845, b = 12.691, c = 10.688 Å, β = 112.34°, Z = 4.   Mo
radiation, R = 0.054 for 1054 reflexions.

The P$_4$O$_{12}$$^{4-}$ rings, located on symmetry centres, are connected by isolated
NdO$_8$ dodecahedra and irregularly shaped Rb polyhedra, to form a layer structure
parallel to (001) (Fig. 1).

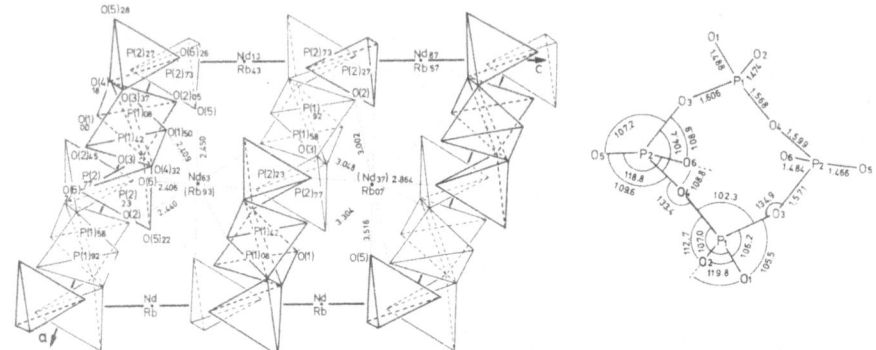

Fig. 1.   Details of the structure of rubidium neodymium tetrametaphosphate.

AMMONIUM PRASEODYMIUM TETRAMETAPHOSPHATE
NH$_4$PrP$_4$O$_{12}$

R. MASSE, J.-C. GUITEL and A. DURIF, 1977.   Acta Cryst., B33, 630-632.

Monoclinic, C2/c, a = 7.916, b = 12.647, c = 10.672 Å, β = 110.34°, Z = 4.
Ag radiation, R = 0.045 for 1068 reflexions.

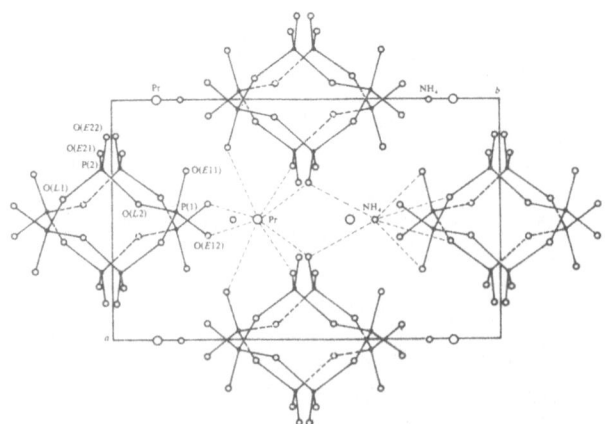

Fig. 1.   Structure of ammonium praseodymium tetrametaphosphate.

The structure (Fig. 1) contains cyclic centrosymmetric tetrametaphosphate anions, P-O = 1.48 (terminal), 1.60(1) Å (ring), P-O-P = 135, 138°. Pr has 8-fold dodecahedral and ammonium has octahedral coordination; Pr-O = 2.44-2.49, N-O = 2.83-3.06 Å.

CHROMIUM(III) HEXAMETAPHOSPHATE
$Cr_2P_6O_{18}$

M. BAGIEU-BEUCHER and J.C. GUITEL, 1977. Acta Cryst., B33, 2529-2533.

Monoclinic, $P2_1/a$, a = 8.311, b = 15.221, c = 6.220 Å, β = 105.85°, Z = 2. Mo radiation, R = 0.046 for 935 reflexions.

Centrosymmetric $P_6O_{18}^{6-}$ cyclic anions have approximately C_{2h} symmetry, and are connected by isolated CrO_6 octahedra (Fig. 1). P-O = 1.46-1.48 (terminal), 1.58-1.59 (bridging), Cr-O = 1.95-1.98 Å, P-O-P = 132-139°.

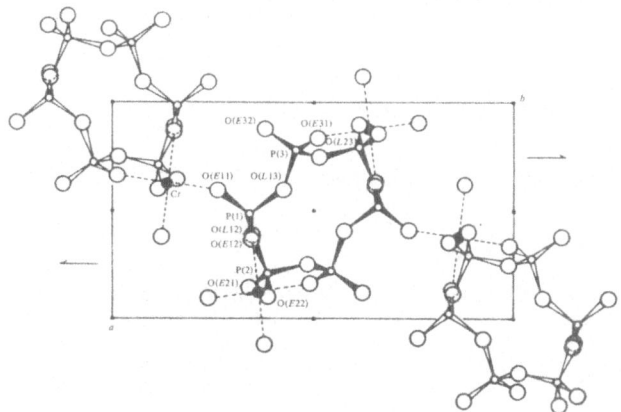

Fig. 1.   Structure of chromium(III) hexametaphosphate.

AMMONIUM ARSENATE TRIHYDRATE
$(NH_4)_3AsO_4 \cdot 3H_2O$

T.H. HSEU and T.H. LU, 1977. Acta Cryst., B33, 3947-3949.

Monoclinic, $P2_1/c$, a = 6.818, b = 6.364, c = 22.811 Å, β = 93.74°, D_m = 1.61, Z = 4. Cu radiation, R = 0.07 for 1127 reflexions.

Isostructural with the corresponding phosphate (1). The structure (Fig. 1) is held together by a system of hydrogen bonds, O-H...O = 2.66-2.95 Å. As-O = 1.67-1.70(1) Å. O(5) and O(7) have large thermal parameters and may be disordered.

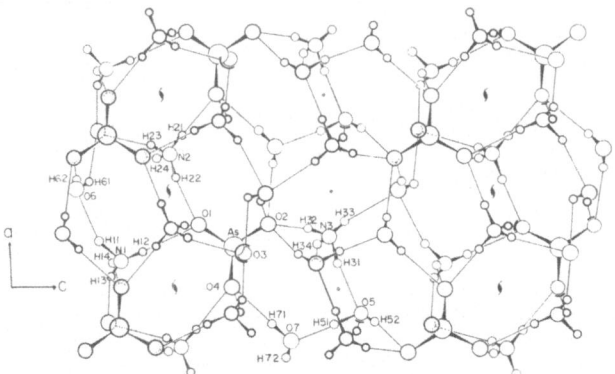

Fig. 1.   Structure of ammonium arsenate trihydrate.

1. Structure Reports, 35A, 331.

SODIUM DIDEUTERIUMARSENATE MONODEUTERATE
$NaD_2AsO_4 \cdot D_2O$

D.F. RENDLE and J. TROTTER, 1977. Acta Cryst., B33, 2684-2686.

Monoclinic, $P2_1/m$, a = 5.865, b = 7.123, c = 5.619 Å, β = 92.58°, Z = 2. Mo radiation, R = 0.049 for 801 reflexions.

The structure (Fig. 1) contains $AsO_2(OD)_2^-$ ions linked into (001) sheets by
O-D...O hydrogen bonds, directly by pairs of bonds along b and via D_2O molecules
along a. The sheets are joined by $NaO_4(OD_2)_2$ coordination polyhedra which share
edges along b. As-OD = 1.725(3), As-O = 1.647, 1.689(4), Na-OD_2 = 2.414, Na-O =
2.379, 2.534, O-D...O = 2.621, 2.781, 3.104 Å.

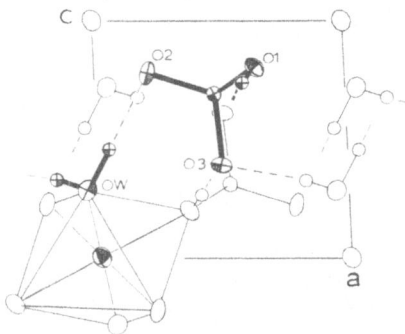

Fig. 1.   Structure of sodium dideuteriumarsenate monodeuterate.

BARIUM TETRAARSENATE
$BaH_6As_4O_{14}$

D. BLUM, A. DURIF and J.C. GUITEL, 1977. Acta Cryst., B33, 3222-3224.

Orthorhombic, Pman, a = 8.496, b = 11.249, c = 5.858 Å, Z = 2. Ag radiation,
R = 0.051 for 1768 reflexions.

Atomic positions

			x	y	z
Ba	in	2(b)	1/2	0	0
As(1)		4(f)	0.1661	0	1/2
As(2)		4(h)	0	0.1903	0.7843
O(L12)		8(i)	0.1656	0.1038	0.7532
O(E21)		4(h)	0	0.3175	0.6419
O(E22)		4(h)	0	0.2398	0.0591
O(L11)		4(h)	0	0.0849	0.3822
O(E1)		8(i)	0.1968	0.4135	0.3457

The structure (Fig. 1) contains a cyclic anion with two AsO_4 tetrahedra and
two AsO_6 octahedra, As-O = 1.656-1.721 (tetrahedral), 1.767-1.887(5) Å (octahedral).
Ba has 8-coordination, Ba-O = 2.800-2.948 Å.

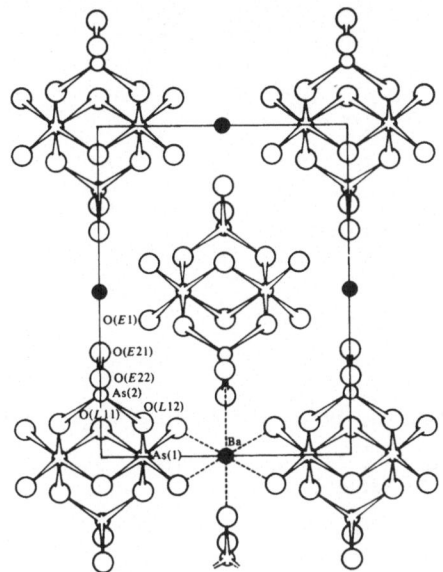

Fig. 1.   Structure of $BaH_6As_4O_{14}$.

STENHUGGARITE
$CaFeSbAs_2O_7$

A. CODA, A. DAL NEGRO, C. SABELLI and V. TAZZOLI, 1977. Acta Cryst., B33,
1807-1811.

Tetragonal, $I4_1/a$, a = 16.144, c = 10.706 Å, Z = 16. Mo radiation, R = 0.041 for
1446 reflexions.

The structure (Fig. 1) contains two types of anion: $As_4O_8^{4-}$ with eight-
membered rings, and $AsSbO_5^{4-}$ with -O-Sb-O-helices and $-OAsO_2$ side groups. Coordination polyhedra are: As, trigonal pyramid with lone-pair completing a tetrahedron; Sb, trigonal bipyramid with lone-pair in one equatorial position; Fe,
trigonal bipyramid ; Ca, between cube and square antiprism.

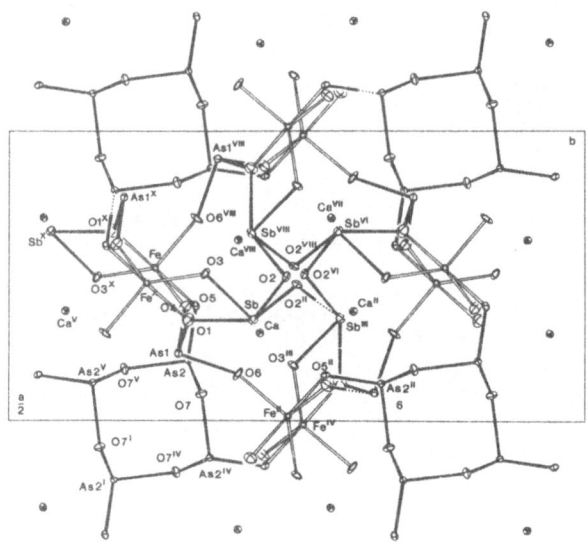

Fig. 1.   Structure of stenhuggarite.

ROSELITE
$Ca_2(Co,Mg)(AsO_4)_2 \cdot 2H_2O$

F.C. HAWTHORNE and R.B. FERGUSON, 1977. Canad. Miner., 15, 36-42.

Monoclinic, $P2_1/c$, a = 5.801, b = 12.898, c = 5.617 Å, β = 107.42°, Z = 2. Mo
radiation, R = 0.058 for 1194 reflexions.

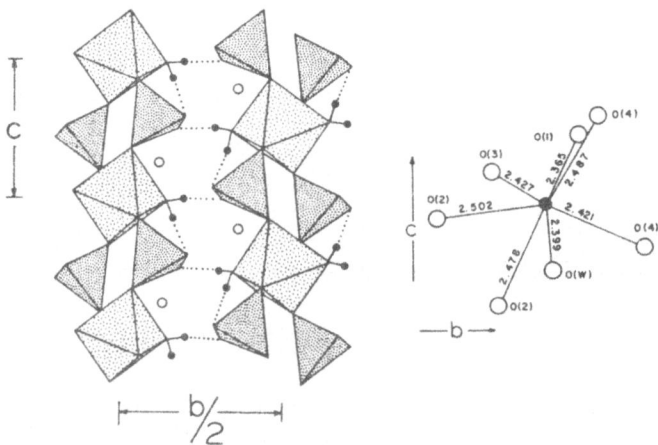

Fig. 1.   Structure of roselite and Ca coordination.

Isostructural with kröhnkite, Na$_2$Cu(SO$_4$)$_2$.2H$_2$O (1). The structure (Fig. 1) contains isolated Mg/Co octahedra linked by corner sharing with As tetrahedra to form chains along c. The chains are linked by Ca ions (7-coordination) and hydrogen bonding. As-O = 1.67-1.70, Mg/Co-O = 2.06-2.20, Ca-O = 2.37-2.50(1) Å.

1. Structure Reports, 41A, 353.

COPPER(II) CHLOROARSENATE   COBALT(II) CHLOROARSENATE
Cu$_2$(AsO$_4$)Cl   Co$_2$(AsO$_4$)Cl

J.R. REA, J.B. ANDERSON and E. KOSTINER, 1977. Acta Cryst., B33, 975-979.

Monoclinic, P2$_1$/m, a = 6.877, 6.880, b = 6.589, 6.593, c = 4.929, 4.932 Å, β = 91.12, 91.12°, Z = 2. Mo radiation, R = 0.054 and 0.055 for 673 and 688 reflexions.

Atomic positions (Cu, similar values for Co compound)

	x	y	y
Cu(1)	0	0	0
Cu(2)	0.2735	1/4	0.3538
As	0.2471	3/4	0.5085
Cl	0.2426	1/4	0.8376
O(1)	0.4496	3/4	0.6697
O(2)	0.0636	3/4	0.7582
O(3)	0.1990	0.5454	0.3062

The structures (Fig. 1) contain chains along b of face-sharing MO$_4$Cl$_2$ (M = Cu, Co) trans-octahedra, which resemble those in F̄e$_2$(PO$_4$)Cl; adjacent chains are linked by sharing Cl corners and by the arsenate tetrahedra. Cu-O = 1.911-2.373(7), Cu-Cl = 2.399-2.550(3), Co-O = 1.911-2.376(8), Co-Cl = 2.403-2.549(3), As-O = 1.644-1.731(8) Å, O-As-O = 105.8-114.1°.

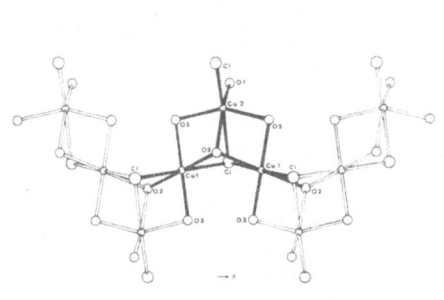 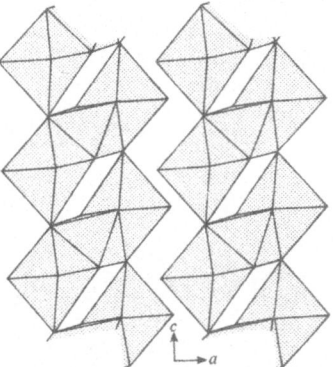

Fig. 1.   The structural unit (left, As bridges the O(3) vertices) and structure (right) of Cu$_2$(AsO$_4$)Cl.

OLIVENITE
Cu$_2$AsO$_4$(OH)

K. TOMAN, 1977. Acta Cryst., B33, 2628-2631.

Monoclinic, P2₁/n (a-axis unique), a = 8.615, b = 8.240, c = 5.953 Å, α = 90.0°,
Z = 4. Mo radiation, R = 0.065 for 488 reflexions (twinned crystal).

Atomic positions

	x	y	z
As	0.2500	0.2622	-0.0115
Cu(1)	-0.1178	0.3627	0.0219
Cu(2)	0.0007	0.0005	0.2511
O(1)	0.111	0.398	-0.058
O(2)	0.419	0.370	0.001
O(3)	-0.099	0.131	0.013
O(4)	0.222	0.161	0.240
O(5)	0.248	0.132	-0.215

The structure deviates only slightly from the previous orthorhombic description (1). It is related to that of adamite (2) and contains AsO_4 tetrahedra, As-O =
1.62-1.73(3) Å, and CuO_5 trigonal bipyramids, Cu-O = 1.99-2.16(equatorial), 1.92
and 1.98(2) Å (axial).

1. Strukturbericht, 6, 22, 109.
2. Ibid., 5, 17, 95; 6, 22; Structure Reports, 42A, 364.

REINERITE
$Zn_3(AsO_3)_2$

S. GHOSE, P. BOVING, W.A. LaCHAPELLE and CHE'NG WAN, 1977. Amer. Min., 62, 1129-
1134.

Orthorhombic, Pbam, a = 6.092, b = 14.407, c = 7.811 Å, D_m = 4.27, Z = 4. Mo
radiation, R = 0.051 for 1289 reflexions.

Atomic positions

	x	y	z
Zn(1)	0	1/2	0.3145
Zn(2)	0.5649	0.7791	0.7843
As(1)	0.9131	0.8744	1/2
As(2)	0.2231	0.9013	0
O(1)	0.3304	0.0608	1/2
O(2)	0.3349	0.2807	0
O(3)	0.1519	0.1997	0.3262
O(4)	0.0904	0.3965	0.1736

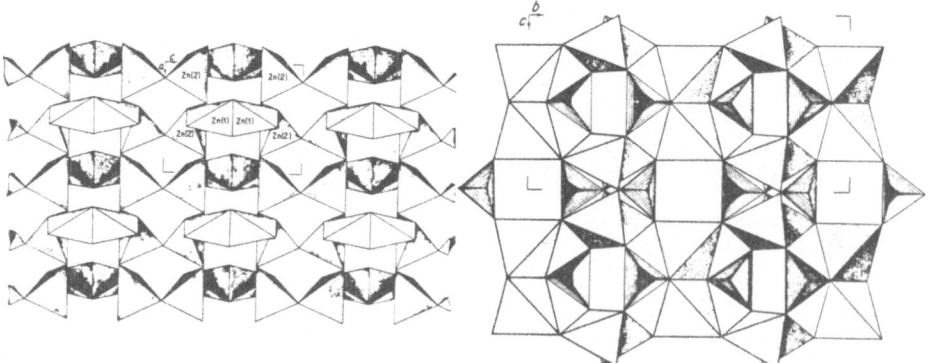

Fig. 1.   Reinerite structure; partial view down b (left, arsenite groups omitted),
and view down a (right, Zn_4O_{10} chains viewed end-on).

The structure (Fig. 1) contains a tetrahedral double chain along $\underline{a}$, $[Zn(2)_4O_{10}]_\infty$, with four-membered rings; these chains are cross-linked by corner sharing with edge-shared tetrahedral dimers, $Zn(1)_2O_6$, and with two types of trigonal pyramidal arsenite groups. Zn-O = 1.933-1.988(3), As-O = 1.753-1.772(4) Å, O-Zn-O = 86.1-127.4, O-As-O = 95.2-100.0°.

SULPHAMIC ACID   (at 78°K)
H_3NSO_3

J.W. BATS, P. COPPENS and T.F. KOETZLE, 1977. Acta Cryst., B33, 37-45 and 1304.

Orthorhombic, Pbca, a = 8.034, b = 8.020, c = 9.236 Å, Z = 8. Mo radiation, R = 0.028 for 4265 reflexions, and neutron diffraction data, R = 0.041 for 1283 reflexions.

The structure is as previously determined (1). S-O = 1.441-1.446, S-N = 1.775 Å, O-S-O = 114.7-116.0, N-S-O = 101.7, 102.9°; N-H = 1.033-1.036 Å, S-N-H = 109.3-111.6°; N-H...O = 2.930-2.964 Å. (X-ray - neutron) difference maps reveal bonding and lone-pair electrons.

1. Structure Reports, 29, 391.

DISILVER SULPHAMIDE
$SO_2(NHAg)_2$

H. GRESCHONIG, E. NACHBAUR and H. KRISCHNER, 1977. Acta Cryst., B33, 3595-3597.

Monoclinic, P2₁/c, a = 7.661, b = 5.719, c = 10.464 Å, β = 93.89°, D_m = 4.53, Z = 4. Cu radiation, R = 0.098 for 925 reflexions.

Atomic positions

	x	y	z
Ag(1)	-0.0033	0.1978	0.1752
Ag(2)	0.2886	0.3916	0.4641
S	0.3162	0.5396	0.1587
N(1)	0.266	0.268	0.146
N(2)	0.268	0.624	0.296
O(1)	0.214	0.663	0.057
O(2)	0.497	0.066	0.351

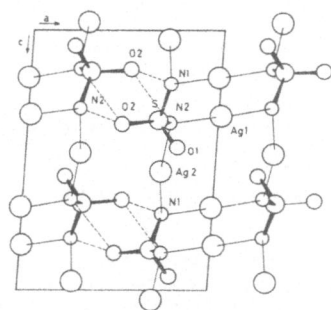

Fig. 1.   Structure of disilver sulphamide.

The structure (Fig. 1) contains SO_2N_2 tetrahedra, with each Ag^+ ion bonded to 2 N atoms of different tetrahedra to form double chains along $\underline{c}$; these double chains are linked by hydrogen bonds. S-O = 1.45, S-N = 1.60, Ag-$\overline{N}$ = 2.12-2.20, N-H...O = 2.93, 3.11 Å.

AMMONIUM SULPHITE MONOHYDRATE
$(NH_4)_2SO_3 \cdot H_2O$

J. DURAND, J.L. GALIGNÉ and L. COT, 1977. Acta Cryst., B33, 1414-1417.

Monoclinic, $P2_1/c$, a = 6.340, b = 8.068, c = 12.340 Å, β = 97.79°, D_m = 1.428, Z = 4. Mo radiation, R = 0.048 for 1590 reflexions.

The structure (Fig. 1) is the same as that of $(NH_4)_2PO_3F \cdot H_2O$ (1), the SO_3^{2-} lone pair replacing the PO_3F^{2-} fluorine atom. S-O = 1.517(3) Å, O-S-O = 104.0-105.5°.

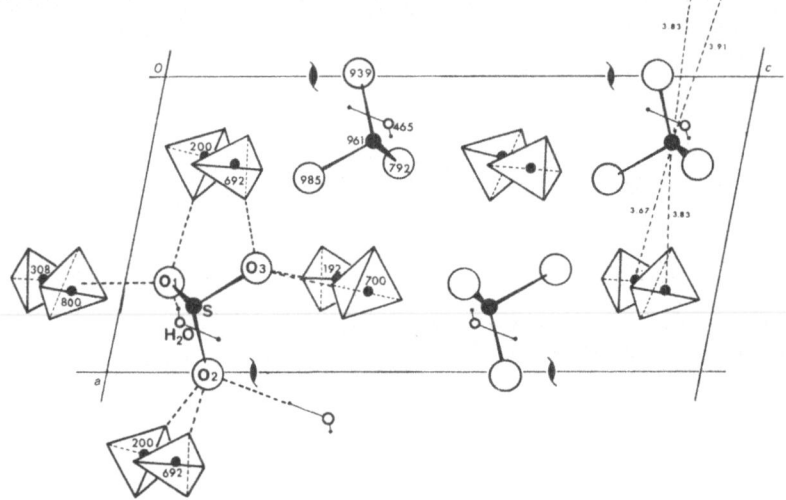

Fig. 1.   Structure of ammonium sulphite monohydrate.

1. Structure Reports, 38A, 317.

POTASSIUM HYDROXYLAMINETRISULPHONATE SESQUIHYDRATE
$K_3[O_3SON(SO_3)_2] \cdot 1 \cdot 5H_2O$

G.M. BROWN and O.A.W. STRYDOM, 1977. Acta Cryst., B33, 1591-1594.

Monoclinic, I2/c, a = 25.587, b = 6.777, c = 13.969 Å, β = 92.08°, Z = 8. Mo radiation, R = 0.048 for 2872 reflexions, and neutron radiation, R = 0.066 for 3065 reflexions.

The structure contains hydroxylaminetrisulphonate anions (Fig. 1), linked via the water molecules by one normal O-H...O hydrogen bond (2.94 Å) and by two long bifurcated contacts (H...O = 2.30-2.49 Å). $K^+...OH_2$ = 2.74-3.35 Å.

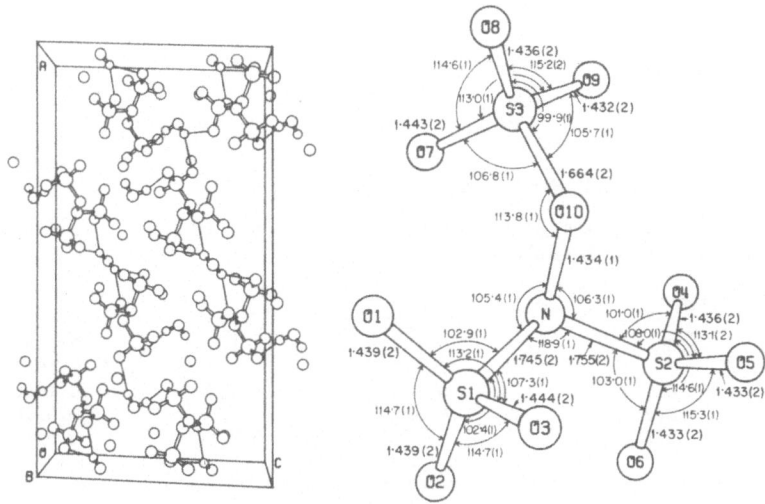

Fig. 1.   Structure of potassium hydroxylaminesulphonate sesquihydrate.

SODIUM THALLIUM(I) SULPHITE
NaTl$_3$(SO$_3$)$_2$

Y. ODDON, G. PEPE and A. TRANQUARD, 1977.   J. Chem. Research (S), 138-139; (M),
1621-1631.

Trigonal, P$\bar{3}$m1, a = 5.723, c = 7.216 Å, D$_m$ = 6.41, Z = 1. Mo radiation, R =
0.041 for 649 reflexions.

Atomic positions

			x	y	z
Tl(1)	in	1(a)	0	0	0
Tl(2)		2(d)	1/3	2/3	0.6131
Na		1(b)	0	0	1/2
S		2(d)	1/3	2/3	0.1975
0		6(i)	0.1956	0.3913	0.2863

The sulphite ion is trigonal pyramidal S-O = 1.51(3) Å, O-S-O = 104°. Na
and Tl(1) have octahedral coordination and Tl(2) 9-coordination; Na-O = 2.48,
Tl-O = 2.73-2.97 Å.

MATTEUCCITE
NaHSO$_4$.H$_2$O

M. CATTI, G. FERRARIS and M. FRANCHINI-ANGELA, 1975. Atti. Accad. Sci. Torino,
Cl. Sci. Fis. Mat. Nat., 109, 531-545.

Monoclinic, Cc, a = 7.811, b = 7.823, c = 8.025 Å, β = 117.49°, D$_m$ = 2.118, Z = 4.
Mo radiation, R = 0.027 for 613 reflexions, synthetic material.

Structure as previously described (1).

1. Structure Reports, 30A, 365; 37A, 306.

BARITE
$BaSO_4$

R.J. HILL, 1977. Canad. Miner., 15, 522-526.

Orthorhombic, Pnma, a = 8.8842, b = 5.4559, c = 7.1569 Å, Z = 4. Mo radiation,
R = 0.0295 for 326 reflexions.

Atomic positions

	x	y	z
Ba	0.1845	1/4	0.1585
S	0.4373	3/4	0.1913
O(1)	0.5890	3/4	0.1066
O(2)	0.3183	3/4	0.0518
O(3)	0.4204	0.9700	0.3116

The structure is as previously described (1).

1. Strukturbericht, 1, 343; 382; Structure Reports, 28, 209; 32A, 327.

CALCIUM ALUMINUM SULPHATE HYDRATE
$[Ca_2Al(OH)_6]_2SO_4.6H_2O$

R. ALLMANN, 1977. Neues Jb. Miner., Mh., 136-144.

Rhombohedral, R$\bar{3}$, a = 5.759, c = 26.795 Å, D_m = 2.03, Z = 3/2. Mo radiation,
R = 0.038 for 938 reflexions.

Atomic positions

			x	y	z
Al	in	3(a)	0	0	0
Ca		6(c)	2/3	1/3	0.0213
OH {O		18(f)	0.2511	-0.0547	0.0373
{H		18(f)	0.211	-0.115	0.0667
$H_2O(1)$		6(c)	2/3	1/3	0.1154
0.25 S		6(c)	0	0	0.4955
0.25 O(1)		6(c)	0	0	0.5486
0.25 O(2)					
0.16 $H_2O(2)$ }		18(f)	0.2688	0.1821	0.4771

The structure of this platy compound, $3CaO.Al_2O_3.CaSO_4.12H_2O$, is as previously
described (1), with the final water molecule, $H_2O(2)$, replacing sulphate in a
disordered manner. It contains a brucite-like $[Ca_2Al(OH)_6]^+$ layer, Al-OH =
1.909(x 6), Ca-OH = 2.357 and 2.455 (each x 3) Å; the Ca ion is shifted 0.57 Å
from the centre of its octahedron, and approaches an interlayer water molecule
at 2.497 Å, resulting in 7-coordination. The remaining part of the interlayer
is disordered, and main and interlayer are connected by O-H...O hydrogen bonds.

1. R. ALLMANN, 1968. Neues Jb. Miner., Mh., 140.

WOODHOUSEITE
$CaAl_3(PO_4)(SO_4)(OH)_6$

T. KATO, 1977. Neues Jb. Miner., Mh., 54-58.

Rhombohedral, $R\bar{3}m$ a = 6.993, c = 16.386 Å, Z = 3. Mo radiation, R = 0.057.

Structure as previously described (1, 2).

1. Structure Reports, 11, 409.
2. T. KATO, 1971. Neues Jb. Miner., Mh., 241.

CHROMIUM(III) HYDROGEN SULPHATE HEPTAHYDRATE
$CrH(SO_4)_2 \cdot 7H_2O$

T. GUSTAFSSON, J.-O. LUNDGREN and I. OLOVSSON, 1977. Acta Cryst., B33, 2373-2376.

Monoclinic, $P2_1/c$, a = 7.371, b = 10.536, c = 16.703 Å, β = 109.80°, Z = 4. Mo
radiation, R = 0.045 for 2815 reflexions.

The structure contains $Cr(H_2O)_6^{3+}$ and H_3O^+ ions (Fig. 1) hydrogen bonded to
the sulphate ions.

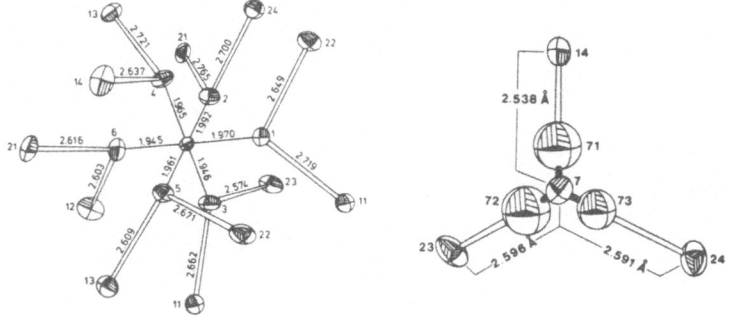

Fig. 1.   Environments of the $Cr(H_2O)_6^{3+}$ and H_3O^+ ions.

IRON(III) SULPHATE (Rhombohedral)
$Fe_2(SO_4)_3$

P.C. CHRISTIDIS and P.J. RENTZEPERIS, 1976. Z. Kristallogr., 144, 341-352.

Rhombohedral, $R\bar{3}$, a = 8.7901 Å, α = 55.87° (hexagonal cell has a = 8.2362, c =
22.1786 Å), D_m = 3.06, Z = 2. Mo radiation, R = 0.057 for 2361 reflexions.

[The structure has been determined independently (1).] It contains corner-
sharing FeO_6 octahedra and SO_4 tetrahedra (Fig. 1), Fe-O = 1.980-1.993, S-O =
1.463-1.468(2) Å. The monoclinic form is described in 2.

Atomic positions (hexagonal axes)

	x	y	z
Fe(1)	0	0	0.1441
Fe(2)	0	0	0.3507
S	0.2904	0.2882	0.2507
O(1)	0.1943	0.2048	0.1938
O(2)	0.2240	0.1456	0.2984
O(3)	0.4922	0.3634	0.2436
O(4)	0.2527	0.4393	0.2659

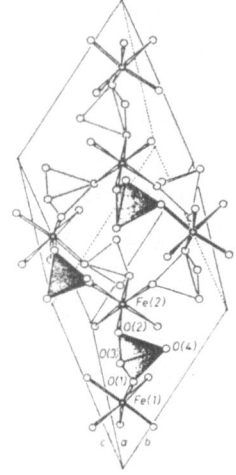

Fig. 1.   Rhombohedral cell of $Fe_2(SO_4)_3$.

1. Structure Reports, 39A, 313.
2. Ibid., 40A, 264; 41A, 349.

FERRINATRITE

$Na_3Fe(SO_4)_3 \cdot 3H_2O$

F. SCORDARI, 1977. Miner. Mag., 41, 375-383.

Trigonal, $P\bar{3}$, a = 15.566, c = 8.69 Å, D_m = 2.56, Z = 6. Mo radiation, R = 0.068 for 2378 reflexions.

The structure (Fig. 1) contains FeO_6 octahedra and SO_4 tetrahedra linked to form infinite Fe-O-S chains along c. These chains are joined by $NaO_5(H_2O)_2$ polyhedra and probably by hydrogen bonds. Fe-O = 1.98-2.01, S-O = 1.45-1.51, Na-O = 2.33-3.01(1) Å.

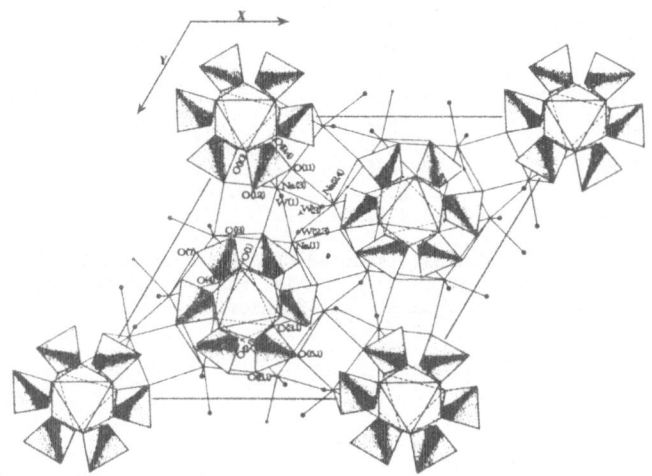

Fig. 1.   Structure of ferrinatrite.

CHLOROTIONITE
$K_2CuCl_2SO_4$

G. GIACOVAZZO, E. SCANDALE and F. SCORDARI, 1976. Z. Kristallogr., <u>144</u>, 226-237.

Orthorhombic, Pnma, a = 7.732, b = 6.078, c = 16.292 Å, D_m = 2.69, Z = 4. Mo
radiation, R = 0.032 for 893 reflexions.

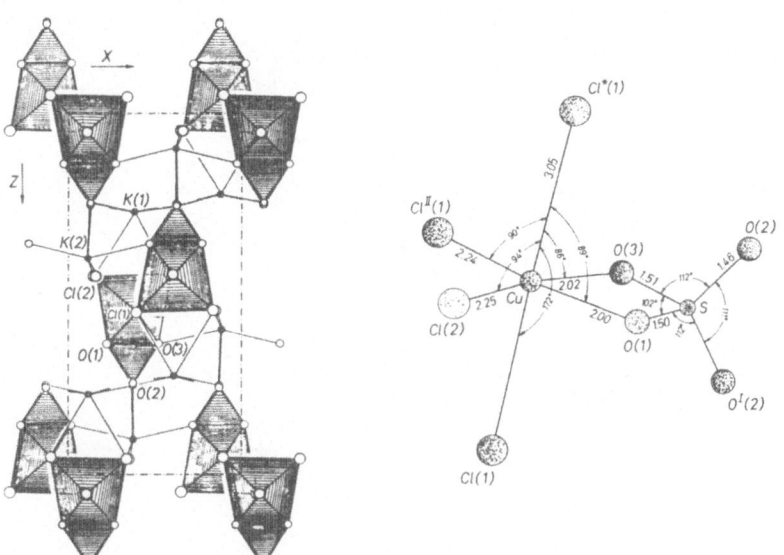

Fig. 1.   Structure of chlorotionite.

Atomic positions

	x	y	z
Cu	0.6212	3/4	0.4615
K(1)	0.3821	1/4	0.2808
K(2)	0.1193	3/4	0.4050
Cl(1)	0.6026	1/4	0.5485
Cl(2)	0.8402	3/4	0.5526
S	0.6296	3/4	0.3024
O(1)	0.7758	3/4	0.3629
O(2)	0.6324	0.9481	0.2518
O(3)	0.4752	3/4	0.3587

The structure (Fig. 1) contains distorted $CuCl_4O_2$ octahedra which share
Cl...Cl edges to form $[Cu_2Cl_4O_4(SO_4)_2^{12-}]_n$ double chains along $\underline{b}$. The chains are
linked by the K^+ ions which have 4 oxygen and 2 chlorine neighbours.

POTASSIUM CADMIUM SULPHATE
$K_2Cd_2(SO_4)_3$

S.C. ABRAHAMS and J.L. BERNSTEIN, 1977.   J. Chem Phys., $\underline{67}$, 2146-2150.

Orthorhombic, $P2_12_12_1$, a = 10.2082, b = 10.2837, c = 10.1661 Å, at 298°K, D_m =
3.70, Z = 4. Mo radiation, R = 0.023 for 3554 reflexions.

The structure (Fig. 1) is similar to that of cubic langbeinite ($\underline{1}$). S-O =
1.460-1.484(3), Cd-O = 2.239-2.327, K-O = 2.642-3.214 Å.

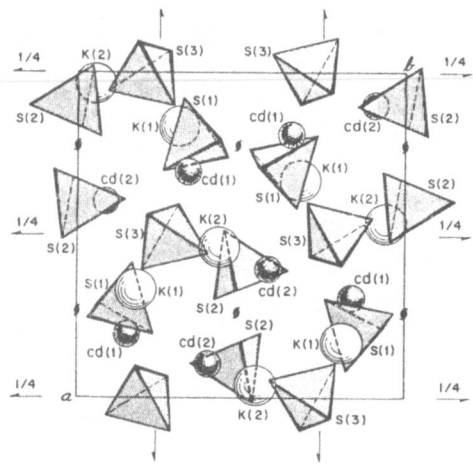

Fig. 1.   Structure of $K_2Cd_2(SO_4)_3$.

$\underline{1}$. Structure Reports, $\underline{21}$, 362.

CERIUM(IV) SULPHATE TETRAHYDRATE
$Ce(SO_4)_2 \cdot 4H_2O$

O. LINDGREN, 1977. Acta Chem. Scand., $A\underline{31}$, 453-456.

Orthorhombic, Pnma, a = 14.599, b = 11.006, c = 5.660 Å, Z = 4. Mo radiation,
R = 0.040 for 1409 reflexions.

Isostructural with the uranium salt (1). The structure (Fig. 1) contains
$Ce(SO_4)_2$ layers parallel to (100), linked by hydrogen bonding via the water molecules. Ce has eight-coordination (square-antiprism) to 4 O and 4 H_2O, Ce-O =
2.29-2.39(1) Å. S-O = 1.45-1.49, O-H...O = 2.69-2.74 Å.

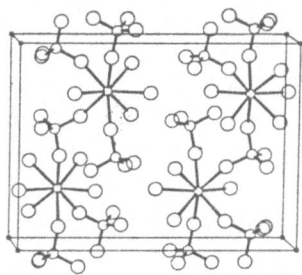

Fig. 1.   Structure of cerium(IV) sulphate tetrahydrate viewed approximately
            along c.

1. Structure Reports, 20, 354.

CERIUM(IV) HYDROXIDE SULPHATE HYDRATE
$Ce_2(SO_4)_3(OH)_2 \cdot 4H_2O$

O. LINDGREN, 1977. Acta Chem. Scand., A31, 163-166.

Monoclinic, A2/a, a = 15.583, b = 13.448, c = 6.748 Å, γ = 95.39°, Z = 4. Mo
radiation, R = 0.043 for 1425 reflexions.

The structure (Fig. 1) contains discrete hydroxy-bridged dimeric cations,
$(H_2O)_2Ce(OH)_2Ce(H_2O)_2{}^{6+}$, linked by sulphate ions to form a three-dimensional
structure. Ce has eightfold distorted-dodecahedral coordination, Ce-O = 2.22-
2.41 Å.

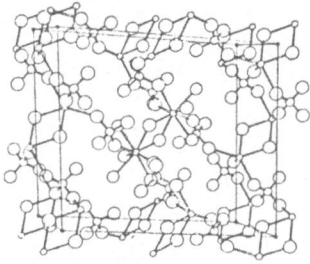

Fig. 1.   Structure of $Ce_2(SO_4)_3(OH)_2 \cdot 4H_2O$; for clarity complete Ce-O coordina-
            tion is shown only for the two polyhedra in the centre of the diagram.

SODIUM THIOSULPHATE PENTAHYDRATE
$Na_2S_2O_3 \cdot 5H_2O$

A. AYDIN URAZ and N. ARMAĞAN, 1977. Acta Cryst., B33, 1396-1399.

Monoclinic, $P2_1/c$, a = 5.941, b = 21.570, c = 7.525 Å, β = 103°55', Z = 4. Cu
radiation, R = 0.082 for 1301 reflexions (films, photometer intensities).

The structure is as previously described (1, 2), except that H(9) is considered to participate in hydrogen bonding (compare 2), O...S = 3.333(6), H...S =
2.42(2) Å, O-H...S = 147°.

1. Structure Reports, 16, 293; 21, 402.
2. Ibid., 37A, 312.

THIOSULPHATOPENTAAMMINECOBALT(III) CHLORIDE MONOHYDRATE
$[(NH_3)_5CoS_2O_3]Cl \cdot H_2O$

R.J. RESTIVO, G. FERGUSON and R.J. BALAHURA, 1977. Inorg. Chem., 16, 167-172.

Orthorhombic, Pnma, a = 13.092, b = 7.995, c = 10.299 Å, Z = 4. Mo radiation, R =
0.031 for 1065 reflexions.

The structure is as previously described (1). In the octahedral cation,
Co-N = 1.960-1.986(4), Co-S = 2.287(1), S-S = 2.048(2), S-O = 1.463(3) Å, and
there is an intra-ion hydrogen bond, N-H...O = 2.923 Å. The cation, chloride
anion, and water molecule are held together by a system of hydrogen bonds.

1. Structure Reports, 35A, 408.

AMMONIUM SELENATE
$(NH_4)_2SeO_4$

R.L. CARTER, C. KOERNTGEN and T.N. MARGULIS, 1977. Acta Cryst., B33, 592-593.

Monoclinic, C2/m, a = 12.152, b = 6.418, c = 7.711 Å, β = 115.50°, D_m = 2.194,
Z = 4. Cu radiation, R = 0.052 for 511 reflexions.

Isostructural with ammonium chromate (1); Se-O = 1.66(1) Å, O-Se-O = 108-
111°, N-H...O = 2.80-3.08(1) Å.

1. Structure Reports, 9, 183; 34A, 274; 35A, 384.

SODIUM TRIHYDROGEN SELENITE
$NaH_3(SeO_3)_2$

S. CHOMNILPAN, R. TELLGREN and R. LIMINGA, 1977. Acta Cryst., B33, 2108-2112.

Monoclinic, $P2_1/n$, a = 10.343, b = 4.837, c = 5.788 Å, β = 91.16°, Z = 2.
Neutron radiation, R = 0.040 for 757 reflexions.

The structure is as previously described (1, 2), except that both H atoms
are considered to be disordered. It contains hydrogen-bonded chains of H_2SeO_3
molecules and $HSeO_3^-$ ions (Fig. 1); Se-O = 1.702-1.718(1) Å, O-Se-O = 99.4-103.2°,

Na-O = 2.409-2.457(1) Å.

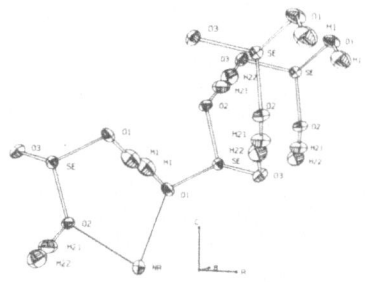

Fig. 1.   Structure of NaH$_3$(SeO$_3$)$_2$.

<u>1</u>. Structure Reports, <u>22</u>, 473; <u>33</u>A, 389; <u>37</u>A, 315; <u>38</u>A, 345; <u>41</u>A, 358.
<u>2</u>. S.F. KAPLAN, M.I. KAY and B. MOROSIN, 1970. Ferroelectrics, <u>1</u>, 31.

RUBIDIUM TRIHYDROGEN SELENITE
RbH$_3$(SeO$_3$)$_2$

I.   R. TELLGREN and R. LIMINGA, 1977. Ferroelectrics, <u>15</u>, 15-20.
II.   Ju.Z. NOZIK, L.E. FYKIN, L.A. MURADJAN and V.A. SARĪN, 1977. Kristallo-
      grafija, <u>22</u>, 69-72 [Soviet Physics - Crystallography, <u>22</u>, 37-39].

I.   Orthorhombic, P2$_1$2$_1$2$_1$, a = 5.919, b = 17.951, c = 6.252 Å (from <u>1</u>), Z = 4.
Neutron radiation, R = 0.028 for 998 reflexions.

II. Orthorhombic, P2$_1$2$_1$2$_1$, a = 17.909, b = 6.243, c = 5.899 Å, Z = 4. Neutron
radiation, R = 0.10 for 687 reflexions.

      The structure is as determined by X-ray methods (<u>1</u>, <u>2</u>), with hydrogen
positions similar to those in the isostructural ammonium compound (<u>3</u>). It
contains Rb$^+$ and HSeO$_3^-$ ions, and H$_2$SeO$_3$ molecules. The SeO$_3$ groups are connected
to form a three-dimensional structure by three almost linear O-H...O hydrogen bonds,
2.51, 2.57, and 2.60 Å.

<u>1</u>. Structure Reports, <u>39</u>A, 319.
<u>2</u>. Ibid., <u>38</u>A, 345.
<u>3</u>. Ibid., <u>39</u>A, 318; <u>40</u>A, 269.

MOLYBDOMENITE   (LEAD SELENITE)
PbSeO$_3$

M. KOSKENLINNA and J. VALKONEN, 1977. Cryst. Struct. Comm., <u>6</u>, 813-816.

Monoclinic, P2$_1$/m, a = 4.552, b = 5.525, c = 6.633 Å, β = 106.40°, Z = 2. Mo
radiation, R = 0.061 for 601 reflexions.

      The structure is essentially as previously described (<u>1</u>, where the cell is
<u>a</u>+<u>c</u>, <u>b</u>, -<u>a</u>). Se-O = 1.68, 1.70(2), Pb-O = 2.53-3.05(2) Å.

Atomic positions

	x	y	z
Pb	0.6389	1/4	0.3018
Se	0.0859	3/4	0.1622
O(1)	0.7068	3/4	0.1267
O(2)	0.1991	0.5144	0.3322

<u>1</u>. Structure Reports, <u>38A</u>, 346.

MANGANESE(III) SELENITE TRIHYDRATE
$Mn_2(SeO_3)_3 \cdot 3H_2O$

M. KOSKENLINNA and J. VALKONEN, 1977. Acta Chem. Scand., A<u>31</u>, 611-614.

Monoclinic, $P2_1/c$, a = 7.751, b = 10.330, c = 13.429 Å, β = 92.74°, D_m = 3.4, Z = 4. Mo radiation, R = 0.043 for 1839 reflexions.

Mn ions have Jahn-Teller-distorted octahedral coordinations; trigonal pyramidal SeO_3 groups bridge the octahedra to form a three-dimensional structure (Fig. 1). Se-O = 1.71-1.74(1) Å, O-Se-O = 97-100°, Mn-O = 1.89-2.26(1) Å.

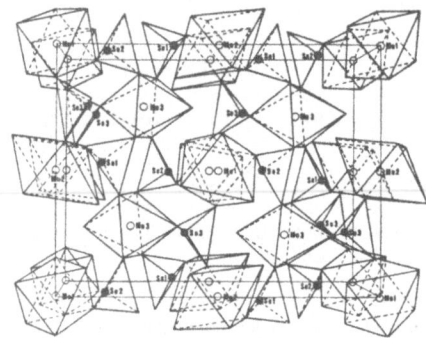

Fig. 1.   Structure of manganese(III) selenite trihydrate viewed along <u>b</u>.

MANGANESE(II) DISELENITE
$MnSe_2O_5$

M. KOSKENLINNA, L. NIINISTÖ and J. VALKONEN, 1976. Acta Chem. Scand., A<u>30</u>, 836-837.

Orthorhombic, Pbcn, a = 6.81, b = 10.63, c = 6.31 Å, D_m = 4.2, Z = 4. Mo radiation, R = 0.074 for 599 reflexions.

Atomic positions

	x	y	z
Se	0.3717	0.1556	0.5386
Mn	1/2	0.5580	1/4
O(1)	0.3324	0.4317	0.0479
O(2)	0.3045	0.7133	0.1558
O(3)	1/2	0.9283	1/4

Isostructural with $ZnSe_2O_5$ (<u>1</u>).   In the Se_2O_5 group, Se-O = 1.68, 1.64, 1.83(1) Å, O-Se-O = 96-103, Se-O-Se = 122°. The MnO_6 octahedron is almost regular, Mn-O = 2.18-2.21(1) Å.

<u>1</u>. Structure Reports, <u>40</u>A, 271.

MANGANESE(III) HYDROGEN SELENITE DISELENITE
$MnH(SeO_3)(Se_2O_5)$

M. KOSKENLINNA and J. VALKONEN, 1977. Acta Chem. Scand., A<u>31</u>, 638-640.

Monoclinic, $P2_1/n$, a = 7.451, b = 12.583, c = 7.575 Å, β = 92.82°, Z = 4. R = 0.035 for 1366 reflexions.

The structure contains trigonal pyramidal SeO_3 and O_2Se-O-SeO_2 groups, which bridge between MnO_6 Jahn-Teller-distorted octahedra (Fig. 1). Se-O = 1.66-1.80(1) Å, O-Se-O = 96-106, Se-O-Se = 126°, Mn-O = 1.92-1.99 (4 bonds), 2.12, 2.18 Å.

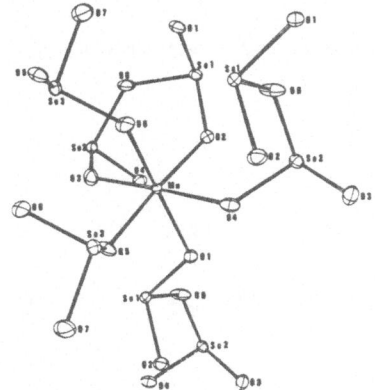

Fig. 1.   Bonding scheme in $MnH(SeO_3)(Se_2O_5)$.

PRASEODYMIUM TRIHYDROGEN BISELENITE DISELENITE
$PrH_3(SeO_3)_2(Se_2O_5)$

M. KOSKENLINNA and J. VALKONNEN, 1977. Acta Chem. Scand., A<u>31</u>, 457-460.

Monoclinic, $P2_1/c$, a = 12.933, b = 7.334, c = 10.811 Å, β = 91.68°, Z = 4. Mo radiation, R = 0.054 for 1517 reflexions.

The structure (Fig. 1) contains double layers parallel to (100), in which Se_2O_5 and one of the two non-equivalent SeO_3 groups link Pr atoms into an infinite two-dimensional network. Pr has 9-coordination, Pr-O = 2.40-2.87(2) Å. Se-O = 1.62-1.85 Å, O-Se-O = 90-105, Se-O-Se = 124°.

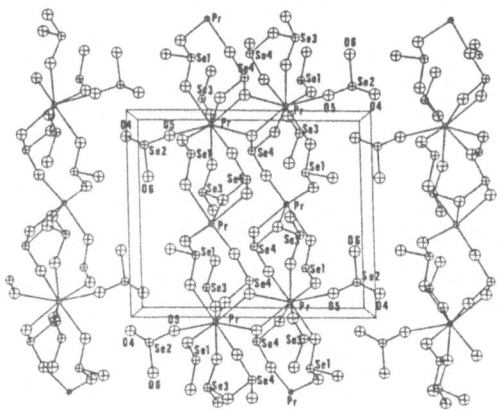

Fig. 1.   Structure of $PrH_3(SeO_3)_2(Se_2O_5)$ viewed along $\underline{b}$.

LITHIUM TELLURATE
Li_2TeO_4

F. DANIEL, J. MORET, E. PHILIPPOT and M. MAURIN, 1977.   J. Solid State Chem.,
22, 113-119.

Tetragonal, $P4_122$, a = 6.045, c = 8.290 Å, D_m = 4.48, Z = 4.   Mo radiation, R =
0.039 for 355 reflexions.

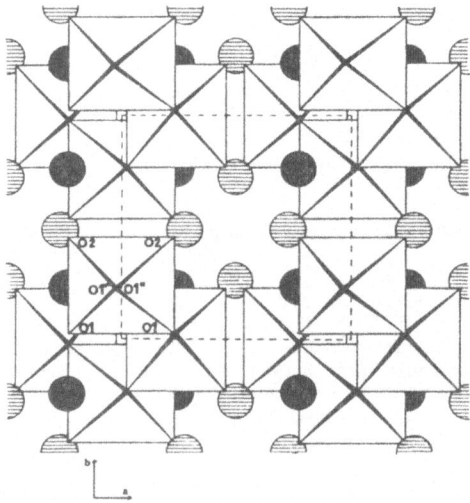

Fig. 1.   Structure of Li_2TeO_4; striped circles are octahedral Li(1), black
            circles are tetrahedral Li(2).

Atomic positions

			x	y	z
Te	in	4(a)	0	0.2686	0
O(1)		8(d)	0.225	0.018	-0.019
O(2)		8(d)	0.232	0.463	0.066
Li(1)		4(b)	1/2	0.26	0
Li(2)		4(c)	0.24	0.24	3/8

The structure (Fig. 1) is a distorted inverse spinel, with a chain of edge-sharing TeO_6 octahedra, and Li ions which have octahedral and tetrahedral coordinations. Te-O = 1.83-2.04(1), Li-O = 2.02-2.24(5) (octahedral), 1.95 and 2.06(4) Å (tetrahedral).

SODIUM METATELLURATE
Na_2TeO_4

I. B. KRATOCHVÍL and L. JENŠOVSKÝ, 1977. Acta Cryst., B<u>33</u>, 2596-2598.

Orthorhombic, Pbcn, a = 5.798, b = 12.24, c = 5.214 Å, D_m = 4.28, Z = 4. Cu radiation, R = 0.071 for 230 reflexions (film data).

Atomic positions

	x	y	z
Te	1/2	0.4284	3/4
Na(1)	1/2	0.7297	3/4
Na(2)	1/2	0.1374	3/4
O(1)	0.3246	0.5509	0.5707
O(2)	0.2977	0.3268	0.6144

TeO_6 octahedra share corners to form infinite chains which are held together by Na^+ ions (also approximately octahedral coordination) (Fig. 1). Te-O = 1.85, 1.97, 2.04(2), Na-O = 2.21-2.70(2) Å, O-Te-O = 77-96, 165, 170°.

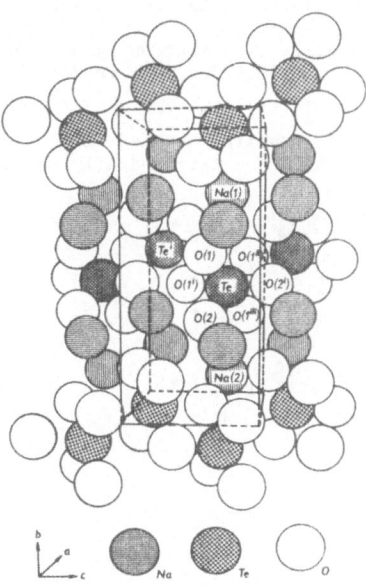

Fig. 1.   Structure of Na_2TeO_4 (orthorhombic).

II. F. DANIEL, M. MAURIN, J. MORET and E. PHILIPPOT, 1977. J. Solid State Chem.,
     $\underline{22}$, 385-391.

Monoclinic, P2$_1$/c, a = 10.632, b = 5.161, c = 13.837 Å, β = 103.27°, Z = 8. Mo
radiation, R = 0.039 for 1199 reflexions.

     The structure (Fig. 1) contains $[TeO_4{}^{2-}]_n$ octahedral chains along $\underline{b}$. Na
ions have distorted octahedral coordinations. Te-O = 1.84-2.05(1), Na-$\overline{O}$ = 2.22-
2.84(2) Å.

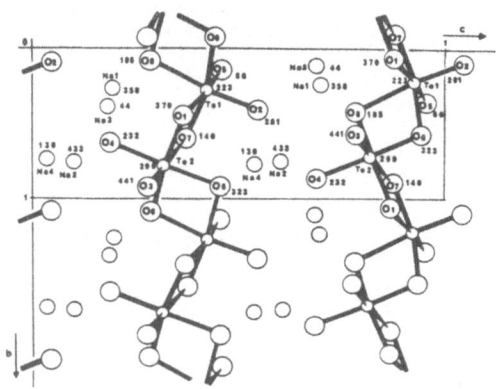

Fig. 1.   Structure of Na$_2$TeO$_4$ (monoclinic).

BERYLLIUM TELLURATE
Be$_4$TeO$_7$

M. TRÖMEL, J. MAETZ and M. MÜLLNER, 1977. Acta Cryst., B$\underline{33}$, 3959-3961.

Cubic, F$\bar{4}$3m, a = 7.5770 Å, D$_m$ = 4.2, Z = 4. Neutron powder data, with profile
analysis. Te in 4(a); Be in 16(e): x = 0.6270; O(1) in 24(f): x = 0.2532; O(2) in
4(d).

     The proposed structure (Fig. 1) contains octahedral TeO$_6$ groups, tetrahedrally-
coordinated Be atoms, and O atoms which are tetrahedrally coordinated by Be.
Te-O = 1.92, Be-O = 1.61 and 1.64 Å.

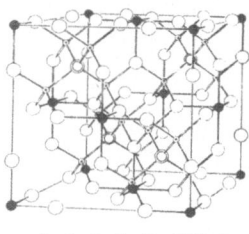

⊖ = Be, ● = Te, ○ = O(1), ◐ = O(2)

Fig. 1.   Structure of Be$_4$TeO$_7$.

MOLYBDENUM TELLURIUM OXIDE
Mo_5TeO_{16}

Y. ARNAUD and J. GUIDOT, 1977. Acta Cryst., B$\underline{33}$, 2151-2155.

Monoclinic, $P2_1/c$, a = 10.038, b = 14.431, c = 8.162 Å, β = 90.85°, Z = 4. Mo
radiation, R = 0.032 for 2003 reflexions.

The structure (Fig. 1) contains double layers of corner-sharing MoO_6 octahedra parallel to (100), separated by a mixed layer of tetrahedral TeO_4 units
alternating with MoO_6 octahedra along c. Mo-O = 1.69-2.39, Te-O = 1.81-2.45(1) Å.

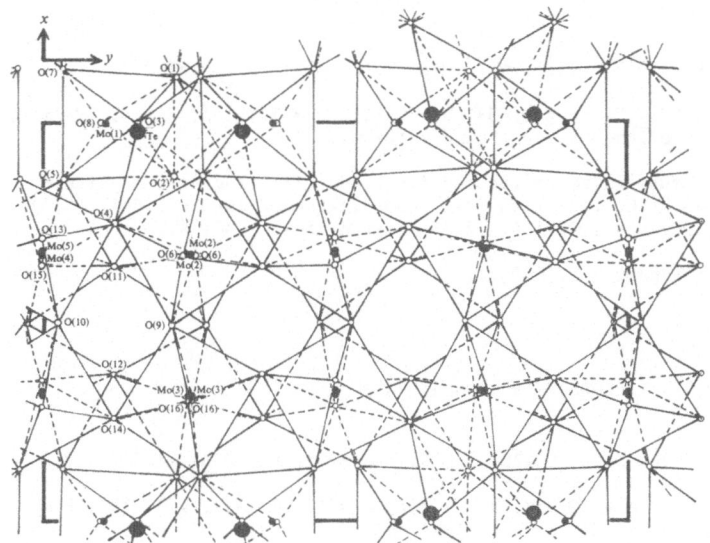

Fig. 1.   Structure of Mo_5TeO_{16}.

MACKAYITE
$FeTe_2O_5(OH)$

F. PERTLIK and A. GIEREN, 1977. Neues Jb. Miner., Mh., 145-154.

Tetragonal, $I4_1/acd$, a = 11.80, c = 15.10 Å, Z = 16. Mo radiation, R = 0.035
for 587 reflexions.

Atomic positions

		x	y	y
Fe in	16(d)	0	1/4	0.2304
Te	32(g)	0.2984	0.3475	0.1114
O(H)	16(f)	0.4294	0.6794	1/8
O(1)	16(f)	0.2608	0.5108	1/8
O(2)	32(g)	0.3824	0.3779	0.0053
O(3)	32(g)	0.3877	0.1744	0.0617

Structure as previously described (1).. Two $FeO_4(OH)_2$ octahedra share the
OH...OH edge to form pairs, which are further joined by corner sharing via O_2Te-O-
TeO_2 groups and by hydrogen bonds. Te has trigonal pyramidal coordination, Te-O =
1.90, 1.92, 1.99 (bridging in Te_2O_5 group) Å, with a fourth (axial) oxygen at
2.41 Å and an (equatorial) lone-pair completing a trigonal bipyramid. Fe-OH =
1.98, Fe-O = 2.01, 2.08 Å.

1. Structure Reports, 41A, 129.

TEINEITE
$CuTeO_3 \cdot 2H_2O$

H. EFFENBERGER, 1977. Tschermaks Min. Petr. Mitt., 24, 287-298.

Orthorhombic, $P2_12_12_1$, a = 6.634, b = 9.597, c = 7.428 Å, Z = 4. Mo radiation, R =
0.047 for 541 reflexions, synthetic material.

The structure is as previously described (1); it contains pyramidal TeO_3
groups, linked by Cu ions with (4+1)-coordination. Te-O = 1.86-1.89(1) Å,
O-Te-O = 96°, Cu-O = 1.93-1.97, 2.35 (next nearest 3.37) Å.

1. Structure Reports, 27, 635.

AMMINECOPPER(II) TELLURATE(IV) MONOHYDRATE
$Cu(NH_3)TeO_3 \cdot H_2O$

G.B. JOHANSSON and O. LINDQVIST, 1977. Acta Cryst., B33, 2418-2421.

Monoclinic, C2/c, a = 12.988, b = 7.333, c = 10.022 Å, β = 97.18°, Z = 8. Mo
radiation, R = 0.040 for 1283 reflexions.

The structure (Fig. 1) contains trigonal-pyramidal TeO_3^{2-} ions and square-
planar $Cu(NH_3)O_3$ groups which share oxygens atoms to form layers; the water molecule is situated between the layers. Te-O = 1.88(1), Cu-O = 1.96, Cu-N = 2.00 Å
(two long Cu-O distances, 2.60 and 3.07 Å, complete a distorted octahedron). N...O
and O...O distances are consistent with hydrogen bonding, but the water hydrogen
atoms do not conform to C2/c.

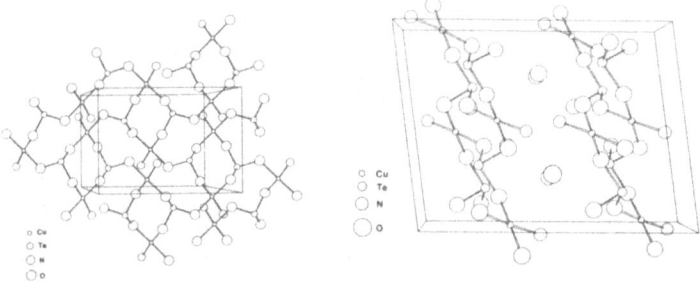

Fig. 1.   Views of the $Cu(NH_3)TeO_3 \cdot H_2O$ structure.

SODIUM CHLORATE   SODIUM BROMATE
$NaClO_3$ $NaBrO_3$

I. M.E. BURKE-LAING and K.N. TRUEBLOOD, 1977. Acta Cryst., B33, 2698-2699.
II. S.C. ABRAHAMS and J.L. BERNSTEIN, 1977. Ibid., B33, 3601-3604.

Cubic, $P2_13$, a = 6.57584, 6.70717 Å (at 298.2°K), D_m = 2.49, 3.34, Z = 4. Mo
radiation, R = 0.045, 0.013, 0.034 for 406, 671, 489 reflexions for chlorate (I,
film data), chlorate (II), bromate (II). Na in 4(a): x = 0.0687, 0.0775; Cl, Br
in 4(a): x = 0.4182, 0.4067; O in 12(b): (0.3035, 0.5924, 0.5047), (0.2882, 0.5964,
0.5085), from II for chlorate and bromate, respectively.

The structure is as previously described (1-4). Crystals of the two compounds
with identical chirality have opposite senses of optical rotation of 6328 Å
laser light (confirming the results of 4). Cl-O = 1.485, Br-O = 1.648 Å (from II,
no libration correction).

1. Strukturbericht, 1, 297, 308; 2, 407; Structure Reports, 21, 356; 23, 460.
2. W.H. ZACHARIASEN, 1965. Acta Cryst., 18, 703.
3. Strukturbericht, 1, 300, 310; 6, 93.
4. G. BEURSKENS-KERSSEN, J. KROON, H.J. ENDEMAN, J. van LAAR and J.M. BIJVOET,
   1963. Crystallography and Crystal Perfection, ed. G.N. RAMACHANDRAN, p. 225.
   London: Academic Press.

LITHIUM PERCHLORATE TRIHYDRATE
$LiClO_4 \cdot 3H_2O$

S. CHOMNILPAN, R. LIMINGA and R. TELLGREN, 1977. Acta Cryst., B33, 3954-3957.

Hexagonal, $P6_3mc$, a = 7.7192, c = 5.4531 Å, D_m = 1.89, Z = 2. Cu radiation, R =
0.017 for 175 reflexions.

Atomic positions

			x	y	z
Li	in	2(a)	0	0	0.2218
Cl		2(b)	1/3	2/3	0
O(1)		2(b)	1/3	2/3	0.2617
O(2)		6(c)	0.4346	-0.4346	-0.0894
O(3)		6(c)	0.1218	-0.1218	0.4726
H		12(d)	0.2290	0.3106	0.4774

Structure as previously described (1).

1. Strukturbericht, 3, 117, 468; Structure Reports, 33A, 450; 41A, 362.

POTASSIUM PERCHLORATE
$KClO_4$

G.B. JOHANSSON and O. LINDQVIST, 1977. Acta Cryst., B33, 2918-2919.

Orthorhombic, Pnma, a = 8.866, b = 5.666, c = 7.254 Å, Z = 4. Mo radiation, R =
0.031 for 533 reflexions.

$BaSO_4$-type structure, as previously described (1). Cl-O = 1.427-1.441(4),
K-O = 2.862-3.453(3) Å (12-coordination), O-Cl-O = 108.3-110.6°.

Atomic positions

	x	y	z
K	0.1809	1/4	0.3374
Cl	0.0679	1/4	0.8105
O(1)	0.1885	1/4	0.9419
O(2)	0.4197	0.5438	0.1952
O(3)	0.4253	1/4	0.5981

1. Strukturbericht, 1, 344, 372; 2, 84, 411, 414; Structure Reports, 21, 358.

ZINC PERCHLORATE HEXAHYDRATE
$Zn(ClO_4)_2 \cdot 6H_2O$

M. GHOSH and S. RAY, 1977. Z. Kristallogr., 145, 146-154.

Hexagonal pseudo-cell, $P6_3mc$, a = 7.715, c = 5.22 Å. Cu radiation, R = 0.10 for 75 reflexions. The arrangement of perchlorate ions and water molecules is similar to that in lithium perchlorate trihydrate (1). The distribution of Zn ions among twice their number of available sites gives rise to orthorhombic symmetry ($Pmn2_1$), and the crystals grow as three-component twins.

1. Strukturbericht, 3, 117, 468; Structure Reports, 33A, 450; 41A, 362.

URANYL DIPERCHLORATE HEPTAHYDRATE
$UO_2(ClO_4)_2 \cdot 7H_2O$

N.W. ALCOCK and S. ESPERÅS, 1977. J. Chem. Soc., Dalton, 893-896.

Orthorhombic, $Pca2_1$, a = 9.302, b = 14.692, c = 10.842 Å, Z = 4. Mo radiation, R = 0.046 for 493 reflexions.

The structure (Fig. 1) contains a pentagonal-bipyramidal $UO_2(H_2O)_5^{2+}$ cation, two ClO_4^- anions, and two additional water molecules, held together by hydrogen bonds. U-O(uranyl) = 1.71, U-OH$_2$ = 2.45 Å (mean values).

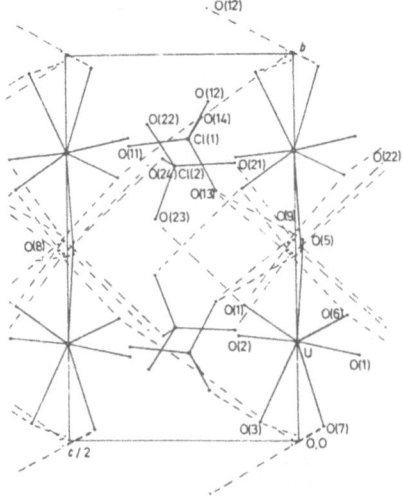

Fig. 1.   Structure of uranyl diperchlorate heptahydrate.

PRASEODYMIUM BROMATE NONAHYDRATE   YTTERBIUM BROMATE NONAHYDRATE
$Pr(BrO_3)_3.9H_2O$ $Yb(BrO_3)_3.9H_2O$

J. ALBERTSSON and I. ELDING, 1977. Acta Cryst., B33, 1460-1469.

Hexagonal, $P6_3/mmc$, a = 11.840, 11.706, c = 6.801, 6.647 Å, D_m = not given, 3.00,
Z = 2. Mo radiation, R = 0.034, 0.040 for 278, 188 reflexions.

    The structures are as previously described for the Sm salt (1); since all the
Ln bromates are isostructural, the Nd structure (2) should be described in $P6_3/mmc$.

1. Structure Reports, 34A, 362.
2. Strukturbericht, 7, 29, 135.

POTASSIUM HYDROGEN IODATE   (Orthorhombic)
$KIO_3.HIO_3$

G. KUNZE and S.A. HAMID, 1977. Acta Cryst., B33, 2795-2803.

Orthorhombic, Fdd2, a = 39.294, b = 8.157, c = 11.580 Å, D_m = 4.20, Z = 24. Mo
radiation, R = 0.059 for 580 reflexions.

    The structure (Fig. 1) contains trigonal pyramidal IO_3 groups, I-O = 1.76-
1.93 Å, O-I-O = 93-102°, with longer I...O contacts (2.40-3.28 Å) completing
octahedral coordination for two iodine atoms and sevenfold capped-octahedral
coordination for the third independent iodine [bond lengths here are from the text
of the paper; the abstract gives different values]. K ions have 8-coordination,
K-O = 2.66-3.23 Å. The monoclinic form is described in 1 [and an apparently different orthorhombic form in 2].

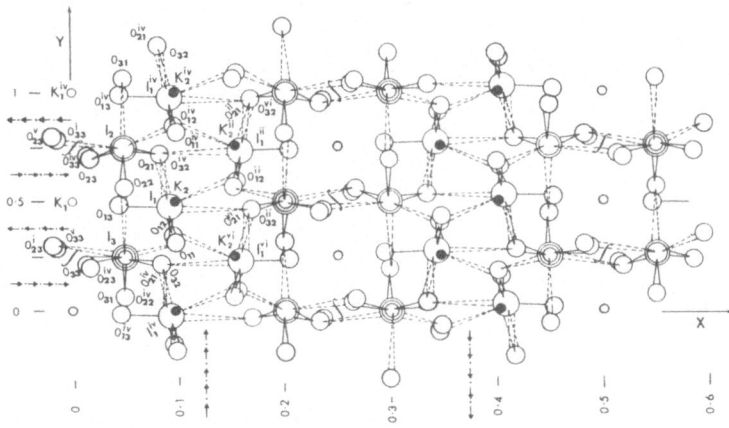

Fig. 1.   Structure of potassium hydrogen iodate.

1. Structure Reports, 37A, 320; 38A, 354.
2. Ibid., 41A, 390.

TETRAPOTASSIUM DIHYDROGENDECAOXODIIODATE(VII) OCTAHYDRATE
$K_4[H_2I_2O_{10}] \cdot 8H_2O$

I. MIKHAIL, 1977. Mater. Res. Bull., 12, 489-495.

Triclinic, PĪ, a = 7.161, b = 10.553, c = 7.081 Å, α = 98°1', β = 117°8', γ = 90°6', D_m = 2.48, Z = 1. Cu radiation, R = 0.066 for 877 reflexions (films, photo-meter intensities). Previous study in 1.

The structure contains $H_2I_2O_{10}{}^{4-}$ ions which consist of two $IO_5(OH)$ distorted octahedra sharing an O...O edge, I-O = 1.78-2.24 Å [values in diagram and table are inconsistent]. K ions have 6- and 8-coordination.

1. Structure Reports, 30A, 350.

ALUMINUM HYDROGEN IODATE HEXAHYDRATE
$AlH_2(IO_3)_5 \cdot 6H_2O$ $Al(IO_3)_3 \cdot 2HIO_3 \cdot 6H_2O$

P.D. CRADWICK and A.S. de ENDREDY, 1977. J. Chem. Soc., Dalton, 146-149.

Hexagonal, P6$_3$, a = 16.107, c = 12.378 Å, D_m > 2.9, Z = 6. Mo radiation, R = 0.029 for 2546 reflexions.

The structure (Fig. 1) may be formulated $[Al(OH_2)_6{}^{3+}][IO_3{}^-]_2[HI_2O_6{}^-] \cdot HIO_3$. It contains octahedral cations and trigonal pyramidal IO_3 groups linked by a system of hydrogen bonds. One very short I...O contact (2.48 Å) results in the $HI_2O_6{}^-$ grouping.

Fig. 1.   Structure of aluminum hydrogen iodate hexahydrate; broken lines indicate the short I...O contact.

GADOLINIUM IODATE
$Gd(IO_3)_3$

R. LIMINGA, S.C. ABRAHAMS and J.L. BERNSTEIN, 1977. J. Chem. Phys., 67, 1015-1023.

Monoclinic, P2$_1$/a, a = 13.4365, b = 8.5226, c = 7.1356 Å, β = 99.717°, at 297°K, D_m = 5.70, Z = 4. Mo radiation, R = 0.033 for 5012 reflexions.

The structure contains trigonal pyramidal iodate ions, linked by Gd ions with distorted dodecahedral coordination. I-O = 1.782-1.819(4) Å, O-I-O = 96.5-103.0° (each I has three further O neighbours at 2.741-3.194 Å); Gd-O = 2.278-2.556(4) Å.

LOW TRIDYMITE
SiO_2

W.H. BAUR, 1977. Acta Cryst., B33, 2615-2619.

Monoclinic, Cc, a = 18.494, b = 4.991, c = 23.758 Å, β = 105.79°, Z = 12. Data of 1 for synthetic material, R = 0.076 for 4117 reflexions.

The results are in agreement with those in 1 and 2, except that the disorder in 1 is not found. Si-O = 1.576-1.622, mean 1.597 Å.

1. Structure Reports, 42A, 393.
2. Ibid., 42A, 394.

COESITE
SiO_2

G.V. GIBBS, C.T. PREWITT and K.J. BALDWIN, 1977. Z. Kristallogr., 145, 108-123.

Monoclinic, C2/c, a = 7.135, b = 12.372, c = 7.174 Å, β = 120.36°, Z = 16. Mo radiation, R = 0.024 for 805 reflexions.

Atomic positions

	x	y	z
Si(1)	0.1403	0.1084	0.0723
Si(2)	0.5066	0.1580	0.5405
O(1)	0	0	0
O(2)	1/2	0.1161	3/4
O(3)	0.2660	0.1233	0.9400
O(4)	0.3110	0.1037	0.3280
O(5)	0.0177	0.2119	0.4784

The structure is as previously determined (1).

1. Structure Reports, 23, 340; 34A, 363.

LITHIUM METASILICATE
Li_2SiO_3

K.-F. HESSE, 1977. Acta Cryst., B33, 901-902.

Orthorhombic, $Cmc2_1$, a = 9.392, b = 5.397, c = 4.660 Å, D_m = 2.54, Z = 4. Mo radiation, R = 0.066 for 275 (of 1459) reflexions. Previous studies in 1.

Atomic positions

	x	y	z
Si in 4(a)	0	0.1703	0.4912
Li 8(b)	0.1737	0.3449	-0.0024
O(1) 8(b)	0.1446	0.3077	0.4108
O(2) 4(a)	0	0.1143	0.8461

The structure (Fig. 1) contains silicate chains along c with a repeat period of two tetrahedra, and tetrahedrally-coordinated Li ions. S̄i-O = 1.59 (terminal), 1.68(2) (bridging), Li-O = 1.96-2.17(1) Å (with a fifth O at 2.76 Å), Si-O-Si = 124°. The Na compound is isostructural (2), but Na has fivefold coordination.

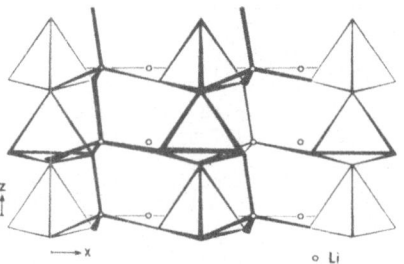

Fig. 1.   Structure of Li_2SiO_3.

1. Structure Reports, 20, 404; 21, 431; 33A, 452.
2. Ibid., 16, 334; 32A, 448.

RUBIDIUM BERYLLOSILICATE
$Rb_2Be_2Si_2O_7$

R.A. HOWIE and A.R. WEST, 1977. Acta Cryst., B33, 381-385.

Orthorhombic, P2nn, a = 8.92, b = 8.32, c = 5.15 Å, D_m = 3.2, Z = 2. Mo radiation,
R = 0.054 for 241 reflexions.

Atomic positions

	x	y	z
Rb(1)	0	0	0
Rb(2)	0.0590	1/2	0
Be	0.6850	0.7414	0.0000
Si	0.3595	0.1965	0.0103
O(1)	0.2380	0.2506	0.2177
O(2)	0.5204	0.2631	0.0697
O(3)	0.3744	0	0
O(4)	0.3050	0.2483	-0.2656

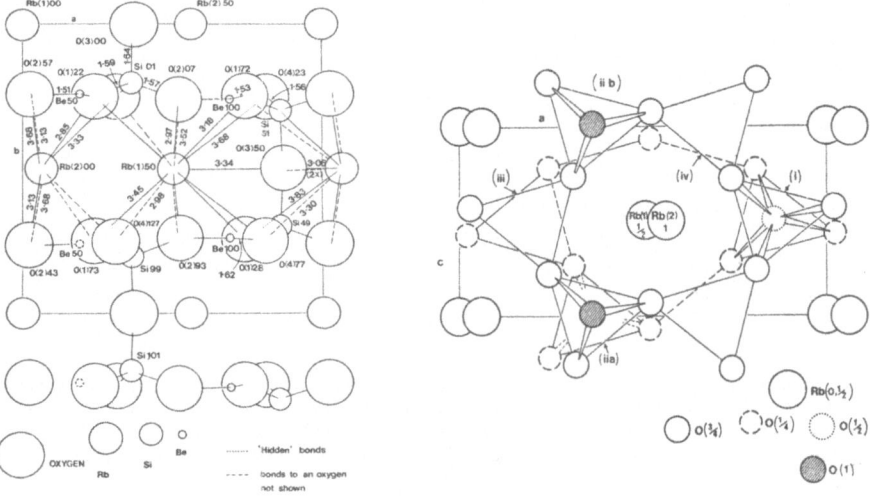

Fig. 1.   Projections of the rubidium beryllosilicate structure.

The structure (Fig. 1) contains an infinite beryllosilicate framework, with large cavities containing 13- and 14-coordinated Rb ions (Rb-O = 2.85-3.83 Å). The framework contains Si_2O_7 units whose non-bridging O atoms form the corners of BeO_3 groups; mean Be-O = 1.55, Si-O = 1.59 Å.

DICALCIUM SILICATE
β-Ca_2SiO_4

K.H. JOST, B. ZIEMER and R. SEYDEL, 1977. Acta Cryst., B<u>33</u>, 1696-1700.

Monoclinic, $P2_1/n$, a = 5.502, b = 6.745, c = 9.297 Å, β = 94.59°, Z = 4. Mo radiation, R = 0.048 for 859 reflexions.

Atomic positions

	x	y	z
Ca(1)	0.2738	0.3428	0.5694
Ca(2)	0.2798	0.9976	0.2981
Si	0.2324	0.7814	0.5817
O(1)	0.2864	0.0135	0.5599
O(2)	0.0202	0.7492	0.6919
O(3)	0.4859	0.6682	0.6381
O(4)	0.1558	0.6710	0.4264

An indexing error in a previous study (<u>1</u>) led to incorrect z parameters, but the structure is roughly as previously described. It contains $Si\bar{0}_4$ tetrahedra and 7- and 8-coordinate calcium ions (Fig. 1).

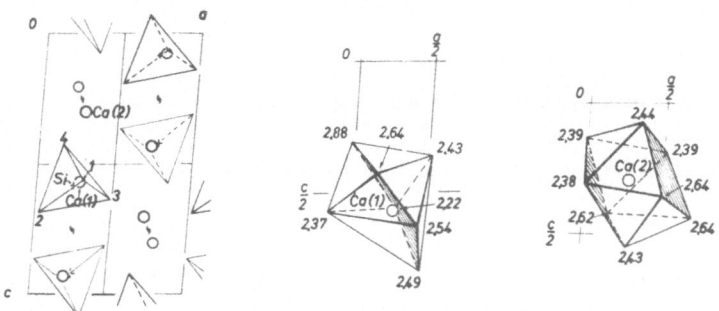

Fig. 1.   Structure of β-Ca_2SiO_4 and Ca coordination polyhedra.

<u>1</u>. Structure Reports, <u>16</u>, 335; <u>29</u>, 394.

POTASSIUM ALUMINUM SILICATE
$KAlSiO_4$

W.A. DOLLASE and W.P. FREEBORN, 1977. Amer. Min., <u>62</u>, 336-340.

Hexagonal, $P6_3mc$, a = 5.153, c = 8.682 Å, Z = 2. Mo radiation, R = 0.098 for 228 reflexions.

Atomic positions

			x	y	z
2 K	in	2(a)	0	0	0.25
2 Si		2(b)	1/3	2/3	0.437
2 Al		2(b)	1/3	2/3	0.056
2 O(1)		12(d)	0.333	0.720	0.258
6 O(2)		12(d)	0.614	0.019	0.993

The structure is the same as that of kalsilite ($\underline{1}$), but with further disordering of atom O(2) resulting in the higher-symmetry space group (numbering of oxygen atoms here is reversed with respect to that in $\underline{1}$).

$\underline{1}$. Structure Reports, $\underline{30}$A, 425.

LITHIUM SCANDIUM SILICATE
$LiScSi_2O_6$

F.C. HAWTHORNE and H.D. GRUNDY, 1977. Canad. Miner., $\underline{15}$, 50-58.

Monoclinic, C2/c, a = 9.803, b = 8.958, c = 5.352 Å, ß = 110.28°, Z = 4. Mo radiation, R = 0.024 for 799 reflexions (twinned crystal).

Atomic positions

	x	y	z
M(1)	0	0.8950	1/4
M(2)	0	0.2574	1/4
Si	0.2990	0.0868	0.2778
O(1)	0.1209	0.0835	0.1582
O(2)	0.3709	0.2486	0.3448
O(3)	0.3545	0.0057	0.0579

The material is a synthetic pyroxene with spodumene structure ($\underline{1}$), but with a different distortion of the tetrahedral chain.

$\underline{1}$. Strukturbericht, $\underline{2}$, 130, 527; Structure Reports, $\underline{39}$A, 355; $\underline{41}$A, 394.

CALCIUM SCANDIUM SILICATE
$Ca_3Sc_2Si_3O_{12}$

CALCIUM SCANDIUM GERMANATE
$Ca_3Sc_2Ge_3O_{12}$

CADMIUM SCANDIUM GERMANATE
$Cd_3Sc_2Ge_3O_{12}$

B.V. MILL', E.L. BELOKONEVA, M.A. SIMONOV and N.V. BELOV, 1977. Ž. Strukt. Khim., $\underline{18}$, 399-402 [J. Struct. Chem., $\underline{18}$, 321-323].

Cubic, Ia3d, a = 12.250, 12.512, 12.458 Å, Z = 8. Mo radiation, R = 0.049, 0.032, 0.054 for 71, 73, 73 reflexions. Garnet structure ($\underline{1}$); oxygen parameters are (-0.0400, 0.0501, 0.1589), (-0.0352, 0.0526, 0.1551), (-0.0344, 0.0530, 0.1556).

$\underline{1}$. Strukturbericht, $\underline{1}$, 363.

POTASSIUM BARIUM SILICOTANTALATES
$K_4BaTa_6Si_4O_{26}$ $K_3Ba_{1.5}Ta_6Si_4O_{26}$

J. CHOISNET, N. NGUYEN and B. RAVEAU, 1977. Mater. Res. Bull., 12, 91-96.

Hexagonal, P$\bar{6}$2m, a = 9.05, 9.04, c = 7.81, 7.79 Å, Z = 1. Powder data, $R(F^2)$ =
0.061 and 0.081 for 104 and 96 reflexions.

     Isostructural with similar materials (1). The data do not distinguish
between possible K/Ba distributions in 3(f) and 3(g) sites.

1. Structure Reports, 42A, 396.

IRON(II) SILICATE   COBALT(II) SILICATE
γ-Fe_2SiO_4 γ-Co_2SiO_4

F. MARUMO, M. ISOBE and S. AKIMOTO, 1977. Acta Cryst., B33, 713-716.

Cubic, Fd3m, a = 8.234, 8.140 Å, Z = 8. Mo radiation, R = 0.014 and 0.015 for
217 and 202 reflexions. x(O) = 0.3659 and 0.3673.

     Normal spinel structures, as previously reported (1), with a small amount
of inversion in the Fe compound (0.024 Fe in tetrahedral site). Final difference
maps indicate trigonally-deformed 3d-electron distributions.

1. Structure Reports, 40A, 280.

SODIUM COPPER SILICATE
$Na_2CuSi_4O_{10}$

K. KAWAMURA and A. KAWAHARA, 1977. Acta Cryst., B33, 1071-1075.

Triclinic, P$\bar{1}$, a = 10.613, b = 7.850, c = 6.944 Å, α = 118.20, β = 116.53, γ =
93.65°, D_m = 2.90, Z = 2. Mo radiation, R = 0.042 for 1741 reflexions.

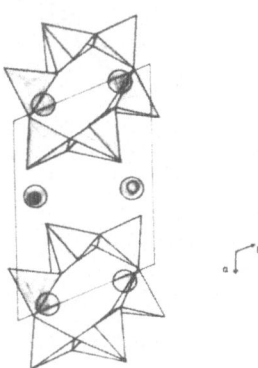

Fig. 1.   Structure of $Na_2CuSi_4O_{10}$.

The material is synthetic and is isostructural with litidionite, $NaKCuSi_4O_{10}$ (1), and fenaksite, $NaKFeSi_4O_{10}$ (2). The structure (Fig. 1) contains silicate pipes with a $Si_8O_{20}^{8-}$ repeat unit, linked by Cu and one Na, with the other Na in the silicate pipe. Cu has square-planar coordination to four O at 1.943-1.981(5) Å, with a fifth oxygen at 2.511(4) Å, and Na ions have 7 oxygen neighbours at 2.34-3.05 Å. Another synthetic phase is described in 3.

1. Structure Reports, 41A, 386.
2. Ibid., 35A, 480.
3. Ibid., 42A, 396.

SILVER DISILICATE
$Ag_6Si_2O_7$

M. JANSEN, 1977. Acta Cryst., B33, 3584-3586.

Monoclinic, P2/n, a = 10.264, b = 5.259, c = 8.052 Å, β = 110.5°, D_m = 6.58, Z = 2. Mo radiation, R = 0.062 for 1761 reflexions.

Atomic positions

	x	y	z
Ag(1)	0.1323	0.2276	0.5800
Ag(2)	0.6000	0.3018	0.2871
Ag(3)	0.3851	0.2425	0.4495
Si	-0.1380	0.2816	0.6628
O(1)	0.6338	0.5915	-0.1801
O(2)	0.5142	0.8395	0.2694
O(3)	0.6882	0.1986	0.0435
O(4)	1/4	0.8313	1/4

The structure (Fig. 1) contains $Si_2O_7^{6-}$ anions and Ag^+ ions with three- and four-coordinations, Ag-O = 2.18-2.76 Å.

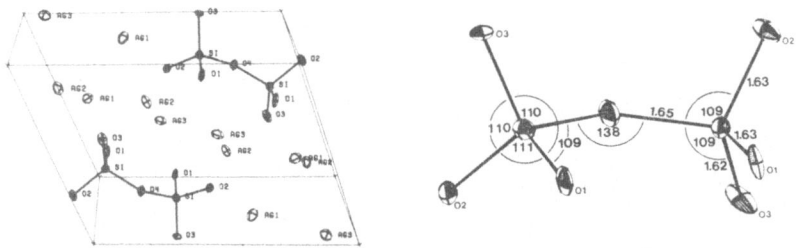

Fig. 1.   Structure of silver disilicate (left) and the $Si_2O_7^{6-}$ anion (right).

SODIUM ZINC SILICATES
$Na_2ZnSi_3O_8$

I. K.-F. HESSE, F. LIEBAU, H. BÖHM, P.H. RIBBE and M.W. PHILLIPS, 1977. Acta Cryst., B33, 1333-1337.

Monoclinic, $P2_1$, a = 6.660, b = 8.629, c = 6.411 Å, β = 103.70°, D_m = 2.94, Z = 2. Mo radiation, R = 0.033 for 1077 reflexions.

The results are in fair agreement with those of an independent study (1). The structure (Fig. 1) contains a framework of SiO_4 and ZnO_4 tetrahedra resembling that in paracelsian, $BaAl_2Si_2O_8$; corner-sharing SiO_4 tetrahedra form chains with a repeat period of three tetrahedra, and these are linked into Si_3O_8 layers, which are joined by ZnO_4 tetrahedra to give a three-dimensional framework. Na ions have seven- or eight-coordination.

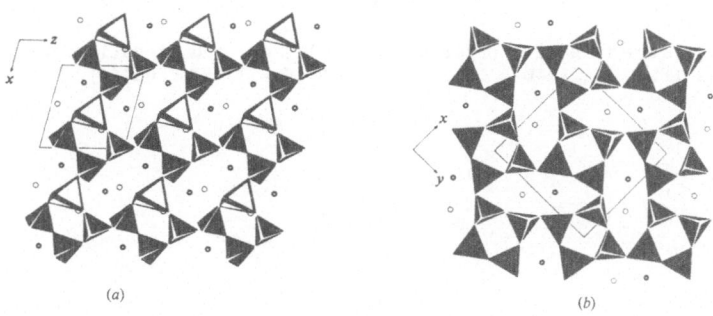

(a)   (b)

Fig. 1.   One tetrahedral layer in (a) $Na_2ZnSi_3O_8$, and (b) $BaAl_2Si_2O_8$; ZnO_4 tetrahedra are white.

$Na_2Zn_2Si_2O_7$

II. M.A. SIMONOV, E.L. BELOKONEVA and N.V. BELOV, 1977. Ž. Strukt. Khim., 18, 203-206 [J. Struct. Chem., 18, 173-175].

Orthorhombic, C2cm, a = 5.100, B = 9.276, c = 13.663 Å, Z = 4 [not 2]. Mo radiation, R = 0.060 for 1150 reflexions. Previous study in 2.

Atomic positions

	x	y	z
Zn	0.5584	0.2152	0.0802
Si	0.0446	0.3806	0.1405
Na(1)	0	0	0
Na(2)	0.0152	0.0374	1/4
O(1)	0.1379	0.3167	1/4
O(2)	0.7386	0.3866	0.1276
O(3)	0.1862	0.2663	0.0617
O(4)	0.6826	0.0365	0.1293

The structure contains a tetrahedral Zn, Si framework with $[Zn_2Si_2O_{11}]_n$ ribbons. Na ions have 8- and 6-coordinations. Zn-O = 1.90-2.05, Si-O = 1.57-1.68, Na-O = 2.37-2.82(1) Å.

1. Structure Reports, 40A, 295.
2. Ibid., 32A, 437.

LITHIUM CADMIUM SILICATE
Li_2CdSiO_4

C. RIEKEL, 1977. Acta Cryst., B33, 2656-2657.

Orthorhombic, Pmnb, a = 6.479, b = 10.715, c = 5.119 Å, Z = 4. Mo radiation, R = 0.06 for 1103 reflexions.

Atomic positions

	x	y	z
Cd	1/4	0.1627	0.1745
Si	1/4	0.3727	0.6676
O(1)	1/4	0.3535	0.3505
O(2)	-0.0466	0.0896	0.2890
O(3)	1/4	0.1929	0.7451
Li	0.499	0.910	0.327

Wurtzite structure with ordering of the Li, Cd, and Si cations in the tetrahedral sites (Fig. 1); Li-O = 1.95-2.00(2), Si-O = 1.627-1.654(6), Cd-O = 2.156-2.234(6) Å.

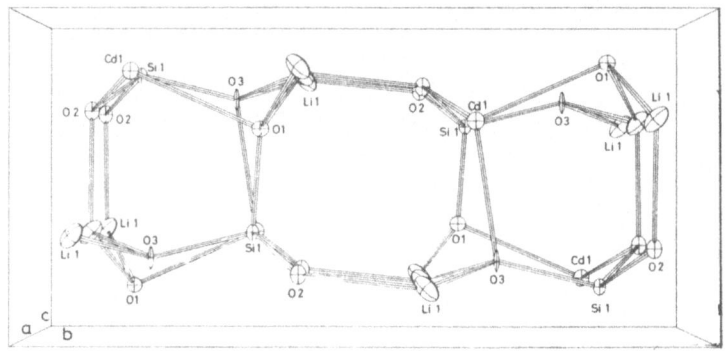

Fig. 1.   Structure of lithium cadmium silicate.

POTASSIUM CERIUM SILICATE
$K_2CeSi_6O_{15}$

I.   E.E. STRELKOVA, O.G. KARPOV, B.N. LITVIN, E.A. POBEDIMSKAJA and N.V. BELOV, 1977.   Kristallografija, 22, 174 [Soviet Physics - Crystallography, 22, 98-99].
II.   O.G. KARPOV, E.A. POBEDIMSKAJA and N.V. BELOV, 1977.   Ibid., 22, 382-384 [Ibid., 22, 215-217].

Monoclinic, I2/b [or B2/b? Both are given], a = 13.059, b = 11.854, c = 8.698 Å, γ = 90.15°, Z = 4. Mo radiation, R = 0.022 for 3793 reflexions.

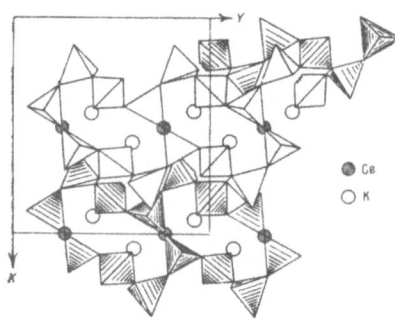

Fig. 1.   Structure of $K_2CeSi_6O_{15}$.

The structure (Fig. 1) contains a Si-O tetrahedral framework, Ce in octahedral coordination, and K in 8-coordination. Si-O = 1.582-1.642(3), Ce-O = 2.189-2.250(3), K-O = 2.717-3.212(3) Å.

CADMIUM NEODYMIUM SILICATE
$CdNd_4(SiO_4)_3O$

P.M. SIDOROV, E.L. BELOKONEVA, N.F. FEDOROV, T.A. TUNIN, M.A. SIMONOV and N.V. BELOV, 1977. Ž. Strukt. Khim., 18, 397-399 [J. Struct. Chem., 18, 319-320].

Hexagonal, $P6_3/m$, a = 9.562, c = 7.075 Å, D_m = 5.86, Z = 2. Mo radiation, R = 0.045 for 1380 reflexions.

Atomic positions

	x	y	z
Nd(1)	-0.0114	0.2317	1/4
Cd,Nd(2)	2/3	1/3	-0.0009
Si	0.3707	0.3988	1/4
O(1)	0.484	0.316	1/4
O(2)	0.473	0.596	1/4
O(3)	0.249	0.339	0.066
O(4)	0	0	1/4

Fluorapatite structure (1).

1. Strukturbericht, 2, 99.

SAMARIUM CHROMIUM SILICON OXYNITRIDE
$Sm_4CrSi_3NO_{12}$

M. MAUNAYE, C. HAMON, P. L'HARIDON and Y. LAURENT, 1976. Bull. Soc. Fr. Miner. Crist., 99, 203-205.

Hexagonal, $P6_3$, a = 9.469, c = 6.890 Å, D_m = 5.7, Z = 2. Mo radiation, R = 0.036 for 717 reflexions.

Apatite-like structure, as for $Sm_5Si_3NO_{12}$ (1), with 2 Cr and 2 Sm disordered in two 2(b) sites.

1. Structure Reports, 41A, 376.

ALBITE (LOW)
$NaAlSi_3O_8$

J.K. WINTER, S. GHOSE and F.P. OKAMURA, 1977. Amer. Min., 62, 921-931.

Triclinic, $C\bar{1}$, a = 8.15-8.28, b = 12.78-12.86, c = 7.17-7.18 Å, α = 94.3-93.3, β = 116.7-116.1, γ = 87.7-87.6°, at 25-970°C, Z = 4. Mo radiation, R = 0.032, 0.035, 0.039 for 1980, 1990, 2002 reflexions at 500, 750, 900°C, respectively.

Structure as previously described (1). The observed high anisotropy of the Na atom is considered to result strictly from thermal motion.

1. Strukturbericht, 3, 164, 546; Structure Reports, 22, 494; 27, 676; 29, 398; 34A, 368; 37A, 328.

BAOTITE
$Ba_4Ti_4(Ti,Nb,Fe)_4(Si_4O_{12})O_{16}Cl$

L.A. MURADJAN and V.I. SIMONOV, 1977. Kristallografija, 22, 486-493 [Soviet Physics - Crystallography, 22, 277-281].

Tetragonal, $I4_1/a$, a = 19.99, c = 5.908 Å, Z = 4. Previous data, R = 0.039 for 800 reflexions. Structure as previously described (1).

1. Structure Reports, 24, 504; 26, 536; 28, 274; 34A, 382.

BARYLITE
$BaBe_2Si_2O_7$

P.D. ROBINSON and J.H. FANG, 1977. Amer. Min., 62, 167-169.

Orthorhombic, Pnma, a = 9.82, b = 11.67, c = 4.69 Å, Z = 4. Mo radiation, R = 0.060 for 1045 reflexions.

Atomic positions

	x	y	z
Ba	0.1516	3/4	0.2470
Be	0.161	0.502	0.703
Si	0.0885	0.3779	0.1995
O(1)	0.4291	0.3846	0.2175
O(2)	0.1874	0.4712	0.3558
O(3)	0.1072	0.3868	0.8592
O(4)	0.1454	1/4	0.2958

The initial determination of the barylite structure (1) had several incorrect z parameters, and a redetermination is described in 2. The present study gives results in agreement with 2. Refinement in Pnma is satisfactory, but a second harmonic generation test indicates that the structure is non-centrosymmetric.

1. Structure Reports, 27, 704; 29, 398.
2. E. CANNILLO, A. DAL NEGRO and G. ROSSI, 1969. Rend. Soc. Ital. Mineral. Petrol., 26, 2.

BERYL (CAESIUM-LITHIUM)
$(Be,Li)_3Al_2Si_6O_{18}(Na,Cs)_{0.5} \cdot 0 \cdot 7H_2O$

F.C. HAWTHORNE and P. ČERNÝ, 1977. Canad. Miner., 15, 414-421.

Hexagonal, P6/mcc, a = 9.212, c = 9.236 Å, D_m = 2.781, Z = 2. R = 0.055 for 381 reflexions.

Beryl structure (1), with Li in the Be and not the Al site, and Na, Cs, and H_2O in 2(a) and 2(b) sites.

Atomic positions

			x	y	z
Si	in	12(ℓ)	0.3892	0.1189	0
Be		6(f)	1/2	0	1/4
Al		4(c)	2/3	1/3	1/4
O(1)		12(ℓ)	0.3048	0.2352	0
O(2)		24(m)	0.4983	0.1473	0.1445
Na		2(b)	0	0	0
Cs		2(a)	0	0	1/4
H_2O		2(a)	0	0	1/4

1. Strukturbericht, 1, 305, 329; Structure Reports, 13, 370; 30A, 427; 31A, 221; 33A, 480; 38A, 362.

BREWSTERITE
$(Sr,Ba)Al_2Si_6O_{16} \cdot 5H_2O$

J.L. SCHLENKER, J.J. PLUTH and J.V. SMITH, 1977. Acta Cryst., B33, 2907-2910.

Monoclinic, $P2_1/m$, a = 6.793, b = 17.573, c = 7.759 Å, β = 94.54°, Z = 2. Mo radiation, R = 0.046 for 2505 reflexions.

Structure as previously described (1).

1. Structure Reports, 29, 399.

BUSTAMITE   WOLLASTONITE
$(Ca,Mn)SiO_3$ $(Ca,Fe)SiO_3$

Y. OHASHI and L.W. FINGER, 1976. Carnegie Inst. Wash. Yearbook, 75, 746-753.

Structures as previously described (1, 2).

1. Structure Reports, 27, 701; 28, 246; 39A, 342.
2. Strukturbericht, 4, 71, 207; Structure Reports, 13, 359; 20, 409; 28, 246; 34A, 377.

CELADONITE (FERROUS)
$K(Fe,Mg)_2Si_4O_{10}(OH)_2$

A.P. ŽUKHLISTOV, B.B. ZVIAGIN, E.K. LAZARENKO and V.I. PAVLIŠIN, 1977.
Kristallografija, 22, 498-504 [Soviet Physics - Crystallography, 22, 284-288].

Monoclinic, C2/m, a = 5.23, b = 9.06, c = 10.15 Å, β = 100.6°, Z = 2. Electron diffraction, R = 0.11 for 163 reflexions.

Atomic positions

	x	y	z
Fe,Mg	1/2	0.1648	0
Si	0.421	0.3326	0.275
K	1/2	0	1/2
O(1)	0.367	0.324	0.111
OH	0.386	0	0.116
O(2)	0.187	0.254	0.338
O(3)	0.445	0	0.338

Structure as previously described (1).

1. Structure Reports, 21, 459.

CHABAZITE

DEHYDRATED CALCIUM-EXCHANGED
$Ca_2Al_4Si_8O_{24}.xH_2O$ (I)

DEHYDRATED CALCIUM-EXCHANGED CARBON MONOXIDE
$Ca_2Al_4Si_8O_{24}.xH_2O.1\cdot4CO$ (II)

DEHYDRATED SODIUM-EXCHANGED
$Na_{15}Al_{15}Si_{33}O_{96}.xH_2O$

I.   W.J. MORTIER, J.J. PLUTH and J.V. SMITH, 1977. Mater. Res. Bull., 12, 97-102.
II.   Idem, 1977. Ibid., 12, 103-108.
III. Idem, 1977. Ibid., 12, 241-249.

I, II
Rhombohedral, R3m, a = 9.442, 9.430 Å, α = 93.09, 93.06°, Z = 1. Mo radiation, R = 0.047 and 0.047 for 852 and 633 reflexions.

III
Monoclinic, C2/m (or lower symmetry), a = 19.319, b = 13.833, c = 11.849 Å, β = 113.48°. Mo radiation, R = 0.097 for 3457 reflexions (crystal may be twinned and only pseudo-monoclinic).

Structures of I and II are as previously described (1). Most Ca atoms (position 2(c)) are displaced into the large cavity from the centre of a 6-ring, Ca-O = 2.33 (x 3), 2.78 (x 3) Å. The centre of the ditrigonal prism contains 0.1 Ca (position 1(a)), apparently with 6 Ca-O = 2.8 Å. Adsorption of CO slightly decreases T-O and increases Ca-O distances; CO is absorbed onto the 2(c) Ca atom (Ca-C = 2.74 Å), is statistically distributed, and is probably canted off the triad axis.

For the Na compound, reduction of the ideal rhombohedral symmetry to monoclinic (or lower) is interpreted to result from near-elliptical distortion of most 8-rings to accommodate 57% of the Na^+ cations. All these cations have four framework oxygens as nearest neighbours, but positional disorder complicates interpretation of the Na-O distances, which range from 2.34 to 2.92 Å. Most other Na^+ cations (38%) are displaced into the large cavity from 6-rings, with three framework oxygens at 2.32-2.49 Å.

CHABAZITE (COPPER-EXCHANGED)
$(Cu,K)_2Al_4Si_8O_{24}.nH_2O$

IV.   J.J. PLUTH, J.V. SMITH and W.J. MORTIER, 1977. Mater. Res. Bull., 12, 1001-1007.

Rhombohedral, R3m, a = 9.411, 9.310 Å, α = 95.31, 92.01°, for hydrated and dehydrated samples, respectively, Z = 1. Mo radiation, R = 0.057 and 0.078 for 606 and 583 reflexions.

Structure as for Ca chabazites (I, II). Cu is near the centre of a 6-ring, 1.97 Å from three O(4) and 2.83 Å from three O(3).

1. Strukturbericht, 3, 151, 526; Structure Reports, 17, 564; 22, 513; 27, 691; 28, 250; 29, 401.

HYDROXYL-CHONDRODITE
$Mg_5(SiO_4)_2(OH)_2$ $Mg(OH)_2 \cdot 2MgSiO_4$

K. YAMAMOTO, 1977. Acta Cryst., B33, 1481-1485.

Monoclinic, $P2_1/b$, a = 4.752, b = 10.350, c = 7.914 Å, α = 108.71°, Z = 2. Cu radiation, R = 0.042 for 714 reflexions.

The material is synthetic, and the structure is as previously described for F-rich natural chondrodite (1). Most of the interatomic distances (Fig. 1) are longer than those in 1; statistical distribution is suggested for the OH orientation.

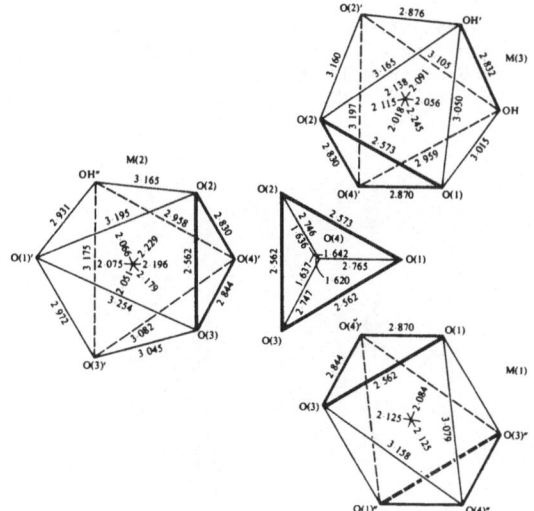

Fig. 1.   Exploded diagram of the cation coordination polyhedra in OH-chondrodite; heavier lines are edges shared between two octahedra, and double lines are edges shared between a tetrahedron and an octahedron.

1. Strukturbericht, 2, 119, 517; Structure Reports, 35A, 469.

CLINOENSTATITE
$(Mg,Ca)SiO_3$

Y. OHASHI and L.W. FINGER, 1976. Carnegie Inst. Wash. Yearbook, 75, 743-746.

Monoclinic, $P2_1/c$, a $\sim$ 9.6, b $\sim$ 8.8, c $\sim$ 5.2 Å, β $\sim$ 108°, for three samples with Ca content 0, 0.056, and 0.084, Z = 8. Mo radiation, R = 0.03-0.06.

Structure as previously described (1), with Ca confined to the M(2) site.

1. Structure Reports, 24, 465.

CLINOHEDRITE
$CaZnSiO_4 \cdot H_2O$

C.C. VENETOPOULOS and P.J. RENTZEPERIS, 1976.   Z. Kristallogr., 144, 377-392.

Monoclinic, Cc, a = 5.090, b = 15.829, c = 5.386 Å, β = 103.26°, D_m = 3.3, Z = 4.
Mo radiation, R = 0.041 for 1294 reflexions.

Atomic positions

	x	y	z
Zn	0	0.2493	0
Ca	0.3480	0.5724	0.6384
Si	0.0139	0.3618	0.5183
O(1)	0.1415	0.9591	0.8564
O(2)	0.1876	0.1443	0.9489
O(3)	0.1410	0.2934	0.3476
O(4)	0.1279	0.3406	0.8155
O(5)	0.1067	0.5459	0.9479

The structure differs substantially from that of 1 (which seems to contain errors of sign in y and z coordinates). It contains $S\overline{i}O_4$ and ZnO_4 tetrahedra, which share corners with each other and with edge-sharing CaO_6 octahedral chains (Fig. 1). Si-O = 1.61-1.65, Zn-O = 1.95-1.97, Ca-O = 2.30-2.45, O-H...O = 2.68, 2.79 Å.

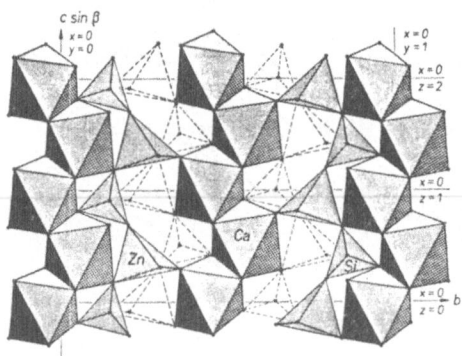

Fig. 1.   Structure of clinohedrite.

1. Structure Reports, 28, 261.

CORDIERITE (LOW)
$Mg_2Al_4Si_5O_{18} \cdot nH_2O$

J.P. COHEN, F.K. ROSS and G.V. GIBBS, 1977. Amer. Min., 62, 67-78.

Orthorhombic, Cccm, a = 17.079, b = 9.730, c = 9.356 Å, D_m = 2.57, Z = 4.   X-ray and neutron diffraction data (no details).

The structure is as previously described (1). Alkali metal ions ($Na_{0.05}K_{0.02}$-$Ca_{0.02}$) are centred about position (0,0,0) and 0.56 water molecules are fourfold disordered around (0,0,1/4).

1. Structure Reports, 8, 222; 9, 258; 31A, 223.

DEERITE
$Fe_9O_3[Si_6O_{17}](OH)_5$

M.E. FLEET, 1977. Amer. Min., <u>62</u>, 990-998.

Monoclinic, $P2_1/a$, a = 10.786, b = 18.88, c = 9.564 Å, β = 107.45°, D_m = 3.84,
Z = 4. Mo radiation, R = 0.07 for 515 reflexions.

The structure of a pseudo-cell, $P2_1/a$, c' = c/3, disordered Si and O positions,
has been determined. This structure contains two structural units continuous along
<u>c</u>: a strip of edge-shared FeO_6 octahedra, six octahedra wide and parallel to {110},
and a hybrid single-double Si_6O_{17} silicate chain.

DIOPTASE
$CuSiO_3 \cdot H_2O$

P.H. RIBBE, G.V. GIBBS and M.M. HAMIL, 1977. Amer. Min., <u>62</u>, 807-811.

Rhombohedral, $R\bar{3}$, a = 14.566, c = 7.778 Å, Z = 18. Mo radiation, R = 0.039
for 969 reflexions.

Atomic positions

	x	y	z
Cu	0.4065	0.4025	0.0630
Si	0.1756	0.2174	0.0413
O(1)	0.0715	0.1809	-0.0827
O(2)	0.2807	0.2995	-0.0641
O(3)	0.1599	0.2678	0.2139
O(W)	0.1422	0.1820	0.5785

The structure is as previously described (<u>1</u>). It contains Si_6O_{18} cyclic
anions, Si-O = 1.645, 1.646 (ring), 1.600, 1.617 Å (terminal), linked by Cu ions
with tetragonally-distorted octahedral coordination, Cu-O = 1.952-1.983, $Cu-OH_2$ =
2.502, 2.648 Å. Puckered trigonal rings of six water molecules with an ice-like
configuration are sandwiched between the anions.

<u>1</u>. Structure Reports, <u>16</u>, 348; <u>19</u>, 465.

ERIONITE   (Dehydrated)
$(Na,Mg,Ca)_2K_2Ca_2Al_9Si_{27}O_{72}$

J.L. SCHLENKER, J.J. PLUTH and J.V. SMITH, 1977. Acta Cryst., B<u>33</u>, 3265-3268.

Hexagonal, $P6_3/mmc$, a = 13.252, c = 14.810 Å, Z = 1. Mo radiation, R = 0.062
for 642 reflexions. Mineral from Mazé, Japan, dehydrated at 300°C under vacuum.

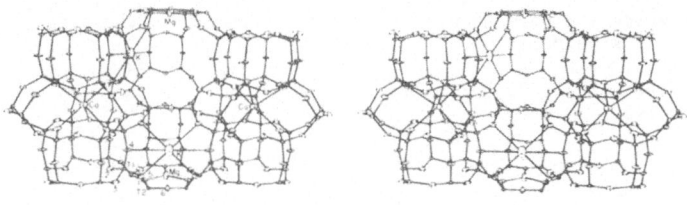

Fig. 1.   Stereoview of the structure of dehydrated erionite.

The framework is as in hydrated erionite (1), but with about 6% of the single six-membered rings stacked as in offretite. The centre of the cancrinite cage is occupied by trigonal-prismatically coordinated Ca (instead of K as in the hydrated variety), with K moved to a planar six-coordinated site at the centre of an eight-membered ring. The Na, Mg and remaining Ca ions are bonded to three oxygens. There is Si/Al disorder. See Fig. 1.

1. Structure Reports, 23, 494; 34A, 375.

FAYALITE   HORTONOLITE
Fe_2SiO_4 $(Fe,Mg,Mn)_2SiO_4$

R.M. HAZEN, 1977. Amer. Min., 62, 286-295.

Orthorhombic, Pbnm, a $\sim$ 4.8, b $\sim$ 10.4, c $\sim$ 6.1 Å, for various temperatures and pressures, Z = 4. Mo radiation, R = 0.05-0.14 for 174-1137 reflexions.

Olivine structure (1). SiO_4 tetrahedra are little affected, but octahedra expand and compress with increasing temperature and pressure, respectively.

1. Strukturbericht, 1, 352.

FELDSPAR
$(K,Na)AlSi_3O_8$

P.M. FENN and G.E. BROWN, 1977. Z. Kristallogr., 145, 124-145.

Monoclinic, C2/m, a = 8.434, b = 13.015, c = 7.172 Å, β = 116.09°, Z = 4. Mo radiation, R (weighted) = 0.029 for 1126 reflexions.

Typical feldspar structure (e.g. sanidine, 1) with two sites for the alkaki metal ions separated by 0.08 Å, M(1) = 0.5K, M(2) = 0.08K + 0.42Na.

1. Strukturbericht, 3, 161, 547; Structure Reports, 28, 242; 38A, 375; 41A, 395.

FERROBUSTAMITE
$(Ca,Fe,Mn)SiO_3$

T. YAMANAKA, R. SADANAGA and Y. TAKÉUCHI, 1977. Amer. Min., 62, 1216-1224.

Triclinic, Al̄, a = 7.862, b = 7.253, c = 13.967 Å, α = 89°44', β = 95°28', γ = 103°29', Z = 12. Mo radiation, R = 0.035 for 3083 reflexions.

Structure as previously described (1, 2). This Ca-rich specimen possesses pseudo-mirrors which are more nearly true mirror planes than those in 2. Fe (Mn) is concentrated in site M(3).

1. Structure Reports, 27, 701; 28, 246.
2. Ibid., 39A, 342.

GERSTMANNITE
$(Mn,Mg)Mg(OH)_2[ZnSiO_4]$

P.B. MOORE and T. ARAKI, 1977. Amer. Min., <u>62</u>, 51-59.

Orthorhombic, Bbcm, a = 8.185, b = 18.65, c = 6.256 Å, D_m = 3.68, Z = 8. Mo
radiation, R = 0.042 for 1091 reflexions.

    The structure (Fig. 1) contains $[MnMgO_3(OH)_2]$ octahedral sheets linked to
$[ZnSiO_4]$ tetrahedral sheets which are parallel to (010). Mean bond distances are
Mn-O = 2.21, Mg-O = 2.08, Zn-O = 1.95, Si-O = 1.64 Å.

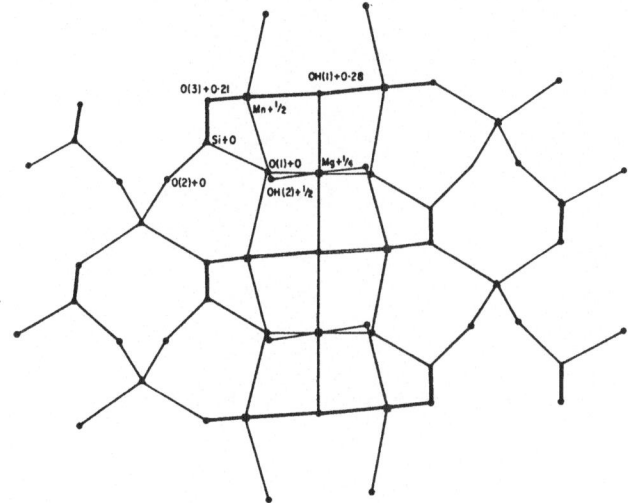

Fig. 1.   Structure of gerstmannite viewed down <u>c</u>.

SUB-SILICIC HASTINGSITE
$NaCa_2(Mg,Fe,Al)_5Si_5Al_3O_{22}(OH)_2$ (idealized)

F.C. HAWTHORNE and H.D. GRUNDY, 1977. Miner. Mag., <u>41</u>, 43-50.

Monoclinic, C2/m, a = 9.866, b = 18.014, c = 5.355 Å, β = 105.08°, Z = 2. Mo
radiation, R = 0.041 for 1263 reflexions.

    Amphibole structure (see e.g. ferrotschermakite, <u>1</u>).

<u>1</u>. Structure Reports, <u>39</u>A, 342.

HELLANDITE
$(Ca_{5.5}Ln_5\square_{1.5})(AlFe)(OH)_4[Si_8B_8O_{40}(OH)_4]$

M. MELLINI and S. MERLINO, 1977. Amer. Min., <u>62</u>, 89-99.

Monoclinic, P2/a, a = 18.99, b = 4.715, c = 10.30 Å, β = 111.4°, D_m = 3.63, Z = 1.
Mo radiation, R = 0.05 for 1480 reflexions.

The structure (Fig. 1) contains infinite silicoborate chains $[Si_4B_4O_{20}(OH)_2]$ parallel to $\underline{c}$ at $y \sim 1/2$. Fe/Al in octahedral coordination and Ca/Y/Ln in square antiprismatic coordination are all at $y \sim 0$, and connect the silicoborate chains.

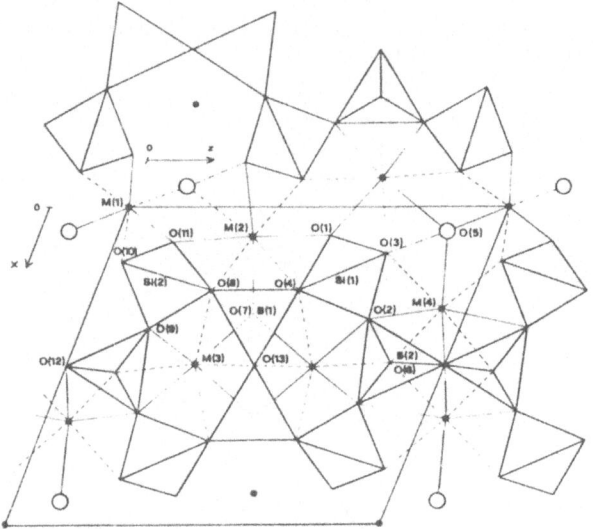

Fig. 1.   Structure of hellandite viewed along $\underline{b}$; dot-dashed line is a hydrogen bond.

HOLDENITE
$(Mn,Mg)_6Zn_3(OH)_8(AsO_4)_2(SiO_4)$

P.B. MOORE and T. ARAKI, 1977. Amer. Min., 62, 513-521.

Orthorhombic, Abma, a = 11.99, b = 31.46, c = 8.697 Å, D_m = 4.11, Z = 8. Mo radiation, R = 0.064 for 3478 reflexions.

An open tetrahedral $[Zn_2SiO_6(OH)_2]_n$ sheet parallel to (010) links to AsO_4 tetrahedra to form thick slabs of composition $[Zn_2As_2SiO_{12}(OH)_2]_n$. Insular $Zn(OH)_4$ tetrahedra also occur and these link to complex aggregates of $Mn(O,OH)_6$ condensed octahedra to form a framework structure. Mean interatomic distances are Mn-O = 2.15-2.22, Zn-O = 1.95, 1.98, As-O = 1.69, Si-O = 1.64 Å. There are five hydrogen bonds, O-H...O = 2.89-3.22 Å.

HYALOPHANE
$(K,Ba)(Al,Si)_4O_8$

R. de PIERI, S. QUARENI and K.M. HALL, 1977. Acta Cryst., B33, 3073-3076.

Monoclinic, C2/m, a = 8.557, 8.556, b = 13.040, 13.045, c = 7.200, 7.189 Å, β = 115.69, 115.63°, for natural and heat-treated specimens, Z = 4. Mo radiation, R = 0.084 and 0.080 for 1311 and 1122 reflexions (film data).

Atomic positions (natural specimen, similar values for heat-treated specimen)

	x	y	z
K,Ba	0.2826	0	0.1325
T(1)	0.0085	0.1832	0.2243
T(2)	0.7044	0.1178	0.3453
O(A1)	0	0.1431	0
O(A2)	0.6265	0	0.2854
O(B)	0.8255	0.1406	0.2253
O(C)	0.0296	0.3103	0.2570
O(D)	0.1847	0.1261	0.4020

Feldspar structure (e.g. 1), with only slight differences between the natural and heat-treated specimens (in spite of large differences in optical properties).

1. Structure Reports, 28, 243.

ILVAITE
$CaFe_2(Fe,Mn)OSi_2O_7(OH)$

N. HAGA and Y. TAKEUCHI, 1976. Z. Kristallogr., 144, 161-174.

Orthorhombic, Pbnm, a = 8.818, b = 13.005, c = 5.853 Å. Z = 4. Neutron radiation, R = 0.06 for 550 reflexions.

The structure is as determined by X-ray methods (1). M(1) contains Fe^{2+} and Mn^{2+}, and M(2) Fe^{2+} and Fe^{3+}. O-H = 0.98(1) Å.

1. Structure Reports, 11, 463; 15, 309; 17, 569; 38A, 369; 41A, 385.

INDIALITE
$(Mg,Fe)_2Al_4Si_5O_{18}$

E.P. MEAGHER and G.V. GIBBS, 1977. Canad. Miner., 15, 43-49.

Hexagonal, P6/mcc, a = 9.800, c = 9.345 Å, Z = 2. Mo radiation, R = 0.061 for 168 reflexions.

Atomic positions

	in	x	y	z
Mg,Fe	4(c)	1/3	2/3	1/4
Al,Si(1)*	6(f)	1/2	1/2	1/4
Al,Si(2)†	12(ℓ)	0.3727	0.2668	0
O(1)	24(m)	0.4851	0.3494	0.1445
O(2)	12(ℓ)	0.2305	0.3093	0

* 0.72Al + 0.28Si
† 0.30Al + 0.70Si

Isostructural with beryl (1), the structure being related to that of ortho-rhombic low-cordierite (2).

1. Strukturbericht, 1, 305, 329; Structure Reports, 13, 370; 30A, 427; 31A, 221; 33A, 480; 38A, 362.
2. Structure Reports, 8, 222; 9, 258; 31A, 223.

KASOLITE
$Pb(UO_2)SiO_4 \cdot H_2O$

A. ROSENZWEIG and R.R. RYAN, 1977. Cryst. Struct. Comm., 6, 617-621.

Monoclinic, $P2_1/c$, a = 6.704, b = 6.932, c = 13.252 Å, β = 104.22°, Z = 4. Mo
radiation, R = 0.034 for 885 reflexions.

Atomic positions

	x	y	z
U	0.4757	0.2299	0.1788
Pb	0.0754	0.0706	0.3821
Si	0.4079	0.7120	0.0734
O(1)	0.205	0.276	0.166
O(2)	0.250	0.697	0.300
O(3)	0.180	0.638	0.026
O(4)	0.451	0.267	0.009
O(5)	0.526	0.559	0.167
O(6)	0.418	0.903	0.147
O(7)	0.940	0.932	0.089

    The structure is essentially as previously described (1, compare 2), with
the water molecule, O(7), now being located, coordinated to Pb.

1. Structure Reports, 29, 405.
2. Ibid.', 28, 278.

KILLALAITE
$Ca_3(HSi_2O_7)(OH)$

H.F.W. TAYLOR, 1977. Miner. Mag., 41, 363-369.

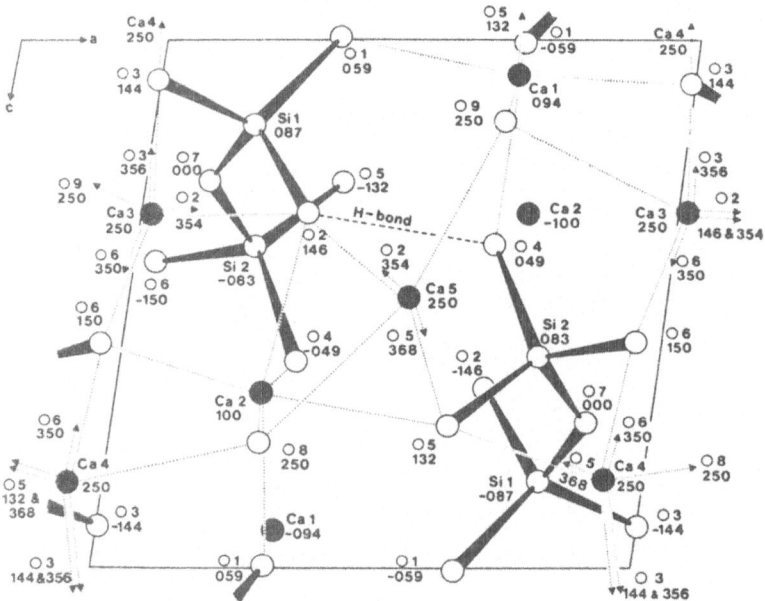

Fig. 1.   Structure of killalaite.

Monoclinic, $P2_1/m$, a = 6.807, b = 15.459, c = 6.811 Å, β = 97.76°, D_m = 2.88, Z = 4 (pseudo-cell; true cell is B-centred with doubled $\underline{a}$ and $\underline{c}$). Mo radiation, R = 0.039 for 1425 reflexions.

The composition is $Ca_{3+x}(H_{1-2x}Si_2O_7)(OH)$, x ∿ 0.2, and site Ca(5) has occupancy 0.43; the true cell may arise from ordering of these Ca ions. The structure (Fig. 1) contains Ca ions (octahedral and seven-coordinated), pyrosilicate groups, and OH^- ions. Si-O = 1.60-1.65(1) Å, Si-O-Si = 138°.

$2M_1$ LEPIDOLITE
$KLi(Al,Li)_2(Si,Al)_4O_{10}(OH,F)_2$

F. SARTORI, 1977. Tschermaks Min. Petr. Mitt., 24, 23-37.

Monoclinic, C2/c, a = 5.209, b = 9.053, c = 20.185 Å, β = 99.125° (from 1), D_m = 2.828, Z = 4. Cu radiation, R = 0.113 for 590 reflexions (films, densitometer intensities), material from Czechoslovakia.

Atomic positions

	x	y	z
T_1	0.4614	0.9244	0.1340
T_2	0.4558	0.2554	0.1341
M_1	0.2550	0.0851	0.0001
M_2	1/4	3/4	0
K	0	0.0906	1/4
O_1	0.4417	0.9281	0.0524
O_2	0.4153	0.2517	0.0529
O_3	0.4396	0.0898	0.1658
O_4	0.2353	0.8217	0.1629
O_5	0.2394	0.3546	0.1670
(OH,F)	0.4435	0.5701	0.0498

$2M_1$-muscovite structure (2). The mica layer is similar to those in $2M_2$ (3) and 1M (4) polytypes; there is ordering of the octahedral cations, distortion of the tetrahedra, and a rather low deviation from hexagonality of the surface oxygen ring.

1. P. ČERNÝ, M. RIEDER and P. POVONDRA, 1970. Lithos, 3, 319.
2. Structure Reports, 37A, 339.
3. Ibid., 39A, 346.
4. Ibid., 42A, 407.

MALAYAITE
$CaSnOSiO_4$

I. J.B. HIGGINS and F.K. ROSS, 1977. Cryst. Struct. Comm., 6, 179-182.
II. J.B. HIGGINS and P.H. RIBBE, 1977. Amer. Min., 62, 801-806.

Monoclinic, A2/a, a = 7.149, b = 8.906, c = 6.667 Å, β = 113.4°, Z = 4. Mo radiation, R = 0.047 for 872 reflexions.

Atomic positions

	x	y	z
Ca	1/4	0.9117	3/4
Sn	1/2	3/4	1/4
Si	3/4	0.9335	3/4
O(1)	3/4	0.8367	1/4

318   INORGANIC COMPOUNDS

	x	y	z
O(2)	0.9125	0.8171	0.9255
O(3)	0.3724	0.9614	0.1414

Isostructural with titanite (1). The structure (Fig. 1) contains infinite chains of corner-sharing SnO_6 octahedra, cross-linked by isolated SiO_4 tetrahedra; Ca occupies a large irregular polyhedron in this framework. Sn-O = 1.95-2.09, Si-O = 1.63, 1.65, Ca-O = 2.24-2.75 Å.

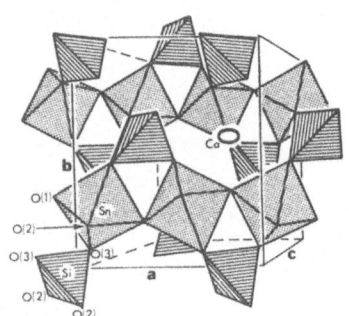

Fig. 1.   Structure of malayaite.

1. Structure Reports, 42A, 417.

ORTHOENSTATITE   ORTHOPYROXENE (SYNTHETIC)
$MgSiO_3$ $(Mg,Mn,Co)SiO_3$

F.C. HAWTHORNE and J. ITO, 1977. Canad. Miner., 15, 321-338.

Orthorhombic, Pbca, a = 18.216, 18.246, b = 8.813, 8.839, c = 5.179, 5.196 Å, Z = 16. Mo radiation, R = 0.033 and 0.041 for 1223 and 1129 reflexions.

Structure as previously described (1). In the synthetic material site occupancies are: M(1) = 0.904Mg + 0.065$\overline{C}$o + 0.031Mn, M(2) = 0.658Mg + 0.198Co + 0.144Mn.

1. Structure Reports, 34A, 365.

PARAGONITE (1M)
$NaAl_2(Si_3Al)O_{10}(OH)_2$

S.V. SOBOLEVA, O.V. SIDORENKO and B.B. ZVIAGIN, 1977. Kristallografija, 22, 510-514 [Soviet Physics - Crystallography, 22, 291-293].

Monoclinic, C2/m, a = 5.14, b = 8.89, c = 9.74 Å, β = 99.6°. Electron diffraction, R = 0.12 for 140 reflexions.

Mica-type structure [see e.g. celadonite (1)].

Atomic positions

	x	y	z
Al	0	0.322	0
Si,Al	0.420	0.330	0.280
Na	1/2	0	1/2
OH	0.391	0	0.104
O(1)	0.365	0.324	0.112
O(2)	0.031	0	0.334
O(3)	0.140	0.302	0.344

<u>1</u>. Structure Reports, <u>21</u>, 459; this volume, p. 307.

PARWELITE

$Mn_{10}Sb_2As_2Si_2O_{24}$ (idealized)

P.B. MOORE and T. ARAKI, 1977. Inorg. Chem., <u>16</u>, 1839-1847.

Monoclinic, Aa, a = 10.048, b = 19.418, c = 9.735 Å, β = 95.83°, D_m = 4.62, Z = 4.
Mo radiation, R = 0.048 for 1665 reflexions.

    The compound is a complex anion-deficient fluorite derivative structure,
containing three non-equivalent polyhedral sheets (Fig. 1). Mean polyhedral distances are: Mn-O = 2.07 - 2.42; Sb-O = 1.99; As-O = 1.68; Si-O = 1.63-1.64 Å.

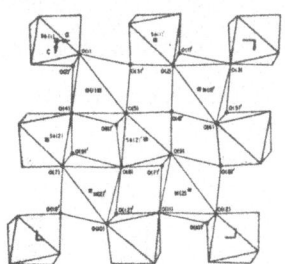

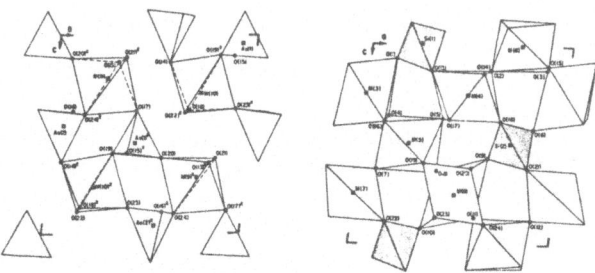

Fig. 1.   Polyhedral sheets in parwelite.

PLAGIOCLASE

$(Ca,Na)(Al,Si)_4O_8$

B. BERKING, 1976. Z. Kristallogr., <u>144</u>, 189-197.

Triclinic, $P\bar{1}$, a = 8.184, b = 12.890, c = 14.214 Å, α = 93.22, β = 115.90, γ = 91.13°, Z = 8. Mo radiation, R = 0.084 for 7048 reflexions. Lunar material.

The average structure of this An_{90} lunar plagioclase is refined and compared with those of An_{100} (1) and An_{94}.

1. Structure Reports, 27, 679; 37A, 328.

PROTOPYROXENE
$(Mg,Li,Sc)SiO_3$

J.R. SMYTH and J. ITO, 1977. Amer. Min., 62, 1252-1257.

Orthorhombic, Pbcn, a = 9.251, b = 8.773, c = 5.377 Å, Z = 8. Mo radiation, R = 0.053 for 826 reflexions.

Atomic positions

	x	y	z
M(1)	0	0.0994	3/4
M(2)	0	0.2639	1/4
Si	0.2935	0.0900	0.0740
O(1)	0.1199	0.0908	0.0805
O(2)	0.3736	0.2504	0.0710
O(3)	0.3493	0.9831	0.3045

M(1) = 0.3 Sc + 0.7 Mg; M(2) = 0.3 Li + 0.7 Mg

The structure is as previously described for protoenstatite (1), with Sc in M(1) and Li in M(2).

1. Structure Reports, 22, 487; 23, 475; 37A, 343.

RANKINITE
$Ca_3Si_2O_7$

I. KUSACHI, C. HEMMI, A. KAWAHARA and K. HEMMI, 1975. Mineral. J., 8, 38-47.

Monoclinic, $P2_1/a$, a = 10.60, b = 8.92, c = 7.89 Å, β = 119.6°, D_m = 2.97, Z = 2. R = 0.12.

The structure contains arrays of Si_2O_7 groups along c, linked by Ca ions which have 7-coordination. Si-O = 1.52-1.65, Ca-O = 2.25-2.90 Å.

LITHIUM-HYDRORHODONITE
$LiMn_4HSi_5O_{15}$

T. MURAKAMI, Y. TAKÉUCHI, T. TAGAI and K. KOTO, 1977. Acta Cryst., B33, 919-921.

Triclinic, $P\bar{1}$, a = 7.530, b = 11.736, c = 6.710 Å, α = 92°58', β = 95°14', γ = 106°16', Z = 2. Mo radiation, R = 0.041 for 1879 reflexions.

Isostructural with the Na analogue, nambulite (1), with Li in one octahedral site (M(5)) and Mn in the other four octahedral sites. In nambulite (1) the Na coordination is better described as eightfold, but in the present structure Li has six O at 2.02-2.80 Å, with the two additional O at 3.06 and 3.07 Å. Si-O = 1.58-1.65(1), Mn-O = 2.07-2.80(1) Å, Si-O-Si = 132-142°.

1. Structure Reports, 41A, 390.

ROSENHAHNITE
$Ca_3Si_3O_8(OH)_2$

CHE'NG WAN, S. GHOSE and G.V. GIBBS, 1977. Amer. Min., 62, 503-512.

Triclinic, P$\bar{1}$, a = 6.955, b = 9.484, c = 6.812 Å, α = 108.64, β = 94.84, γ = 95.89°, D_m = 2.89, Z = 2. Mo radiation, R = 0.035 for 3071 reflexions.

The structure contains $Si_3O_8(OH)_2{}^{6-}$ groups, linked by Ca ions with 6- and 7-co-ordinations   (Fig. 1), and by hydrogen bonds. Si-O = 1.584-1.677 Å, Si-O-Si = 128 and 146°, Ca-O = 2.290-2.917 Å.

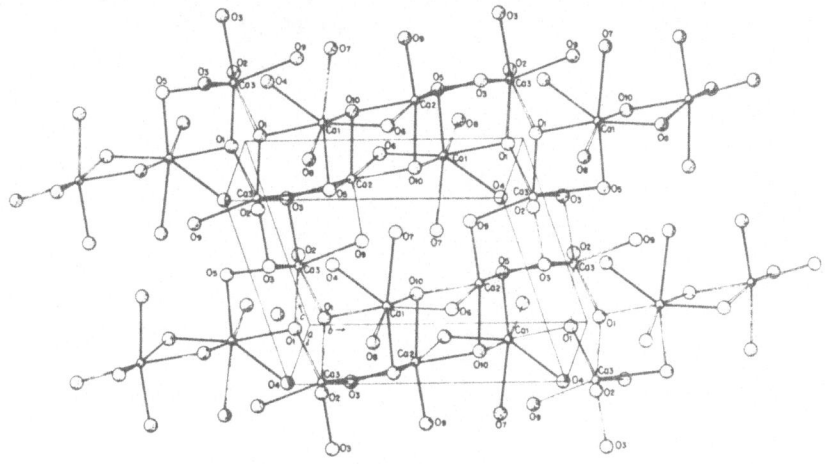

Fig. 1.   Part of the structure of rosenhahnite.

SARCOLITE
$Na_{1.7}Ca_6Al_4Si_6O_{23}(OH)(H_2O)_{0.4}[(Si,P)O_4]_{0.5}[CO_3,Cl]_{0.5}$

G. GIUSEPPETTI, F. MAZZI and C. TADINI, 1977. Tschermaks Min. Petr. Mitt., 24, 1-21.

Tetragonal, I4/m, a = 12.343, c = 15.463 Å, Z = 4. Mo radiation, R = 0.054 for 1637 reflexions.

The structure contains an aluminosilicate tetrahedral framework (Fig. 1), mean Si-O = 1.616, Al-O = 1.763 Å, with isolated $(Si,P)O_4$, CO_3, OH, H_2O, and Cl species disordered in cavities in the framework. Two independent Ca ions coordinate with 5 and 6 framework oxygens, with further contacts to available anions (Fig. 2); Ca-O = 2.34-2.69 Å. Na ions coordinate with 4 framework oxygens and one from CO_3; Na-O = 2.42 (x 4), 2.37 Å.

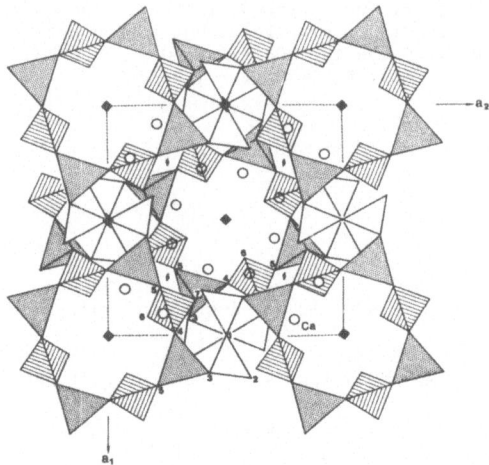

Fig. 1.   Tetrahedral framework of sarcolite. Si and Al tetrahedra are lined
           and dotted, respectively; unshaded groups are Si_2O_7.

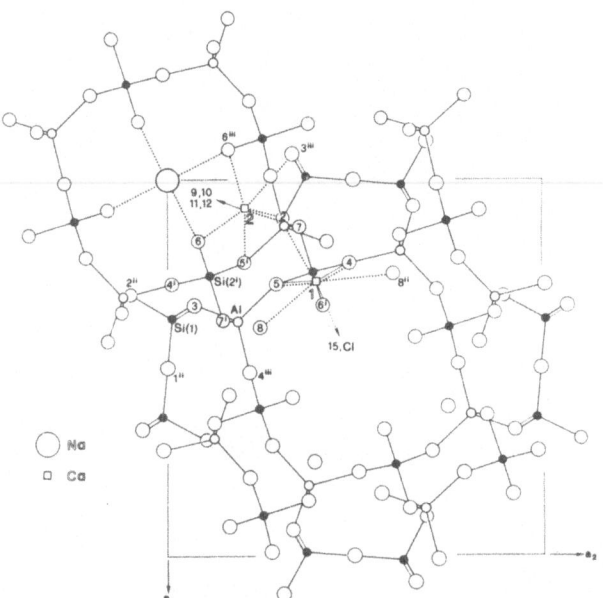

Fig. 2.   A portion of the sarcolite structure, showing Ca and Na coordination.

SHATTUCKITE   PLANCHEITE
$Cu_5(SiO_3)_4(OH)_2$ $Cu_8(Si_4O_{11})_2(OH)_4 \cdot xH_2O$

I. H.T. EVANS and M.E. MROSE, 1977. Amer. Min., <u>62</u>, 491-502.

Shattuckite
Orthorhombic, Pcab, a = 9.885, b = 19.832, c = 5.383 Å, D_m = 4.11, Z = 4. Mo
radiation, R = 0.056 for 1980 reflexions.

Plancheite
Orthorhombic, Pcnb, a = 19.043, b = 20.129, c = 5.269 Å, D_m = 3.65-3.80, Z = 4. R =
0.129 for 429 reflexions.

The z-coordinates for shattuckite differ from those given previously ($\underline{1}$).
The structure (Fig. 1) consist of brucite-like $(CuO_2)_n$ layers with pyroxene-type
$(SiO_3)_n$ chains joined to their surfaces. Adjacent complex sheets are linked by
additional Cu atoms in square coordination, forming ladder-like ribbons along the
chain axis parallel to $\underline{c}$.

Poor quality of the asbestos-like plancheite crystals allowed only a semi-
quantitative analysis. The structure is very similar to that of shattuckite,
except that pyroxene-type chains are replaced by amphibole-type chains. Water
content (x < 0.43) is probably variable.

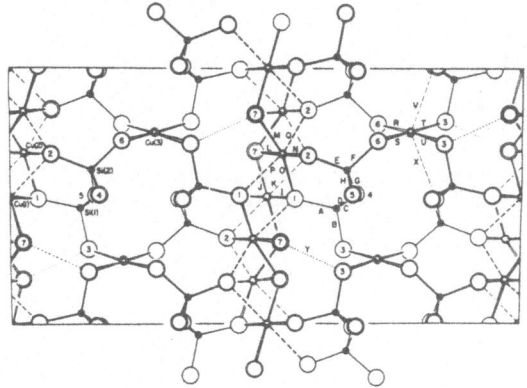

Fig. 1.   Structure of shattuckite.

II. A. KAWAHARA, 1976. Mineral. J. (Japan), $\underline{8}$, 193-199.

Results for shattuckite (R = 0.128 for 251 reflexions) agree with those in I.

$\underline{1}$. Structure Reports, $\underline{31}$A, 233; $\underline{32}$A, 460.

SKLODOWSKITE
$(H_3O)_2Mg[(UO_2)_2(SiO_4)_2]\cdot 4H_2O$ $MgO\cdot 2UO_3\cdot 2SiO_2\cdot 7H_2O$

R.R. RYAN and A. ROSENZWEIG, 1977. Cryst. Struct. Comm., $\underline{6}$, 611-615.

Monoclinic, C2/m, a = 17.382, b = 7.047, c = 6.610 Å, β = 105.9°, Z = 2. Mo
radiation, R = 0.022 for 1146 reflexions.

The structure is essentially as previously determined ($\underline{1}$). Hydrogen atoms
have not been located, but the presence of hydroxonium ions is proposed.

Atomic positions

	x	y	z
U	0.2576	0	0.1384
Si	0.2836	1/2	0.3578
Mg	1/2	1/2	1/2
O(1)	0.3639	0	0.1619
O(2)	0.1509	0	0.1065
O(3)	0.2263	1/2	0.5119
O(4)	0.3779	1/2	0.5016
O(5)	0.2673	0.3243	0.1917
$H_2O(6)$	0.5196	0.3029	0.7301
$H_3O^+(7)$	0.5760	0	0.2275

1. Structure Reports, 23, 471; 27, 711; 29, 410.

SLAWSONITE
$Sr_{0.87}Ca_{0.13}Al_2Si_2O_8$

D.T. GRIFFEN, P.H. RIBBE and G.V. GIBBS, 1977. Amer. Min., 62, 31-35.

Monoclinic, $P2_1/a$, a = 8.888, b = 9.344, c = 8.326 Å, β = 90.33°, Z = 4. Mo
radiation, R = 0.048 for 1617 reflexions.

    Isostructural with paracelsian (1), with an ordered Al/Si distribution, mean
Al-O = 1.748, Si-O = 1.624 Å. Sr has 7-coordination, mean Sr-O = 2.630 Å.

1. Structure Reports, 17, 556; 24, 492.

TIRODITE (ZINCIAN)
$(Mg,Mn,Fe,Zn)_7Si_8O_{22}(OH)_2$

F.C. HAWTHORNE and H.D. GRUNDY, 1977. Canad. Miner., 15, 309-320.

Monoclinic, C2/m, a = 9.606, b = 18.126, c = 5.317 Å, β = 102.63°, D_m = 3.24,
Z = 2. Mo radiation, R = 0.037 for 1383 reflexions.

    C-centred cummingtonite amphibole structure (1). There is some ordering of
the cations.

1. Structure Reports, 26, 526; 31A, 224; 39A, 339.

TRIMERITE
$CaMn_2(BeSiO_4)_3$

K.H. KLASKA and O. JARCHOW, 1977. Z. Kristallogr., 145, 46-65.

Monoclinic, $P2_1/n$, a = 8.098, b = 7.613, c = 14.065 Å, β = 90°, Z = 4. Mo
radiation, R = 0.033 for 2250 reflexions.

    The structure is derived from that of beryllonite (1). It contains an ordered
Be,Si tetrahedral framework with two channels along b (Fig. 1) which contain the
Mn (octahedral coordinations) and Ca cations (9-coordination). Si-O = 1.623-
1.643(3), Be-O = 1.612-1.650(6), Mn-O = 2.113-2.513(3), Ca-O = 2.449-2.876(3) Å.

Fig. 1.   Structure of trimerite.

<u>1</u>. Structure Reports, <u>26</u>, 456; <u>27</u>, 580; <u>39</u>A, 283.

TUSCANITE
$K(H_2O)Ca_6(Si,Al)_{10}O_{22}[SO_4,CO_3OH,O_4H_4]_2$

I.   P. ORLANDI, L. LEONI, M. MELLINI and S. MERLINO, 1977. Amer. Min., <u>62</u>,
      1110-1113.
II.   M. MELLINI, S. MERLINO and G. ROSSI, 1977. Ibid., <u>62</u>, 1114-1120.

Monoclinic, $P2_1/a$, a = 24.03, b = 5.11, c = 10.88 Å, β = 106.94°, D_m = 2.83, Z = 2.
Mo radiation, R = 0.034 for 2420 reflexions.

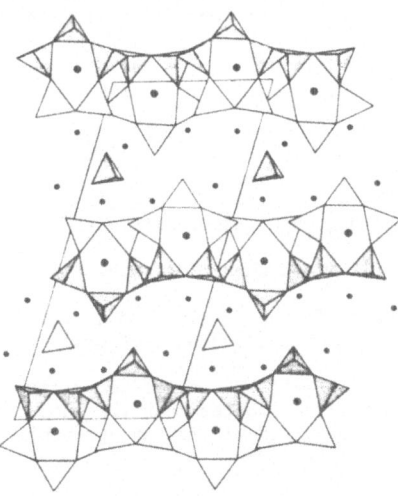

Fig. 1.   Structure of tuscanite, viewed along <u>b</u>.

The structure (Fig. 1) contains double layers of (Si,Al) tetrahedra parallel to (100), with five-membered rings in which three tetrahedra point upward and two downward (compare latiumite (1), in which all five tetrahedra point upward). The layers are connected by calcium and sulphate ions, with some substitution of sulphate by carbonate and O_4H_4. K is located in the cavity between two successive five-membered rings, and is partially substituted by water molecules.

1. Structure Reports, 39A, 346.

WADEITE
$K_2ZrSi_3O_9$

V.A. BLINOV, N.G. ŠUMJACKAJA, A.A. VORONKOV, V.V. ILJUKHIN and N.V. BELOV, 1977. Kristallografija, 22, 59-65 [Soviet Physics - Crystallography, 22, 31-35].

Hexagonal, $P\bar{6}$, a = 6.926, c = 10.177 Å, Z = 2. R = 0.030 for 1268 reflexions.

The structure is essentially as previously described (1), but the observation of some odd-index 00ℓ reflexions suggests small deviations from $P6_3/m$. Si-O = 1.59-1.64, Zr-O = 2.07, 2.09 Å (octahedral).

1. Structure Reports, 19, 463.

ZEOLITE A (RUBIDIUM-EXCHANGED)
$Rb_{11}NaAl_{12}Si_{12}O_{48}\cdot xH_2O$

R.L. FIROR and K. SEFF, 1977. J. Amer. Chem. Soc., 99, 1112-1117.

Cubic, Pm3m, a = 12.261 (dehydrated), 12.321 Å (hydrated), Z = 1. Mo radiation, R = 0.087 and 0.134 for 300 and 226 reflexions.

The results for the dehydrated sample are identical to those previously reported by the same authors (1), with one Rb ion not coordinated (4.35 Å from the nearest atom or ion). In the hydrated sample, five Rb are in the large cavity, three in the 8-oxygen ring, two inside the sodalite unit, and one in the large cavity near a 4-oxygen ring.

1. Structure Reports, 42A, 419.

ZEOLITE A (THALLIUM-EXCHANGED)
$Tl_{12}Al_{12}Si_{12}O_{48}$ $Tl_{13}Al_{12}Si_{12}O_{48}\cdot 9H_2O$ (approximately)

R.L. FIROR and K. SEFF, 1977. J. Amer. Chem. Soc., 99, 4039-4044.

Cubic, Pm3m, a = 12.180, 12.380 Å, for dehydrated and hydrated samples, Z = 1. Mo radiation, R = 0.057 and 0.062 for 312 and 321 reflexions.

Structure as described for Tl_{11}-A (1). In the dehydrated sample one Tl^+ is 3.21 Å from the nearest oxygen (0.42 Å greater than the sum of the ionic radii). The hydrated sample contains a thirteenth Tl^+ ion and probably a hydroxide anion.

1. Structure Reports, 38A, 376.

ZEOLITE A   (CAESIUM AND POTASSIUM EXCHANGED, DEHYDRATED)
$Cs_7K_5Al_{12}Si_{12}O_{48}$

R.L. FIROR and K. SEFF, 1977. J. Amer. Chem. Soc., 99, 6249-6253.

Cubic, Pm3m, a = 12.266 Å, Z = 1. Mo radiation, R = 0.047 for 314 reflexions.

Typical zeolite A structure. One K^+ ion is zero-coordinate.

ZEOLITE A   (SILVER-EXCHANGED DEHYDRATED)
$Ag_{12}Al_{12}Si_{12}O_{48}$

ZEOLITE A   (SILVER-EXCHANGED AMMONIA DEHYDRATED)
$Ag_{12}Al_{12}Si_{12}O_{48} \cdot 3N_3H_3 \cdot 4N_3H_5$

ZEOLITE A   (EUROPIUM(II)-EXCHANGED DEHYDRATED)
$Eu_6Si_{12}Al_{12}O_{48}$ (approximately)

I.   Y. KIM and K. SEFF, 1977. J. Amer. Chem. Soc., 99, 7055-7057.
II.   Y. KIM, J.W. GILJE and K. SEFF, 1977. Ibid., 99, 7057-7059.
III. R.L. FIROR and K. SEFF, 1977. Ibid., 99, 7059-7061.

Cubic, Pm3m, a = 12.295, not given, 12.296 Å, Z = 1. Mo radiation, R = 0.110, 0.076, 0.127, and 0.114 for 416, 441 (I, two crystals), 169 and 370 reflexions.

Typical zeolite A structures. The first structure contains an octahedral Ag_6 cluster (Ag-Ag = 2.93 Å) inside a cube of Ag^+ ions. In the ammonia-treated sample the Ag_6 clusters have been displaced and N_3H_3 (three-membered ring, N-N = 1.6 Å) and N_3H_5 ($H_2N.NH.NH_2$, N-N = 1.5 Å) molecules are bonded to Ag^+ ions (π-bonded and σ-bonded, respectively). In the europium-exchanged sample one Eu(II) ion has near-zero coordination (closest O at 3.38 Å).

ZINNWALDITE-1M
$K(Al,Fe,Li)_3(Si,Al)_4O_{10}(OH)F$ (idealized)

S. GUGGENHEIM and S.W. BAILEY, 1977. Amer. Min., 62, 1158-1167.

Monoclinic, C2, a = 5.296, b = 9.140, c = 10.096 Å, β = 100.83°, Z = 2. Mo radiation, R = 0.057 for 1550 reflexions.

1M-Mica structure, with lower symmetry resulting from occupancy of the M(2) octahedral site completely by Al, with Fe and Li distributed over M(1) and M(3). (F,OH) is off the pseudo-mirror plane to coordinate more closely with Al in M(2).

TABLE I

Some structural information has also been given for the following materials (listed with abbreviated 1977 reference).

Compound	Structure	Reference
Silver iodide, AgI	12H-polytype, Ždanov notation 1211111121	Acta Cryst., A33, 25
$Fe_{1-x}O$	Model for a superstructure proposed from electron diffraction and microscopy studies	Ibid., A33, 268
Hollandite	3 Å electron microscope study	Ibid., A33, 672
Cadmium iodide, CdI_2	Faulted 4H polytype	Ibid., A33, 687
$Bi_4Ti_4NbO_{21}$	Electron microscope study	Ibid., A33, 701
Sodium fluoride, NaF	NaCl type; electron distribution studied	Ibid., A33, 776
Graphite	Valence-charge density studied	Ibid., A33, 823
V_4O_9	Structure proposed	Ibid., A33, 834
Barium hectorite	Ba position deduced from diffraction pattern of oriented films	Ibid., B33, 633 and 2349
K_2ReCl_6, K_2ReBr_6	K_2PtCl_6-type, x = 0.239, 0.242. Solid solution studied	Ibid., B33, 2124
Plagioclases	Periodic antiphase structure	Amer. Min., 62, 932
Paraspurrite, $Ca_5(SiO_4)_2CO_3$	Structure probably similar to that of spurrite	Ibid., 62, 1003
$Li_{0.5}Fe_{2.5}O_4$-$CuGa_2O_4$	Spinel	Ann. Chim., 2, 25
PbFI	BiOCl	Chem. Scripta, 10, 206
CrO_2	Rutile	C.R. Acad. Sci. Paris, C, 285, 225
$K_2Pt(CN)_4Br_{0.3}\cdot3H_2O$ $K_2Pt(CN)_4Cl_{0.3}\cdot3H_2O$ $K_{1.75}Pt(CN)_4\cdot1.5H_2O$	Mainly a review of previous work	Ferroelectrics, 16, 135 and 145
$EuM_2M'O_{5.5}$ $Eu_2MM'O_{5.5}$ M = Ca, Sr, Ba M' = Nb, Ta	Cubic perovskites, with oxygen deficiency	Inorg. Chem., 16, 328
KVF_3, $RbVF_3$	Cubic perovskites	Ibid., 16, 646
$NaVF_3$	Orthorhombic, probably $GdFeO_3$-type	

TABLE I

329

Compound	Structure	Reference
Ammonium tetrachloro-zincate, $(NH_4)_2ZnCl_4$	Orthorhombic, 12.620, 7.211, 9.275 Å; super-structure, with c = 4c', $P2_1cn$	Inst. Fiz. Tech. Jadrow. A.G.H., Rap., no. 105, 1-18
$Ti(N,O)$ $Ti(C,N,O)$ $Zr(C,N,O)$	NaCl	Izv. Akad. Nauk SSSR, Neorg. Mater., 12, 1396 and 1581 (1976)
$EuF_{2.5}$	CaF_2	Ibid., 13, 175
Pb_2NbFeO_6 Pb_2TaFeO_6 $Pb_3WFe_2O_9$	Pyrochlores	Ibid., 13, 893
Lithium zinc ferrites	Spinels	Izv. Vyssh. Učebn. Zaved., Fiz., 20, 111
Potassium cyanide, KCN	Neutron diffraction powder data at 6°K indicate 'antiferroelectric' ordering of CN^- ions	J. Chem Phys., 66, 5147
$AB(PO_4)_2$ $AB(AsO_4)_2$ A = Sr, Ba B = Ge, Ti, Zr, Hf	Yavapaiite structure	J. Inorg. Nucl. Chem., 39, 11
$Pb(Fe,W,Mg,Ta)O_3$	Perovskite	J. Phys. Soc. Japan, 41, 542
$Ge_2Zn_{1-x}Fe_2O_4$	Spinel	Ibid., 41, 1522
Scandium disilicate, $Sc_2Si_2O_7$ (high-pressure) Indium disilicate, $In_2Si_2O_7$ (high-pressure)	Pyrochlore, x = 0.4313 Pyrochlore, x = 0.4272	J. Solid State Chem., 20, 219
$Mg_2(OH)_2CO_3$	Structural model proposed	Ibid., 21, 211
$BaPtO_3$	12-layer perovskite	Ibid., 21, 277
$SrLnMnO_4$ $BaLnMnO_4$ $(Sr,La)MnO_4$ $(Ba,La)MnO_4$	K_2NiF_4-type	Ibid., 22, 121
$(La,Sr)_2NiO_4$	K_2NiF_4-type	Ibid., 22, 145
Tutton's salts	Review of octahedral distortions and hydrogen bonding	Ibid., 22, 379
$Rb_{12}Nb_{30}W_3O_{90}$	Pyrochlore-related structure (see this volume, p. 194)	Ibid., 22, 393
$Ba_3(Fe,Ho)_2UO_9$	Perovskite	Ibid., 22, 411
$Cu_{0.5}TaO_3$	Perovskite	Kristallografija, 21, 1030 (1976) [Soviet Physics - Crystallography, 21, 593]

Compound	Structure	Reference
Ba_2NiReO_4	Hexagonal, $BaTiO_3$ type	Magnet Letters, 1, 23
$KBiF_4$	Fluorite	Mater. Res. Bull., 12, 145
$RbBiF_4$	Fluorite	
β-$TlBiF_4$ (high-temp.)	Fluorite	
α-$TlBiF_4$	Ordered α-PbF_2	
$Ba_3In_2(OH)_{12}$	Hydrogarnets	Ibid., 12, 161
$Ba_3Sc_2(OH)_{12}$		
$Cs_2NaM(O,F)_6$	Various structure	Ibid., 12, 803
Rb_2LiMOF_5	types were found:	
M = Ti, V, Nb	HT-K_2LiAF_6	
Mo, W	K_2NaCrF_6-elpasolite	
	$K_2VO_2F_3$	
$Ca_3MTa_2O_9$	Monoclinic perovskites	Ibid., 12, 815
M = Ca, Cd, Mn, Fe,		
Zn, Co, Mg, Ni		
$SrO.(Ln,Sr)VO_3$	K_2NiF_4-type	Ibid., 12, 831
Fe_2TeO_5	Structure as previously determined (Structure Reports, 42, 387). Magnetic structure now described	Ibid., 12, 1063
$Ln_2V_2O_7$	Pyrochlore	Ibid., 12, 1149
Ln = Tm, Yb, Lu		
$CsAlSiO_4$	$RbAlSiO_4$ [Structure Reports, 41A, 369]	Ibid., 12, 1183
Plagioclase (Na-rich $An_{16.5}$)	Study of superstructure	Neues Jb. Miner., Mh., 132
MgO-V_2O_3	Metastable spinel with cation vacancies	Nippon Kagaku Kaishi, 1539 (1976)
$MgMn_2O_4$	Spinel	Phys. Status Solidi, A, 37, 719 (1976)
$(Mn,Cr)_3O_4$	Spinel, $x(O) = 0.389$	Ibid., A, 37, K47 (1976)
$Cs_2AgLnCl_6$	Elpasolites	Rev. Chim. Minér., 14, 52
Ln = Sc, Y, Ce-Nd, Sm-Lu		
$Sr_2(Fe,Nb)O_4$	K_2NiF_4	Rev. Int. Hautes Temp. Refract., 14, 97
$Sr_2(Fe,Ta)O_4$		
$Sr_2(Fe,Sb)O_4$		
TbO_2	[CaF_2]	Solid State Comm., 18, 557 (1976)

TABLE I

331

Compound	Structure	Reference
$CsZnGaF_6$ $CsZnInF_6$ $CsZnTlF_6$ $CsZnScF_6$ $CsZnTiF_6$ $CsZnVF_6$ $CsZnMnF_6$ $CsZnCuF_6$ $CsZnRhF_6$ $CsPdInF_6$ $CsPdScF_6$ $CsPdFeF_6$ $CsPdRhF_6$ $CsNiInF_6$ $CsNiTlF_6$ $CsNiScF_6$ $CsNiRhF_6$ $CsMnGaF_6$ $CsMnFeF_6$ $CsMnRhF_6$	$RbNiCrF_6$-type (Structure Reports, 38A, 205), a = 10.2- 10.9 Å, powder data, x(F) = 0.31-0.32	Z. anorg. Chem., 428, 83, 91, 97
$Ba_2Y_{0.67}UO_6$ $Ba_2Y_{0.67}WO_6$ $Ba_2Y_{0.67}TeO_6$ $Ba_2Gd_{0.67}UO_6$	Perovskites, Fm3m, Ba in 8(c), 0.67 Y or Gd in 4(b), U or W or Te in 4(a), O in 24(e): x = 0.25. Powder data	Ibid., 429, 181
$Ba_2(Y,Ca)(U,W)O_6$	Deformed perovskite	Ibid., 429, 185
$(Ba,Sr)_2CdUO_6$ $(Ba,Sr)_2ZnUO_6$	Deformed perovskites	Ibid., 429, 198
$AgB(CN)_4$ $CuB(CN)_4$	Fd3m, a = 11.3, 10.7 Å. No atomic positional parameters given, but structures said not to contain $B(CN)_4^-$ ions	Ibid., 430, 38
$NaScO_2$ $NaYO_2$ $NaDyO_2$ $NaTmO_2$ $NaYbO_2$ $NaLuO_2$ K_2CeO_3 K_2PrO_3 K_2ThO_3 Rb_2CeO_3 Rb_2PrO_3 Rb_2ThO_3 Cs_2ThO_3	NaCl-type structures, with disorder of metal ions	Ibid., 430, 144
$LiMgInF_6$ $LiCaInF_6$ $LiMnInF_6$ $LiCoInF_6$ $LiNiInF_6$ $LiZnInF_6$ $LiCdInF_6$ $LiMnTiF_6$	Na_2SiF_6-type	Ibid., 430, 161

Compound	Structure	Reference
$LiMgTiF_6$ $LiMnTiF_6$ (H.T.)	Trirutile-type	Z. anorg. Chem., <u>430</u>, 161
$LiCaTiF_6$ $LiCdTiF_6$	Li_2ZrF_6-type super-structure	
$LiMnVF_6$ $LiFeGaF_6$	Na_2SiF_6 and trirutile phases	
Yb_3F_7 $Yb_{14}F_{33}$ $Yb_{27}F_{64}$ $Yb_{13}F_{32}$ $TmF_{2.4}$ $(Tm_{13}F_{32})$	Anion-excess fluorite-superstructures	Ibid., <u>430</u>, 175 and <u>434</u>, 89
$Ba_2(Nd,Y)_{0.67}WO_6$ $Ba_2Nd_{0.67}(W,U)O_6$	Perovskites	Ibid., <u>431</u>, 134
Ba_2MgWO_6-$Ba_2Y_{0.67}WO_6$ Ba_2CaWO_6-$Ba_2Y_{0.67}WO_6$	Perovskites	Ibid., <u>431</u>, 144
Ba_2CaUO_6-$Ba_2Lu_{0.67}UO_6$ Ba_2SrUO_6-$Ba_2Lu_{0.67}UO_6$	Distorted perovskites	Ibid., <u>431</u>, 153
$Ba_2Gd_{0.67}(U,Ta)O_6$	Perovskites	Ibid., <u>431</u>, 239
Lead iodide, PbI_2	$48R_1$ polytype, $[(11)_51212]_3$	Z. Kristallogr., <u>144</u>, 409 (1976)
Nickel tungstate $NiWO_4$	Structure as determined by X-rays (Structure Reports, <u>21</u>, 305); magnetic structure now determined	Ibid., <u>145</u>, 96
Cobalt tungstate $CoWO_4$	Isostructural with $FeWO_4$ and $NiWO_4$	

Cs_2MLnBr_6,
  M = Na, K
  Ln = Sc, Tm, Sm
	Elpasolites	Z. Naturforsch., <u>32B</u>, 594

LnOCl	Normal pressure, SmSI-type; high pressure, PbFCl-type	Ibid., <u>32B</u>, 1015
Ln_3O_4Cl	Eu_3O_4Br-type	

$AgLnF_4$
  Ln = Nd, Sm, Eu
     Gd, Tb, Dy, Ho
  Ln = Er, Tm, Yb
     Lu, Y
	
	$NaNdF_4$-type
	
	$KErF_4$-type
		Ibid., <u>32B</u>, 1093

$KBiO_3$	$KSbO_3$ [see Structure Reports, <u>11</u>, 443, 445]	Ibid., <u>32B</u>, 1340
$YbIO_5.4H_2O$	$HoIO_5.4H_2O$	Z. Neorg. Khim., <u>22</u>, 833
Lanthanum perrhenate tetrahydrate $La(ReO_4)_3.4H_2O$	Determination of H positions	Ibid., <u>22</u>, 1511

The following compounds have been studied by electron diffraction of the vapours (listed with abbreviated 1977 reference). Bond lengths are in Å, angles in degrees.

Molybdenum tetrafluoride oxide $MoOF_4$	Square pyramid, with O apical	Bull. Chem. Soc. Japan, 50, 373
	Mo-O 1.650	
	Mo-F 1.836	
	O-Mo-F 104	
	F-Mo-F 87	

Iminosulphur oxydifluoride, $HNSOF_2$	H is trans to O	Inorg. Chem., 16, 2959
	S-N 1.466	
	S-O 1.420	
	S-F 1.549	
	N-H 1.023	
	N-S-O 119.5	
	N-S-F 112.9	
	F-S-F 93.7	
	H-N-S 115.5	

Monochlorodiborane, B_2H_5Cl

ClHB⟨H/H⟩BH₂

	B-B 1.775	Ibid., 16, 3230
	B-Cl 1.775	
	B-H 1.21 (terminal)	
	B-H 1.33 (bridging)	

Potassium tetrafluoro-aluminate, $KAlF_4$	Al-F 1.69	Inorg. Chim. Acta, 25, L143
	K...F 2.51 (x 2)	
	3.89, 4.53	

Diaminodifluoro-phosphorane, $PF_2H(NH_2)_2$	Trigonal bipyramid with F atoms axial	J. Chem. Soc., Dalton, 585
	P-F 1.643	
	P-N 1.640	
	N-P-H 118.8	

Tungsten oxide tetra-fluoride, WOF_4	Square pyramid, C_{4v} symmetry	J. Molec. Struct., 37, 105
	W=O 1.666	
	W-F 1.847	
	O-W-F 104.8	
	F-W-F 86.2	

Aluminum bromide ammonia, $Br_3Al.NH_3$	Al-N 2.00	Ibid., 39, 225
	Al-Br 2.26	

Gallium bromide ammonium, $Br_3Ga.NH_3$	Ga-N 2.08	
	Ga-Br 2.29	

Lead(II) chloride, $PbCl_2$	Pb-Cl 2.45	Ibid., 42, 147
	Cl-Pb-Cl 98	

Copper(II) nitrate, $Cu(NO_3)_2$	Cu has approximately tetrahedral coordination to two bidentate nitrate groups.		Koordin. Khim., 2, 1203 (1976)
	Cu-O	1.95(1)	
	N-O	1.31, 1.21(1)	
	O-Cu-O	70 (O atoms of same NO_3 group)	
	Previous study in Structure Reports, 28, 361		
Scandium fluoride, ScF_3	Sc-F	1.926	Ž. Strukt. Khim., 17,
	F-Sc-F	110.0	797 (1976)
Lutetium chloride, $LuCl_3$	Lu-Cl	2.417	
	Cl-Lu-Cl	111.5	

MICROWAVE SPECTRA

Trichlorosilane, $SiHCl_3$	Si-Cl	2.020	Bull. Chem. Soc. Japan,
	Si-H	1.464	50, 1633
	Cl-Si-Cl	109.4	
	H-Si-Cl	109.5	
Pentaborane(9), B_5H_9	C_{4v} symmetry assumed		Inorg. Chem., 16, 3219
	B(1)-B(2)	1.690	
	B(2)-B(3)	1.803	
	B-H (terminal)	1.18	
	B-H (bridging)	1.35	

PAPERS REFERRED TO LATER YEARS

Many preliminary notes have not been reported, since fuller accounts will appear at a later date. The compounds studied, and abbreviated 1977 references, are listed below.

Gadolinium monochloride, GdCl Terbium monochloride, TbCl	Angew. Chem., 88, 685 (1976)
$CuCl.S_4N_4$	Ibid., 88, 807 (1976)
$Te_5O_4F_{22}$	Ibid., 88, 846 (1976) [see this volume p. 145]
N_6P_4S	Ibid., 88, 853 (1976)
$Ba_3[Mo_2S_2(CN)_8].14H_2O$ $K_7[Mo_2S(CN)_{12}].0·5[MoO_4].5H_2O$	Ibid., 88, 855 (1976)
Sodium oxoferrate, $Na_{14}Fe_6O_{16}$	Ibid., 89, 45
Arsenic(V) oxide, As_2O_5	Ibid., 89, 326
Boron sulphide, B_2S_3	Ibid., 89, 327 [see this volume, p. 37]

Antimony trichloride difluoride, $SbCl_3F_2$ — Chem. Comm., 653 (1976) [see this volume, p. 120]

Iodocycloheptasulphur hexafluoroantimonate(V), $[S_7I][SbF_6]$ — Ibid., 689 (1976)

$Er_6Pb_3(SiO_4)_6$ — Ibid., 706 (1976)

$Te_6(AsF_6)_4.2AsF_3$ — Ibid., 791 (1976)

$H_2Os_6(CO)_{18}$ — Ibid., 883 (1976)

Deuterium β-alumina, $D_2O.11Al_2O_3$ — Ibid., 895 (1976)

5,10-Dibromo-nido-decaborane(14) — Ibid., 121

Cyclotetrathiazyl bis(hexachloroantimonate(V)), $(S_4N_4)(SbCl_6)_2$ — Ibid., 253

Cyclotetrathiazyl hexafluoroantimonate(V) tetradecafluorotriantimonate, $(S_4N_4)-(SbF_6)(Sb_3F_{14})$

$Os_7(CO)_{21}$ — Ibid., 385

Sarabanite, $CaSb_{10}O_{10}S_6$ — Chem. letters, 275

Potassium zinc molybdate, $K_4Zn(MoO_4)_3$ — C.R. Acad. Sci. Paris, C, $\underline{283}$, 533 (1976)

3PbO.Nb_2O_5
3PbO.Ta_2O_5
2PbO.Nb_2O_5 — Ibid., C, $\underline{284}$, 179 and 331

Cs_2AgF_6 — Ibid., C, $\underline{284}$, 231

$NaAl(CrO_4)_2.2H_2O$
$KAl(CrO_4)_2.2H_2O$ — Ibid., C, $\underline{284}$, 565

$MoRh_2O_6$ — Ibid., C, $\underline{284}$, 921

$NaCd_4(PO_4)_3$
$NaCd_4(VO_4)_3$ — Ibid., C, $\underline{284}$, 963

$β-NaFe_2O_3$ — Ibid., C, $\underline{285}$, 129

Shungite — Dokl. Akad. Nauk SSSR, $\underline{232}$, 1189

Strontium dodecahydrodecaborate heptahydrate — Ibid., $\underline{232}$, 1366

$Sm_4Ge_3O_9(OH)_6$ — Ibid., $\underline{233}$, 362

Strontium metagermanate — Ibid., $\underline{233}$, 1086

Manganese milarite, $K_2Mn_5Si_{12}O_{30}.H_2O$ — Ibid., $\underline{233}$, 1090

Rhodonite, $CaMn_4Si_5O_{15}$
$NaHCd_4Ge_5O_{15}$
$LiHCd_4Ge_5O_{15}$ — Ibid., $\underline{234}$, 586

Zirconium polyphosphate, $Zr(PO_3)_4$ — Ibid., $\underline{234}$, 628

Uralborite, $Ca_2B_4O_4(OH)_8$ — Ibid., $\underline{234}$, 822

Potassium neodymium silicate, $K_3NdSi_6O_{15}$	Dokl. Akad. Nauk SSSR, $\underline{234}$, 1323
Triphylite, $LiFePO_4$	Ibid., $\underline{235}$, 93
Sodium uranyl metaphosphate, $NaUO_2(PO_3)_3$	Ibid., $\underline{235}$, 394
Sodium indium selenate hexahydrate, $NaIn(SeO_4)_2.6H_2O$	Ibid., $\underline{235}$, 575
Yttrium decavanadate, $Y_2V_{10}O_{28}.24H_2O$	Ibid., $\underline{235}$, 578
Potassium lead molybdate, $K_2Pb(MoO_4)_2$	Ibid., $\underline{235}$, 820
Lomonosovite	Ibid., $\underline{235}$, 1064
Hilgardite, $Ca_2B_5O_9Cl.H_2O$	Ibid., $\underline{236}$, 91
Hydrazinium(2+) pentafluorooxoniobate hydrate	Ibid., $\underline{236}$, 393
$NaNdSi_6O_{13}(OH)_2.nH_2O$	Ibid., $\underline{236}$, 593
Zinc hydrogen phosphate hydrate, $ZnHPO_4.H_2O$	Ibid., $\underline{236}$, 597
Tienshanite	Ibid., $\underline{236}$, 863
$Na_2Cd_3Si_3O_{10}$	Ibid., $\underline{236}$, 866
Alluaudite (synthetic), $Na_2Fe_3(PO_4)_3$	Ibid., $\underline{236}$, 1123
Clinohedrite, $CaZnSiO_4.H_2O$	Ibid., $\underline{237}$, 334 [see this volume, p. 310]
Willemite, Zn_2SiO_4	Ibid., $\underline{237}$, 581
$Na_8Al_6Ge_6O_{24}(CO_3).2H_2O$	Ibid., $\underline{237}$, 585
Ammonium 12-molybdotellurate hydrate, $(NH_4)_2TeMo_{12}O_{40}.15H_2O$	Indian J. Chem., A, $\underline{14}$, 694 (1976)
Holmium chloride, Ho_5Cl_{11}	Inorg. Chem., $\underline{16}$, 2134
Cm_2O_2Sb Cm_2O_2Bi Am_2O_2Bi $Pu_2(O,N)_2Sb$	Inorg. Nucl. Chem. Letters, $\underline{13}$, 161
Lutetium hydroxide, $Lu(OH)_3$	Ibid., $\underline{13}$, 173
Caesium hexabromoneptunate, Cs_2NpBr_6	Ibid., $\underline{13}$, 529
$M_2Pt(CN)_4(FHF)_{0.39}.xH_2O$, M = Rb, Cs	J. Amer. Chem. Soc., $\underline{99}$, 1668
Iridium tricarbonyl chloride, $Ir(CO)_3Cl$	Ibid., $\underline{99}$, 4184
$Ce_6Mo_{10}O_{39}$ $K_2Mo_2O_7.H_2O$ $(NH_4)_4Mo_8O_{26}.4H_2O$	J. Less-Common Metals, $\underline{54}$, 283
$CsTb(PO_3)_4$	Koordin. Khim., $\underline{3}$, 275
Mitridatite, $Ca_2Fe_3O_2(PO_4)_3.3H_2O$	Miner. Mag., $\underline{41}$, 527 and M8 [see this volume, p. 256]

$YSiO_2N$	Mater. Res. Bull., __12__, 251
YBO_3	
$KGaGe_3O_8$	Naturwissenschaften, __64__, 92
Sulphate hydrocancrinite, $Na_8(OH)_2$-$(AlSiO_4)_6.5H_2O$	Ibid., __64__, 93
Bicchulite (synthetic), $Ca_2Al_2SiO_7.H_2O$	Ibid., __64__, 94
Sodium diferrate(III), $Na_4Fe_2O_5$	Ibid., __64__, 271
Sodium dimanganate(III), $Na_4Mn_2O_5$	Ibid., __64__, 272
Gianellaite, $Hg_4N_2(SO_4)$	Neues Jb. Miner., Mh., 119
Deerite, $Fe_9O_3(Si_6O_{17})(OH)_5$	Trans. Amer. Geophys. Union, __58__, 821 [see this volume, p. 311]
$ZrBr_2$	Vest. Moskov. Univ., __18__, 82
Potassium hexachlorostannate(IV) (high-temperature), K_2SnCl_6	Z. Kristallogr., __144__, 426 (1976)
$K_2Pt(CN)_4Cl_{0.3}.3H_2O$	Ibid., __144__, 427 (1976)
$K_2Pt(CN)_4Br_{0.3}.3H_2O$	
$Cs_2Au_2Cl_6$ (high-pressure)	Ibid., __144__, 428 (1976)
Plagioclase	Ibid., __144__, 433 (1976)
Rubidium strontium octaborate, $Rb_2O.SrO.4B_2O_3.12H_2O$	Ž. Strukt. Khim., __17__, 950 (1976)

ADDITIONAL PAPERS

The following reports were prepared too late for inclusion in the main text.

cis-TETRACARBONYL-BIS(CHLOROMERCURY)IRON
$(ClHg)_2Fe(CO)_4$

C.L. RASTON, A.H. WHITE and S.B. WILD, 1976. Aust. J. Chem., __29__, 1905-1911.

Monoclinic, $C2/c$, a = 36.81, b = 11.181, c = 20.369 Å, β = 95.28°, Z = 32. Cu radiation, R = 0.071 for 2647 reflexions.

There are four cis-octahedral molecules in the asymmetric unit; in one of these one of the carbonyl sites is occupied by a more substantial moiety, possibly partial occupancy by HgCl as a result of disorder or decomposition. Mean Fe-Hg = 2.50, Fe-C = 1.82, C-O = 1.13, Hg-Cl = 2.42 Å; Fe-Hg-Cl = 165-174°.

POTASSIUM AMIDOZINCATE
$K_2Zn(NH_2)_4$

L. GUÉMAS and P. PALVADEAU, 1977. Rev. Chim. Minér., __14__, 381-386.

Triclinic, $P\bar{1}$, a = 6.76, b = 7.56, c = 7.99 Å, α = 102.7, β = 95.4, γ = 117.4°, D_m = 1.98, Z = 2. Mo radiation, R = 0.07 for 2982 reflexions.

The structure contains tetrahedral $Zn(NH_2)_4{}^{2-}$ ions, Zn-N = 2.05-2.07 Å, and K^+ ions with nearly-planar four-coordination, K-N = 2.84-3.01 Å.

POTASSIUM CALCIUM AMIDE
$KCa(NH_2)_3$

H. JACOBS and U. FINK, 1977. Z. anorg. Chem., **435**, 137-145.

Monoclinic, $P2_1/c$, a = 6.767, b = 11.68, c = 6.624 Å, β = 106.7°, D_m = 1.70, Z = 4. Mo radiation, R = 0.028 for 589 reflexions.

Atomic positions

	x	y	z
Ca	0.1753	0.2597	0.4080
K	0.2618	0.5910	0.2783
N(1)	0.9225	0.1688	0.5680
N(2)	0.2022	0.0899	0.1734
N(3)	0.4201	0.2252	0.2279

H atom parameters are also given.

New type of ternary amide structure which contains chains of face-sharing $Ca(NH_2)_6$ octahedra (Fig. 1) linked by K ions. Ca-N = 2.46-2.56(1) Å; K has 4 N at 2.98-3.06, 2 N at 3.16 and 3.42, and 4 N at 3.74-4.30 Å.

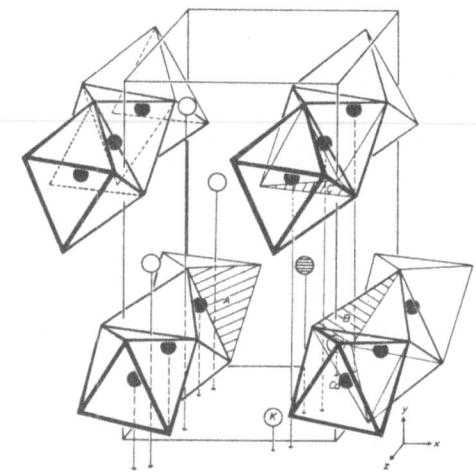

Fig. 1.   Structure of $KCa(NH_2)_3$.

POTASSIUM HEXAAMIDOSAMARATE
$K_3Sm(NH_2)_6$

POTASSIUM HEPTAAMIDODISAMARATE
$KSm_2(NH_2)_7$

H. JACOBS and K. KISTRUP, 1977. Z. anorg. Chem., **435**, 127-136.

$K_3Sm(NH_2)_6$. Monoclinic, C2/m, a = 6.685, b = 11.57, c = 7.095, β = 108.6°, D_m = 2.30, Z = 2. Powder data, parameters of $K_3La(NH_2)_6$ (**1**) assumed.

$KSm_2(NH_2)_7$. Orthorhombic, Pnnm, a = 6.654, b = 10.060, c = 15.66 Å, D_m = 2.89, Z = 4. Mo radiation, R = 0.025 for 1273 reflexions.

Isostructural with the corresponding La compounds (1).

1. Structure Reports, 40A, 127.

TIN BROMIDE FLUORIDE
$Sn_5Br_4F_6$

C. GENEYS and S. VILMINOT, 1977. Rev. Chim. Minér, 14, 395-403.

Orthorhombic, Pnma, a = 22.563, b = 4.315, c = 14.934 Å, D_m = 4.66, Z = 4. Mo radiation, R = 0.076 for 702 reflexions.

The structure contains $(SnBr_3^-)_n$ and $(Sn_4F_6^{2+})_n$ chains (Fig. 1) and Br^- ions. Sn-Br = 2.66-3.02(1) Å (square-pyramidal coordination), Sn-F = 2.01-2.55(3) Å (square-pyramids, with Sn at the apex); additional neighbours complete SnX_8E and SnX_9E coordinations, E = lone pair. One F atom is disordered over two positions.

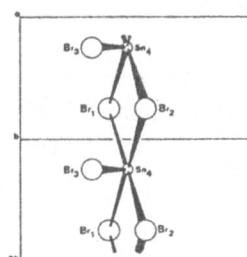 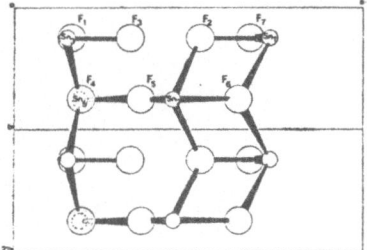

Fig. 1. $(SnBr_3)_n$ and $(Sn_4F_6)_n$ chains in $Sn_5Br_4F_6$.

LEAD FLUORIDE CHLORIDE
$Pb_7F_{12}Cl_2$ $6PbF_2 \cdot PbCl_2$

B. AURIVILLIUS, 1976. Chem. Scripta, 10, 206-209.

Hexagonal, $P\bar{6}$, a = 10.274, c = 3.988 Å, D_m = 8.19, Z = 1. Mo radiation, R = 0.045 for 1140 reflexions.

Atomic positions

	x	y	z
Pb(1)	0	0	1/2
Pb(2)	0.8931	0.5842	1/2
Pb(3)	0.1140	0.4007	0
Cl(1)	2/3	1/3	0
Cl(2)	1/3	2/3	1/2
F(1)	0.3894	0.4270	0
F(2)	0.1650	0.2766	1/2
F(3)	0.4284	0.4600	1/2
F(4)	0.2100	0.0803	0

The structure (Fig. 1) contains Pb ions with 9-coordination. The bromide is probably isostructural.

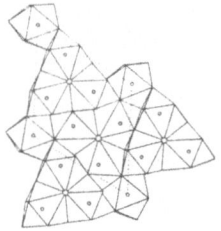

Fig. 1.   Structure of $6PbF_2 \cdot PbCl_2$.

BISMUTH(III) FLUORIDE (ORTHORHOMBIC)
BiF_3

O. GREIS and M. MARTINEZ-RIPOLL, 1978. Z. anorg. Chem., **436**, 105-112.

Orthorhombic, Pnma, a = 6.561, b = 7.015, c = 4.841 Å, Z = 4. Mo radiation, R = 0.050 for 553 reflexions.

Atomic positions

	x	y	z
Bi	0.3547	1/4	0.0349
F(1)	0.5361	1/4	0.6271
F(2)	0.1652	0.0577	0.3528

Isostructural with YF_3, as previously reported (1). A small amount of oxygen produces a tysonite-type structure (2).

1. Structure Reports, **40A**, 134.
2. Ibid., **19**, 342.

ZINC PENTAFLUOROINDATE(III) HEPTAHYDRATE
$ZnInF_5 \cdot 7H_2O$

CADMIUM PENTAFLUOROGALLATE(III) HEPTAHYDRATE
$CdGaF_5 \cdot 7H_2O$

B.V. BUKVECKIJ, S.A. POLIŠČUK and V.I. SIMONOV, 1977. Koordin. Khim., **3**, 926-938.

$ZnInF_5 \cdot 7H_2O$, triclinic, $P\bar{1}$, a = 6.638, b = 11.812, c = 9.039 Å, α = 120.72, β = 98.21, γ = 111.55°, Z = 2. Mo radiation, R = 0.016 for 2212 reflexions.

$CdGaF_5 \cdot 7H_2O$, triclinic, $P\bar{1}$, a = 5.949, b = 9.362, c = 10.033 Å, α = 107.45, β = 102.43, γ = 94.70°, Z = 2. Mo radiation, R = 0.024 for 3040 reflexions.

Both structures contain octahedral cations and anions (Fig. 1). In $ZnInF_5 \cdot 7H_2O$, there are two $Zn(H_2O)_6^{2+}$ cations, and InF_6^{3-} and $InF_4(H_2O)_2^-$ anions, all located on symmetry centres. $CdGaF_5 \cdot 7H_2O$ contains $Cd(H_2O)_6^{2+}$ and $GaF_5(H_2O)^{2-}$ ions, both in general positions. Zn-O = 2.04-2.16, In-F = 2.04-2.09, In-O = 2.15; Cd-O = 2.24-2.32, Ga-F = 1.87-1.90, Ga-O = 2.00 Å. The ions in both structures are linked by hydrogen bonds (Fig. 1).

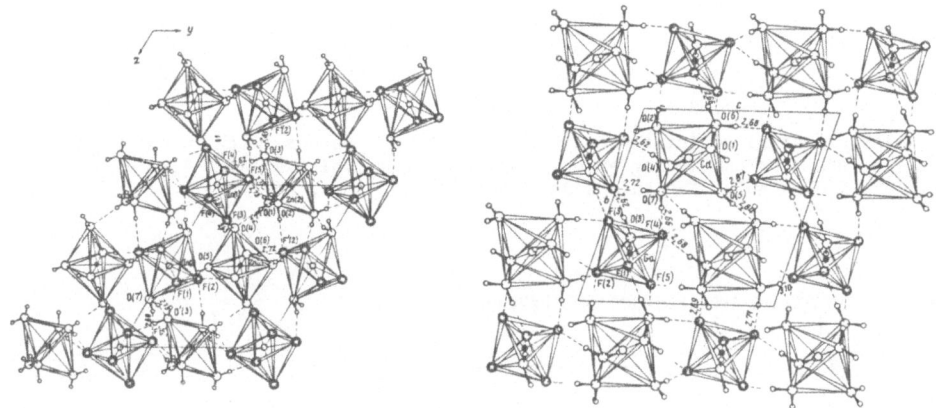

Fig. 1.   Structures of $ZnInF_5 \cdot 7H_2O$ and $CdGaF_5 \cdot 7H_2O$.

RUBIDIUM BROMOTRIFLUOROANTIMONATE(III)
$RbSbBrF_3$

B. DUCOURANT, J.C. JUMAS, R. FOURCADE and G. MASCHERPA, 1977. Rev. Chim. Minér.,
14, 76-82.

Triclinic, $P\bar{1}$, a = 7.720, b = 4.448, c = 8.906 Å, α = 100.50, β = 109.96, γ =
103.68°, D_m = 4.23, Z = 2. Mo radiation, R = 0.076 for 693 reflexions.

Atomic positions

	x	y	z
Sb	0.1626	0.3091	0.2099
Br	0.7687	0.4343	0.0704
Rb	0.6864	0.0097	0.3447
F(1)	0.236	0.764	0.301
F(2)	0.125	0.306	0.415
F(3)	0.435	0.403	0.320

The structure contains trigonal-pyramidal SbF_3 groups, Sb-F = 1.91 Å. One
F from a neighbouring group, 4 Br and the lone-pair complete 9-coordination at
Sb (trigonal prism of 4 Br and 2 F, with the lone-pair capping the 4 Br face,
and two F atoms capping the other two rectangular faces); Sb...F = 2.81, Sb-Br =
3.08-4.02 Å. Rb has 6 F and 3 Br neighbours, Rb-F = 2.93-3.25, Rb-Br = 3.48-3.89 Å.

CAESIUM PENTAFLUOROTELLURATE(IV)
$CsTeF_5$

J.C. JUMAS, M. MAURIN and E. PHILIPPOT, 1977. J. Fluor. Chem., 10, 219-230.

Orthorhombic, Pnma, a = 10.221, b = 6.651, c = 8.330 Å, D_m = 4.12, Z = 4. Mo
radiation, R = 0.046 for 568 reflexions.

Atomic positions

			x	y	z
Cs	in	4(c)	0.1682	1/4	0.6358
Te	in	4(c)	0.0702	1/4	0.1399
F(1)	in	8(d)	0.138	0.049	-0.010
F(2)	in	8(d)	0.070	0.046	0.302
F(3)	in	4(c)	0.242	1/4	0.192

The structure contains square-pyramidal TeF_5^- anions, with Te below the base and the lone pair completing an octahedron; Te-F = 1.81 (apical), 1.93(2) Å (basal), F(ap)-Te-F(b) = 80°. The anions are linked by Cs^+ ions, Cs-12F = 3.07-3.60 Å.

TRIFLUOROXENON(IV) HEXAFLUOROBISMUTHATE(V)
$[XeF_3][BiF_6]$

R.J. GILLESPIE, D. MARTIN, G.J. SCHROBILGEN and D.R. SLIM, 1977. J. Chem. Soc., Dalton, 2234-2237.

Triclinic, P$\bar{1}$, a = 5.698, b = 7.811, c = 8.854 Å, α = 99.45, β = 110.09, γ = 92.84°, Z = 2. Mo radiation, R = 0.097 for 1238 reflexions.

The structure contains XeF_3^+ (T-shaped) and BiF_6^- (octahedral) ions, with a close Xe...F interionic contact, 2.25 Å (Fig. 1).

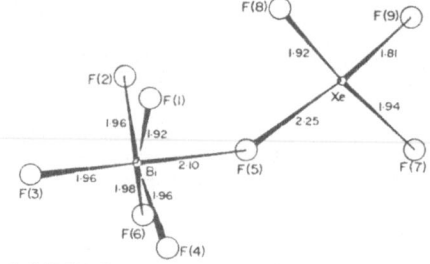

Fig. 1.   The $[XeF_3][BiF_6]$ structural unit.

AMMONIUM CHROMIUM(III) FLUORIDE HEXAHYDRATE
$(NH_4)_2CrF_5 \cdot 6H_2O$

W. MASSA, 1977. Z. anorg. Chem., 436, 29-38.

Monoclinic, C2/c, a = 11.997, b = 6.928, c = 13.574 Å, β = 90.0°, D_m = 1.70, Z = 4. R = 0.042 for 791 reflexions.

The structure (Fig. 1) contains octahedral $Cr(H_2O)_6^{3+}$ ions, NH_4^+, and F^- ions, linked by O-H...F and N-H...F hydrogen bonds; Cr-O = 1.965(2), O...F 2.53-2.56, N...F 2.72-2.73 Å.

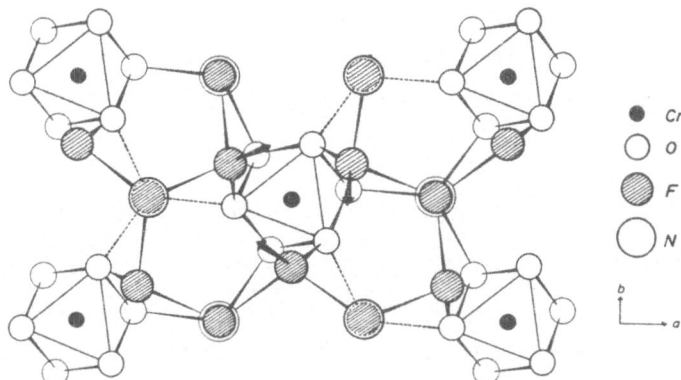

Fig. 1.   Structure of ammonium chromium(III) fluoride hexahydrate.

STRONTIUM TRICHLOROSTANNATE(II) PENTAHYDRATE
$Sr(SnCl_3)_2 \cdot 5H_2O$

H.J. HAUPT, F. HUBER and H.-W. SANDBOTE, 1977. Z. anorg. Chem., **435**, 191-196.

Monoclinic, C2/c, a = 14.945, b = 12.006, c = 9.100 Å, β = 116.10°, Z = 4. Mo
radiation, R = 0.075 for 2650 reflexions.

The structure contains trigonal pyramidal $SnCl_3^-$ ions, Sn-Cl = 2.58, 2.61,
2.63 Å; three more distant Cl atoms (3.22, 3.32, 3.42 Å) complete distorted octahedra which share edges to form a three-dimensional structure. Sr has 2 Cl and 7 O
neighbours, Sr-Cl = 3.06, Sr-O = 2.58-2.74 Å; these 9-coordinate polyhedra share
oxygen atoms to form chains along c̲.

BARIUM PHOSPHORUS HALIDES
$Ba_{1.6}PX_{0.2}$, Ba_2PX (X = Cl, Br, I)

C. HADENFELDT, 1977. Z. anorg. Chem., **436**, 113-121.

$Ba_{1.6}PX_{0.2}$, cubic, I4̄3d, a = 9.76, 9.80, 9.86 Å, D_m = 4.58, 4.68, 4.70, for X = Cl,
Br, I, Z = 10. Cu radiation, powder data for Br compound. Anti-Th_3P_4 type structure
(1̲), as for Ba_3P_2 (2̲); 16 Ba in 16(c): x,x,x, x = 0.063; (10 P + 2 Br) in 12(a):
3̲/8,0,1/4.

Ba_2PX, rhombohedral, R3̄m, a = 4.63, 4.67, 4.68, c = 22.71, 23.38, 24.69 Å, D_m = 4.01,
4.33, 4.48, for X = Cl, Br, I, Z = 3. Cu radiation, powder data for Cl compound.
Anti-α-$NaFeO_2$ type structure (3̲); Ba in 6(c): 0,0,z, z = 0.245; P in 3(b): 0,0,1/2;
Cl in 3(a): 0,0,0.

1̲. Strukturbericht, **7**, 15, 112.
2̲. Structure Reports, 33A, 40; 35A, 28.
3̲. Strukturbericht, 3, 75, 392.

CAESIUM TETRACHLOROFERRATE(III)
CsFeCl$_4$

G. MEYER, 1977. Z. anorg. Chem., **436**, 87-94.

Orthorhombic, Pbnm, a = 9.42, b = 11.71, c = 7.14 Å, D$_m$ = 2.74, Z = 4. R = 0.104
for 724 reflexions.

Atomic positions

	x	y	z
Cs	0.1654	0.1800	1/4
Fe	0.695	0.070	1/4
Cl(1)	0.594	-0.101	1/4
Cl(2)	0.550	0.218	1/4
Cl(3)	0.829	0.083	0.004

BaSO$_4$-type structure (1); Fe-Cl = 2.17-2.22 Å. CsAlCl$_4$ is isostructural.

1. Strukturbericht, **1**, 343.

GERMANIUM DIBROMIDE
GeBr$_2$

R.C. ROUSE, D.R. PEACOR and B.R. MAXIM, 1977. Z. Kristallogr., **145**, 161-171.

Monoclinic, P2$_1$/c, a = 11.68, b = 9.12, c = 7.02 Å, β = 101.9°, Z = 8. Mo
radiation, R = 0.12 for 901 reflexions.

Atomic positions

	x	y	z
Ge(1)	0.0121	0.1254	0.2693
Ge(2)	0.4500	0.3435	0.2197
Br(1)	0.3369	0.1201	0.2622
Br(2)	0.8884	0.3331	0.3970
Br(3)	0.1402	0.4441	0.1747
Br(4)	0.5968	0.2056	0.0583

The structure (Fig. 1) contains GeBr$_3$ trigonal pyramids which share corners
to give a layer parallel to (100) of very distorted corner-sharing octahedra, with
layers linked by further Br bridging.

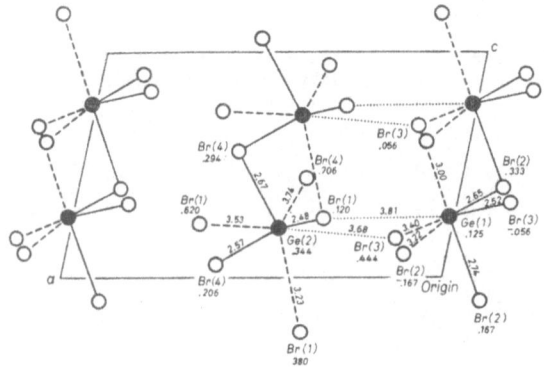

Fig. 1.   Structure of germanium dibromide.

CAESIUM CADMIUM BROMIDE
$CsCdBr_3$

C.K. MØLLER, 1977. Acta Chem. Scand., A31, 669-672.

Hexagonal, $P6_3/mmc$, a = 7.68, c = 6.73 Å, Z = 2. Cu radiation, R = 0.16 for some
film data. Cd in 2(a); Cs in 2(d): Br in 6(h): x = 1/6. Isostructural with
$RbNiCl_3$ (1).

1. Structure Reports, 34A, 211.

SILVER IODIDE
α-AgI

A.F. WRIGHT and B.E.F. FENDER, 1977. J. Phys., C, 10, 2261-2267.

Cubic, Im3m, a = 5.062-5.106 Å, at 180-450°C, Z = 2. Neutron powder data. 2 I in
2(a): 0,0,0; 2 Ag in 12(d): 1/4,0,1/2.

The Ag ions are confined to the tetrahedral sites, rather than distributed over
several sites as previously described (1), but have large anharmonic thermal vibra-
tions.

1. Strukturbericht, 3, 8, 232; Structure Reports, 18, 350; 21, 216.

THALLIUM LEAD IODIDES
$TlPbI_3$ Tl_4PbI_6

W. STOEGER, 1977. Z. Naturforsch., 32B, 975-981.

$TlPbI_3$
Orthorhombic , Cmcm, a = 4.625, b = 14.885, c = 11.857 Å, D_m = 6.60, Z = 4. Mo
radiation, R = 0.044 for 646 reflexions.

Tl_4PbI_6
Orthorhombic, Pbam, a = 19.117, b = 9.877, c = 4.586 Å, D_m = 6.70, Z = 2. Mo
radiation, R = 0.042 for 944 reflexions.

Atomic positions

$TlPbI_3$	x	y	z
Tl	0	0.2489	1/4
Pb	0	0	0
I(1)	0	0.6436	0.0565
I(2)	0	0.9224	1/4
Tl_4PbI_6			
Tl(1)	0.0807	0.6028	1/2
Tl(2)	0.2046	0.2930	1/2
Pb	0	0	0
I(1)	0.0538	0.3055	0
I(2)	0.1155	0.9396	1/2
I(3)	0.2861	0.0929	0

Both structures (Fig. 1) contain PbI_6 octahedra and elements of the TlI structure.

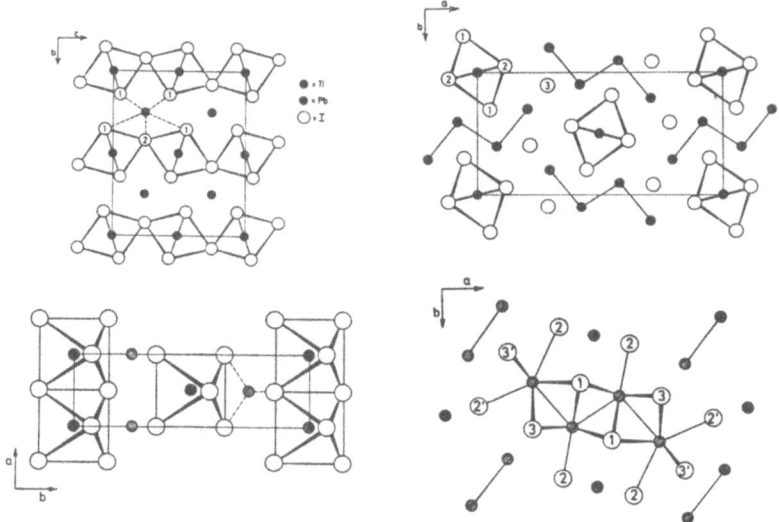

Fig. 1.   Projections of the structures of $TlPbI_3$ (left) and Tl_4PbI_6 (right).

TRIANTIMONY NONACHLORIDE TETRAFLUORIDE OXIDE
$Sb_3Cl_9F_4O$

A.J. EDWARDS and G.R. JONES, 1977.   J. Chem. Soc., Dalton, 1968-1971.

Triclinic, P$\bar{1}$, a = 9.79, b = 13.24, c = 6.94 Å, α = 93.1, β = 105.3, γ = 91.1°,
Z = 2. Mo radiation, R = 0.067 for 3379 reflexions.

The structure contains discrete molecules (Fig. 1), with three Sb atoms,
bridged by two F atoms and by a double disordered O/F bridge. Each Sb has octahedral coordination, Sb-Cl = 2.256-2.273(4), Sb-F = 2.02-2.12(1) (bridging), 1.94(1)
(terminal), Sb-O/F = 1.99-2.01(1) Å.

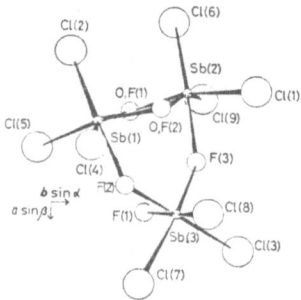

Fig. 1.   The $Sb_3Cl_9F_4O$ molecule.

AMMONIUM PENTAFLUOROPEROXOTITANATE(IV)
$(NH_4)_3[TiF_5(O_2)]$

R. STOMBERG and I.-G. SVENSSON, 1977. Acta Chem. Scand., A31, 635-637.

Cubic, Fm3m, a = 9.231 Å, Z = 4. Cu radiation, R = 0.058 for 68 reflexions (films, visual intensities). Previous study in 1.

Atomic positions

			x	y	z
Ti	in	4(a)	0	0	0
N(1)	in	4(b)	1/2	1/2	1/2
N(2)	in	8(c)	1/4	1/4	1/4
0.208 F(1)	in	96(j)	0	0.046	0.205
0.083 F(2)	in	96(j)	0	0.112	0.180

The data are consistent with a statistically oriented pentagonal-bipyramidal complex anion, with the H atoms of N(1) also disordered.

1. Structure Reports, 8, 163.

POTASSIUM HYDROGEN OXYFLUOROTUNGSTATE
$K_3H[W_2O_4F_6]$

AMMONIUM OXYFLUOROTUNGSTATE HYDRATE
$(NH_4)_5[W_3O_4F_9] \cdot NH_4F \cdot H_2O$

R. MATTES and K. MENNEMANN, 1977. Z. anorg. Chem., 437, 175-182.

Potassium salt, monoclinic, C2/c, a = 15.14, b = 11.88, c = 8.34 Å, β = 138.2°, Z = 4. Mo radiation, R = 0.064 for 1091 reflexions (0.058 for Cc).

Ammonium salt, monoclinic, P2₁/c, a = 13.589, b = 8.216, c = 15.44 Å, β = 93.82°, Z = 4. Mo radiation, R = 0.073 for 2087 reflexions.

The structures contain bi- and trinuclear anions (Fig. 1), W...W = 2.62 and 2.51 Å, linked by the cations, with hydrogen bonding in the ammonium compound.

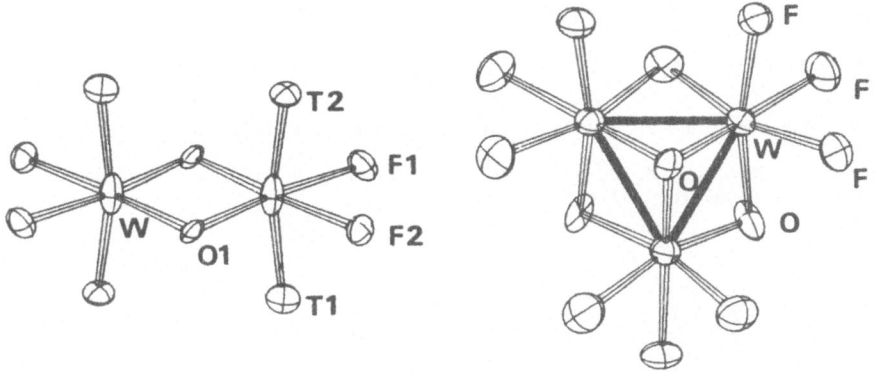

Fig. 1.   The $W_2O_4F_6^{4-}$ and $W_3O_4F_9^{5-}$ anions; T1 and T2 are disordered F and O atoms.

ALUMINA
$\eta-Al_2O_3$

K. SHIRASUKA, H. YANAGIDA and G. YAMAGUCHI, 1976. J. Ceram. Soc. Japan, $\underline{84}$, 610-613.

Cubic, Fd3m, a = 7.906 Å, [Z = 32/3]. Cu radiation, powder data. 32 O in 32(e):
x = 0.370; 4 Al in 8(a); 4 Al in 48(f): x = 0.250; 9 Al in 16(d); 4.3 Al in 16(c)
(origin at $\bar{4}$3m). Spinel-type structure, $Al_{2/3}\square_{1/3}[Al_2]O_4$.

STRONTIUM CALCIUM INDATE
$SrCa_2In_2O_6$

W. MUSCHICK and H. MÜLLER-BUSCHBAUM, 1977. Z. anorg. Chem., $\underline{435}$, 56-60.

Orthorhombic, Pbam, a = 11.046, b = 16.630, c = 3.279 Å, Z = 4. R = 0.064 for
672 reflexions.

    Isostructural with $Ca_3In_2O_6$ ($\underline{1}$). Disordered Ca/In sites have octahedral
coordination, Ca/In-O = 2.12-2.98 Å, and Sr has 9-coordination (trigonal prism
Sr-O = 2.49-2.56 Å, with three further O at 2.77, 3.49, 3.90 Å).

$\underline{1}$. Structure Reports, $\underline{39}$A, 212.

POTASSIUM NIOBATE GERMANATE
$K_6Nb_6Ge_4O_{26}$

J. CHOISNET, N. NGUYEN and B. RAVEAU, 1977. Rev. Chim. Minér., $\underline{14}$, 311-317.

Hexagonal, P$\bar{6}$2m, a = 9.19, c = 8.12 Å, D_m = 4.19, Z = 1. Cu radiation, powder data.

    Isostructural with the Si compound ($\underline{1}$).

$\underline{1}$. Structure Reports, $\underline{42}$A, 396.

LITHIUM PLUMBATE(IV)
Li_4PbO_4

K.-P. MARTENS and R. HOPPE, 1977. Z. anorg. Chem., $\underline{437}$, 105-115.

Orthorhombic, Cmcm, a = 8.32, b = 7.30, c = 6.52 Å, D_m = 4.98, Z = 4. Mo
radiation, R = 0.067 for 197 reflexions.

Atomic positions

			x	y	z
Pb	in	4(c)	0	0.3533	1/4
O(1)	in	8(f)	0	0.795	0.022
O(2)	in	8(g)	0.276	0.981	1/4
Li(1)	in	8(g)	0.145	0.738	1/4
Li(2)	in	8(e)	0.148	0	0

    The structure contains isolated PbO_4^{4-} tetrahedra (Pb-O = 2.08 Å, O-Pb-O = 104-
127°), linked by Li^+ ions which also have tetrahedral coordination (Li-O = 2.42-
2.51 Å, O-Li-O = 98-129°).

POTASSIUM DIPLUMBATE(II)
$K_2Pb_2O_3$

K.-P. MARTENS and R. HOPPE, 1977. Z. anorg. Chem., 437, 116-122.

Cubic, $I2_13$, a = 8.42 Å, D_m = 5.98, Z = 4. Mo radiation, R = 0.060 for 237 reflexions. K and Pb in 8(a): x,x,x, x = 0.5232 and 0.2562; O in 12(b): 0.2887,0,1/4.

Perovskite superstructure with oxygens omitted. Pb-3 O = 2.18, K-6 O = 2.80, 2.95 Å.

RUBIDIUM PLUMBATE(IV)
Rb_2PbO_3

R. HOPPE and H.-D. STÖVER, 1977. Z. anorg. Chem., 437, 123-126.

Orthorhombic, $Cmc2_1$, a = 10.84, b = 7.49, c = 6.01 Å, Z = 4. Mo radiation, R = 0.12 for 227 reflexions.

Atomic positions

			x	y	z
Pb	in	4(a)	0	0.0924	1/4
Rb	in	8(b)	0.3378	0.144	1/4
O(1)	in	8(b)	0.123	0	0
O(2)	in	4(a)	0	0.39	0.19

Isostructural with K_2PbO_3 (1) and Cs_2PbO_3 (2), contrary to a previous description (3).

1. R. HOPPE and H.J. RÖHRBORN, 1964. Naturwissenschaften, 51, 103.
2. Structure Reports, 38A, 251.
3. Ibid., 35A, 223.

PALLADIUM BISMUTH OXIDE
$PdBi_2O_4$

P. CONFLANT, J.-C. BOIVIN and D. THOMAS, 1977. Rev. Chim. Minér., 14, 249-255.

Tetragonal, P4/ncc, a = 8.622, c = 5.907 Å, D_m = 8.7, Z = 4. Mo radiation, R = 0.041 for 228 reflexions. Pd in 4(c): x = 0.0819; Bi in 8(f): x = 0.0794; O in 16(g): (0.458, 0.140, 0.089).

Isostructural with $CuBi_2O_4$ (1) [note, however, that both $CuBi_2O_4$ and $PdBi_2O_4$ have previously been reported, with different structures from that given here (2)].

1. This volume, p. 187.
2. Structure Reports, 42A, 251, 447.

RUBIDIUM SCANDATE
β-$RbScO_2$

H. WIENCH, G. BRACHTEL and R. HOPPE, 1977. Z. anorg. Chem., 436, 169-172.

Hexagonal, $P6_3/mmc$, a = 3.25, c = 12.79 Å, Z = 2. Mo radiation, R = 0.043 for 83 reflexions. Rb in 2(d); Sc in 2(a); O in 4(f): z = 0.0837.

The structure was previously described in $P\bar{6}m2$ (1). In this new description the structure is of the $AlCCr_2$ type (2).

1. Structure Reports, 41A, 420.
2. Ibid., 28, 3.

BARIUM HEXATITANATE
$Ba_2Ti_6O_{13}$ $2BaO.Ti_2O_3.4TiO_2$

J. SCHMACHTEL and H. MÜLLER-BUSCHBAUM, 1977. Z. anorg. Chem., 435, 243-246.

Monoclinic, C2/m, a = 15.004, b = 3.953, c = 9.085 Å, β = 98.01°, Z = 2. R = 0.048 for 1078 reflexions.

Isostructural with $Na_2Ti_6O_{13}$ (1), with statistical distribution of Ti^{3+} and Ti^{4+} in the octahedra, Ti-O = 1.77-2.21 Å; Ba-O = 2.73-3.00 Å.

1. Structure Reports, 27, 528.

CALCIUM COPPER DITITANATE   COPPER TANTALATE TITANATE
$Ca_{0.5}Cu_{1.5}Ti_2O_6$ $Cu_{1.5}TaTiO_6$
COPPER DITANTALATE
$CuTa_2O_6$

V. PROPACH, 1977. Z. anorg. Chem., 435, 161-171.

$Ca_{0.5}Cu_{1.5}Ti_2O_6$, $Cu_{1.5}TaTiO_6$
Cubic, Im3, a = 7.37, 7.43 Å, Z = 4. Neutron powder data. Ca (or vacancy) in 2(a); Cu in 6(b); Ti (or Ta+Ti) in 8(c); O in 24(g): y = 0.3040 (0.3040), z = 0.1793 (0.1859).

$CuTa_2O_6$
Cubic, Pm3, a = 7.487 Å, Z = 4. Neutron powder data. 2 Cu(1) in 3(c); 2 Cu(2) in 3(d); Ta in 8(i): x = 0.2500; O(1) in 12(j): y = 0.2978, z = 0.1812; O(2) in 12(k): y = 0.1850, z = 0.3049. The material may have lower symmetry with ordered Cu positions (compare 1).

The structures are related to the perovskite type, and consist of a framework of corner-sharing MO_6 octahedra (M = Ti or Ta), with Cu ions occupying strongly distorted 12-coordinate sites.

1. J.M. LONGO, 1975. Mater. Res. Bull., 10, 1273.

SODIUM VANADATE HYDRATE
$NaVO_3.1·89H_2O$

A. BJORNBERG and B. HEDMAN, 1977. Acta Chem. Scand., A31, 579-584.

Report not available. See volume 44A.

CAESIUM NIOBATE
$CsNbO_3$

G. MEYER and R. HOPPE, 1977. Z. anorg. Chem., **436**, 75-86.

Monoclinic, $P2_1/c$, a = 5.15, b = 15.89, c = 9.14 Å, β = 93.3°, D_m = 4.83, Z = 8.
R = 0.072 for 613 reflexions.

The structure contains Nb_4O_{12} groups of four edge-sharing tetragonal pyramids
(Fig. 1), Nb-O = 1.76-2.09 Å. Cs ions have 6 and 8 coordinations.

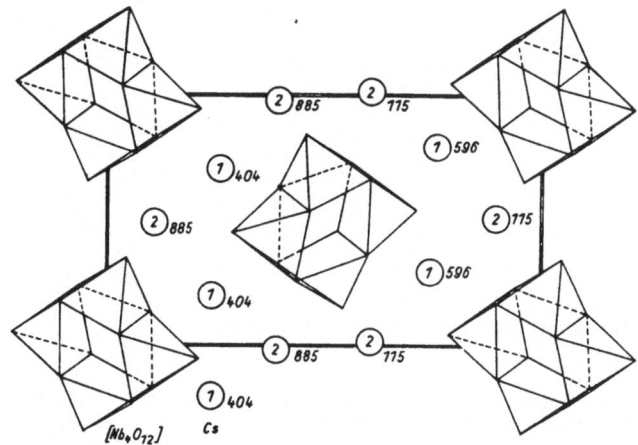

Fig. 1.   Structure of $CsNbO_3$.

AMMONIUM OCTAMOLYBDATE HYDRATE
$(NH_4)_4Mo_8O_{26} \cdot 4H_2O$

H. VIVIER, J. BERNARD and H. DJOMAA, 1977. Rev. Chim. Minér., **14**, 584-604.

Triclinic, $P\bar{1}$, a = 10.051, b = 10.603, c = 7.881 Å, α = 101.06, β = 105.67, γ =
113.41°, Z = 1. Mo radiation, R = 0.064 for 5135 reflexions.

The structure contains the $Mo_8O_{26}^{4-}$ anion, consisting of condensed MoO_6 octahedra as previously described (1). Ammonium cations and water molecules are
distinguished from chemical criteria, and a probable network of hydrogen bonds is
suggested.

1. Structure Reports, **13**, 267.

POTASSIUM DIIODODIMOLYBDATE HYDRATE
$K_6[Mo_2I_2O_{16}] \cdot 10H_2O$

R. MATTES, C. MATZ and E. SICKING, 1977. Z. anorg. Chem., **435**, 207-213.

Triclinic, $P\bar{1}$, a = 8.383, b = 9.878, c = 8.287 Å, α = 97.01, β = 95.57, γ = 104.67°,
D_m = 2.84, Z = 1. Mo radiation, R = 0.042 for 2955 reflexions.

The structure contains tetranuclear heteropolyanions in which the Mo and I
atoms have distorted octahedral coordinations (Fig. 1); Mo-O = 1.72-2.27, I-O =

1.80-2.01 Å. The anions are linked by the K ions (10-coordination) and by hydrogen bonding via the water molecules.

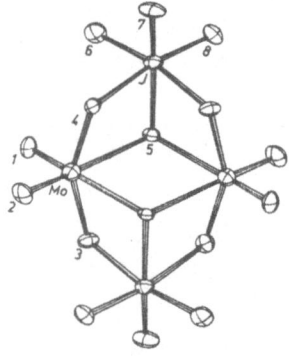

Fig. 1.   Structure of $K_6[Mo_2I_2O_{16}] \cdot 10H_2O$.

SODIUM DECATUNGSTOURANATE(IV) HYDRATE
$Na_8[UW_{10}O_{36}] \cdot 30H_2O$

A.M. GOLUBEV, L.A. MURADJAN, L.P. KAZANSKIJ, E.A. TORČENKOVA, V.I. SIMONOV and
V.I. SPICYN, 1977. Koordin. Khim., 3, 920-925.

Monoclinic, B2/b, a = 18.139, b = 18.509, c = 18.602 Å, γ = 96.0°, D_m = 3.42, Z = 4.
R = 0.064 for 1612 reflexions.

      The structure contains $UW_{10}O_{36}^{8-}$ anions, which consist of condensed WO_6 octa-
hedra and a central 8-coordinate U atom (Fig. 1). The anions are linked by Na
ions and water molecules.

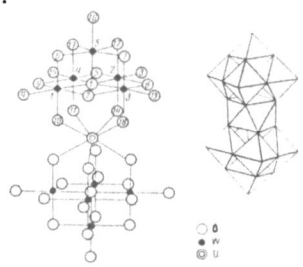

Fig. 1.   The $UW_{10}O_{36}^{8-}$ anion.

SODIUM OXOFERRATE(II)
Na_4FeO_3

R. HOPPE and H. RIECK, 1977. Z. anorg. Chem., 437, 95-104.

Monoclinic, Cc, a = 10.97, b = 5.82 [5.83 also given], c = 8.22 Å, β = 114.0°
[113.1° also given], D_m = 2.75, Z = 4. Mo radiation, R = 0.045 for 382 reflexions.

Atomic positions

	x	y	z
Na(1)	0.7072	0.1606	0.631
Na(2)	0.3062	0.7879	0.351
Na(3)	0.1127	0.1620	0.445
Na(4)	0.8856	0.841	0.532
Fe	0.9785	0.6464	0.2441
O(1)	0.3159	0.995	0.109
O(2)	0.6466	0.989	0.843
O(3)	0.4391	0.444	0.296

The structure contains nearly planar FeO_3 groupings (Fe-O = 1.87, 1.88, 1.89 Å, O-Fe-O = 108, 125, 126°), linked by Na ions with 3- and 4-coordinations (Na-O = 2.28-2.64 Å).

POTASSIUM SILVER OXIDE
K_3AgO_2

B. DARRIET, M. DEVALETTE and B. LECART, 1977. Rev. Chim. Minér., 14, 423-428.

Orthorhombic, $P2_12_12_1$, a = 12.30, b = 12.30, c = 13.48 Å, D_m = 3.45, Z = 16. R = 0.076 for 1375 reflexions.

The structure can be derived from the antifluorite structure of K_2O by doubling the cell edges, and substituting Ag for K (K-K-Ag-Ag-K-K along c, and alternation of K-Ag rows and K only rows along a and b). Ag has linear coordination, Ag-O = 1.95-2.12 Å, and K tetrahedral coordination, K-O = 2.33-3.50 Å.

STRONTIUM URANATE
γ-$SrUO_4$

T. FUJINO, N. MASAKI and H. TAGAWA, 1977. Z. Kristallogr., 145, 299-309.

Rhombohedral, R3̄m, a = 6.551 Å, α = 34.82°, Z = 1 (hexagonal cell has a = 3.921, c = 18.443 Å), for $SrUO_{3.95}$, data given also for $SrUO_{3.78}$ and $SrUO_{3.60}$. Cu radiation, powder data. U in 1(a): 0,0,0; Sr in 1(b): 1/2,1/2,1/2; O(1) and O(2) in 2(c): x,x,x, x = 0.112 and 0.357.

Structure as previously described (1, 2); in the oxygen-deficient samples the vacancies are found exclusively at the O̅(2) site. The orthorhombic α-form is described in 3.

1. Structure Reports, 11, 324.
2. K.L. REŠETOV and L.M. KOVBA, 1966. Ž. Strukt. Khim., 7, 625.
3. Structure Reports, 34A, 289; 38A, 286.

MOLYBDENUM OXIDE HYDROXIDE
$MoO_{2.7}(OH)_{0.3}$

F.A. SCHRÖDER and H. WEITZEL, 1977. Z. anorg. Chem., 435, 247-256.

Although the material has been described previously as $Mo_4O_{10}(OH)_2$ [i.e. $MoO_{2.5}$-$(OH)_{0.5}$] the composition obtained here is $MoO_{2.7}(OH)_{0.3}$. The structure is essentially as previously described (1), except that the H atom is disordered in two positions between bridging oxygen atoms from corner sharing Mo-O octahedra (neutron powder data, H parameters: 0, 0.461, 0.496).

1. Structure Reports, 26, 354; 34A, 236.

SELENIUM SULPHUR

$Se_{2.9}S_{5.1}$ $Se_{3.3}S_{4.7}$ $Se_{4.7}S_{3.3}$

J. WEISS, 1977. Z. anorg. Chem., 435, 113-118.

$Se_{2.9}S_{5.1}$, $Se_{3.3}S_{4.7}$
Monoclinic, P2/c, a = 8.578, 8.579, b = 13.386, 13.405, c = 9.368, 9.354 Å, β = 124.3, 124.1°, Z = 4. Mo radiation, R = 0.069 and 0.074 for 1266 and 1597 reflexions.

$Se_{4.7}S_{3.3}$
Monoclinic, $P2_1/n$, a = 8.737, b = 9.104, c = 11.316 Å, β = 90.8°, Z = 4. Mo radiation, R = 0.114 for 1729 reflexions.

The first two compounds have the γ-sulphur structure (1), and the last has the α-selenium structure (2). All three structures contain eight-membered rings with random Se/S occupancy, bond lengths 2.08-2.28 Å.

1. Structure Reports, 40A, 119.
2. Ibid., 15, 132; 16, 156; 38A, 141.

AMMONIUM HEPTATHIOTETRAANTIMONATE
$(NH_4)_2Sb_4S_7$

G. DITTMAR and H. SCHÄFER, 1977. Z. anorg. Chem., 437, 183-187.

Orthorhombic, Pbca, a = 11.330, b = 26.252, c = 9.940 Å, D_m = 3.26, Z = 8. R = 0.075 for 2117 reflexions.

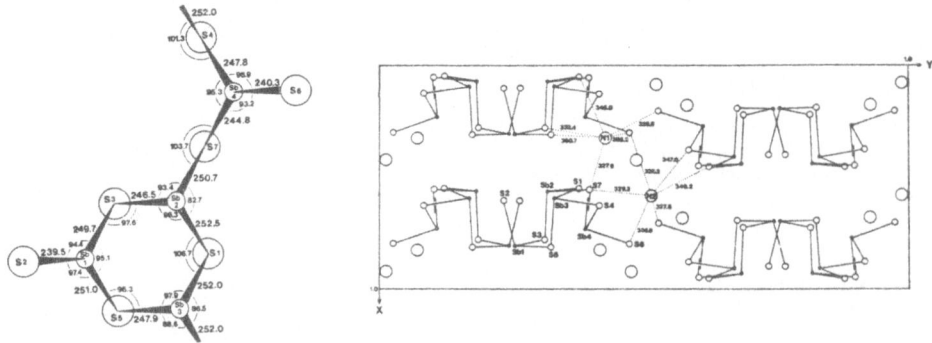

Fig. 1.   The unit of the chain (left, distances in Å x 10^2) and the structure (right) of $(NH_4)_2Sb_4S_7$.

The structure (Fig. 1) contains a six-membered ring of three SbS_3 trigonal pyramids, linked into chains along c by a fourth SbS_3 pyramid. The chains are linked by the ammonium ions, N...S = 3.28-3.65 Å.

AMMONIUM PALLADIUM POLYSULPHIDE DIHYDRATE
$(NH_4)_2PdS_{11}\cdot2H_2O$

P.S. HARADEM, J.L. CRONIN, R.A. KRAUSE and L. KATZ, 1977.   Inorg. Chim. Acta, 25, 173-179.

Tetragonal, P4/mnc, a = 11.065, c = 6.862 Å, D_m = 2.09, Z = 2. Mo radiation, R = 0.113 for 243 reflexions.

Atomic positions

			x	y	z
Pd	in	2(a)	0	0	0
0.5 S(1)	in	16(i)	0.1481	0.1483	0.0231
0.5 S(2)	in	16(i)	0.3157	0.0799	0.0496
0.4 S(3)	in	16(i)	0.0046	0.3878	0.1892
0.5 O/N	in	16(i)	0.3420	0.4124	0.0444

The structure contains Pd atoms (square-planar coordination) linked by S_6 chains, with deficiency at the central S atoms, and with fourfold disorder. Pd-S = 2.32, S-S = 2.01-2.06 Å, S-S-S = 107 and 109°.

LITHIUM EUROPIUM(III) BORATE
$Li_3Eu_2(BO_3)_3$

G.K. ABDULLAEV, K.S. MAMEDOV and G.G. DŽAFAROV, 1977. Azerbajdž. Khim. Ž., no. 2, 115-119.

Monoclinic, $P2_1/n$, a = 8.66, b = 13.95, c = 5.69 Å, β = 103.7°, Z = 4. R = 0.12 for 570 reflexions (film data).

The structure contains triangular BO_3^{3-} ions, Li ions with tetrahedral coordination, and Eu ions with 9-coordination. B-O = 1.33-1.39, Li-O = 1.85-2.19, Eu-O = 2.35-2.77 Å.

SODIUM PERCARBONATE
$Na_2CO_3\cdot1\cdot5H_2O_2$

M.A.A.F.deC.T. CARRONDO, W.P. GRIFFITH, D.P. JONES and A.C. SKAPSKI, 1977. J. Chem. Soc., Dalton, 2323-2327.

Orthorhombic, Aba2, a = 9.183, b = 15.745, c = 6.730 Å, D_m = 2.15, Z = 8. Cu radiation, R = 0.029 for 486 reflexions.

The structure contains sodium and carbonate ions, and H_2O_2 molecules disordered in fourfold and eightfold sites. [Independent study in 1, where only the fourfold H_2O_2 sites are considered to be disordered.]

1. This volume, p. 232.

MALACHITE
$Cu_2(OH)_2CO_3$

F. ZIGAN, W. JOSWIG, H.D. SCHUSTER and S.A. MASON, 1977. Z. Kristallogr., 145, 412-426.

Monoclinic, $P2_1/a$, a = 9.502, b = 11.974, c = 3.240 Å, β = 98.75°, Z = 4. Neutron radiation, R = 0.021 for 635 reflexions.

The structure is as previously described (1). O-H = 0.969, 0.975(2), O...O = 2.716, 2.737(1) Å, O-H...O = 148.1, 153.7°.

1. Strukturbericht, 2, 397; 15, 280; 32A, 419.

SODIUM OXIDE NITRITE
$Na_3O(NO_2)$

M. JANSEN, 1977. Z. anorg. Chem., 435, 13-20.

Cubic, Pm3m, a = 4.605 Å, D_m = 2.20, Z = 1. Mo radiation, R = 0.05 for 54 reflexions. Na in 3(d); O in 1(a); disordered NO_2 group.

Anti-perovskite type structure, with a disordered NO_2^- anion.

TELLURIUM(IV) OXIDE PHOSPHATE
$Te_8O_{10}(PO_4)_4$

H. MAYER and G. PUPP, 1977. Z. Kristallogr., 145, 321-333.

Monoclinic, $P2_1/n$, a = 20.076, b = 4.650, c = 11.220 Å, β = 92.99°, D_m = 4.90, Z = 2. Mo radiation, R = 0.049 for 1932 reflexions.

The structure contains a triple chain, $Te_6O_{10}^{4+}$ (mean Te-O = 1.95 Å), linked via a fourth Te atom which is strongly bonded to three PO_4 groups (Te-O = 1.95, 2.09, 2.10 Å). This gives $[Te_6O_{10}][Te(PO_4)_2]_2$ structural units, which are connected into puckered layers, which are further held together by weak Te...O interactions (2.65, 2.87 Å). The environments of all four independent Te atoms indicate stereochemically active lone pairs.

TALMESSITE
$Ca_2(Mg,Co)(AsO_4)_2 \cdot 2H_2O$

M. CATTI, G. FERRARIS and G. IVALDI, 1977. Bull. Soc. Fr. Minér. Crist., 100, 230-236.

Triclinic, $P\bar{1}$, a = 5.874, b = 6.943, c = 5.537 Å, α = 97.3, β = 108.7, γ = 108.1°, D_m = 3.57, Z = 1. Mo radiation, R = 0.081 for 979 reflexions.

The structure contains layers parallel to ($\bar{1}$10) of CaO_8 polyhedra and $(Mg,Co)O_6$ distorted octahedra; there are two very short hydrogen bonds, O-H...O = 2.56, 2.61 Å. As-O = 1.66-1.70, Ca-O = 2.36-2.77, (Mg,Co)-O = 2.04-2.14 Å.

SODIUM CERIUM(III) SULPHATE MONOHYDRATE
$NaCe(SO_4)_2 \cdot H_2O$

O. LINDGREN, 1977. Acta Chem. Scand., A31, 591-594.

Report not available. See volume 44A.

POTASSIUM TRISULPHATOURANYLATE
$K_4UO_2(SO_4)_3$

Ju.N. MIKHAJLOV, L.A. KOKH, V.G. KUZNECOV, T.G. GREVCEVA, S.K. SOKOL and
G.V. ÈLLERT, 1977. Koordin. Khim., 3, 508-513.

Orthorhombic, Pnma, a = 13.053, b = 23.200, c = 9.379 Å, D_m = 3.3, Z = 8. Mo
radiation, R = 0.099 for 1290 reflexions (film data).

     The structure (Fig. 1) contains $[(SO_4)_2UO_2(SO_4)_2UO_2(SO_4)_2]$ groupings and K^+
ions. U has pentagonal bipyramidal coordination, U-O = 1.89 (uranyl), 2.33-2.55(4)
Å (sulphate oxygen).

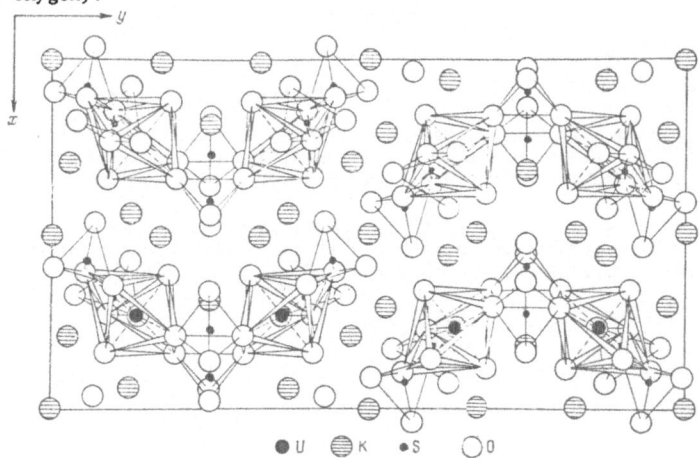

Fig. 1.   Structure of $K_4UO_2(SO_4)_3$.

MANGANESE(II) SELENITE MONODEUTERATE
$MnSeO_3 \cdot D_2O$

M. KOSKENLINNA and J. VALKONEN, 1977. Acta Chem. Scand., A31, 752-754.

Orthorhombic, Pnma, a = 13.179, b = 5.826, c = 4.933 Å, D_m = 3.5, Z = 4. Mo
radiation, R = 0.066 for 500 reflexions.

Atomic positions

	x	y	z
Mn	0.7683	1/4	0.3985
Se	0.4018	3/4	0.0247
O(1)	0.3632	3/4	0.3489
O(2)	0.1671	0.0303	0.3824
$H_2O(3)$	0.8908	1/4	0.0918

The structure contains trigonal pyramidal SeO_3 groups (Se-O = 1.68; 1.72(1) Å, O-Se-O = 96, 103°) linked by $MnO_5(H_2O)$ octahedra which share four O corners with neighbouring octahedra; Mn-O = 2.14-2.31(1) Å.

CHAPMANITE   BISMUTHOFERRITE
$SbFe_2Si_2O_8OH$ $BeFe_2Si_2O_8OH$

A.P. ŽUKHLISTOV and B.B. ZVJAGIN, 1977. Kristallografija, 22, 731-738 [Soviet Physics - Crystallography, 22, 419-423].

Monoclinic, Cm, a = 5.19, 5.21, b = 8.99, 9.02, c = 7.70, 7.74 Å, β = 100°40', 100°40', Z = 2. Electron diffraction.

A structure is derived with two-stage kaolinite-like layers of composition $Fe_2Si_2O_8(OH)$, with Sb (Bi) between the layers.

CLINOPTILOLITE
$(Ca,Na,K,Mg)(Al,Si)_9O_{18} \cdot 6H_2O$

K. KOYAMA and Y. TAKÉUCHI, 1977. Z. Kristallogr., 145, 216-239.

Monoclinic, C2/m, a = 17.660, 17.662, b = 17.963, 17.911, c = 7.400, 7.407 Å, β = 116.47, 116.40°, for specimens from Kurama Pass, Japan, and Agoura, U.S.A., respectively, Z = 4. Cu radiation, R = 0.080 and 0.058 for 1701 and 1951 reflexions.

Isostructural with heulandite (1), with two new partially-occupied cation sites: K in M(3), which is too close to M(1) for simultaneous occupancy, and Mg (or Al) in M(4) (0,0,1/2). The K location differs from that in 2.

1. Structure Reports, 33A, 484; 38A, 368.
2. Ibid., 41A, 380.

SUBJECT INDEX

This index contains the names of substances printed
at the head of the reports, and some additional general
entries. Greek letter and numerical prefixes, and prefixes
such as cis, trans etc. are disregarded in fixing the
alphabetical order.

Aikinite, 31
Albite (low), 305
Alkaline-earth antimonides, 3
Alkaline-earth beryllium
  germanides, 3
Alkaline-earth beryllium
  silicides, 3
Alkaline-earth bismuthides, 3
Alkaline-earth group IV ternary
  compounds, 98
Alkaline-earth pnictides, 4
Alloclasite, 20
Alluaudite (synthetic), 336
beta-Alumina, 174, 179, 335
beta-Alumina (alkali-free), 174
Alumina (eta), 348
Aluminum barium germanium, 4
Aluminum barium iron sulphur, 5
Aluminum bismuth oxide, 187
Aluminum boron, 6
Aluminum bromide ammonia (gas),
  333
Aluminum cerium, 8
Aluminum chromate dichromate
  tetrahydrate (alpha), 198
Aluminum chromium, 8
Aluminum cobalt lanthanum, 10
Aluminum hydrogen iodate hexa-
  hydrate, 296
Aluminum iridium niobium (A'-
  phase), 10
Aluminum magnesium selenium, 11
Aluminum orthoborate, 228
Aluminum oxide hydrate, 174
Aluminum palladium, 12
Aluminum praseodymium, 8
Aluminum rhenium silicon, 12
Aluminum vanadium, 8
Americium silicon, 12
Amminecopper(II) tellurate(IV)
  monohydrate, 292
Ammonium arsenate trihydrate, 269
Ammonium cadmium phosphate
  monohydrate (phase B), 262
Ammonium chromium(III) fluoride

hexahydrate, 342
Ammonium cobalt(II) tetrafluoro-
  beryllate hexahydrate, 123
Ammonium decabromodibismuth-
  ate(III) tetrahydrate, 140
Ammonium dilead chloride, 136
Ammonium heptathiotetraantimonate
  354
Ammonium hexaammine-tri-mu-
  hydroxo-dicobalt tetrachloride,
  149
Ammonium hexachlorotungstate(V),
  137
Ammonium hydrogen aluminum
  tripolyphosphate, 264
Ammonium iron(III) chromate
  (alpha), 200
Ammonium metavanadate, 190
Ammonium octamolybdate hydrate,
  351
Ammonium oxyfluorotungstate hydr-
  ate, 347
Ammonium palladium polysulphide
  dihydrate, 355
Ammonium pentafluoroperoxotitan-
  ate(IV), 347
Ammonium phosphoberyllate, 245
Ammonium praseodymium tetra-
  metaphosphate, 268
Ammonium selenate, 284
Ammonium sodium cadmium
  pyrophosphate trihydrate, 263
Ammonium sulphite monohydrate,
  276
Ammonium tetrachloropalladate(II)
  138
Ammonium tetrachlorozincate, 329
Ammonium thiocyanate, 171
Ammonium 12-molybdotellurate
  hydrate, 336
Ankerite, 235
Antimony arsenic lead sulphur, 13
Antimony arsenic uranium, 13
Antimony beryllium, 13
Antimony beryllium sodium, 14

Antimony chloride fluoride, 120
Antimony chloride fluoride oxide,
    346
Antimony gold yttrium, 14
Antimony(III) iodide sulphide,
    157
Antimony lithium, 15
Antimony lithium zinc, 15
Antimony manganese zinc, 16
Antimony oxide (alpha), 174
Antimony palladium, 16
Antimony rhodium, 17
Antimony ruthenium, 17
Antimony silver, 17
Antimony thallium, 17
Antimony titanium, 18
Antimony(V) trichloride
    difluoride, 120, 335
Antimony trifluoride - antimony
    pentafluoride (1:1), 119
Antimony vanadium, 18
Argyrodite (synthetic), 66
Aristarainite, 225
Armalcolite, 189
Arsenian ullmannite, 19
Arsenic beryllium lithium, 19
Arsenic beryllium sodium, 14
Arsenic cobalt iron sulphur, 20
Arsenic europium, 20
Arsenic magnesium, 21
Arsenic mercury sulphur thallium,
    22
Arsenic molybdenum nickel, 22
Arsenic(V) oxide, 334
Arsenic sodium sulphur, 23
Arsenic strontium, 23
Arsenic sulphur thallium, 24
Arsenic vanadium, 25
Artinite, 233

Baotite, 306
Barite, 278
Barium aluminum germanate, 180
Barium antimony titanium oxide,
    188
Barium bismuthate, 186
Barium cobaltate(IV), 214
Barium copper(II) oxybromide, 151
Barium copper oxychloride, 150
Barium copper(II) oxychloride,
    151
Barium dibromotetracyanoplatin-
    ate(IV) hydrate, 168
Barium dihydrogen phosphate, 250
Barium ferrotantalate, 197
Barium fluoride bromide, 118
Barium fluoride iodide, 118

Barium gadolinium uranate, 331
Barium germanium, 26
Barium hectorite, 328
Barium hexafluorocuprate(II), 130
Barium hexatitanate, 350
Barium iron(II) germanate, 180
Barium iron sulphur, 5, 26
Barium lead, 27
Barium lead magnesium, 28
Barium nickel antimonate, 185
Barium oxocuprate(II), 215
Barium phosphochromate monohydr-
    ate, 198
Barium phosphochromate trihydrate
    , 198
Barium phosphorus bromide, 343
Barium phosphorus chloride, 343
Barium phosphorus halides, 343
Barium phosphorus iodide, 343
Barium platinum oxide, 214
Barium silicon, 29
Barium silver trimetaphosphate
    tetrahydrate, 265
Barium tetraarsenate, 270
Barium tetracyanoplatinate(II)
    tetrahydrate, 168
Barium tin zinc, 28
Barium titanotantalate, 197
Barium yttrium tellurate, 331
Barium yttrium tungstate, 331
Barium yttrium uranate, 331
Barylite, 306
Beryl (caesium-lithium), 306
Beryllium calcium germanium, 28
Beryllium nitrate tetrahydrate,
    238
Beryllium phosphorus, 30
Beryllium polyphosphate (form II)
    244
Beryllium tellurate, 290
Bicchulite (synthetic), 337
Bismuth cerium sulphur, 30
Bismuth chromium niobium oxide,
    196
Bismuth copper lead sulphur, 30
Bismuth(III) fluoride
    (orthorhombic), 340
Bismuth nickel selenium, 33
Bismuth sulphur, 33
Bismuth tellurium iodide, 157
Bismuthinite, 33
Bismuthoferrite, 358
Boehmite, 219
Boracite (lithium), 223
Boron (beta-rhombohedral), 34
Boron germanium molybdenum, 34
Boron lithium nickel, 35

Boron lutetium ruthenium, 35
Boron magnesium nickel, 35
Boron molybdenum thorium, 36
Boron ruthenium tungsten, 37
Boron sulphide, 334
Boron sulphur, 37
Boron thorium tungsten, 36
Brewsterite, 307
Bromopentacarbonylrhenium(I), 110
Bukovite, 103
Bustamite, 307, 312

Cadmium apatites, 261
Cadmium arsenate bromide, 261
Cadmium bromapatite, 261
Cadmium chlorapatite, 261
Cadmium gold, 38
Cadmium heptagermanate, 181
Cadmium hexacyanocobaltate(III)
  dodecahydrate, 164
Cadmium hexahydroxoplumbate(IV),
  220
Cadmium iodide, 328
Cadmium neodymium silicate, 305
Cadmium pentafluorogallate(III)
  heptahydrate, 340
Cadmium scandium germanate, 300
Cadmium selenium, 40
Cadmium vanadate bromide, 261
Cadmium vanadate iodide, 261
Caesium barium hexacyanoferr-
  ate(II) dihydrate, 162
Caesium cadmium bromide, 345
Caesium cadmium trimetaphosphate,
  266
Caesium chlorotrifluoroantimon-
  ate(III), 126
Caesium divanadate, 190
Caesium divanadate bronze, 191
Caesium enneabromodibismuth-
  ate(III), 140
Caesium hexabromoneptunate, 336
Caesium hexachlorotungstate(V),
  137
Caesium lithium hexacyanochrom-
  ate(III) (tetragonal), 160
Caesium lithium hexacyanoferr-
  ate(III), 162
Caesium manganese gallium
  fluoride, 331
Caesium manganese iron fluoride,
  331
Caesium manganese nickel fluoride
  , 129
Caesium manganese rhodium
  fluoride, 331
Caesium metavanadate, 190

Caesium nickel indium fluoride,
  331
Caesium nickel rhodium fluoride,
  331
Caesium nickel scandium fluoride,
  331
Caesium nickel thallium fluoride,
  331
Caesium niobate, 351
Caesium octachlorodirhenate(III)
  hydrate, 138
Caesium palladium indium fluoride
  331
Caesium palladium iron fluoride,
  331
Caesium palladium rhodium
  fluoride, 331
Caesium palladium scandium
  fluoride, 331
Caesium pentafluorotellurate(IV),
  341
Caesium pentavanadate, 190
Caesium potassium hexacyanoferr-
  ate(III), 162
Caesium praseodymate, 217
Caesium sodium hexacyanoferr-
  ate(III), 162
Caesium suboxide, 173
Caesium sulphur, 40
Caesium tetrachloroferrate(III),
  344
Caesium tetracyanoplatinate(II)
  monohydrate, 167
Caesium tetrafluoromanganate(III)
  dihydrate, 128
Caesium titanoniobate, 196
Caesium titanotantalate, 196
Caesium trimetaphosphate
  monohydrate, 263
Caesium trivanadate bronze, 191
Caesium vanadates, 190
Caesium vanadium bronzes, 191
Caesium zinc copper fluoride, 331
Caesium zinc gallium fluoride,
  331
Caesium zinc indium fluoride, 331
Caesium zinc manganese fluoride,
  331
Caesium zinc rhodium fluoride,
  331
Caesium zinc scandium fluoride,
  331
Caesium zinc thallium fluoride,
  331
Caesium zinc titanium fluoride,
  331
Caesium zinc vanadium fluoride,

331
Calcite (magnesian), 235
Calcium aluminum sulphate hydrate
, 278
Calcium ammonium hydrogenpyro-
phosphate, 262
Calcium borate, 226
Calcium bromide hexahydrate, 131
Calcium chloride dihydrate, 131
Calcium chloride hexahydrate, 131
Calcium chromate monohydrate, 198
Calcium copper dititanate, 350
Calcium copper germanate, 181
Calcium copper oxybromide, 150
Calcium copper oxychloride, 150
Calcium deuterium, 41
Calcium fluoride chloride, 118
Calcium hexahydroxoplumbate(IV),
220
Calcium holmium oxide, 218
Calcium iron aluminum oxide, 179
Calcium lutetium sulphur, 41
Calcium magnesium phosphate
(beta), 249
Calcium nitrate tetrahydrate
(alpha), 239
Calcium phosphate (alpha), 248
Calcium phosphate (beta), 249
Calcium scandium germanate, 300
Calcium scandium silicate, 300
Calcium silicate (beta), 299
Calcium silicon, 42
Calcium thorium niobate, 196
Calcium thulium oxide, 218
Calcium tin, 42
Cancrinite, 337
Carbon cobalt tungsten, 43
Carbon molybdenum, 44
Celadonite (ferrous), 307
Cerium, 44
Cerium(IV) chromate dihydrate,
203
Cerium dysprosium sulphur, 45
Cerium(IV) hydroxide sulphate
hydrate, 283
Cerium rhenium silicon, 46
Cerium(IV) sulphate tetrahydrate,
282
Cervantite, 174
Chabazite (calcium-exchanged de-
hydrated), 308
Chabazite (calcium-exchanged de-
hydrated carbon monoxide), 308
Chabazite (copper-exchanged), 308
Chabazite (copper-exchanged de-
hydrated), 308
Chabazite (sodium-exchanged de-

hydrated), 308
Chapmanite, 358
Chlorotionite, 281
Chloroxiphite, 151
Chondrodite (hydroxyl), 309
Christite, 22
Chromium(III) hexametaphosphate,
269
Chromium(III) hydrogen sulphate
heptahydrate, 279
Chromium lanthanum sulphur, 47
Chromium manganese oxide, 330
Chromium niobium silicon, 47
Chromium nitrogen vanadium, 48
Chromium(III) oxide hydroxide
(alpha), 221
Chromium(III) oxide hydroxide
(beta), 221
Chromium phosphorus sulphur, 48
Clinoenstatite, 309
Clinohedrite, 310, 336
Clinoptilolite, 358
Cobalt(II) chloroarsenate, 273
Cobalt germanium sulphur, 48
Cobalt indium, 49
Cobalt iron yttrium, 49
Cobalt(II) niobate, 211
Cobalt niobium phosphorus, 50
Cobalt(II) phosphate hydroxide,
257
Cobalt(II) silicate (gamma), 301
Cobalt silicon uranium, 50
Cobalt(II) tungstate, 211
Cobaltite, 20
Coesite, 297
Copper bismuth oxide, 187
Copper(II) chloroarsenate, 273
Copper ditantalate, 350
Copper gallium selenium, 51
Copper iron sulphur tin, 51
Copper(II) manganese(IV) oxide,
212
Copper mercury sulphur tin, 52
Copper mercury titanium, 52
Copper molybdenum sulphur, 53
Copper nickel tin, 53
Copper(II) nitrate (gas), 334
Copper oxide phosphate, 258
Copper oxychloride, 149
Copper(II) phosphate, 257
Copper(II) phosphate hydroxide,
259
Copper potassium sulphur, 54
Copper sodium silicate, 301
Copper sulphur tin, 55
Copper tantalate titanate, 350
Copper tin, 53

Copper titanate (alpha), 189
Copper titanate (beta), 189
Cordierite (low), 310
Cyclotetrathiazyl bis(hexachloro-
antimonate(V)), 335
Cyclotetrathiazyl hexafluoro-
antimonate(V) tetradecafluoro-
triantimonate, 335
Cyclotriazane, 327
Cyclo-tri-mu-hydroxo-tris(cis-
diammineplatinum(II)) nitrate,
242

Dawsonite, 235
Deerite, 311, 337
Deuterium beta-alumina, 335
Deuterium thorium, 56
Diaminodifluorophosphorane (gas),
333
Diborane, 109
5,10-Dibromo-nido-decaborane(14),
335
Dicalcium silicate, 299
Dicyanotrisulphane, 159
Di-mu-fluoro-hexafluorohexaaquo-
dizirconium(IV), 121
Di-mu-hydrido-decacarbonyltri-
osmium, 111
Di-mu-hydroxo-bis(diammineplatin-
um(II)) nitrate, 242
Diiron pentafluoride dihydrate,
122
cis-Dinitratodiammineplatinum(II)
, 241
Dioptase, 311
Disilver sulphamide, 275
Disodium di-mu-hydroxo-dizinc
molybdate, 208
Disodium hydrogen phosphate
dihydrate, 247
Ditellurium tetraselenide hexa-
fluoroantimonate(V), 155
Ditellurium tetraselenide hexa-
fluoroarsenate(V), 155
Ditin(II) chloride trifluoride,
118
Dodecacarbonyldiiodotriosmium,
112
Dodecacarbonyltriosmium, 111
Dodecacarbonyltriruthenium, 110
Dodecatungstophosphoric acid
hexahydrate, 210
Dolomite, 235
Dyscrasite, 17
Dysprosium hydroxide, 222
Dysprosium iron silicon, 56

Erbium hydroxide, 222
Erbium hydroxide nitrate, 244
Erbium oxyiodide, 155
Erbium platinum silicon, 57
Erbium selenium silver, 58
Erionite (dehydrated), 311
Europium arsenic oxide, 183
Europium silicon, 42
Europium sulphur tin, 58

Fayalite, 312
Feldspar, 312
Ferrinatrite, 280
Ferrobustamite, 312
Ferrous celadonite, 307
Fluorapatite, 262
mu-Fluorosulphato-bis(fluoro-
xenon(II)) hexafluoroarsenate(V)
145

Gadolinium hydroxide, 222
Gadolinium iodate, 296
Gadolinium monochloride, 334
Gadolinium oxocuprate, 216
Gadolinium phosphorus sulphur, 59
Gallium bromide ammonia (gas),
333
Gallium oxide (beta), 174
Gallium oxide deuteroxide, 219
Gallium oxide hydroxide (alpha),
219
Gallium selenium thallium, 60
Gallium tellurium, 61
Geocronite, 13
Germanium dibromide, 344
Germanium gold sodium, 61
Germanium iron lithium, 62
Germanium nickel palladium, 64
Germanium palladium, 64
Germanium platinum selenium, 65
Germanium silver sulphur, 66
Gerstmannite, 313
Gianellaite, 337
Gladite, 31
Gold indium, 66
Gold sodium tin, 61
Graphite, 328
Group VIII plutonium compounds,
67

Hafnium nickel silicon, 68
Hammarite, 32
Harkerite, 228
Hastingsite (sub-silicic), 313
Heazlewoodite, 84
Hectorite, 328
Hellandite, 313

Heptascandium decachloride, 133
Heteromorphite, 99
Hexaamminecobalt(III) hexacyano-
  chromate(III), 161
Hexaaquoiron(III) nitrate tri-
  hydrate, 241
B-Hexachlorocyclotriborazane, 116
Hexadecacarbonylpentaosmium, 113
Hexakis(ammonia-cyanoborane)-
  sodium iodide, 109
Hilgardite, 336
Holdenite, 314
Hollandite, 328
Holmium aluminate, 179
Holmium chloride, 336
Holmium hydroxide, 222
Holmium oxocuprate, 217
Hortonolite, 312
Hubnerite, 211
Hungchaoite, 225
Hyalophane, 314
Hydrazinium(2+) pentafluoro-
  oxoniobate hydrate, 336
Hydrogen lithium palladium, 69
Hydrogen lithium platinum, 69
Hydrogen ruthenium ytterbium, 69
Hydromagnesite, 234
mu-Hydroxo-bis(pentaamminechrom-
  ium(III)) chloride dihydrate,
  147
Hydroxyl-chondrodite, 309

Ilvaite, 315
Imhofite, 24
Iminosulphur oxydifluoride (gas),
  333
Indialite, 315
Indium disilicate (high-pressure)
  , 329
Indium samarium sulphur, 70
Indium selenium, 70
Indium silver, 66
Indium sulphur terbium, 70
Iodocycloheptasulphur hexafluoro-
  antimonate(V), 335
Iridium plutonium, 67, 71
Iridium tricarbonyl chloride, 336
Iron(III) chromate dichromate
  tetrahydrate (alpha), 199
Iron(III) dihydrogen tripolyphos-
  phate monohydrate, 265
Iron fluoride hydrate, 122
Iron lanthanum phosphorus, 71
Iron lanthanum sulphur, 72
Iron manganese silicon, 73
Iron(II) niobate, 211
Iron(II) nitrate nonahydrate, 241

Iron scandium silicon, 56
Iron(II) silicate (gamma), 301
Iron silicon tungsten, 73
Iron(III) sulphate (rhombohedral)
  , 279
Iron sulphur, 75
Iron sulphur tin, 76
Iron sulphur ytterbium, 76
Iron tellurate, 330
Iron(III) tungstate, 212

K-phase, 73
Kaliborite, 225
Kasolite, 316
Killalaite, 316
Krupkaite, 31
Krutaite, 103

Lanthanite, 236
Lanthanon hydroxides, 222
Lanthanum borate, 230
Lanthanum chromite, 202
Lanthanum hydroxide, 222
Lanthanum magnesium nitrate hydr-
  ate, 244
Lanthanum nickelate, 214
Lanthanum osmium oxide, 214
Lanthanum oxocuprate, 216
Lanthanum oxyiodide, 155
Lanthanum perrhenate tetrahydrate
  , 332
Lanthanum phosphate, 262
Lanthanum rhenium silicon, 77
Lanthanum ruthenium oxide, 214
Lanthanum strontium chromite, 202
Lead azide (alpha), 115
Lead(II) chloride (gas), 333
Lead(II) fluoride chloride, 339
Lead fluoride oxide, 144
Lead metaniobate, 195
Lead selenite, 285
Lepidolite (2M1), 317
Likasite, 242
Lindstromite, 32
Lithium aluminate (beta), 178
Lithium boracite (alpha form),
  223
Lithium boracite (beta form), 223
Lithium boracite (gamma form),
  223
Lithium cadmium silicate, 303
Lithium europium(III) borate, 355
Lithium-hydrorhodonite, 320
Lithium lead(II) polyphosphate,
  246
Lithium magnesium zinc, 78
Lithium metasilicate, 297

Lithium neodymium borate, 231
Lithium nickel molybdate, 207
Lithium nitrate trihydrate, 237
Lithium perchlorate trihydrate,
 293
Lithium plumbate(IV), 348
Lithium praseodymium borate, 231
Lithium scandium molybdate, 207
Lithium scandium silicate, 300
Lithium stannate, 183
Lithium tantalate (M-form), 197
Lithium tellurate, 288
Lithium tellurium, 79
Lithium tetrachloroaluminate, 135
Lithium tungstate(VI) (high-
 pressure), 135
Lithium ytterbium borate, 232
Lithium yttrium molybdate, 207
Lithium zinc ferrites, 329
Lomonosovite, 336
Lutetium chloride (gas), 334
Lutetium hydroxide, 336
Lutetium oxocuprate, 217

Mackayite, 291
Magnesian calcite, 235
Magnesium dimolybdate, 205
Magnesium germanate (spinel
 phase), 180
Magnesium molybdate pentahydrate,
 204
Malachite, 356
Malayaite, 317
Manganese(II) diselenite, 286
Manganese(II) hexacyanocobalt-
ate(III) dodecahydrate, 164
Manganese(III) hydrogen selenite
 diselenite, 287
Manganese milarite, 335
Manganese(II) niobate, 211
Manganese(II) nitrate hexahydrate
 , 240
Manganese(II) nitrate monohydrate
 , 240
Manganese nitrogen, 79
Manganese nitrogen silicon, 80
Manganese oxychloride, 148
Manganese(II) pentafluorochrom-
ate(III), 127
Manganese(II) phosphate hydroxide
 , 257
Manganese(II) selenite
 monodeuterate, 357
Manganese(II) selenite monohydr-
ate, 357
Manganese(III) selenite trihydr-
ate, 286

Manganese silicon uranium, 80
Manganese(II) tantalate, 211
Marcasite-type compounds, 81
Maricite, 255
Matteuccite, 277
Mawsonite, 51
Melonjosephite, 255
Mercury(II) fluoride dihydrate,
 122
Mercury nickel, 81
Milarite (manganese), 335
Minyulite, 251
Mitridatite, 256, 336
Molybdenum nickel phosphorus, 82
Molybdenum oxide hydroxide, 353
Molybdenum tellurium oxide, 291
Molybdenum tetrafluoride oxide
 (gas), 333
Molybdomenite, 285
Monetite, 250
Monochlorodiborane (gas), 333

Neodymium bromate nonahydrate,
 295
Neodymium cobalt metaborate, 231
Neodymium hydroxide, 222
Neodymium silicon, 82
Neptunium oxyiodide, 154
Nickel hexaborate octahydrate,
 230
Nickel scandium silicon, 83
Nickel sulphur, 84
Nickel tungstate, 211, 332
Nickel yttrium, 85
Niobite structures, 211
Niobium dioxide, 175
Niobium dioxide (high-temp), 176
Niobium nitrogen, 86
Niobium oxynitride, 109
Niobium selenide iodide, 158
Niobium sulphur tin, 86
Niobium tetrachloride, 133
Nitrogen silicon, 86
Nitrogen tantalum, 87
Nitrogen tellurium uranium, 88
Nitrogen titanium, 88
Nitrogen uranium, 88
Nyerereite, 234

Olivenite, 273
Orthoenstatite, 318
Orthopyroxene (synthetic), 318
Osmium(IV) chloride (high-temp),
 134
Overite, 252

Palladium bismuth oxide, 349

Palladium phosphorus, 89
Palladium plutonium, 90
Palladium selenium, 91
Palladium tellurium, 91
Palladseite, 91, 105
Paragonite (1M), 318
Paraspurrite, 328
Parkerite, 33, 102
Parwelite, 319
Pekoite, 30
Pentaborane(9) (gas), 334
Pentadecacarbonylhydridotri-
  osmiumrhenium, 113
Pentahydroborite, 227
Pentlandites, 91
Phosphophyllite, 261
Phosphorus silver sulphur, 92
Plagioclases, 319, 328, 330, 337
Plancheite, 322
Platinum oxide, 176
Platinum plutonium, 92
Platinum selenium, 93
Plutonium rhodium, 92, 94
Potassium beta-alumina, 179
Potassium aluminum silicate, 299
Potassium amidozincate, 337
Potassium aquotetrabromonitrido-
  osmate monohydrate, 147
Potassium azide, 114
Potassium barium silicotantalate,
  301
Potassium bismuth oxide hydroxide
  , 221
Potassium bismuthate(V), 332
Potassium boroniobate, 229
Potassium cadmium sulphate, 282
Potassium cadmium vanadate, 193
Potassium calcium amide, 338
Potassium cerium silicate, 304
Potassium cyanide, 329
Potassium decabromodibismuth-
  ate(III) tetrahydrate, 140
Potassium dihydrogendecaoxodiiod-
  ate(VII) octahydrate, 296
Potassium diiododimolybdate hydr-
  ate, 351
Potassium diplumbate(II), 349
Potassium ferrotantalate, 197
Potassium fluoride diperoxyhydr-
  ate, 118
Potassium germanium niobium oxide
  , 195
Potassium heptaamidodisamarate,
  338
Potassium heptacyanomolybdate(II)
  monohydrate, 161
Potassium heptafluorodiindate,

123
Potassium hexaamidosamarate, 338
Potassium hexaaminostannate(IV),
  115
Potassium hexabromorhenate(IV),
  328
Potassium hexachlororhenate(IV),
  328
Potassium hexachlorostannate(IV),
  337
Potassium hexachlorotungstate(V),
  137
Potassium hexacyanonitrosovanad-
  ate(I) hydroxide hydrate, 159
Potassium hexatungstate, 208
Potassium hydrogen iodate
  (orthorhombic, Fdd2), 295
Potassium hydrogen oxyfluoro-
  tungstate, 347
Potassium hydroxylaminetrisulph-
  onate sesquihydrate, 276
Potassium iron(III) chromate, 201
Potassium lead(II) hexanitrocupr-
  ate(II) (orthorhombic), 237
Potassium lead molybdate, 336
Potassium lead(II) polyphosphate,
  246
Potassium lithium iron sulphide
  chloride, 156
Potassium magnesium chloride, 134
Potassium magnesium chloride
  (high-temp), 134
Potassium mercury(II) cyanate,
  171
Potassium metaborate hydrate, 224
Potassium metavanadate, 190
Potassium molybdenum bronze (red)
  , 204
Potassium neodymium silicate, 336
Potassium niobate germanate, 348
Potassium niobium sulphide hydr-
  ate, 222
Potassium nonamolybdomangan-
  ate(IV) hexahydrate, 207
Potassium octatungstate, 208
Potassium pentachlorooxomolybd-
  ate(V), 147
Potassium pentafluoroferrate(III)
  , 129
Potassium pentanitrocuprate(II),
  237
Potassium perchlorate, 293
Potassium selenium, 94
Potassium silver oxide, 353
Potassium sodium tetracyano-
  platinate trihydrate, 165
Potassium sulphur, 94

Potassium tantalum oxyfluoride, 147
Potassium tantalum sulphide hydrate, 222
Potassium terbium fluoride, 130
Potassium tetracyanoplatinate bromide trihydrate, 328
Potassium tetracyanoplatinate chloride trihydrate, 328
Potassium tetracyanoplatinate hydrate, 328
Potassium tetrafluoroaluminate (gas), 333
Potassium titanotantalate, 197
Potassium triantimonate(III), 185
Potassium trisulphatouranylate, 357
Potassium uranyl oxychloride hydrate, 152
Potassium ytterbium fluoride (gamma), 130
Potassium zinc molybdate, 335
Praseodymium bromate nonahydrate, 295
Praseodymium oxyiodide, 154
Praseodymium selenite, 287
Praseodymium silicon, 82
Praseodymium trihydrogen biselenite diselenite, 287
Preobrazhenskite, 225
Protopyroxene, 320
Prussian Blue, 163
Pseudomalachite, 259
Pyrite, 75
Pyroxene, 318, 320
Pyrrhotite, 99

Quartz, 262

Rankinite, 320
Raspite, 210
Reinerite, 274
Rhodium samarium, 95
Rhodonite, 320, 335
Roscherite (triclinic), 248
Roselite, 272
Rosenhahnite, 321
Rubidium beryllosilicate, 298
Rubidium bromotrifluoroantimonate(III), 341
Rubidium cadmium bromide, 141
Rubidium cadmium iodide monohydrate, 143
Rubidium cerate, 217
Rubidium decafluorotriindate, 124
Rubidium dibromotetracyanoplatinate(IV), 167

Rubidium dysprosium tungstate, 212
Rubidium hexaaminostannate(IV), 115
Rubidium hexachlorotungstate(V), 137
Rubidium iodide triiodide hexaiodobismuthate(III) dihydrate, 142
Rubidium metavanadate, 190
Rubidium neodymium tetrametaphosphate, 268
Rubidium niobate, 193
Rubidium plumbate(IV), 349
Rubidium polytungstate, 209
Rubidium scandate (beta), 349
Rubidium silver iodide, 142
Rubidium strontium octaborate, 337
Rubidium suboxide, 172
Rubidium terbate, 217
Rubidium tetrachloroaurate(III), 139
Rubidium tetrachloromanganate(II), 137
Rubidium tetracyanoplatinate bifluoride, 166
Rubidium tetracyanoplatinate(II) sesquihydrate, 166
Rubidium tetrafluorovanadate(III) dihydrate, 126
Rubidium tetraiodocobaltate(II), 142
Rubidium titanoniobate, 196
Rubidium titanotantalate, 196
Rubidium trichloromanganate(II), 137
Rubidium trihydrogen selenite, 285
Rubidium uranyl oxychloride hydrate, 152

Samarium chromium silicon oxynitride, 305
Samarium hydroxide, 222
Samuelsonite, 253
Sarabanite, 335
Sarcolite, 321
Scandium chlorides, 132, 133
Scandium disilicate (high-pressure), 329
Scandium fluoride (gas), 334
Scandium monochloride, 132
Scandium oxocuprate, 217
Schoonerite, 260
Seqelerite, 252
Selenium sodium vanadium, 95

Selenium sulphur, 354
Shattuckite, 322
Shungite, 335
Silicon oxynitride, 109
Silicon titanium, 96
Silver bromide nitrate, 243
Silver dicyanamide, 170
Silver disilicate, 302
Silver iodide, 328, 345
Silver iodide tetratungstate, 154
Silver sulphamide, 275
Silver(I) tetrachloroaurate(III),
139
Sklodowskite, 323
Slawsonite, 324
Sodium azide (beta), 114
Sodium borate, 223
Sodium bromate, 293
Sodium bromotrifluoroantimon-
ate(III) monohydrate, 125
Sodium cerium(III) sulphate
monohydrate, 357
Sodium chlorate, 293
Sodium copper silicate, 301
Sodium cyanide dihydrate, 158
Sodium decatungstouranate(IV)
hydrate, 352
Sodium dibromotetracyanoplatin-
ate(IV) dihydrate, 165
Sodium dicyanocuprate(I) dihydr-
ate, 169
Sodium dideuteriumarsenate mono-
deuterate, 270
Sodium diferrate(III), 337
Sodium dihydrogenarsenate mono-
hydrate, 270
Sodium dimanganate(III), 337
Sodium dodecamolybdogermanate
octahydrate, 205
Sodium fluoride, 328
Sodium heptacyanomolybdate(II)
decahydrate, 161
Sodium hexacyanochromate(II)
decahydrate, 160
Sodium hydrogen phosphate dihydr-
ate, 247
Sodium hydrogen sulphate mono-
hydrate (matteuccite), 277
Sodium indium selenate hexahydr-
ate, 336
Sodium lanthanum hydroxygermanate
, 182
Sodium manganese(II) hexafluoro-
chromate(III), 127
Sodium metatellurate (monoclinic)
, 290
Sodium metatellurate (ortho-

rhombic), 289
Sodium oxide nitrite, 356
Sodium oxoferrate, 334
Sodium oxoferrate(II), 352
Sodium pentaantimonate(III), 184
Sodium pentaborate monohydrate,
224
Sodium pentamolybdodiphosphate
tetradecahydrate, 206
Sodium percarbonate, 232, 355
Sodium tantalate, 197
Sodium tetraantimonate(III), 184
Sodium tetrachloroaluminate, 135
Sodium tetrachlorocobaltate(II),
138
Sodium tetrachlorozincate, 138
Sodium tetracyanoplatinate(II)
trihydrate, 164
Sodium thallium(I) sulphite, 277
Sodium thiocyanate, 172
Sodium thiosulphate pentahydrate,
284
Sodium titanium zinc germanate,
181
Sodium titanium zinc oxyfluoride,
146
Sodium trihydrogen selenite, 284
Sodium uranyl metaphosphate, 336
Sodium vanadate hydrate, 350
Sodium zinc hydroxide molybdate,
208
Sodium zinc silicate, 302, 303
Sodium zinc tripolyphosphate
nonahydrate, 266
Solonqoite, 227
Stenhuggarite, 271
Strontium calcium indate, 348
Strontium copper oxybromide, 153
Strontium dodecahydrodecaborate
heptahydrate, 335
Strontium gadolinium gallate, 180
Strontium metagermanate, 335
Strontium niobium titanium oxide,
188
Strontium nitroprusside tetra-
hydrate, 163
Strontium silicon, 42
Strontium tantalum titanium oxide
, 188
Strontium trichlorostannate(II)
pentahydrate, 343
Strontium uranate (gamma), 353
Sulphamic acid, 275
Sulphate hydrocancrinite, 337
Sulphur tantalum tin, 86
Sulphur thallium vanadium, 96
Sulphur ytterbium, 97

mu-Supersulphido-bis(pentaammineruthenium) chloride
dihydrate, 156

Talmessite, 356
Tantalates, 197
Tantalum oxide, 176
Tantalum oxynitride, 109
Teineite, 292
Tellurium(IV) oxide phosphate,
356
Tellurium oxybromide, 153
Tellurium oxyfluoride, 145
Tellurium selenides, 155
Tellurium sulphide, 155
Tellurium(VI) tetrakis(oxopenta-
fluorotellurate(VI)) difluoride,
145
Terbium chromate(V), 203
Terbium dioxide, 330
Terbium hydroxide, 222
Terbium iron garnet, 218
Terbium monochloride, 334
cis-Tetracarbonyl-bis(chloro-
mercury)iron, 337
Tetrachlorophosphonium penta-
chlorovanadate(IV), 136
Tetrametaphosphimic acid dihydr-
ate, 267
Tetrapotassium dihydrogendecaoxo-
diiodate(VII) octahydrate, 296
Tetrasulphur tetraimide, 117
Thallium beta-alumina, 179
Thallium lead chloride, 136
Thallium lead iodide, 141, 345
Thallium niobate, 194
Thallium triniobate, 194
Thiosulphatopentaamminecobalt-
(III) chloride monohydrate, 284
Thiotrithiazyl trichloromercur-
ate(II), 116
Thorium silicides, 99
Thulium oxyiodide, 155
Tienshanite, 336
Tin(II) bis(difluorooxcstann-
ate(II)), 143
Tin bromide fluoride, 118, 339
Tin(II) chloride dihydrate, 132
Tin chloride fluoride, 118
Tin difluoride - arsenic penta-
fluoride (1:1), 124
Tin(II) oxyfluoride, 143
Tin(II) phosphate, 254
Tirodite (zincian), 324
Titanium aluminum oxide, 189
Titanium(III) oxide, 175
Titanium scandium oxide, 189

Titanium vanadium oxide, 189
Titanoniobates, 196
Titanotantalates, 196
Transition-metal thorium
silicides, 99
Triantimony nonachloride tetra-
fluoride oxide, 346
Triazane, 327
Trichlorosilane (gas), 334
fac-Trichlorotriammineruthenium-
(III), 133
Tridymite (low), 297
Trifluoroxenon(IV) hexafluoro-
bismuthate(V), 342
Trimerite, 324
Triphylite, 336
Triruthenium dodecacarbonyl, 110
Tritellurium trisulphide hexa-
fluoroarsenate(V), 155
Tritin(II) bromide pentafluoride,
118
Tungsten(VI) oxide (monoclinic),
176
Tungsten(VI) oxide (orthorhombic)
, 176
Tungsten oxide tetrafluoride
(gas), 333
Tuscanite, 325
Tutton's salts, 329

Ullmannite (arsenian), 19
Undecacarbonyldihydridotriosmium,
111
Uralborite, 335
Uranium oxide (alpha), 177
Uranium oxide tetrafluoride
(alpha), 147
Uranium(V) oxybromide, 153
Uranium trioxide (gamma), 177
Uranium trioxide (gamma-high-
temp), 177
Uranyl bis(hydroxylaminate)
tetrahydrate, 113
Uranyl bis(hydroxylaminate)
trihydrate, 113
Uranyl diperchlorate heptahydrate
, 294

Vanadate pyroxenes, 190
Vanadium bronzes, 191
Vanadium dioxide (M2), 175
Vanadium dioxide (tetragonal),
175
Vanadium dioxide (triclinic), 175
Vanadium(III) metaphosphate, 247
Vandenbrandeite, 221
Variscite, 251

Velikite, 52

Wadeite, 326
Willemite, 336
Wolframite structures, 211
Wollastonite, 307
Woodhouseite, 279

Ytterbium bromate nonahydrate,
 295
Ytterbium perrhenate tetrahydrate
 , 213
Yttrium decavanadate, 336
Yttrium fluoroselenide (8M), 156
Yttrium fluoroselenide (10M), 156
Yttrium hydroxide, 222

Zeolite A (caesium and potassium
 exchanged, dehydrated), 327
Zeolite A (europium(II)-exchanged
 dehydrated), 327
Zeolite A (rubidium-exchanged),
 326
Zeolite A (rubidium-exchanged
 dehydrated), 326
Zeolite A (silver-exchanged
 ammonia dehydrated), 327
Zeolite A (silver-exchanged
 dehydrated), 327
Zeolite A (thallium-exchanged),
 326
Zeolite A (thallium-exchanged
 dehydrated), 326
Zinc hydrogen phosphate hydrate,
 336
Zinc magnesium phosphate, 259
Zinc pentafluoroindate(III)
 heptahydrate, 340
Zinc perchlorate hexahydrate, 294
Zinc thiocyanate (beta), 172
Zinnwaldite-1M, 327
Zirconium(IV) fluoride trihydrate
 , 121
Zirconium hydrogen phosphate
 monohydrate, 255
Zirconium monobromide, 139
Zirconium polyphosphate, 335
Zirconium yttrium oxide, 188

FORMULA INDEX

The entries are in alphabetical order by formula. Compounds in Table I of the Metals Section are excluded from this index, and that Table (pp. 100-105), which serves as its own index, should be consulted for additional metallic structures.

Ag12 Al12 H29 N21 O48 Si12, 327
Ag12 Al12 O48 Si12, 327
Ag Au Cl4, 139
Ag B C4 N4, 331
Ag Ba H8 O13 P3, 265
Ag2 Ba Sn2, 98
Ag2 Br N O3, 243
Ag C2 N3, 170
Ag Cl6 Cs2 Ln, 330
Ag Cs2 F6, 335
Ag Er Se2, 58
Ag F4 Ln, 332
Ag8 Ge S6, 66
Ag2 H2 N2 O2 S, 275
Ag I, 328, 345
Ag26 I18 O16 W4, 154
Ag I3 Rb2, 142
Ag9 In4, 66
Ag K3 O2, 353
Ag6 O7 Si2, 302
Ag4 P2 S7, 92
Ag3.15 Sb0.85, 17
Ag2 Si2 Sr, 98
Al3(Al, Fe, Mg)5 Ca2 H2 Na O24 Si5, 313
(Al, Fe, Li)3(Al, Si)4 F H K O11, 327
(Al, Li)2(Al, Si)4(F, O H)2 K Li O10, 317
Al B12, 6
Al2 B8 C8 Ca24(Cl, H2 O) Mg8 O48(O, O H)32 Si8, 228
Al B8(Ca, Ln)12 Fe H8 O48 Si8, 313
Al B C3, 228
Al2 Ba0.5 Ca9(Fe, Mn, Na)4 H2 O42 P10, 253
Al0.4 Ba3.6 Fe2 S7.4, 5
Al3 Ba10 Ge7, 4
Al2 Ba Ge2 O8, 180
Al2(Ba, Sr) H10 O21 Si6, 307
(Al, Si)4(Ba, K) O8, 314
(Al, Fe, Mn) Be2 Ca H8 Mn2 O17 P3, 248
Al2(Be, Li)3(Cs, Na)0.5 O18

Si6. 0.7 H2 O, 306
Al4 Bi2 O9, 187
Al Br3 H3 N, 333
Al4 C1.4 Ca2 O25.4 Si8.x H2 O, 308
Al6 C Ge6 H4 Na8 O29, 336
Al C H2 Na O5, 235
Al9 Ca2(Ca, Mg, Na)2 K2 O72 Si27, 311
Al6 Ca Fe6 O19, 179
Al Ca H9 Mg O13 P2, 252
(Al, Si)9(Ca, K, Mg, Na) H12 O24, 358
Al3 Ca H6 O14 P S, 279
Al2 Ca4 H24 O22 S, 278
Al2 Ca2 H2 O8 Si, 337
(Al, Si)4(Ca, Na) O8, 319
Al4 Ca2 O24 Si8.x H2 O, 308
Al4 Ca2 O24 Si8.x H2 O. 1.4 C O, 308
Al2 Ca0.13 O8 Si2 Sr0.87, 324
Al2 Ca Si2, 38
Al Ce, 8
Al Cl4 Li, 135
Al Cl4 Na, 135
Al4 Co La, 10
Al8 Cr5, 8
Al Cr2 H4 K O10, 335
Al Cr2 H4 Na O10, 335
Al2 Cr4 H8 O19, 198
Al12 Cs7 K5 O48 Si12, 327
Al Cs O4 Si, 330
Al4(Cu, K)2 O24 Si8.n H2 O, 308
Al22 D2 O34, 335
Al12 Eu6 O48 Si12, 327
Al2 F H8 K O12 P2, 251
Al F4 K, 333
Al4(Fe, Mg)2 O18 Si5, 315
Al H14 I5 O21, 296
Al H K O10 P3, 264
Al H5 N O10 P3, 264
Al3 H2 Na O12 Si3, 318
Al6 H12 Na8 O31 Si6, 337
Al H O2, 219
Al22 H2 O34, 335

Al22 H6 O36, 174
Al H4 O6 P, 251
Al12 H18 O57 Si12 Tl13, 326
Al Ho O3, 179
Al Ir Nb, 10
Al22 K2 O34, 179
Al K O4 Si, 299
Al(K, Na) O8 Si3, 312
Al Li5 O4, 178
Al4 Mg2 O18 Si5, 315
Al4 Mg2 O18 Si5.n H2 O, 310
Al2 Mg5 Se8, 11
Al112 Na O48 Rb11 Si12.x H2 O, 326
Al Na O8 Si3, 305
Al2 O3, 348
(Al, Ti)2 O3, 189
Al2 O8 Si2 Sr, 324
Al12 O48 Si12 Tl12, 326
Al22 O34 Tl2, 179
Al10.015 O2 V0.985, 175
Al2 Pb2 Sr, 98
Al Pd, 12
Al Pr, 8
Al1.2 Re Si0.8, 12
Al8 V5, 8
Am2 Bi O2, 336
Am Si2, 12
As4 Ba H6 O14, 270
As Be Li, 19
As Be Na, 14
As0.10 Bi0.01 Co0.03 Ni0.97 S0.98
  Sb0.91, 19
As3 Br Cd5 O12, 261
As2 Ca Cd2, 38
As2 Ca2(Co, Mg) H4 O10, 272, 356
As2 Ca Fe O7 Sb, 271
As2 Ca Mg2, 4
As2 Ca Zn2, 38
As Cl Co2 O4, 273
As Cl Cu2 O4, 273
As(Co, Fe) S, 20
As Cu2 H O5, 273
As D4 Na O5, 270
As2 Eu2, 20
As3 Eu, 21
As2 Eu4 O, 183
As F9 O3 S Xe2, 145
As2 F12 S3 Te3, 155
As2 F12 Se4 Te2, 155
As F7 Sn, 124
As6 F30 Te6, 335
As2 H8(Mg, Mn)6 O20 Si Zn3, 314
As H18 N3 O7, 269
As H4 Na O5, 270
As Hg S3 Tl, 22
As4 Mg, 21
As2 Mn10 O24 Sb2 Si2, 319

As3 Mo2 Ni0.83, 22
As Na3 S3, 23
(As, Sb) Ni S, 19
As2 O5, 334
As2 O6 Zn3, 274
As2 Os, 81
As5.5 Pb28 S46 Sb6.5, 13
As2 Ru, 81
As15 S25.3 Tl5.6, 24
As Sb U, 13
As4 Sr3, 23
As2 V3, 25
As3 V5, 25
Au2(Au, Ge)2 Ba, 98
Au Br4 Rb, 139
Au2 Ca Ge2, 98
Au Cd, 38
Au Cd3, 39
Au2 Cl6 Cs2, 337
Au Cl4 Rb, 139
Au Ge Na2, 61
Au2 Ge2 Sr, 98
Au9 In4, 66
Au Na Sn, 61
Au3 Sb4 Yb3, 14

B, 34
B C4 Cu N4, 331
B6 C6 H30 I N12 Na, 109
B3 Ca2 Cl H4 O8, 227
B5 Ca2 Cl H2 O10, 336
B2 Ca H10 O9, 227
B4 Ca2 H8 O12, 335
B4 Ca O7, 226
(Ba, Sr)2 Cd O6 U, 331
B2 Cl H5, 333
B3 Cl6 H6 N3, 116
B7 Cl Li4 O12, 223
B5 Co Nd O10, 231
B3 Eu2 Li3 O9, 355
B Ge0.3 Mo1.7, 34
B2 H6, 109
B5 H9, 334
B3 H8 K3 O10, 224
B12 H16 Mg Na2 O28, 225
B4 H18 Mg O16, 225
B5 H4 Na3 O11, 224
B6 H22 Ni O21, 230
B8 H24 O26 Rb2 Sr, 337
B2 K3 Nb3 O12, 229
B La O3, 230
B3 Li3 Nd2 O9, 231
B2 Li1.2 Ni2.5, 35
B3 Li3 O9 Pr2, 231
B3 Li6 O9 Yb, 232
B4 Lu Ru4, 35
B2 Mg Ni2.5, 35

B4 Mo Th, 36
B Na3 O3, 223
B O3 Y, 337
B2 Ru1.25 W1.75, 37
B2 S3, 37, 334
B4 Th W, 36
Ba Be2 O7 Si2, 306
Ba Be0.75 Si1.25, 3
Ba5 Bi3, 3
Ba Bi2 Mg2, 4
Ba Bi O3, 186
Ba Br2 C4 H9 N4 O4.5 Pt, 168
Ba Br2 C4 N4 Pt. 4.5 H2 O, 168
Ba88 Er2 Cu88 O175, 151
Ba Br F, 118
Ba1.6 Br0.2 P, 343
Ba2 Br P, 343
Ba C6 Cs2 Fe H4 N6 O2, 162
Ba3 C8 H28 Mo2 N8 O14 S2, 334
Ba C4 H8 N4 O4 Pt, 168
Ba2 Ca O6 U, 332
Ba2(Ca, Y) O6(U, W), 331
Ba2 Ca O6 W, 332
Ba2 Cl Cu O2, 150
Ba44 Cl4 Cu45 O87, 151
Ba4 Cl(Fe, Nb, Ti)4 O28 Si4 Ti4, 306
Ba1.6 Cl0.2 P, 343
Ba2 Cl P, 343
Ba Co O3, 214
Ba Cr2 H3 O11 P, 198
Ba Cr2 H7 O13 P, 198
Ba2 Cu F6, 130
Ba Cu O2, 215
Ba F I, 118
Ba3 Fe2 O21 Ta6, 197
Ba3(Fe, Ho)2 O9 U, 329
Ba4 Fe2 S7.33, 5
Ba9 Fe16 S32, 26
Ba2 Gd0.67 O6(Ta, U), 332
Ba2 Gd0.67 O6 U, 331
Ba Ge2, 26
Ba2 Ge2 Fe O7, 180
Ba3 H12 In2 O12, 330
Ba H4 O8 P2, 250
Ba3 H12 O12 Sc2, 330
Ba1.6 I0.2 P, 343
Ba2 I P, 343
Ba1.5 K3 O26 Si4 Ta6, 301
Ba K4 O26 Si4 Ta6, 301
Ba Ln Mn O4, 329
Ba2 Lu0.67 O6 U, 332
Ba2 Mg O6 W, 332
Ba Mg2 Pb2, 28
Ba Mg2 Sb2, 4
(Ba, La) Mn O4, 329
Ba2 Nd0.67 O6(U, W), 332

Ba2(Nd, Y) 0.67 O6 W, 332
Ba2 Ni O4 Re, 330
Ba3 Ni O9 Sb2, 185
Ba O3 Pt, 329
Ba3 O7 Pt2, 214
Ba O4 S, 278
Ba3 O21 Sb4 Ti4, 188
Ba2 O6 Sr U, 332
Ba3 O21 Ta4 Ti4, 197
Ba2 O6 Te Y0.67, 331
Ba2 O13 Ti6, 350
Ba2 O6 U Y0.67, 331
(Ba, Sr)2 O6 U Zn, 331
Ba2 O6 W Y0.67, 331, 332
Ba3 Pb5, 27
Ba2 Sb, 3
Ba5 Sb3, 3
Ba Si2, 29
Ba Sn2 Zn2, 28
Be0.75 Ca Ge1.25, 3
Be2 Ca Ge2, 28
Be3 Ca Mn2 O12 Si3, 324
Be0.75 Ca Si1.25, 3
Be2 Co F8 H20 N2 O6, 123
Be0.75 Ge1.25 Sr, 3
Be H8 N2 O10, 238
Be2 H4 N O10 P3, 245
Be Na Sb, 14
Be O6 P2, 244
Be2 O7 Rb2 Si2, 298
Be4 O7 Te, 290
Be P2, 30
Be0.75 Si1.25 Sr, 3
Be13 Sb, 13
Bi2 Br9 Cs3, 140
Bi2 Br10 H8 K4 O4, 140
Bi2 Br10 H24 N4 O4, 140
Bi2 Ca Mg2, 4
Bi3.78 Ce1.25 S8, 30
Bi Cm2 O2, 336
Bi1.34 Cr Nb O6, 196
Bi2 Cu O4, 187
Bi Cu Pb S3, 31
Bi1.25 Cu0.75 Pb0.75 S3, 31
Bi1.5 Cu0.5 Pb0.5 S3, 31
Bi1.667 Cu0.33 Pb0.33 S3, 31
Bi1.8 Cu0.2 Pb0.2 S3, 31
Bi4 Cu2 Pb2 S9, 32
Bi7 Cu3 Pb3 S15, 32
Bi11 Cu Pb(S, Se)18, 30
Bi F3, 340
Bi F4 K, 330
Bi F4 Rb, 330
Bi F4 Tl, 330
Bi F9 Xe, 342
Bi Fe2 H O9 Si2, 358
Bi H4 I10 O2 Rb5, 142

Bi2.37 H2 K1.14 O7, 221
Bi I Te, 157
Bi K O3, 332
Bi4 Nb O21 Ti4, 328
Bi2 Ni3 S2, 33
Bi2 Ni3 Se2, 33
Bi2 O4 Pd, 349
Bi2 S3, 33
Bi Sr2, 3
Bi3 Sr5, 3
Br0.3 C4 H6 K2 N4 O3 Pt, 328, 337
Br2 C4 H4 N4 Na2 O2 Pt, 165
Br2 C4 N4 Pt Rb2, 167
Br C5 O5 Re, 110
Br2 Ca2 Cu O2, 150
Br2 Ca H12 O6, 131
Br3 Cd Cs, 345
Br Cd5 O12 P3, 261
Br Cd5 O12 V3, 261
Br3 Cd Rb, 141
Br6 Cs2 K Sc, 332
Br6 Cs2 K Sm, 332
Br6 Cs2 K Tm, 332
Br6 Cs2 Na Sc, 332
Br6 Cs2 Na Sm, 332
Br6 Cs2 Na Tm, 332
Br6 Cs2 Np, 336
Br2 Cu O2 Sr2, 153
Br F3 H2 Na O Sb, 125
Br F3 Rb Sb, 341
Br F5 Sn3, 118
Br4 F6 Sn5, 339
Br3 Ga H3 N, 333
Br2 Ge, 344
Br4 H4 K N O2 Os, 147
Br3 H18 Nd O18, 295
Br3 H18 O18 Pr, 295
Br3 H18 O18 Yb, 295
Br6 K2 Re, 328
Br Na O3, 293
Br2 O11 Te6, 153
Br O2 U, 153
Br Zr, 139

C2 Ca(Fe, Mg) O6, 235
C2 Ca(K, Na)2 O6, 234
C Ca0.9 Mg0.1 O3, 235
C2 Ca Mg O6, 235
C(Ca, Mg) O3, 235
C Ca5 O11 Si2, 328
C12 Cd3 Co2 H24 N12 O12, 164
C3(Ce, La)2 H16 O17, 236
C4 Cl2 Fe Hg2 O4, 337
C4 Cl0.3 H6 K2 N4 O3 Pt, 328, 337
C3 Cl Ir O3, 336
C6 Co Cr H18 N12, 161
C12 Cc2 H24 Mn3 N12 O12, 164

C3.4 Co3 W10, 43
C6 Cr Cs2 Li N6, 160
C6 Cr H20 N6 Na4 O10, 160
C4 Cs2 F0.78 H0.39 N4 Pt.x H2 O, 336
C6 Cs2 Fe K N6, 162
C6 Cs2 Fe Li N6, 162
C6 Cs2 Fe N6 Na, 162
C4 Cs2 H2 N4 O Pt, 167
C2 Cu H4 N2 Na O2, 169
C Cu2 H2 O5, 356
C4 F0.8 H0.4 N4 Pt Rb2, 166
C4 F0.78 H0.39 N4 Pt Rb2.x H2 O, 336
C18 Fe7 H30 N18 O15, 163
C5 Fe H8 N6 O5 Sr, 163
C7 H2 K5 Mo N7 O, 161
C12 H10 K7 Mo2.5 N12 O7 S, 334
C4 H6 K N4 Na O3 Pt, 165
C4 H3 K1.75 N4 O1.5 Pt, 328
C6 H1.5 K4.5 N7 O2 V, 159
C H2 Mg2 O5, 329
C H8 Mg2 O8, 233
C4 H10 Mg5 O18, 234
C7 H20 Mo N7 Na5 O10, 161
C H4 N Na O2, 158
C4 H6 N4 Na2 O3 Pt, 164
C4 H3 N4 O1.5 Pt Rb2, 166
C H4 N2 S, 171
C H3 Na2 O6, 232, 355
C10 H2 O10 Os3, 111
C11 H2 O11 Os3, 111
C18 H2 O18 Os6, 335
C15 H O15 Os3 Re, 113
C Mo2, 44
C N Na S, 172
C4 N4 Pt Rb2. 0.4(F2 H), 166
C4 N4 Pt Rb2. 1.5 H2 O, 166
C2 N2 S3, 159
C2 N2 S2 Zn, 172
C Na2 O3. 1.5 H2 O2, 232, 355
C12 O12 Os3, 111
C16 O16 Os5, 113
C21 O21 Os7, 335
C12 O12 Ru3, 110
C3 Hg K N3 O3, 171
C12 I2 O12 Os3, 112
C K N, 329
C6 K4 N7 O V. 0.5 K O H. 0.5 H2 O, 159
(C, N, O) Ti, 329
(C, N, O) Zr, 329
Ca Cd2 P2, 38
Ca2 C12 Cu O2, 150
Ca Cl F, 118
Ca Cl2 H4 O2, 131
Ca Cl2 H12 O6, 131

Ca Cr H2 O5, 198
Ca Cu3 Ge4 O12, 181
Ca0.5 Cu1.5 O6 Ti2, 350
Ca D2, 41
Ca F6 In Li, 331
Ca F6 Li Ti, 332
Ca5 F O12 P3, 262
Ca Fe2(Fe, Mn) H O9 Si2, 315
Ca Fe H9 Mg O13 P2, 252
Ca Fe2 H O9 P2, 255
Ca2 Fe3 H6 O17 P3, 256, 336
Ca Fe3 H O9 Si2, 315
Ca3 Ge3 O12 Sc2, 300
Ca H6 K N3, 338
Ca H10 N Na O10 P2, 263
Ca H8 N2 O10, 239
Ca H5 N O7 P2, 262
Ca H O4 P, 250
Ca H6 O6 Pb, 220
Ca3 H2 O8 Si2, 316
Ca3 H2 O10 Si3, 321
Ca H2 O5 Si Zn, 310, 336
Ca0.07 Ho1.86 O2.86, 218
Ca Ho O2.5, 218
Ca2 In2 O6 Sr, 348
Ca Lu2 S4, 41
Ca2.71 Mg0.29 O8 P2, 249
Ca2.89 Mg0.11 O8 P2, 249
Ca Mg2 Sb2, 4
Ca Mn4 O15 Si5, 335
Ca Nb2 O8 Th, 196
Ca Nb2 O8 U, 197
Ca3 O8 P2, 248, 249
(Ca, Mg)3 O8 P2, 249
Ca O10 S6 Sb10, 335
Ca3 O12 Sc2 Si3, 300
(Ca, Fe) O3 Si, 307
(Ca, Fe, Mn) O3 Si, 312
(Ca, Mg) O3 Si, 309
(Ca, Mn) O3 Si, 307
Ca2 O4 Si, 299
Ca3 O7 Si2, 320
Ca O5 Si Sn, 317
Ca0.5 O2.75 Tm1.5, 218
Ca P2 Zn2, 38
Ca Si2, 42
Ca31 Sn20, 42
Cd5 Cl O12 P3, 261
Cd Cs O9 P3, 266
Cd F5 Ga H14 O7, 340
Cd F6 In Li, 331
Cd F6 Li Ti, 332
Cd4 Ge5 H Li O15, 335
Cd4 Ge5 H Na O15, 335
Cd2 Ge7 O16, 181
Cd3 Ge3 O12 Sc2, 300
Cd2 Ge2 Sr, 98

Cd H2 I3 O Rb, 143
Cd H10 N Na O10 P2, 263
Cd H6 N O5 P, 262
Cd H6 O6 Pb, 220
Cd I2, 328
Cd5 I O12 V3, 261
Cd2 K2 O12 S3, 282
Cd4 K O12 V3, 193
Cd Li2 O4 Si, 303
Cd4 Na O12 P3, 335
Cd3 Na2 O10 Si3, 336
Cd4 Na O12 V3, 335
Cd Nb2 O8 Th, 197
Cd Nd4 O13 Si3, 305
Cd Se, 40
Ce, 44
Ce Cs2 O3, 218
Ce Cr2 H4 O10, 203
Ce Dy S3, 45
Ce H2 Na O9 S2, 357
Ce H8 O12 S2, 282
Ce2 H10 O18 S3, 283
Ce K2 O3, 331
Ce K2 O15 Si6, 304
Ce6 Mo10 O39, 336
Ce O3 Rb2, 217, 331
Ce Re4 Si2, 46
C14 Co2 H25 N7 O3, 149
Cl Co H17 N5 O4 S2, 284
C14 Co Na2, 138
C15 Cr2 H35 N10 O3, 147
Cl Cs F3 Sb, 126
C14 Cs Fe, 344
C18 Cs2 H2 O Re2, 138
C16 Cs W, 137
Cl2 Cu H2 O4 Pb3, 151
Cl2 Cu K2 O4 S, 281
Cl Cu N4 S4, 334
Cl2 Cu2 O, 149
C14 D8 N2 Pd, 138
Cl2 D4 O2 Sn, 132
C19 F4 O Sb3, 346
Cl2 F12 Pb7, 339
Cl3 F2 Sb, 120, 335
Cl11 F4 Sb3, 120
Cl F3 Sn2, 118
Cl Fe24 K6 Li S26, 156
Cl Gd, 334
C14 H14 K2 O18 U4, 152
Cl H6 Li O7, 293
C14 H34 N10 O2 Ru2 S2, 156
C15 H4 N Pb2, 136
C14 H8 N2 Pd, 138
Cl3 H9 N3 Ru, 133
C14 H8 N2 Zn, 329
C16 H4 N W, 137
C14 H4 O7 Rb2 U2, 152

Cl2 H4 O2 Sn, 132
Cl6 H10 O5 Sn2 Sr, 343
Cl2 H14 O17 U, 294
Cl2 H12 O14 Zn, 294
Cl3 H Si, 334
Cl Hf, 139
Cl3 Hg N3 S4, 116
Cl11 Ho5, 336
Cl3 K Mg, 134
Cl5 K2 Mo O, 147
Cl K O4, 293
Cl6 K2 Re, 328
Cl6 K2 Sn, 337
Cl6 K W, 137
Cl Ln O, 332
Cl Ln3 O4, 332
Cl3 Lu, 334
Cl3 Mn8 O10, 148
Cl3 Mn Rb, 137
Cl4 Mn Rb2, 137
Cl112 N4 S4 Sb2, 335
Cl Na O3, 293
Cl4 Na2 Zn, 138
Cl4 Nb, 133
Cl4 Os, 134
Cl9 P V, 136
Cl2 Pb, 333
Cl5 Pb Tl3, 136
Cl6 Rb W, 137
Cl Sc, 132, 139
Cl1.43 Sc, 133
Cl10 Sc7, 133
Cl Tb, 334
Cm2 O2 Sb, 336
Co F6 In Li, 331
Co0.07 Fe3.97 Ni4.84 S8, 91
Co5.60 Fe1.63 Ni1.87 S8, 91
Co5.95 Fe13.25 Y2, 49
Co13.25 Fe5.15 Y2, 49
Co2 Ge3 S3, 48
Co5 H4 O12 P2, 257
Co I4 Rb2, 142
Co In3, 49
Co Nb2 O6, 211
Co Nb4 P, 50
(Co, Mg, Mn) O3 Si, 318
Co2 O4 Si, 301
Co O4 W, 211, 332
(Co, Fe, Ni)9 S8, 91
Co2 Si2 Th, 99
Co3 Si5 U2, 50
Cr D O2, 221
Cr F5 H20 N2 O6, 342
Cr F5 Mn, 127
Cr F6 Mn Na, 127
Cr2 Fe H4 N O8, 200
Cr4 Fe2 H8 O19, 199

Cr2 Fe K O8, 201
Cr H O2, 221
Cr H15 O15 S2, 279
Cr La O3, 202
Cr(La, Sr) O3, 202
Cr La0.75 O3 Sr0.25, 202
Cr La S3, 47
Cr N O12 Si3 Sm4, 305
Cr0.875 N V1.25, 48
Cr2 Nb4 Si5, 47
Cr O2, 328
(Cr, Mn)3 O4, 330
Cr2 O18 P6, 269
Cr O4 Tb, 203
Cr P S4, 48
Cr2 Si2 Th, 99
Cs Cu F6 Zn, 331
Cs F6 Fe Mn, 331
Cs F6 Fe Pd, 331
Cs F6 Ga Mn, 331
Cs F6 Ga Zn, 331
Cs F4 H4 O2 Mn, 128
Cs F6 In Ni, 331
Cs F6 In Pd, 331
Cs F6 In Zn, 331
Cs F4 Mn, 128
Cs2 F6 Mn Ni, 129
Cs F6 Mn Rh, 331
Cs F6 Mn Zn, 331
Cs F6 Ni Rh, 331
Cs F6 Ni Sc, 331
Cs F6 Ni Tl, 331
Cs F6 Pd Rh, 331
Cs F6 Pd Sc, 331
Cs F6 Rh Zn, 331
Cs F6 Sc Zn, 331
Cs F5 Te, 341
Cs F6 Ti Zn, 331
Cs F6 Tl Zn, 331
Cs F6 V Zn, 331
Cs3 H2 O10 P3, 263
Cs Nb O3, 351
Cs2 Nb6 O18 Ti, 196
Cs11 O3, 173
Cs O12 P4 Tb, 336
Cs2 O3 Pr, 217
Cs2 O18 Ta6 Ti, 196
Cs2 O3 Tb, 218
Cs2 O3 Th, 331
Cs0.3 O5 V2, 191
Cs0.35 O7 V3, 191
Cs O3 V, 190
Cs O5 V2, 190
Cs2 O13 V5, 190
Cs2 S, 40
Cu6 Fe2 S8 Sn, 51
Cu Ga2 O4, 328

Cu Ga Se2, 51
Cu Gd2 O4, 216
Cu3 H7 N O6 P2, 242
Cu H5 N O4 Te, 292
Cu5 H4 O12 P2, 259
Cu H2 O4 Si, 311
Cu5 H2 O14 Si4, 322
Cu8 H4 O26 Si8.x H2 O, 322
Cu H4 O5 Te, 292
Cu H4 O6 U, 221
Cu3.75 Hg1.75 S8 Sn2, 52
Cu Hg2 Ti, 52
Cu2 Ho2 O5, 217
Cu K3 N5 O10, 237
Cu K2 N6 O12 Pb, 237
Cu3 K S2, 54
Cu La2 O4, 216
Cu2 Lu2 O5, 217
Cu2 Mn3 O8, 212
Cu2-x Mo3 S4, 53
Cu N2 O6, 334
Cu Na2 O10 Si4, 301
Cu9 Ni Sn3, 53
Cu3 O8 P2, 257
Cu5 O10 P2, 258
Cu2 O5 Sc2, 217
Cu0.5 O3 Ta, 329
Cu O6 Ta2, 350
Cu1.5 O6 Ta Ti, 350
Cu3 O4 Ti, 189
Cu4 S4 Sn, 55
Cu2 Si2 Th, 99
Cu41 Sn11, 53

D Ga O2, 219
D0.70 Li Pd, 69
D0.66 Li Pt, 69
D2 Mn O4 Se, 357
D15 Th4, 56
Dy2 Fe3 Si5, 56
Dy H3 O3, 222
Dy Na O2, 331
Dy O8 Rb W2, 212

Er4 H9 N O13, 244
Er H3 O3, 222
Er I O, 155
Er6 O24 Pb3 Si6, 335
Er Pt2 Si2, 57
Eu F2.5, 329
Eu3 S7 Sn2, 59
Eu5 S12 Sn3, 58
Eu Si2, 42

F6 Fe Ga Li, 332
F5 Fe2 H4 O2, 122
F5 Fe K2, 129

F2 H4 Hg O2, 122
F5 H14 In O7 Zn, 340
F H4 K O4, 118
F6 H K3 O4 W2, 347
F2 H N O S, 333
F5 H12 N3 O2 Ti, 347
F10 H26 N6 O5 W3, 347
F2 H5 N2 P, 333
F4 H4 O2 Rb V, 126
F4 H6 O3 Zr, 121
F I Pb, 328
F6 I S7 Sb, 335
F7 In2 K, 123
F6 In Li Mg, 331
F6 In Li Mn, 331
F6 In Li Ni, 331
F6 In Li Zn, 331
F10 In3 Rb, 124
F6 K2 O3 Ta2, 147
F10 K Tb3, 130
F3 K V, 328
F10 K Yb3, 130
F6 Li Mg Ti, 332
F6 Li Mn Ti, 331
F6 Li Mn V, 332
F4 Mo O, 333
F20 N4 S4 Sb4, 335
F Na, 328
(F, O)7 Na0.76(Ti, Zn)4, 146
F3 Na V, 328
F2 O Pb2, 144
F2 O Sn2, 143
F22 O4 Te5, 145, 334
F4 O U, 147
F4 O W, 333
F3 Rb V, 328
F16 Sb4, 119
F12 Sb2 Se4 Te2, 155
F3 Sc, 334
F Se Y, 156
F2.4 Tm, 332
F32 Tm13, 332
F7 Yb3, 332
F32 Yb13, 332
F33 Yb14, 332
F64 Yb27, 332
Fe6 Ge4 Li, 62
Fe6 Ge5 Li, 62
Fe6 Ge6 Li, 62
Fe2 Ge2 O4 Zn, 329
(Fe, Mg)2 H2 K O12 Si4, 307
Fe3 H20 Mn O23 P3 Zn, 260
Fe H6 Na3 O15 S3, 280
Fe H18 N3 O18, 241
Fe H4 O11 P3, 265
Fe H8 O12 P2 Zn2, 261
Fe2 H O9 Sb Si2, 358

(Fe, Mg, Mn, Zn)7 H2 O24 Si8, 324
Fe9 H5 O25 Si6, 311, 337
Fe H O6 Te2, 291
Fe0.5 K3 O21 Ta7.5, 197
Fe4 La P12, 71
Fe1.87 La2 S5, 72
Fe2 La2 S5, 72
Fe2.5 Li0.5 O4, 328
Fe Li O4 P, 336
Fe4 Mn77 Si19, 73
Fe Na4 O3, 352
Fe2 Na O3, 335
Fe2 Na4 O5, 337
Fe6 Na14 O16, 334
Fe Na O4 P, 255
Fe3 Na2 O12 P3, 336
Fe Nb2 O6, 211
Fe Nb O6 Pb2, 329
Fe1-x O, 328
(Fe, Mg, Ta, W) O3 Pb, 329
Fe O6 Pb2 Ta, 329
Fe2 O9 Pb3 W, 329
Fe2 O12 S3, 279
Fe2 O4 Si, 301, 312
(Fe, Mg, Mn)2 O4 Si, 312
(Fe, Nb) O4 Sr2, 330
(Fe, Sb) O4 Sr2, 330
(Fe, Ta) O4 Sr2, 330
Fe5 O12 Tb3, 218
Fe2 O5 Te, 330
(Fe, Mg) O5 Ti2, 189
Fe2 O6 W, 212
Fe S2, 75
Fe11 S12, 99
Fe2 S4 Sn, 76, 99
Fe S4 Yb2, 76
Fe3 Sc2 Si5, 56
Fe2 Si2 Th, 99
Fe Si W2, 73

Ga Gd O5 Sr2, 180
Ga Ge3 K O8, 337
Ga H O2, 219
Ga2 O3, 174
Ga Se2 Tl, 60
Ga2 Te5, 61
Gd H3 O3, 222
Gd I3 O9, 296
Gd Li Mo3 O8, 207
Gd P S, 59
Ge H La Na2 O5, 182
Ge H16 Mo12 Na4 O48, 205
Ge3 H6 O15 Sm4, 335
Ge4 K6 Nb6 O26, 348
Ge4 K10 Nb22 O68, 195
Ge Mg2 O4, 180

Ge Na2 O7 Ti Zn2, 181
Ge Ni Pd, 64
Ge Pd2, 64
Ge8 Pd21, 64
Ge Pt Se, 65

H3 Ho O3, 222
H20 I2 K6 Mo2 O26, 351
H I2 K O6, 295
H18 I2 K4 O18, 296
H8 I O9 Yb, 332
H12 In Na O14 Se2, 336
H12 K6 Mn Mo9 O38, 207
H2 K2 Mn5 O31 Si12, 335
H2 K2 Mo2 O8, 336
H3 K3 N O11.5 S3, 276
H12 K3 N6 Sm, 338
H14 K N7 Sm2, 338
H12 K2 N6 Sn, 115
H8 K2 N4 Zn, 337
H0.8 K0.5 Nb O0.4 S2, 222
H1.2 K0.33 O0.6 S2 Ta, 222
H48 La2 Mg3 N12 O60, 244
H3 La O3, 222
H8 La O16 Re3, 332
H Li Mn4 O15 Si5, 320
H6 Li N O6, 237
H0.70 Li Pd, 69
H0.66 Li Pt, 69
H3 Ln O3, 222
H3 Lu O3, 336
H2 Mg(Mg, Mn) O6 Si Zn, 313
H10 Mg Mo O9, 204
H2 Mg5 O10 Si2, 309
H14 Mg O18 Si2 U2, 323
H2 Mn N2 O7, 240
H12 Mn N2 O12, 240
H4 Mn5 O12 P2, 257
H Mn O8 Se3, 287
H2 Mn O4 Se, 357
H6 Mn2 O12 Se3, 286
H24 Mo8 N4 O30, 336, 351
H38 Mo12 N2 O55 Te, 336
H28 Mo5 Na6 O37 P2, 206
H Mo Na O5 Zn, 208
H0.3 Mo O3, 353
H2 Mo4 O12, 354
H3 N3, 327
H5 N3, 327
H12 N4 O10 P4, 267
H4 N O12 P4 Pr, 268
H12 N2 O2 Pd S11, 355
H6 N4 O6 Pt, 241
H14 N6 O8 Pt2, 242
H21 N9 O12 Pt3, 242
H3 N O3 S, 275
H10 N2 O4 S, 276

H8 N2 O4 Se, 284
H10 N2 O7 U, 113
H12 N2 O8 U, 113
H4 N O3 V, 190
H12 N6 Rb2 Sn, 115
H4 N4 S4, 117
H8 N2 S7 Sb4, 354
H2 Na Nd O15 Si6.n H2 O, 336
H5 Na2 O6 P, 247
H18 Na O19 P3 Zn2, 266
H3 Na O5 S, 277
H10 Na2 O8 S2, 284
H3 Na O6 Se2, 284
H60 Na8 O66 U W10, 352
H3.78 Na O4.89 V,
H3 Nd O3, 222
H15 O46 P W12, 210
H3 O5 P Zn, 336
H4 O9 P2 Zr, 255
H2 O7 Pb Si U, 316
H3 O11 Pr Se4, 287
H3 O6 Rb Se2, 285
H8 O16 Re3 Yb, 213
H3 O3 Sm, 222
H3 O3 Tb, 222
H48 O52 V10 Y2, 336
H3 O3 Y, 222
H6 Ru Yb2, 69
H15 Th4, 56
Hf3 Ni2 Si3, 68
Hg4 N2 O4 S, 337
Hg Ni, 81

I La O, 155
I0.33 Nb Se4, 158
I Np C, 154
I O Pr, 154
I2 Pb, 332
I3 Pb Tl, 345
I6 Pb Tl4, 345
I10 Pb Tl6, 141
I S Sb, 157
In Li Mo3 O8, 207
In2 O7 Si2, 329
In S6 Sm3, 70
In5 S12 Tb3, 70
In Se, 70
Ir3 Pu5, 67, 71

K0.33 Mo O3, 204
K2 Mo2 O8 Pb, 336
K4 Mo3 O12 Zn, 335
K N3, 114
K3 N O10 S3. 1.5 H2 O, 276
K3 Nd O15 Si6, 336
K2 O12 P4 Pb, 246
K2 O3 Pb2, 349

K2 O3 Pr, 331
K4 O14 S3 U, 357
K O5 Sb3, 185
K2 O9 Si3 Zr, 326
K3 O21 Ta7 Ti, 197
K2 O3 Th, 331
K O3 V, 190
K2 O19 W6, 208
K2 O25 W8, 208
K2 S3, 94
K2 Se3, 94

(La, Sr) Mn O4, 329
La2 Ni O4, 214
(La, Sr)2 Ni O4, 329
La4 O19 Os6, 214
La O4 P, 262
La4 O19 Ru6, 214
La Re2 Si2, 77
Li Lu Mo3 O8, 207
Li0.11 Mg Zn1.89, 78
Li Mg2 Zn3, 78
Li2 Mo3 Ni2 O12, 207
Li Mo3 O8 Sc, 207
Li Mo3 O8 Sm, 207
Li Mo3 O8 Y, 207
Li Mo3 O8 Yb, 207
Li O9 P3 Pb, 246
Li4 O4 Pb, 348
Li O6 Sc Si2, 300
(Li, Mg, Sc) O3 Si, 320
Li2 O3 Si, 297
Li8 O6 Sn, 183
Li O8 Ta3, 197
Li2 O4 Te, 288
Li2 O4 W, 208
Li2 Sb, 15
Li2 Sb Zn, 15
Li Te3, 79
Ln Mn O4 Sr, 329
(Ln, Sr) O4 Sr V, 330
Ln2 O7 V2, 330
Lu Na O2, 331
Lu2 O7 V2, 330

Mg Mn2 O4, 330
Mg Mo2 O7, 205
Mg O8 P2 Zn2, 259
Mg O3 Si, 318
Mg O4 V2, 330
Mg2 Sb2 Sr, 4
Mn2 N0.86, 79
Mn N2 Si, 80
Mn2 Na4 O5, 337
Mn Nb2 O6, 211
Mn O5 Se2, 286
Mn O6 Ta2, 211

Mn O4 W, 211
Mn2 S4 Sn, 99
Mn2 Si2 Th, 99
Mn3 Si5 U2, 80
Mn Sb Zn, 16
Mo Ni P, 82
Mo O6 Rh2, 335
Mo5 O16 Te, 291

N3 Na, 114
N Na3 O3, 356
N Nb, 86
N Nb O, 109
N2 O Si2, 109
N O2 Si Y, 337
N O Ta, 109
N6 P4 S, 334
N6 Pb, 115
(N, C)2 Pu2 Sb, 336
N4 Si3, 86
N Ta2, 87
N5 Ta3, 87
N Te U, 88
N0.61 Ti, 88
(N, O) Ti, 329
N3 U2, 88
Na O11 P3 U, 336
Na O6 S2 Tl3, 277
Na O8 Sb5, 184
Na2 O7 Sb4, 184
Na O2 Sc, 331
Na2 O7 Si2 Zn2, 303
Na2 O8 Si3 Zn, 302
Na O3 Ta, 197
Na2 O4 Te, 289
Na O2 Tm, 331
Na O3 V. 1.89 H2 O,
Na O2 Y, 331
Na O2 Yb, 331
Na0.6 Se2 V, 95
Na Se2 V, 95
Nb O2, 175
Nb2 O6 Pb, 195
Nb2 O7 Pb2, 335
Nb2 O8 Pb3, 335
Nb11 C30 Rb4, 193
Nb6 O18 Rb2 Ti, 196
Nb30 C90 Rb12 W3, 329
Nb2 O8 Sr Th, 197
Nb4 O21 Sr3 Ti4, 188
Nb3 O9 Tl, 194
Nb22 O59 Tl8, 194
Nb S2 Sn, 86
Nd C12 P4 Rb, 268
Nd Si, 82
Ni O4 W, 211, 332
Ni3 S2, 84

Ni S Sb, 19
Ni Sc Si3, 83
Ni2 Si2 Th, 99
Ni2 Y3, 85

O8 P2 Sn3, 254
O26 P4 Te8, 356
O9 P3 V, 247
O12 P4 Zr, 335
O3 Pb Se, 285
O8 Pb3 Ta2, 335
O4 Pb W, 210
O3 Pr Rb2, 331, 349
O4 Pt3.4, 176
O2 Rb9, 172
O2 Rb Sc, 349
O18 Rb2 Ta6 Ti, 196
O3 Rb2 Tb, 217
O3 Rb2 Th, 331
O3 Rb V, 190
O107 Rb22 W32, 209
O4 Sb2, 174
O3(Sc, Ti)2, 189
O7 Sc2 Si2, 329
O2 Si, 262, 297
O4 Si Zn2, 336
O21 Sr3 Ta4 Ti4, 188
O4 Sr U, 353
O3 Ta, 176
O9 Ta5, 176
O2 Tb, 330
O3 Ti2, 175, 189
O3(Ti, V)2, 189
O7 Tm2 V2, 330
O3 U, 177
O8 U3, 177
O2 V, 175
O9 V4, 328
O7 V2 Yb2, 330
O3 W, 176
O12 Y4 Zr3, 188
Os P2, 81
Os3 Pu5, 67
Os Sb2, 81

P2 Pd15, 89
P3 Pd7, 90
P2 Ru, 81
Pb7 Sb8 S19, 99
Pd5 Pu3, 90
Pd20 Sb7, 16
Pd17 Se15, 91
Pd3 Te2, 91
Pr Si, 82
Pt3 Pu5, 67
Pt20 Pu31, 92
Pt5 Se4, 93

Pu5 Rh3, 67
Pu5 Rh4, 94
Pu31 Rh20, 92
Pu5 Ru3, 67

Rh Sb, 17
Rh2 Sm3, 95
Ru Sb, 17
Ru Sb2, 81

S3.3 Se4.7, 354
S4.7 Se3.3, 354
S5.1 Se2.9, 354
S2 Sn Ta, 86
S8 Tl V5, 96
S1.7 Yb, 98
S3 Yb2, 97
Sb3 Ti5, 18
Sb2 Tl7, 17
Sb2 V3, 18
Si2 Sr, 42
Si2 Ti, 96

382

AUTHOR INDEX

Names beginning with a separated prefix are listed
before single-word names beginning with the same letters;
accents are omitted.

Abdullaev, G.K., 230, 231, 232, 355
Abraham, F., 214, 221
Abrahams, S.C., 65, 282, 293, 296
Ackerman, R.J., 177
Adams, J.M., 232
Adrian, H.W.W., 113
Ahtee, M., 197
Akao, M., 233, 234
Akimoto, S., 301
Aksel'rud, L.G., 50, 80
Alapini, F., 61
Alasafi, K.M., 38, 39
Albertsson, J., 295
Alcock, N.W., 294
Aleonard, S., 123, 130
Ali, E.M., 137
Ali, S.Z., 70
Allmann, R., 278
Althoff, P.L., 235
Amiraslanov, I.R., 231
Anderson, J.B., 257, 259, 273
Anderson, M.R., 244
Andersson, Y., 89
Andresen, A.F., 17, 41, 81
Araki, T., 252, 253, 256, 313, 314, 319
Armagan, N., 284
Arnaud, Y., 291
Arpe, R., 149, 186, 187
Aslanov, L.A., 172
Asmat, H., 8
Atoda, T., 6
Atovmjan, L.O., 147
Attig, R., 267
Aubry, J., 21
Aurivillius, B., 144, 339
Averbuch-Pouchot, M.T., 198, 244, 245, 263, 264, 266
Averbuch, M.T., 265
Avilov, A.S., 91
Axe, J.D., 175
Aydin Uraz, A., 284

Baert, F., 135
Bagieu-Beucher, M., 269
Bailey, S.W., 327

Bakhareva, E.D., 213
Balahura, R.J., 284
Baldwin, K.J., 297
Ballard, J.G., 120
Ban, Z., 52, 82, 99
Barbara, B., 8
Barbier, P., 135
Barnea, Z., 40
Barnighausen, H., 139
Bars, O., 198, 204
Bassi, G., 218
Bats, J.W., 158, 159, 171, 172, 275
Baur, W.H., 297
Bayliss, P., 19, 75
Beall, G.W., 162, 164, 222, 244
Beattie, J.K., 241
Belokoneva, E.L., 300, 303, 305
Belov, N.V., 52, 180, 181, 227, 262, 300, 303, 304, 305, 326
Beran, A., 235
Berg, R.W., 138
Berger, R., 18, 25
Berking, B., 319
Bernal, I., 162, 164, 222, 237
Bernard, J., 351
Bernstein, J.L., 65, 282, 293, 296
Bertaut, E.F., 48, 79
Besrest, F., 72
Beyerlein, R.A., 56, 109
Beznosikova, A.V., 67
Bierstedt, P.E., 223
Billiet, Y., 220
Billy, M., 109
Birchall, T., 120
Birnie, R.W., 13
Birtill, J.J., 219
Bither, T.A., 223
Bjornberg, A., 350
Blinov, V.A., 326
Bloembergen, J.R., 95
Blum, D., 270
Bodak, O.I., 46, 56, 77, 83
Bohm, H., 302
Boilot, J.P., 179
Boivin, J.-C., 187, 349

Bonnet, B., 125
Bonnet, M., 218
Bonnin, A., 199, 200
Booth, M.H., 53
Borel, M.M., 131
Borisov, S.V., 130, 154
Bottcher, P., 94
Bottomley, F., 133
Boutonnet, R., 130
Boving, P., 274
Bowman, A.L., 180
Brachtel, G., 349
Brandon, J.K., 8, 53, 66
Braun, D., 71
Braun, R.M., 183
Brice, J.F., 21
Britton, D., 170
Brizard, R.Y., 53, 66
Bronger, W., 54, 69
Brovkin, A.A., 226
Brown, D., 154
Brown, D.B., 122
Brown, G.E., 312
Brown, G.M., 210, 276
Brown, I.D., 141, 142, 143
Brown, K.L., 22
Brown, W.E., 248, 249
Brun, T.O., 56
Brunel-Lauqt, M., 246, 258
Brunn, H., 217
Bruzzone, G., 27
Brynestad, J., 134
Buehler, E., 65
Buisson, G., 148, 203
Bukovec, P., 128
Bukveckij, B.V., 122, 126, 340
Burke-Laing, M.E., 293
Burnham, C.W., 13
Burschka, C., 54
Buser, H.J., 163
Busing, W.R., 210

Calabrese, J.C., 133
Calleri, M., 180
Callmer, B., 34
Calvert, L.D., 20, 183
Calvo, C., 190, 247
Cano, F.H., 228
Carpentier, C.-D., 48
Carre, D., 70, 70
Carrondo, M.A.A.F.DeC.T., 355
Carter, R.L., 284
Cartz, L., 109
Castellano, E.E., 163
Caton, R., 56
Catti, M., 247, 250, 277, 356
Cavin, O.B., 79

Cebotarev, N.T., 67
Cemodina, T.N., 227
Ceolin, R., 30
Cernji, A.V., 67
Cerny, P., 306
Chadwick, B.M., 160
Chailleux, J.M., 197
Chamberland, B.L., 214
Chaminade, J.-P., 147
Champarnaud-Mesjard, J.C., 123, 124
Chan, L.Y.Y., 154
Chang, A.T., 177
Che'ng Wan, see Wan, C.
Chenavas, J., 181
Chevalier, P., 115
Chevrel, R., 53
Chieh, C., 8, 17, 53
Choi, C.S., 115
Choisnet, J., 195, 229, 301, 348
Chomnilpan, S., 284, 293
Chow, Y.H., 170
Chowdhury, M.R., 160
Christ, C.L., 224
Christensen, A.N., 44, 86, 87, 221
Christidis, P.C., 279
Churchill, M.R., 110, 111, 113
Clark, A.M., 91
Clark, J.R., 224
Claus, A., 33
Clearfield, A., 208, 255
Coda, A., 271
Coffey, C.C., 166
Cohen, J.P., 310
Coing-Boyat, J., 245
Collin, G., 48, 72, 179
Comes, R., 179
Conflant, P., 349
Conroy, L.E., 87
Cook, N., 112
Coppens, P., 114, 171, 172, 275
Corazza, E., 235
Corbett, J.D., 132, 133, 139
Cornish, T.F., 167
Cornish, T.L., 168
Cot, L., 276
Cotton, F.A., 134, 138
Couldwell, M.C., 110
Courbion, G., 127
Courtois, A., 21, 21
Cousseins, J.C., 130
Cox, D.E., 202
Cradwick, P.D., 296
Criddle, A.J., 91
Cromer, D.T., 71, 90, 92, 94
Cronin, J.L., 355

Cudennec, Y., 198
Culum, Z., 240

Daake, R.L., 139
Dal Negro, A., 236, 271
Dance, J.-M., 129
Daniel, F., 288, 290
Darriet, B., 143, 353
David, J., 30
Davidovic, R.L., 126
Davis, R.J., 91
de Endredy, A.S., 296
de Pape, R., 127
de Pieri, R., 314
de Roy, A., 146
DeBenedittis, J., 207
DeBoer, B.G., 111
Declercq, J.-P., 242
Delapalme, A., 218
Deller, K., 3, 4, 23
Deschanvres, A., 197
Desgardin, G., 196
Despotovic, Z., 13, 88
Devalette, M., 353
Dickens, B., 248, 249
Dickens, P.G., 219
Dickson, F.W., 22
Diehl, R., 48
Diercks, H., 37
Dittmar, G., 354
Divjakovic, V., 24, 238
Djomaa, H., 351
Djuric, S., 240
Dobrynina, T.A., 118
Dollase, W.A., 299
Donaldson, J.D., 118
Donnay, G., 255
Dotzel, K.-P., 11
Drai, S., 193
Drew, M.G.B., 159, 161
Dubler, E., 128
Ducourant, B., 125, 341
Dudarev, V.Ja., 118
Durand, J., 276
Durif, A., 198, 244, 245, 264,
  265, 266, 268, 270
Dwight, A.E., 14
Dzafarov, G.G., 230, 355

Eddine, M.N., 48, 79
Edenharter, A., 238
Edwards, A.J., 346
Effenberger, H., 292
Egorov-Tismenko, Ju.K., 227
Eichler, W., 137
Eisenmann, B., 3, 4, 23, 28
Elder, R.C., 156

Elding, I., 295
Eliseev, A.A., 97, 98
Ellert, G.V., 357
Ellinger, F.H., 44
Endresen, K., 17
Engel, P., 147
Eppinga, R., 86
Esperas, S., 294
Eulenberger, G., 66
Evans, H.T., 322
Evdokimova, V.V., 98
Evers, J., 26, 29, 42
Evstigneeva, T.L., 181

Faggiani, R., 141, 143, 242
Fallon, G.D., 193
Fanfani, L., 248
Fang, J.H., 306
Farkas, L., 219
Fava, J., 180
Fedorov, N.F., 305
Fender, B.E.F., 345
Fenn, P.M., 312
Ferey, G., 127
Ferguson, G., 284
Ferguson, R.B., 272
Ferraris, G., 247, 250, 277, 356
Filhol, A., 250
Finger, L.W., 307, 309
Fink, U., 338
Finney, J.J., 151
Firor, R.L., 326, 327
Fischer, P., 49
Fleet, M.E., 84, 311
Fletcher, S.R., 162
Flukiger, R., 53
Fonteneau, G., 196
Fornasini, M.L., 42
Forslund, B., 190, 191
Fourcade, R., 125, 341
Fournes, L., 96
Franceschi, E., 27, 42
Franchini-Angela, M., 247, 277
Frankis, E.J., 234
Freeborn, W.P., 299
Freeman, D.K., 40
Freund, H.-R., 217
Frey, M., 195
Frit, B., 123, 124
Fuess, H., 218
Fujino, T., 353
Fujita, T., 210
Furuseth, S., 17
Fykin, L.E., 118, 285

Gabe, E.J., 20, 183
Gabela, F., 121

Gado, P., 219
Galigne, J.L., 276
Galy, J., 143
Garcia-Blanco, S., 228
Garrett, W.L., 115
Gasperin, M., 194, 229
Gatehouse, B.M., 193
Gatilov, Ju.V., 130
Gazzoni, G., 180
Geller, S., 154, 174
Geneys, C., 339
Gerardin, R., 21
Gerault, Y., 199
Germain, G., 242
Gettas, B., 182
Ghedira, M., 175, 181
Ghose, S., 225, 274, 305, 321
Ghosh, M., 294
Giacovazzo, G., 281
Gibb, T.C., 162
Gibbs, G.V., 297, 310, 311, 315,
  321, 324
Gieren, A., 291
Gignoux, D., 8
Gilbert, J.D., 250
Gilje, J.W., 327
Gillespie, R.J., 119, 145, 155,
  342
Giuseppetti, G., 228, 321
Gladysevskij, E.I., 46, 50, 56,
  68, 77, 80, 83
Gleizes, A., 214
Gliemann, G., 167
Golic, L., 124
Golubev, A.M., 352
Goodyear, J., 137
Gopal, R., 208
Graeber, E.J., 151
Graf, H.A., 222
Grande, B., 150, 153, 176, 214,
  216
Grandjean, D., 127, 198, 204
Grannec, J., 129
Gravereau, P., 200, 201
Greis, O., 340
Greschonig, H., 275
Grevceva, T.G., 357
Griffen, D.T., 324
Griffith, W.P., 355
Grin', Ju.N., 68
Groult, D., 188, 195, 196, 197,
  229
Grundy, H.D., 300, 313, 324
Gudel, H.U., 147
Guemas, L., 337
Guerin, R., 23, 82
Guggenheim, S., 327

Guidot, J., 291
Guitel, J.C., 198, 245, 246, 258,
  263, 264, 265, 266, 268, 269,
  270
Guittard, M., 76
Gustafsson, T., 279

Haange, R.J., 95
Haase, A., 13
Haber, J., 205
Hadenfeldt, C., 343
Haga, N., 315
Hair, N.J., 241
Hall, K.M., 314
Hall, L., 154
Hall, W., 122
Hall, W.T., 138
Hamid, S.A., 295
Hamil, M.M., 311
Hamilton, R.D., 151
Hamilton, W.C., 237
Hammann, J., 179
Hamon, C., 305
Hampshire, M.J., 51
Hansen, P., 221
Haradem, P.S., 214, 355
Harder, M., 179
Hardy, A., 127, 200, 201
Hargrave, K.D., 109
Harsta, A., 43
Hasegawa, K., 196
Haupt, H.J., 343
Hawthorne, F., 247
Hawthorne, F.C., 190, 272, 300,
  306, 313, 318, 324
Hazen, R.M., 312
Hedman, B., 206, 350
Hellner, E., 109
Hemmi, C., 320
Hemmi, K., 320
Hermansson, K., 237
Hervieu, M., 195
Hess, H., 116
Hesse, K.-F., 297, 302
Higashi, I., 6
Higgins, J.B., 317
Hilfrich, P., 171
Hill, R.J., 261, 278
Hoch, G., 136
Hoggins, J.T., 5, 26
Hollander, F.J., 110, 111, 113
Holmberg, B., 243
Holt, E., 193
Hope, H., 114
Hoppe, R., 23, 40, 178, 183, 217,
  223, 348, 349, 351, 352
Horiuchi, H., 32

Horyn, R., 10
Howard-Lock, H.E., 142
Howie, R.A., 298
Hseu, T.H., 269
Huber, F., 343
Hugel, R.P., 169
Hulliger, F., 59
Hursthouse, M.B., 117
Hurtgen, C., 154
Hutchinson, J.P., 110, 111

Ievin's, A.F., 230
Ijdo, D.J.W., 136
Iljukhin, V.V., 213, 326
Imamov, R.M., 91
Inoue, Z., 86
Ionov, V.M., 172
Ipatova, E.N., 212
Isobe, M., 301
Ito, J., 318, 320
Itoh, K., 157
Ivaldi, G., 356
Ivanov, Ju.A., 262
Iwai, S., 209, 233, 234
Iwata, M., 161
Iwata, Y., 157

Jacobini, C., 127
Jacobs, H., 338
Jamnova, N.A., 227
Jansen, M., 302, 356
Jarchow, O., 324
Jarmoljuk, Ja.P., 10, 50, 68, 73,
   80
Jarovec, V.I., 56
Jaulmes, S., 55, 58, 59, 61
Jeitschko, W., 16, 71, 96, 223
Jenkin, G.T., 244
Jensovsky, L., 289
Joesten, M.D., 237
Johansson, G.B., 292, 293
Johansson, T., 43
Johnson, P.L., 164, 165, 166,
   167, 168
Johnson, V., 16
Johnston, D.C., 35
Jones, D.P., 355
Jones, G.R., 346
Jordan, T.H., 254
Jorgensen, J.D., 56, 109
Jost, K.H., 299
Joswig, W., 356
Joubert, J.C., 181
Julien-Pouzol, M., 58, 59, 61
Jumas, J.C., 76, 341
Junq, W., 35

Kahn, A., 179
Kampf, A.R., 251, 255, 260
Kanamaru, F., 214
Kaplunnik, L.N., 52
Kappenstein, C., 169
Karl, R., 76
Karpov, O.G., 304
Kasper, J.S., 6
Kato, K., 47, 86, 174, 210
Kato, T., 279
Katric, M.V., 49
Katz, L., 207, 214, 355
Kaucic, V., 128
Kawada, I., 47, 86, 210
Kawahara, A., 301, 320, 323
Kawamura, K., 301
Kawano, S., 157
Kazanskaja, E.V., 227
Kazanskij, L.P., 352
Keller, H.-L., 136
Ketterl, F., 189
Khattak, C.P., 202
Khitrova, V.I., 176
Khodadad, P., 30, 92, 153
Kijima, K., 86
Kim, S., 122
Kim, Y., 327
Kipka, R., 150, 151, 215
Kiriyama, H., 132
Kistrup, K., 338
Kitahama, K., 132
Kjekshus, A., 17, 81
Klanderman, K.A., 237
Klaska, K.H., 324
Kleckovskaja, V.V., 176
Klevcova, R.F., 212
Klufers, P., 38
Klug, A., 208
Kniep, R., 251
Koch, T.R., 164, 165, 166, 167,
   168
Koerntgen, C., 284
Koetzle, T.F., 275
Kohl, P., 185
Koizumi, H., 268
Koizumi, M., 214
Kojic-Prodic, B., 121
Kokh, L.A., 357
Koniq, H., 178, 223
Korenstein, R., 48
Korp, J., 162, 164, 222
Koskenlinna, M., 285, 286, 287,
   357
Kostiner, E., 257, 259, 273
Koto, K., 320
Kotur, B.Ja., 56, 83
Koyama, K., 358

Koyano, N., 157
Kozlowski, R., 205
Krasocka, O.N., 147
Kratochvil, B., 289
Krause, R.A., 355
Krebs, B., 37
Kripjakevic, P.I., 73, 78
Krischner, H., 275
Krstanovic, I., 240
Kudrjavceva, O.V., 182
Kuhn, D., 69
Kunze, G., 295
Kupcik, V., 33
Kusachi, I., 320
Kusuncki, T., 88
Kuz'ma, Ju.B., 12, 34, 50
Kuz'miceva, G.M., 97, 98
Kuznecov, V.G., 357
Kvick, A., 172
Kynev, K., 172

L'Haridon, P., 30, 305
L'Helgoualch, H., 196
LaChapelle, W.A., 274
Labbe, Ph., 195
Labeau, M., 130
Lanq, J., 30
Larsen, E.M., 133
Larsen, F.K., 138
Larson, A.C., 92
Laruelle, P., 55, 58, 70, 156
Lauqhlin, D.R., 118
Launay, J.C., 175
Laurent, Y., 305
Lazarenko, E.K., 307
Lazarini, F., 140, 142
Le Flem, G., 180
Le Marouille, J.Y., 152, 153, 198, 204
Le Paqe, Y., 255
Le Roy, J., 85, 95
Leban, I., 124
Lecart, B., 353
Lecerf, A., 212
Leclaire, A., 131, 239
Lee, G.C., 166
Lehmann, M.S., 221
Lenhert, P.G., 237, 250
Leoni, L., 325
Lerf, A., 222
Levet, J.-C., 153
Levy-Clement, C., 220
Levy, H.A., 210
Levy, J.H., 147
Liebau, F., 302
Liebich, B.W., 118
Liminqa, R., 284, 285, 293, 296

Lindqren, O., 203, 282, 283, 357
Lindqvist, O., 292, 293
Lindsay, R., 69
Linowsky, L., 128
Lippert, B., 241, 242
Litvin, B.N., 304
Ljunqstrom, E., 160
Lock, C.J.L., 241, 242
Lofqren, P., 208
Loopstra, B.O., 176, 177
Lu, T.H., 269
Lucas, J., 196
Ludi, A., 163
Luk, W., 155
Luk'janov, A.S., 67
Lundqren, J.-O., 279
Lux, D., 116

Mackie, P.E., 262
Maeland, A.J., 41
Maetz, J., 290
Maffly, R.L., 165, 168
Maqerramov, A.I., 231
Maharajh, E., 155
Mahlberq, G., 60
Mair, S.L., 40
Mairesse, G., 135
Malaman, B., 18
Malik, K.M.A., 117
Malinovskij, Ju.A., 180
Mamedov, K.S., 230, 231, 232, 355
Mandel, G.S., 149
Mandel, L., 51
Mandel, N.S., 149
Maraine, P., 184, 185
Marchand, R., 80
Marezio, M., 175, 181
Marqulis, T.N., 284
Marko, M.A., 34
Marquart, R., 12
Marsh, R.E., 149
Martens, K.-P., 348, 349
Martin, D., 342
Martinez-Ripoll, M., 13, 340
Marumo, F., 209, 301
Masaki, N., 88, 353
Mascherpa, G., 125, 341
Mason, S.A., 356
Masonkin, V.P., 213
Massa, W., 342
Masse, R., 264, 268
Matejka, G., 129
Mathew, M., 248, 254, 262
Matjusenko, N.N., 49
Matkovic, P., 64, 91, 93
Matkovic, T., 12, 90
Matsunaga, H., 157

Mattes, R., 347, 351
Matthieu, J.-P., 128
Matz, C., 351
Maunaye, M., 79, 80, 305
Maurin, M., 76, 288, 290, 341
Maxim, B.R., 344
May, N., 3, 28, 98
Mayer, H., 356
Mayer, I., 57
Mayer, M., 146
Mazzi, F., 228, 321
McCarroll, W.H., 207
McKie, D., 234
McMullan, R.K., 162
McPhail, A.T., 109
Meagher, E.P., 315
Meerschaut, A., 158
Mel'nik, E.V., 78
Melamud, M., 212
Mellini, M., 313, 325
Menchetti, S., 224
Mennemann, K., 347
Mercurio, D., 124
Merlino, S., 313, 325
Messain, D., 70
Mewis, A., 38
Meyer, G., 344, 351
Middlemiss, N., 247
Mikhail, I., 296
Mikhajlov, Ju.N., 357
Milinski, N., 240
Miljan, V.V., 12
Mill', B.V., 300
Milligan, W.O., 162, 164, 222, 244
Minor, D., 197
Mitchell, P.C.H., 161
Moller, C.K., 345
Monier, J.C., 195, 239
Moore, P.B., 252, 253, 255, 256, 260, 313, 314, 319
Mootz, D., 251, 267
Moreau, J.M., 85, 95
Moret, J., 288, 290
Mortier, W.J., 308
Moseley, P.T., 154
Moyer, R.O., 69
Mrose, M.E., 322
Mueller, M.H., 56
Mullen, D., 109
Muller-Buschbaum, H., 149, 150, 151, 153, 176, 179, 186, 187, 214, 215, 216, 217, 218, 348, 350
Muller, W., 3, 15, 28
Mullner, M., 290
Mumme, W.G., 30

Muradjan, L.A., 126, 285, 306, 352
Murakami, T., 320
Muschick, W., 218, 348
Musselman, R.L., 165
Mys'kiv, M.G., 83

Nabi, S.N., 117
Nachbaur, E., 275
Nacken, B., 69
Nagakura, S., 88
Naqpal, K.C., 70
Nakamura, E., 157
Nakano, J., 268
Naslain, R., 6
Natarajan Iyer, M., 141, 143
Natarajan, M., 142
Navrotsky, A., 180
Needham, G.F., 167
Nguyen-Huy-Dung, 156
Nguyen-Trut-Dinh, 180
Nguyen, N., 301, 348
Nguyen, V.N., 82
Nicollin, D., 118
Niinisto, L., 286
Noe-Spirlet, M.-R., 210
Nord, A.G., 259
Novoksonov, V.I., 98
Nowacki, W., 20, 24, 238
Nowotny, H., 36, 37
Nozik, Ju.Z., 285
Nuber, B., 136

Obenland, S., 60
Ocio, M., 179
Oddon, Y., 277
Oehlinger, G., 26, 29, 42
Ohashi, Y., 307, 309
Okada, K., 209
Okada, T., 157
Okamura, F.P., 305
Olazcuaga, R., 193
Olovsson, I., 237, 279
Orlandi, P., 325
Oswald, H.-R., 128
Otto, H.H., 167
Ozaki, Y., 181
Ozima, M., 207

Paccard, D., 85, 95
Palfij, Ja.F., 50
Palvadeau, P., 158, 337
Paoli, A., 53
Paris, J., 48
Parthe, E., 30, 95
Pavlisin, V.I., 307
Peacor, D.R., 344

Pearson, W.B., 8, 17, 53, 66
Pecarskij, V.K., 46, 77
Pepe, G., 277
Perez, G., 146, 184, 185
Perkins, R.S., 49
Perrin, A., 152
Perrin, M., 129
Persson, K., 243
Pertlik, F., 291
Petrovic, D., 240
Petter, W., 163
Philippot, E., 76, 288, 290, 341
Phillips, M.W., 302
Pinsker, Z.G., 176
Pinto, H., 212
Piret, P., 242
Piro, O.E., 163
Plattner, E., 181
Pluth, J.J., 307, 308, 311
Pobedimskaja, E.A., 52, 180, 182, 304
Podberezskaja, N.V., 130
Poeppelmeier, K.R., 132, 133
Poliscuk, S.A., 122, 340
Potapova, O.G., 130, 154
Potel, M., 23, 153
Pouchard, M., 147
Poulain, M., 127
Prewitt, C.T., 91, 297
Prince, E., 115
Pritchard, R.G., 232
Pritzkow, H., 145
Propach, V., 350
Protas, J., 21
Pupp, G., 356
Puscarovskij, D.Ju., 182
Puselj, M., 52, 82
Puxley, D.C., 118
Pye, M.F., 219
Pygall, C.F., 159, 161
Pynn, R., 175

Quareni, S., 314

Rabenau, A., 141
Rackwitz, R., 76
Radtke, A.S., 22
Rafalko, J.J., 162
Rajamani, V., 91
Rakke, T., 17, 81
Range, K.J., 60, 189
Ras, F.G., 136
Raston, C.L., 337
Raveau, B., 188, 195, 196, 197, 229, 301, 348
Ray, M., 294
Rea, J.R., 273

Reichert, B.E., 113
Reinen, D., 130, 185
Reitveld, H.M., 176
Rendle, D.F., 270
Rendon-Diazmiron, L.E., 5
Rentzeperis, P.J., 279, 310
Restivo, R.J., 284
Ribar, B., 238, 240
Ribbe, P.H., 302, 311, 317, 324
Rice, C.E., 134, 175, 189
Rieck, H., 352
Riedel, E., 76
Riekel, C., 303
Riley, P.W., 8
Riou, A., 212
Ritsma, J., 115
Rivero, B.E., 163
Rivet, J., 55
Robert, C., 196
Robinson, P.D., 306
Robinson, W.R., 175, 189
Rodier, N., 30, 41, 45, 92, 153
Roql, P., 36, 37
Roques, B., 18, 47
Rosenberg, B., 241, 242
Rosenzweig, A., 151, 221, 316, 323
Ross, F.K., 310, 317
Rossat-Mignod, J., 82
Rossi, G., 236, 325
Roth, R.S., 197
Roubin, M., 48
Rouse, R.C., 344
Rouxel, J., 115, 158
Rundqvist, S., 43
Ruszala, F.A., 257, 259
Ruzic-Toros, Z., 121
Ryan, R.R., 221, 316, 323
Rykhal, R.M., 10
Rys, J., 114

Saakjan, L.S., 34
Saalfeld, H., 174
Sabelli, C., 224, 235, 271
Sadanaga, R., 312
Sakurai, T., 6
Saldarriaga-Molina, C.H., 208
Salje, E., 176
Sandbote, H.-W., 343
Sandomipskij, P.A., 181
Santoro, A., 197
Sarin, V.A., 118, 285
Sartori, F., 317
Sato, S., 207
Satterthwaite, C.B., 56
Saurel, C., 188
Sauvage, J.P., 184, 185

Saux, M., 96
Sayetat, F., 203
Scandale, E., 281
Schaefer, W.P., 149
Schafer, H., 3, 4, 11, 28, 98, 354
Scheinert, W., 135
Scheunemann, K., 203
Schlenker, J.L., 307, 311
Schmachtel, J., 350
Schmelczer, R., 59
Schollhorn, R., 222
Schrobilgen, G.J., 145, 342
Schroder, F.A., 204, 353
Schroeder, G., 15
Schroeder, L.W., 248, 249, 254, 262
Schubert, K., 12, 16, 38, 39, 64, 90, 91, 93
Schuckmann, W., 204
Schultz, A.J., 166
Schulz, H., 141, 167
Schuster, H.-U., 14, 15, 19, 61, 62
Schuster, H.D., 356
Schwarz, W., 116
Schwarzenbach, D., 59, 163
Schweizer, M., 216
Scordari, F., 280, 281
Scott, H.G., 188
Scott, J.D., 17, 20
Seethanen, D., 265
Seff, K., 326, 327
Seifert, H.J., 137, 142
Selte, K., 17
Seppelt, K., 145
Sergent, M., 23, 82
Seydel, R., 299
Shaked, H., 212
Sheldrick, G.M., 113
Shibuya, I., 157
Shimada, M., 214
Shirasuka, K., 348
Shoemaker, C.B., 73
Shoemaker, D.P., 73
Shoemaker, G.L., 257, 259
Sicking, E., 351
Sidorenko, O.V., 318
Sidorov, P.M., 305
Sikirica, M., 99
Silin', E.Ja., 230
Simon, A., 172, 173
Simonov, M.A., 181, 227, 262, 300, 303, 305
Simonov, V.I., 122, 126, 306, 340, 352
Simpson, J., 110

Skapski, A.C., 355
Slack, J.F., 22
Slim, D.R., 119, 120, 145, 155, 342
Sljukic, M., 121
Slotfeldt-Ellingsen, D., 41
Smart, L., 112
Smirnova, E.A., 67
Smith, G.P., 134
Smith, J.V., 307, 308, 311
Smyth, J.R., 320
Soboleva, S.V., 318
Sokol, S.K., 357
Soled, S., 48
Solov'eva, L.P., 212
Sommer, H., 23, 40
Sorrell, C.A., 177
Spicyn, V.I., 352
Spielvogel, B.F., 109
Srinivasa, S.R., 109
Stadnicka, K., 205
Staudel, L., 142
Steigmann, G.A., 137
Steinfink, H., 5, 26
Steinmetz, J., 18, 47
Stevens, E.D., 114
Stoeger, W., 141, 345
Stokhuyzen, R., 8, 17
Stomberg, R., 347
Stover, H.-D., 349
Strahle, J., 109, 139
Strandberg, R., 205
Strelkova, E.E., 304
Strydom, O.A.W., 276
Sudarsanan, K., 261
Sumjackaja, N.G., 326
Svensson, I.-G., 347
Swanson, B.I., 162
Szymanski, J.T., 51

Tadini, C., 228, 321
Tagai, T., 320
Tagawa, H., 88, 353
Taguchi, H., 214
Takagi, S., 237
Takeda, Y., 214
Takeuchi, Y., 312, 315, 320, 358
Tanaka, H., 86
Tani, B.S., 156
Taylor, D.R., 133
Taylor, H.F.W., 316
Taylor, J.B., 20, 183
Taylor, J.C., 147, 177
Tazzoli, V., 236, 271
Tcheou, F., 82, 203
Tellgren, R., 284, 285, 293
Terao, N., 87

Thery, J., 179
Thiele, G., 171
Thiemann, K.H., 167
Thomas, D., 187, 214, 221, 349
Thomas, J.O., 43, 237
Thomas, R., 175
Thompson, J.S., 69
Thornton, G., 174
Tiburtius, C., 14, 19
Tien, V., 41, 45
Titov, Ju.G., 49
Tkacev, V.V., 147
Toffoli, P., 30, 92
Toman, K., 273
Tomas, A., 76
Tomlinson, R.D., 51
Tomokiyo, A., 157
Torcenkova, E.A., 352
Tordjman, I., 244, 264
Torii, Y., 196
Tozer, D.J.N., 66
Tranquard, A., 277
Tranqui, D., 123
Trehoux, J., 187, 214, 221
Tressaud, A., 129
Trkula, M., 156
Trojko, R., 13, 88
Tromel, M., 290
Trotter, J., 270
Troup, J.M., 255
Trueblood, K.N., 293
Tunin, T.A., 305

Udovenko, A.A., 126
Unonius, L., 197

Valentine, D.Y., 79
Valkonen, J., 285, 286, 287, 357
van Leon, C.J.J., 138
van Tets, A., 113
Vannucci, S., 235
Varfolomeev, M.B., 213
Vasil'eva, I.G., 154
Vedrine, A., 130
Vegas, A., 228, 251
Vekris, J.E., 119
Venetopoulos, C.C., 310
Verschoor, G.C., 136
Vesela-Novakova, L., 33
Vicat, J., 123
Vignacourt, J.P., 135
Vilminot, S., 339
Vincent, H., 175
Visser, D., 138
Vivier, H., 351
Vlasse, M., 6, 96, 129, 147, 193
Volkova, L.M., 126

Vollenkle, H., 181
von Dreele, R.B., 180, 262
Voronkov, A.A., 326

Waltersson, K., 190, 191, 208
Walton, E.G., 122
Wan, C., 225, 274, 321
Wang, Y., 20, 183
Wanklyn, B.M., 129
Washecheck, D.M., 168
Watts, J.A., 30
Waugh, A.B., 177
Weakley, T.J.R., 207
Weber, K., 33
Wechsler, B.A., 189
Wedgewood, F.A., 160
Weidenhammer, K., 116, 136
Weigel, F., 12
Weishaupt, M., 109
Weiss, A., 26, 29, 42, 135
Weiss, J., 354
Weitzel, H., 130, 211, 353
Welk, E., 62
Werner, P.-E., 208, 219
Werner, W., 139
West, A.R., 298
Westerbeck, E., 173
White, A.H., 337
White, J.W., 244
Widera, A., 4
Wiegers, G.A., 86, 95
Wiench, H., 349
Wild, S.B., 337
Wilde, H.J., 160
Wilhelmi, K.-A., 208
Williams, J.M., 164, 165, 166,
   167, 168
Wilson, A.J.C., 261
Wilson, L.K., 250
Wilson, P.W., 147
Wintenbergen, M., 80
Winter, J.K., 305
Wisian-Neilson, P., 109
Wittmann, F.D., 12
Wolcott, H.A., 244
Wold, A., 48
Woodward, P., 112
Wopersnow, W., 16, 64
Worlton, T.G., 109
Wright, A.F., 345
Wrobel, G., 61
Wuensch, B.J., 32

Yakel, H.L., 79, 134
Yamaguchi, G., 348
Yamamoto, K., 309
Yamanaka, T., 312

Yamane, T., 86
Yanaqida, H., 348
Yersin, H., 167
Yetor, P.D., 57
Young, R.A., 261, 262
Yvon, K., 53

Zachariasen, W.H., 44
Zajakina, N.V., 226
Zanazzi, P.F., 248
Zanne, M., 21
Zanzari, A.R., 248
Zarecnjuk, O.S., 10
Zavodnik, V.E., 118
Zemann, J., 235
Zemnukhova, L.A., 126
Zieqler, M.L., 116, 136
Ziemer, B., 299
Zigan, F., 356
Zoltai, T., 207
Zubieta, J., 122
Zukhlistov, A.P., 307, 358
Zvaqulis, M., 241
Zvjaqin, B.B., 307, 318, 358

Change required on	From	To
37A, 4 and 373	R.A. MURADJAN	L.A. MURADJAN
39A, 38 and 391	L.-A. TERGENIUS	L.-E. TERGENIUS
40A, 266	$Hg(OH)_2.2HgSO_4.2H_2O$	$Hg(OH)_2.2HgSO_4.H_2O$
331	$H_6Hg_3O_{12}S_2$	$H_4Hg_3O_{11}S_2$
41A, 21 and 470	L.N. KAPLUNIK	L.N. KAPLUNNIK
191 and 473	E. PINTCHOVSKI	F. PINTCHOVSKI
206, bottom report	1974	1975
	z(S) 1.3337	1.1137
42A, 95 and 482	J.C.W. FOLMAN	J.C.W. FOLMER
95 and 490	C.A. WIEGERS	G.A. WIEGERS
160 and 485	A.G. MacDAIRMID	A.G. MacDIARMID
169 and 484	C. JACOBINI	C. JACOBONI
193 and 484	D.W.J. IJDO	D.J.W. IJDO
246	2. Ibid., 24, 191.	2. Ibid., 24, 491.
424, Cs_2UO_4	24, 403	26, 403
430	S_2Cl_2	SCl_2
449, $Mn(NO_3)_2.$	39A, 376	39A, 276
$4H_2O$		
473	Cl_2S_2	Cl_2S
481	Corrazo	Corraza
485	Martinego	Martinengo
487	Pickhardt	Pickardt
488	Selte, K., 28, 28, 29	Selte, K., 28, 29